Mapping and Monitoring of Geohazards with Remote Sensing Technologies

Mapping and Monitoring of Geohazards with Remote Sensing Technologies

Editors

Constantinos Loupasakis
Ioannis Papoutsis
Konstantinos G. Nikolakopoulos

MDPI

Basel • Beijing • Wuhan • Barcelona • Belgrade • Novi Sad • Cluj • Manchester

Editors

Constantinos Loupasakis
The National Technical
University of Athens (NTUA)
Athens, Greece

Ioannis Papoutsis
National Observatory of
Athens
Athens, Greece

Konstantinos G.
Nikolakopoulos
University of Patras
Rio, Greece

Editorial Office
MDPI
St. Alban-Anlage 66
4052 Basel, Switzerland

This is a reprint of articles from the Special Issue published online in the open access journal *Remote Sensing* (ISSN 2072-4292) (available at: https://www.mdpi.com/journal/remotesensing/special_issues/Geohazard_RS).

For citation purposes, cite each article independently as indicated on the article page online and as indicated below:

Lastname, A.A.; Lastname, B.B. Article Title. *Journal Name* **Year**, *Volume Number*, Page Range.

ISBN 978-3-0365-9216-9 (Hbk)
ISBN 978-3-0365-9217-6 (PDF)
doi.org/10.3390/books978-3-0365-9217-6

Cover image courtesy of Constantinos Loupasakis

Contents

About the Editors

Constantinos Loupasakis

Dr. Constantinos Loupasakis is a Professor at the school of Mining and Metallurgical Engineering of the National Technical University of Athens (NTUA), and a contracted professor at the at the School of Science and Technology of the Hellenic Open University (HOU). He is the precedent of the Greek National Group of the International Association of Engineering Geology (IAEG) and the Director of the post graduate program of the HOU "Natural Catastrophic Events under climate Change impact". Also he is a member at the Land Subsidence International Initiative (LaSII) of UNESCO. Dr. Loupasakis has received a Ph.D. in Engineering Geology – Geotechnical Engineering (2002), a Diploma in Civil Engineering, (2001), a MSc Diploma in Applied and Environmental Geology (1998) and a Degree in Geology, (1995), from the Aristotle University of Thessaloniki (AUTH), Greece. He has an extensive research experience since 1996 working, with overlaps, in several positions, and he has participated more than 50 research projects and studies as coordinator or principal investigator. He has developed numerous international collaborations with institutes and universities from Germany, Italy, Spain, UK, United Arabs Emirates, Qatar, China etc. He also has an extensive teaching experience since 1996. He has published more than 100 Papers in scientific magazines and congress proceedings, 6 Scientific– educational books, 2 International Conference Proceedings' Volumes and 55 Technical reports.

Ioannis Papoutsis

Dr. Ioannis Papoutsis is an electrical and computer engineer with a research focus on the field of Earth Observation and artificial intelligence through the processing of big volumes of satellite and geospatial data. He holds a PhD on the use of advanced satellite interferometry techniques. He has a deep understanding of the Copernicus flagship programme, being the Operations Manager of the Greek node of European Space Agency Hubs that distribute Sentinel data, as well as a Copernicus Emergency Management Services Manager for Risk and Recovery. He leads the AI4EO research group Orion Lab and his current research focuses on the exploitation, management and processing of big satellite data, and the use of artificial intelligence/deep learning for knowledge extraction using Earth Observation and geo-information based on novel information technologies, with emphasis on disasters management. He has participated in 40 research projects funded mainly from EC (H2020 and FP7), ESA and Copernicus initiatives. Currently, he is the coordinator of the H2020 project DeepCube that focuses on AI pipelines for big Copernicus data, and of the ESA-funded project SeasFire on seasonal wildfire forecasting with deep learning. He has co-authored more than 40 research articles in peer-reviewed journals He was part of the team that won the first prize at the 2014 Copernicus Masters Competition for platform FireHub.

Konstantinos G. Nikolakopoulos

Dr. Konstantinos G. Nikolakopoulos is a Professor at the Department of Geology, University of Patras. He earned a BSc degree in geology and a PhD in Remote Sensing. He made a Post Doc research in the field of hyperspectral data processing. His areas of specialty are remote sensing and geographic information systems (GIS) for geological applications. Since 2012, he is the Chairman of the Special Interest Group Geological Applications of the European Association of Remote Sensing laboratories. He also acts as Associate Editor in European Journal of Remote Sensing. Dr. Konstantinos Nikolakopoulos has been involved in R&D projects funded from organizations such as the European Union, the European Space Agency and the Greek Goverment. He has considerable experience in the area of Earth Observation, GIS and GNSS measurements. His main research interests include: geological mapping, engineering geology, risk analysis, 3D mapping and coastal area mapping. He has more than 200 publications (167 in Scopus) in peer-review journals and conference proceeding with with h index 19 and more than 1600 references.

remote sensing MDPI

Editorial

Special Issue "Mapping and Monitoring of Geohazards with Remote Sensing Technologies"

Constantinos Loupasakis [1,*], Ioannis Papoutsis [2] and Konstantinos G. Nikolakopoulos [3]

1 Department of Geological Sciences, School of Mining and Metallurgical Engineering, National Technical University of Athens (NTUA), 157 72 Athens, Greece
2 Institute of Astronomy, Astrophyiscs, Space Applications and Remote Sensing, National Observatory of Athens, 152 36 Athens, Greece; ipapoutsis@noa.gr
3 GIS and Remote Sensing Laboratory, Division of Applied Geology and Geophysics, University of Patras, 265 04 Patra, Greece; knikolakop@upatras.gr
* Correspondence: cloupasakis@metal.ntua.gr

Geohazard monitoring is crucial for building resilient communities. By leveraging remote sensing technologies, we can assess hazards, implement early warning systems, and evaluate impacts effectively. These cutting-edge tools enable proactive monitoring and real-time analysis, minimizing the impact of geohazards, protecting lives, and fortifying our society against adversity.

Earth observation (EO) techniques have proven to be reliable and accurate for monitoring land surface deformations that occur naturally (landslides, earthquakes, and volcanoes) or due to anthropogenic activities (ground water overexploitation and extraction of oil and gas).

In cases where mitigation methods must be put into practice, the detailed mapping, characterization, monitoring, and simulation of the geocatastrophic phenomena have to precede their design and implementation. EO techniques possess high potential and suitability as alternative, cost-efficient methods for the management of geohazards, and have been proven to be a valuable tool for verifying and validating the spatial extent and the evolution of the deformations.

To this extent, this Special Issue covers innovative applications and case studies on the mapping and monitoring of all kinds of geohazards with remote sensing technologies. It incorporates articles that make use of new tools and methodologies, including the use of data-driven machine learning methods. Machine learning in earth observation have revolutionized geohazard monitoring. By leveraging advanced algorithms, machine learning can analyze vast amounts of satellite imagery and sensor data to detect subtle changes in terrain, identify precursors to hazards, and forecast their evolution, enabling proactive risk mitigation strategies and bolstering societal resilience.

In particular, Orellana et al. [1] focused on the study of the ground deformations taking place in the Santiago basin, combining multi-temporal differential interferometric synthetic aperture radar (DInSAR) with data coming from GNSS stations. The GNSS datasets showed a constant regional uplift in the metropolitan area, while the DInSAR allows for the identification of areas with anomalous local subsidence due to the overexploitation of the aquifers as well as mountainous areas affected by landslides. Overall, the results are fundamental for urban territorial planning in the city of Santiago and demonstrate the importance of geodetic measurements in assessing the impact of climate change on groundwater storage and how this affects the ground surface elevation.

Liu et all [2] also studied the subsidence phenomena taking place due to the long-term excessive extraction of groundwater resources by means of the progressive small baseline subset (SBAS) InSAR time series analysis method. The study was conducted at the eastern Beijing Plain, providing significant information on the deformation mechanisms of land

Citation: Loupasakis, C.; Papoutsis, I.; Nikolakopoulos, K.G. Special Issue "Mapping and Monitoring of Geohazards with Remote Sensing Technologies". *Remote Sens.* **2023**, *15*, 4145. https://doi.org/10.3390/rs15174145

Received: 25 June 2023
Revised: 9 August 2023
Accepted: 10 August 2023
Published: 24 August 2023

subsidence, establishing hydrogeological models, and supporting decision making, early warning, and hazard relief for the urban environment.

Investigating mining geohazards, Ma et al. [3] and Chen et al. [4] studied subsidence phenomena taking place at the perimeter of coal mines. Specifically, Ma et al. [3] proposed an approach for predicting mine subsidence that leverages Interferometric Synthetic Aperture Radar (InSAR) technology and a long short-term memory network (LSTM). Chen et al. [4] investigated the surface deformation by means of the DInSAR-PS-Stacking and SBAS-PS-InSAR methods. The results were verified by means of GPS, indicating the subsidence location, range, distribution, and space–time subsidence law of surface deformation.

Tzampoglou et al. [5] investigated the seasonal ground swelling/settlement of an urban area in Cyprus. The study area is occupied by highly expansive bentonitic clays giving the opportunity to combine the extensive database of geotechnical parameters with the Persistent Scattering Interferometry (PSI) InSAR datasets produced within the framework outcomes of the European Union's research project "PanGeo".

The contributions of Tsironi et al. [6], Ma et al. [7], Chen et al. [8], Tan et al. [9], and Kyriou et al. [10] focused on the study of landslide invents. Specifically, Tsironi et al. [6] studied the kinematics of active landslides in mountain areas of Achaia prefecture, Greece, by processing LiCSAR interferograms using the SBAS tool. The results also suggested a correlation between rainfall and landslide motion. Ma et al. [7] created an inventory map of 2665 rainfall-induced landslides triggered from 5 to 10 May 2016 in Fujian Province, China, by using high-resolution satellite imagery. Numerical simulations proved that the temporal evolution of the landslides could be accurately reproduced by using the MAT.TRIGRS tool. Chen et al. [8] proposed an improved multi-source data-driven landslide prediction method that combines a spatio-temporal knowledge graph and machine learning models. This framework could effectively organize multi-source remote sensing data and generate unified prediction workflows. The proposed workflow can alleviate the problem of poor prediction performance caused by limited data availability in county-level predictions. Tan et al. [9] proposes a landslide time prediction method based on the time series monitoring data of micro-deformation monitoring radar. Deformation displacement, coherence and deformation volume, and the parametric degree of deformation (DOD) are calculated and combined with the use of the tangent angle method. Finally, the effectiveness of the method was verified by using measured data of a landslide in a mining area. Finally, Kyriou et al. [10] used multi-dated data obtained by Unmanned Aerial Vehicle (UAV) campaigns and Terrestrial Laser Scanning (TLS) surveys for the accurate and immediate monitoring of a landslide located in a steep and V-shaped valley. They demonstrated that point clouds arising from a UAV or a TLS sensor can be effectively utilized for landslide monitoring with comparable accuracies. Furthermore, the outcomes were validated using measurements acquired by the Global Navigation Satellite System (GNSS).

Foroughnia et al. [11] proposed a stepwise sequence of unsupervised and supervised classification methods for the delineation of flooding areas using synthetic aperture radar (SAR) and multi-spectral (MS) data. Furthermore, a new unsupervised classification approach based on a combination of thresholding and segmentation (CThS) was developed to deal with the heterogeneity and fragmentation of water patches. The new approach was tested successfully in two flood events in Italy, achieving high precision and accuracy and making it appropriate for rapid flood mapping due to its ease of implementation.

The identification of Fossil Mass Movements is an intriguing subject. Popit et al. [12] conducted a geomorphometric analysis using a high-resolution lidar-derived DEM for the quantification and the visualization of fossil landslides. The proposed methodology was applied at Vipava Valley (SW Slovenia).

Exploiting the recently launched European Ground Motion Service (EGMS) products, Festa and Del Soldato [13] presented a desktop app, the so-called "EGMStream", that enables users to systematically store, customize, and convert ground movement data into geospatial databases, burst per burst or for an area of interest that is directly selectable on the app interface. EGMStream is a value-adding tools for optimal dissemination of radar data from the Copernicus Sentinel-1 satellite mission.

Taking Minqin County, Gansu Province, China as the study area, Yang et al. [14] propose a decision tree model combining four spectral indices for the identification of saline–alkaline areas. The spectral indices are the $NDSI_{34}$ (Normalized Difference Spectral Index of Band 3 and Band 4), the $NDSI_{25}$ (Normalized Difference Spectral Index of Band 2 and Band 5), the $NDSI_{237}$ (Normalized Difference Spectral Index of Band 3 and Band 4), and finally, the NDSInew (New Normalized Difference Salt Index). It was found that this model can be applied for the quick identification of saline–alkaline areas in large regions.

In conclusion, the value of almost any type of remote sensing data, such as radar (SAR), multispectral imagery, data collected by Unmanned Aerial Vehicles and Terrestrial Laser Scanners, and data acquired from airborne Lidar systems, for the mapping and monitoring of geohazards has been demonstrated. Different geohazards, like landslides, ground subsidence in coal mines or urban areas, flooding, and salinization have been addressed in this Special Issue. Most of the researchers that processed SAR data preferred the Sentinel-1 mission. Multispectral data from Sentinel-2, Geoeye, diverse Chinese satellites along with Google Earth data were processed in a fully automatic or semi-automatic way. Different band ratios, supervised and unsupervised classification, diverse spectral indexes, principal component analysis, and soft computing techniques like machine learning were among the methods that were presented for the remote sensing data processing.

Acknowledgments: The Guest Editors of this Special Issue would like to thank all authors who have contributed to this volume for sharing their scientific results and experiences. We would also like to thank the journal editorial team and reviewers for conducting the review process.

Conflicts of Interest: The authors declare no conflict of interest.

References

1. Orellana, F.; Moreno, M.; Yáñez, G. High-Resolution Deformation Monitoring from DInSAR: Implications for Geohazards and Ground Stability in the Metropolitan Area of Santiago, Chile. *Remote Sens.* **2022**, *14*, 6115. [CrossRef]
2. Liu, Y.; Yan, X.; Xia, Y.; Liu, B.; Lu, Z.; Yu, M. Characterizing Spatiotemporal Patterns of Land Subsidence after the South-to-North Water Diversion Project Based on Sentinel-1 InSAR Observations in the Eastern Beijing Plain. *Remote Sens.* **2022**, *14*, 5810. [CrossRef]
3. Ma, F.; Sui, L.; Lian, W. Prediction of Mine Subsidence Based on InSAR Technology and the LSTM Algorithm: A Case Study of the Shigouyi Coalfield, Ningxia (China). *Remote Sens.* **2023**, *15*, 2755. [CrossRef]
4. Chen, Y.; Dong, X.; Qi, Y.; Huang, P.; Sun, W.; Xu, W.; Tan, W.; Li, X.; Liu, X. Integration of DInSAR-PS-Stacking and SBAS-PS-InSAR Methods to Monitor Mining-Related Surface Subsidence. *Remote Sens.* **2023**, *15*, 2691. [CrossRef]
5. Tzampoglou, P.; Loukidis, D.; Koulermou, N. Seasonal Ground Movement Due to Swelling/Shrinkage of Nicosia Marl. *Remote Sens.* **2022**, *14*, 1440. [CrossRef]
6. Tsironi, V.; Ganas, A.; Karamitros, I.; Efstathiou, E.; Koukouvelas, I.; Sokos, E. Kinematics of Active Landslides in Achaia (Peloponnese, Greece) through InSAR Time Series Analysis and Relation to Rainfall Patterns. *Remote Sens.* **2022**, *14*, 844. [CrossRef]
7. Ma, S.; Shao, X.; Xu, C. Landslides Triggered by the 2016 Heavy Rainfall Event in Sanming, Fujian Province: Distribution Pattern Analysis and Spatio-Temporal Susceptibility Assessment. *Remote Sens.* **2023**, *15*, 2738. [CrossRef]
8. Chen, L.; Ge, X.; Yang, L.; Li, W.; Peng, L. An Improved Multi-Source Data-Driven Landslide Prediction Method Based on Spatio-Temporal Knowledge Graph. *Remote Sens.* **2023**, *15*, 2126. [CrossRef]
9. Tan, W.; Wang, Y.; Huang, P.; Qi, Y.; Xu, W.; Li, C.; Chen, Y. A Method for Predicting Landslides Based on Micro-Deformation Monitoring Radar Data. *Remote Sens.* **2023**, *15*, 826. [CrossRef]
10. Kyriou, A.; Nikolakopoulos, K.; Koukouvelas, I. Timely and Low-Cost Remote Sensing Practices for the Assessment of Landslide Activity in the Service of Hazard Management. *Remote Sens.* **2022**, *14*, 4745. [CrossRef]
11. Foroughnia, F.; Alfieri, S.; Menenti, M.; Lindenbergh, R. Evaluation of SAR and Optical Data for Flood Delineation Using Supervised and Unsupervised Classification. *Remote Sens.* **2022**, *14*, 3718. [CrossRef]
12. Popit, T.; Rožič, B.; Šmuc, A.; Novak, A.; Verbovšek, T. Using a Lidar-Based Height Variability Method for Recognizing and Analyzing Fault Displacement and Related Fossil Mass Movement in the Vipava Valley, SW Slovenia. *Remote Sens.* **2022**, *14*, 2016. [CrossRef]

13. Festa, D.; Del Soldato, M. EGMStream, a Desktop App for EGMS Data Downstream. *Remote Sens.* **2023**, *15*, 2581. [CrossRef]
14. Yang, J.; Wang, Q.; Chang, D.; Xu, W.; Yuan, B. A High-Precision Remote Sensing Identification Method on Saline-Alkaline Areas Using Multi-Sources Data. *Remote Sens.* **2023**, *15*, 2556. [CrossRef]

remote sensing

MDPI

Article

High-Resolution Deformation Monitoring from DInSAR: Implications for Geohazards and Ground Stability in the Metropolitan Area of Santiago, Chile

Felipe Orellana [1,*], Marcos Moreno [2] and Gonzalo Yáñez [3]

1 Department of Civil, Building and Environmental Engineering, Sapienza University of Rome, 00184 Rome, Italy
2 Department of Geophysics, University of Concepcion, Concepcion 4070386, Chile
3 Department of Structural and Geotechnical Engineering, Pontificia Universidad Católica de Chile, Santiago 7820436, Chile
* Correspondence: felipe.orellana@uniroma1.it

Abstract: Large urban areas are vulnerable to various geological hazards and anthropogenic activities that affect ground stability—a key factor in structural performance, such as buildings and infrastructure, in an inherently expanding context. Time series data from synthetic aperture radar (SAR) satellites make it possible to identify small rates of motion over large areas of the Earth's surface with high spatial resolution, which is key to detecting high-deformation areas. Santiago de Chile's metropolitan region comprises a large Andean foothills basin in one of the most seismically active subduction zones worldwide. The Santiago basin and its surroundings are prone to megathrust and shallow crustal earthquakes, landslides, and constant anthropogenic effects, such as the overexploitation of groundwater and land use modification, all of which constantly affect the ground stability. Here, we recorded ground deformations in the Santiago basin using a multitemporal differential interferometric synthetic aperture radar (DInSAR) from Sentinel 1, obtaining high-resolution ground motion rates between 2018 and 2021. GNSS stations show a constant regional uplift in the metropolitan area (~10 mm/year); meanwhile, DInSAR allows for the identification of areas with anomalous local subsistence (rates < −15 mm/year) and mountain sectors with landslides with unprecedented detail. Ground deformation patterns vary depending on factors such as soil type, basin geometry, and soil/soil heterogeneities. Thus, the areas with high subsidence rates are concentrated in sectors with fine sedimentary cover and a depressing shallow water table as well as in cropping areas with excess water withdrawal. There is no evidence of detectable movement on the San Ramon Fault (the major quaternary fault in the metropolitan area) over the observational period. Our results highlight the mechanical control of the sediment characteristics of the basin and the impact of anthropogenic processes on ground stability. These results are essential to assess the stability of the Santiago basin and contribute to future infrastructure development and hazard management in highly populated areas.

Keywords: geohazard; ground deformation; landslides; DInSAR; GNSS

Citation: Orellana, F.; Moreno, M.; Yáñez, G. High-Resolution Deformation Monitoring from DInSAR: Implications for Geohazards and Ground Stability in the Metropolitan Area of Santiago, Chile. *Remote Sens.* **2022**, *14*, 6115. https://doi.org/10.3390/rs14236115

Academic Editors: Constantinos Loupasakis, Ioannis Papoutsis and Konstantinos G. Nikolakopoulos

Received: 30 October 2022
Accepted: 25 November 2022
Published: 2 December 2022

1. Introduction

Santiago basin is the homonymous capital of Chile with more than 7 million inhabitants [1]. Along with a high population density, the capital and its surroundings comprise the majority of the economic, political, and social activities of the country. Therefore, knowledge of the geohazards of the territory where the city of Santiago is located is essential for land use planning and sustainable growth. The Santiago forearc basin has been formed in a compressive tectonic environment in which the uplift of the Andes during the Miocene conditions caused the basin development in its eastern flank. The contact between the basin and the Andean Cordillera is formed by reverse faulting (San Ramon Fault). Previous geological/geophysical studies in the Santiago Metropolitan Area aimed at evaluating geohazards have focused on geology and hydrogeology [2–4], seismic potential and micro-zoning [5,6], and tectonics [7–10]. Despite advancement in the knowledge and assessment of geohazards in the Santiago basin, few studies have been carried out to analyze the ground surface stability and deformation of the whole basin. In this sense, this research focuses on studying the ground deformations of the metropolitan area using advanced DInSAR technology and ground deformation time series.

The seismic cycle of megathrust earthquakes is the main driver of ground deformation in subduction zones. This process induces long wavelength deformations with regional vertical patterns that oscillate between uplift and subsidence depending on the stage of the seismic cycle and its distance from the trench [11]. Upper plate faults define a secondary earthquake cycle system that has a direct impact on ground deformation. These faults have a long period of elastic energy accumulation (>1000 years) [12] (much longer than the recurrence of earthquakes in the megathrust), during which they induce a very low magnitude deformation of a few mm/year; this was described by [13] in the Andean region close to those faults. To date, it is unclear under which conditions upper plate faults are reactivated and by which mechanism they interact with the plate interface.

It is difficult to estimate upper plate contribution to the ground deformation due to the lack of observables at broad temporal and spatial scales and the fact that active continental faults can be blind in the sense that their location and recent activity may be unknown. Other natural drivers of subsidence and uplift include glacial isostatic adjustment [14], sediment compaction [15], and seasonal hydrological loading [16]. On the other hand, human-induced processes can cause surface level changes on a smaller spatial scale but with a faster response time. Anthropogenic ground deformations can be associated with the exploitation of groundwater [17,18] and hydrocarbons [19–21], which implies subsidence from local to regional scales. Furthermore, much more rapid anthropogenic causes may be related to urbanization, such as building load or removal of materials typically in unconsolidated alluvial deposits, which are linked to surface processes [22,23].

Land surface deformations caused by anthropogenic processes and hydrogeological phenomena, such as the compaction of aquifers, are slower processes that induce low rates of subsidence; therefore, they do not involve situations of immediate risk as their effects are observed after several years. However, over a period of several years, their effects can change the topography of the land surface, causing damage to the population and civil infrastructure. Quantitative evaluation of ground deformation can be performed based on ground instrumentation and traditional measurement techniques (related to topography/geodesy), i.e., leveling and GNSS [24]; however, these are limited in terms of producing high spatial resolution surface displacement maps over wide areas. DInSAR technology is an alternative solution totally assimilable to terrestrial monitoring [25]. Ground deformation monitoring with DInSAR takes advantage of the amount of available data that is acquired more frequently and accurately at low cost, and such characteristics make it an attractive source of information [26]. One of the main attractions of satellite-based DInSAR is its ability to cover areas at a systematic and continuous rate remotely, which makes it suitable as a structural monitoring and control tool as we can detect near-vertical deformations in the structures along the LOS (line of sight). In recent years, the DInSAR time-series technique has emerged as an essential tool for measuring slow surface displacement [27]. This technology has been widely exploited in a wide variety of contexts such as seismic deformation [28,29], volcanic and landslide monitoring [30–32], infrastructure stability studies [26,33,34], and water overexploitation surface effects [35–37].

DInSAR technology is essentially based on two approaches associated with the selection of coherent pixels: the Persistent Scatterers (PS) technique, developed by Ferretti et al. [38], and the SBAS (Small BAseline Subset), developed by Bernardino et al. [39]. These processing techniques involve multiple time-dependent acquisitions to provide characteristic displacement patterns of surface motion over a period of time, thus providing measurement of surface uplift or subsidence to identify cyclical patterns (due to seasonal variations), trends, and anomalous variations. Such techniques can be applied to collect time series of movements of the earth's surface over wide areas with millimeter precision [40–42]. The evolution of DInSAR techniques observed in recent years is mainly related to the development of advanced computational algorithms [43]. However, it is also the result of greater possibilities for the acquisition of radar images by satellite missions and their high review frequency of 6 to 12 days for the entire world. Of particular note is the European Space Agency (ESA) Copernicus mission material that we use in this work, consisting of two twin SAR satellites, Sentinel 1A and Sentinel 1B [44].

This research quantifies the temporal evolution of ground motion through the analysis of the SAR interferometry with geophysical and geological data characterizing the features of the Santiago basin. The methodology of this study uses DInSAR differential interferometry with ground-based information. We use a C-band Sentinel 1 to process multiple SAR images with the P-SBAS (Parallel Small BAseline Subset) algorithm [43,45,46], an evolution of the traditional SBAS (Small BAseline Subset) method [39] developed by the CNR—IREA (Institute for Electromagnetic Sensing of the Environment). For the processing of SAR images, we used the iCloud platform (GEP) [47–50]. The results presented here provide new information that identifies the stability of the city of Santiago and constrains subsistence-inducing factors, such as hydrogeological phenomena, due to the overexploitation of water and stability of slopes in areas susceptible to landslides.

2. Present Day Regional Deformation and Geological Setting

The subduction of the Nazca Plate beneath the South American Plate controls the seismic cycle of large earthquakes and contributes to the permanent deformation that shapes the main morphological features on the Chilean margin. The Santiago basin is located at about 520 m above sea level at the piedmont of the Andean Precordillera, which reaches over 4000 m in altitude (Figure 1b). Santiago lies in the central segment of the Chilean subduction zone, between the rupture zones of the Maule 2010 [51] and Illapel 2015 [52] earthquakes. This segment is considered a seismic gap, in which the last major earthquake that ruptured the entire segment, both along and across the megathrust, was in 1730 [53]. Velocities derived from GNSS observations between 2018 and 2021 show typical interseismic deformation patterns (Figure 1). In this period, Central Chile moved northeastward with magnitudes of ~20 mm/year near the coast, decreasing to ~10 mm/year in Santiago (Figure 1c). Vertical velocities show subsidence near the coastline, suggesting that the offshore megathrust zone is locked, as previous studies have shown [54]. The Santiago basin and its surroundings show a regional uplift trend of about 10 mm/year. The velocity field shows no horizontal or vertical gradients near the San Ramon fault, so no activity or movement related to this fault are observed.

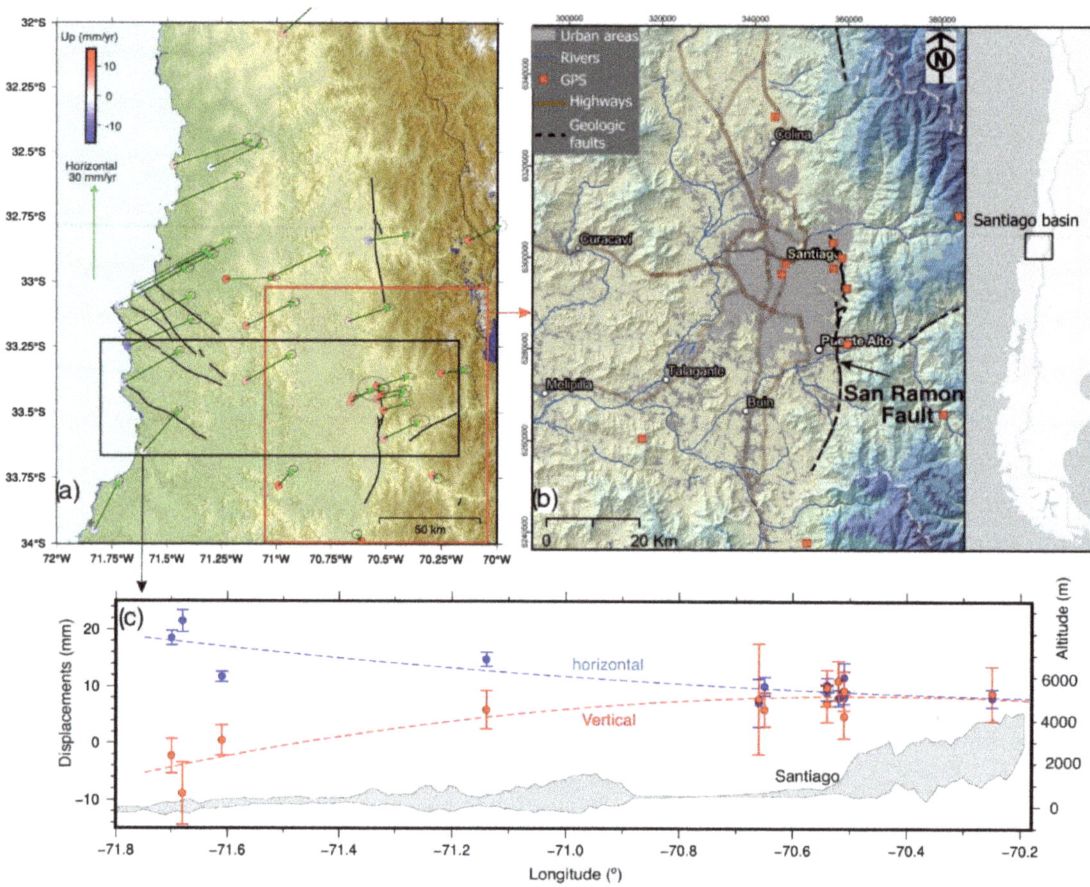

Figure 1. (**a**) Central Chile region showing the GNSS velocity field and traces of active faults (from Maldonado et al. [55]). The red and blue rectangles indicate the corresponding areas of Figures (**b**,**c**), respectively. (**b**) Geographical location of the metropolitan area of Santiago de Chile, ALOS DEM 30 mts, was used as background. (**c**) Cross-section of horizontal (blue) and vertical (red) velocities derived from GNSS data (from Donoso et al. [56]). A swath profile of the topography is shown in gray.

The Santiago forearc basin developed during successive tectonic events divided into two stages. The first stage of the Middle Eocene to the Oligocene-early Miocene was characterized by an extensional setting, which created depocenters where the Abanico Formation accumulated [57]. A second stage of compression linked to an increase in the rate of plate convergence velocity [58] dominated the evolution of the Abanico Basin until the Miocene. This produced the change to a compressive regime, with a partial tectonic inversion of the Abanico basin from the late Oligocene to early Miocene, inverting the larger NS normal fault systems generated in the first stage [59–63]. The Coastal Cordillera is located at the western flank of the basin, which is composed of Jurassic to Late Cretaceous volcanic and sedimentary sequences and Jurassic to Cretaceous intrusive rocks. At the eastern border of the basin, the foothill of the Andean Cordillera corresponds to a thrust deformation front that causes the uplift of the Andean Cordillera [64] and at the same time, the erosional process that provides the sedimentary supply for the basin infill.

Given the high-energy mountain erosional process, most of the basin deposits correspond to coarse gravel material (Figure 2a); however, in the northern part of the basin, some fine sediments associated with low-energy lacustrine-type deposits are still observed (Figure 2b). Based on gravity measurements constrained by geological observations at wells, it has been possible to estimate the thickness of the sedimentary cover as being in the range of 100–400 m with an irregular morphology [3]. The role of the eastern deformation front, whose westernmost branch correspond to the San Ramon Fault System (SRFS), has been a source of large scientific debate in terms of its relevance as a seismic hazard source [7–10,65]; maximum seismic events and recurrence times are among the most relevant unresolved questions. Finally, the basin water table is controlled by surface topography (tilted to the west), asymmetric basin (shallowing to the west), the overexploitation of the water resource, and the limited recharge due to a prolonged drought [66]. Rates of water table descent in the last 10 years range between 1–0.3 m/year, probably due to a combined effect of the lack of recharge and overexploitation. These first order geological characteristics of the Santiago Basin are used in this study to explain the surface deformation derived from the DInSAR observations.

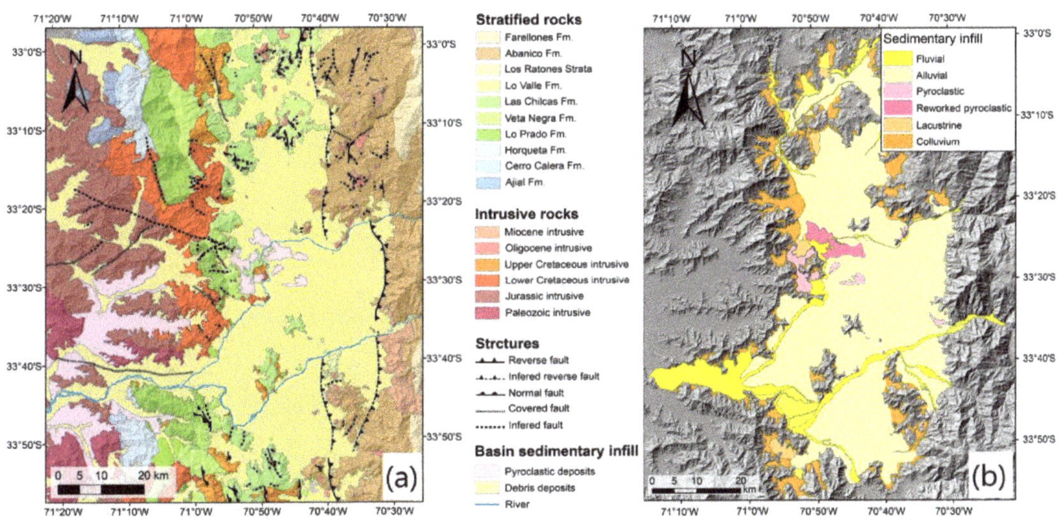

Figure 2. (**a**) Geological map and (**b**) Sediment map of Santiago Basin (From Yanez et al. [3]).

3. Materials and Methods

The methodology describes the study of the deformation in the Santiago Metropolitan Area during an observation period of 3 years (2018–2021), using the multitemporal differential interferometry technique (DInSAR). The data obtained with the interferometric radar is complemented by a contextualization of the deformation obtained with the GNSS stations and, furthermore, by available geological and hydrogeological studies to propose a plausible interpretation of the observed anomalous domains. We integrate the results of the DInSAR analysis with the geological background, which allows us to describe the processes responsible for the detected anomalies. Thus, we present detailed high-resolution maps of the entire Santiago Basin, where records of ground deformation were previously lacking. In addition, the long time series with DInSAR and the high spatial density of PSI (Persistent Scattering Interferometry) can undoubtedly contribute to urban planning in Chile.

The main phases of the workflow include: (1) SAR image processing to generate interferograms, (2) SAR time series analysis using (3) vertical and slope velocity estimation in the post-processing, (4) overlaying with geological and hydrogeological information, and Digital Elevation Models (DTM) (5) to interpret the processes responsible for the anomalies detected.

3.1. Data Set

We have used 204 interferometric C-band VV polarized wide fringe (IW) SAR scenes acquired by the Copernicus Sentinel-1 mission [67] along ascending and descending orbits (Table 1). The review time for each stack is 12 days, with a resolution of 5 (terrestrial range) by 20 (azimuth) and angles of incidence (θ) of 31 to 46°. We use the IW-2 swath width and bursts 2–4 to cover the Santiago Basin area of interest. Through the GEP platform, we retrieved updated SAR images from the ESA Open Access Hub repositories as a Single Look Complex (SLC) interferometric product [68]. The advantages of Sentinel-1 come from its wide range coverage (250 km swath in wide interferometric mode) and sufficient spatial resolution for large areas (90 m × 90 m range vs. azimuth, for this case). The Wide Interferometric Fringe (IW) acquisition mode is based on the ScanSAR terrain observation mode and the use of interferometric fringes with progressive scans (TOPS).

Table 1. Data set including the main features of Sentinel 1.

Orbit	Ascending	Descending
Sensor	1B	1B
N° acquisitions	104	98
Date of measurement start	10 September 2016	10 September 2016
Date of measurement end	30 December 2021	30 December 2021
Relative orbit	156	156
Polarization	VV	VV
Swath	IW-2	IW-3
Bursts	2–3	5–6

3.2. Data Processing

The processing has been based on the Advanced Earth Sciences Cluster operated by Terrafirma with a duration of approximately 48 h of run-time. We adopt SAR processing running on ESA's next-generation GEP iCloud platform, "CNR-IREA P-SBAS Sentinel-1 on-demand processing" service v.1.0.0, implemented in the computer-based operating environment ESA GRID [43]. The processing approach is based on the SBAS technique [39], applied along the descending orbit of Sentinel-1B (C-band SAR sensor wavelength = 5.6 cm). The algorithm was adapted to run efficiently on high-performance distributed computing and configured for Sentinel-1 IW TOPS data processing [45].

The main processing steps of the SBAS method [39] consist of the generation of differential interferograms from the SAR image pairs formed with a small orbital separation (spatial baseline) to reduce spatial decorrelation and topographic effects. The Shuttle Radar Topography Mission (SRTM) [69] with 1 arcsecond DEM from NASA (~30 m pixel size) and precise orbits from the European Space Agency (ESA) were used for the joint registration and removal of the topographic signal from the interferometric phase in each of the computational interferograms.

Each SLC data stack was co-recorded at the single burst level, ensuring very high co-registration accuracy (on the order of 1/1000 azimuth pixel size), as required for TOPS data due to the great Doppler centroid along the path variations [70]. The temporal consistency and the threshold of the minimum temporal consistency was set at 0.85. Atmospheric phase components were identified and removed. The control point for the P-SBAS processing was established in the same place in the city of Santiago center with coordinates lat: −62.555, log: −35.172, where the annual LOS velocity values were taken, and the time series were referenced accordingly. The use of a common reference point allowed internal calibration of the two output data sets.

3.3. Post-Processing

With the persistent scatterer interferometry (PSI) dataset, we project the vertical velocity and the velocity along the steepest slope estimates using the principles mentioned below.

The estimates of the vertical velocity component have been made with the combined ascending and descending data sets to obtain the vertical displacement V_u. A 90-m square element network was used to resample the point data sets on a regular grid and link the output data sets into a single layer. Both P-SBAS outputs (ascending and descending) were available at the same location, *i*; the combination was achieved under the assumption of negligible north−south velocity, $V_N = 0$. This assumption is typically used in DInSAR studies to consider the relatively poor visibility of the north−south horizontal motions that the LOS sensor can achieve [68]. The horizontal E−W displacement components in the Santiago basin region are much lower than the vertical ones; therefore, they were not considered for our case.

Given the known values of the deformation velocity LOS in the ascending (V_{asc}) and descending (V_{dsc}) mode at each location *i*, the V_U is estimated as follows:

$$V_{ui} = \frac{E_{Di} * V_{Ai} - E_{Ai} * V_{Di}}{E_{Di} * U_{Ai} - E_{Ai} * U_{Di}} \quad (1)$$

The estimation of velocity deformation along the Steepest Slope Direction defines zones potentially affected by landslides, considering the geomorphological principle upon which deformation is more likely to occur along the slope direction [71]. This is true for translational landslides and other types of movements present in some sectors of the study area, such as the debris flow. Under the assumption that the displacements occur along the direction of the steepest slope, they were projected considering the local values of slope (β) and aspect (γ) from the DTM ALOS PALSAR 30 mts. These data were used to identify the orientation of the steepest slope of the Andes mountain range, where the value represents the conversion factor of the LOS to slope values. *E*, *N*, and *U* are the directional cosines of the LOS and the SLOPE vectors in the east, north, and zenith directions, respectively, and they are defined as follows:

$E_{Si} = sin\ \gamma_i\ * cos\ \beta_i\ ,\ N_{Si} = cos\ \gamma_i\ *\ cos\ \beta_i\ ,\ and\ U_{Si} = -sin\ \beta_i$. The velocity along such direction (V_{Si}) was then estimated as follows (Cigna et al. [72]):

$$V_{Si} = \frac{V_{LOSi}}{E_{LOSi} * E_{Si} + N_{LOSi} * N_{Si} + U_{LOSi} * U_{Si}} \quad (2)$$

4. Results and Discussion

GNSS velocities allow us to characterize the regional deformation field in Central Chile. These data show widespread uplift in and around the Santiago Basin. However, the distribution of GNSS stations does not allow the identification of local uplift and subsidence. Hence, DInSAR data are an excellent complement to improve the spatial resolution of land-level changes and explore features of local deformation. The results of the SAR image processing provide detailed information to represent the phenomenon of ground deformation over the study area. The data set contains the Persistent Scatterers Interferometry (PSI) data, where each PSI contains a LOS displacement time series; average LOS velocity; temporal Coherence > 0.85; average elevation of the scatterer (topography); and the unit vectors that are the directional cosines of the LOS in the east (*E*), north (*N*), and zenith (*U*) directions, respectively. These were then used for the vertical and velocity along the steepest slope projections in the post-processing phase.

The data set output format is a CSV file, according to the specifications of the European System of Plate Observation—Phase Implementation (EPOS—IP), where the metadata corresponding to the LOS velocity in raster (.png) and Google Earth (.kml) are standardized. With these contents, we georeferenced GIS-based cartography (WGS84—UTM zone 19S) to represent and provide evidence for the phenomenon of deformation over the study area. Our results present an overview of the ground deformation for both orbits, selecting the

PSI of the Santiago Basin area and identifying the local subsidence areas. In addition, we present the displacement map of the estimated vertical components based on the projections of the ascending and descending LOS and the displacement along the maximum slope in the Andean Cordillera area, resampling the LOS velocity.

4.1. PSI Measurement and Classifications

PSI measurements from 2018–2021 have been classified according to deformation velocity rate (mm/year), with a continuous color scale that varies from dark green to dark red. Negative velocity values indicate motion away from the satellite sensor (orange to dark red PSI), while positive values indicate motion toward the sensor (light to dark green PSI). Although we have not used the maximum resolution of the Sentinel 1 (IW) images, the PSI are spaced in 90 × 90 grids, the interferometric data deliver a high density of points, reaching 100 PSI/km^2 in urban areas and PSI 60 PSI/km^2 in rural areas where it is usually covered by more vegetation, which decreases SAR detection. For both cases, we have appointed an optimal data set to represent the phenomenon of deformation in the study area. With the Persistent Scatterer Interferometry (PSI) dataset, we project the vertical velocity and the velocity along the steepest slope estimates using the principles mentioned below.

We obtained the PSI LOS deformation maps for the ascending and descending orbits, covering the entire study area, the Santiago basin, and the western flank of the Andean Mountain range (see Figure 3). An indicator of the compatibility between acquisition geometries is the standard deviation of the mean LOS strain rates, which correspond to 0.47 mm/year and 0.55 mm/year for the ascending and descending orbits, respectively. Given the difference in the area covered by each orbit, the relatively greater number of PSI points in the ascending solution could be explained by the morphology of the area and the angle of incidence on the mountainous cover. Apart from the effect of area coverage, it appears that the total number of PSI targets is comparable. This similarity can be attributed to the common observation period and the ascending and descending data sets. PSI LOS velocities in 2018–2021 ranged from −25.56 to +2.73 cm/year in the ascending dataset and from −28.79 to +3.05 mm/year in the descending dataset (see Table 2).

Table 2. Basic PSI statistical comparison for ascending and descending orbits.

Orbit	18 (Ascending)	156 (Descending)
N° PSI	240.066	259.012
Min. (mm/year)	−25.56	−28.79
Max. (mm/year)	+2.73	+3.05
Media (mm/year)	0.3	0.2
Standard deviation	0.47	0.55

4.2. Deformation Overview and Vertical Displacement

The deformation overview in the Santiago Metropolitan Area, was calculated with the vertical projection using ascending and descending LOS. Thus, we resampled the data considering the similarity of both data sets from both orbits. Vertical velocity component estimates were calculated after combining the two data sets where a 90 square meter grid was used to resample the point data sets on a regular grid and link the output data sets. This single digital layer in SHP format allowed us to create a new vertical deformation map (see Figure 4a).

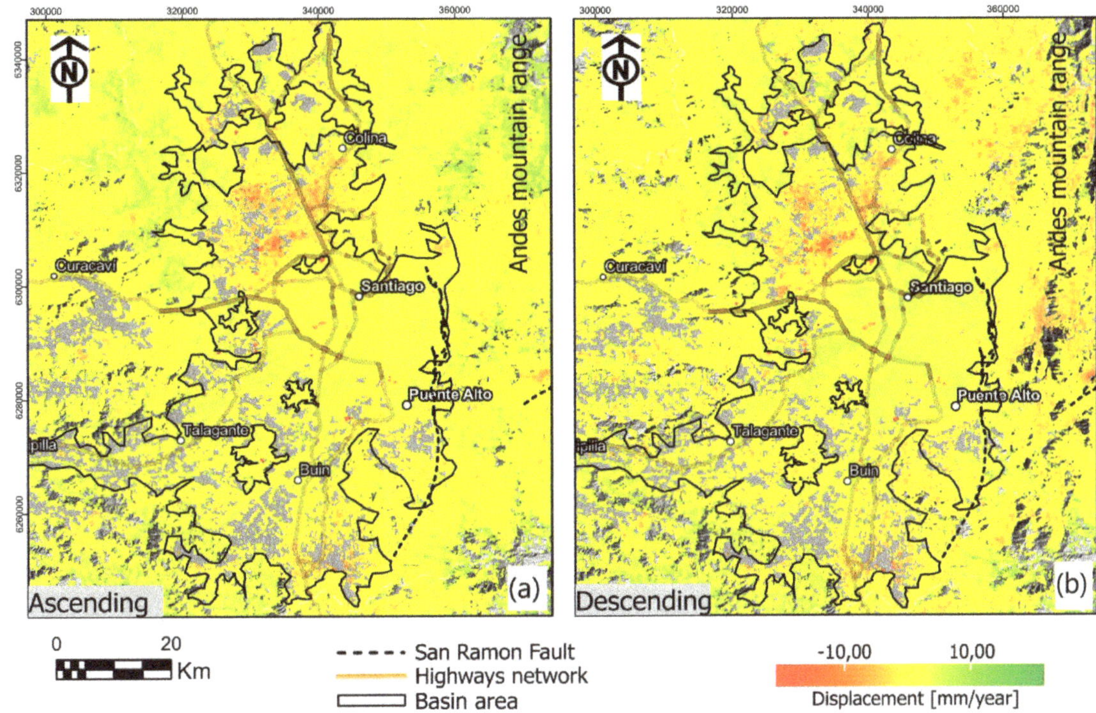

Figure 3. (**a**) LOS displacement maps for the study area, (**a**) Ascending orbits (A18), and (**b**) Descending orbits (D156), with ALOS DEM 30 mts used as background.

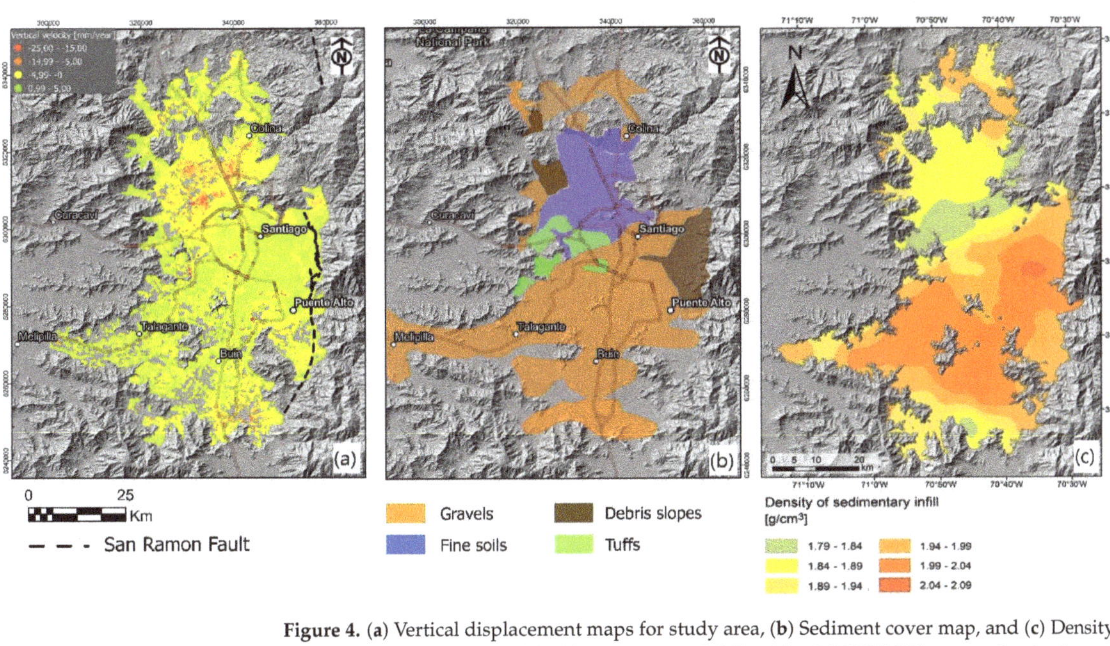

Figure 4. (**a**) Vertical displacement maps for study area, (**b**) Sediment cover map, and (**c**) Density of sedimentary infill map (modified from Yanez et al. [3]), with ALOS DEM 30 mts used as background.

The median ground stability is represented by the yellow PSI and varies between −4.99–0.00 (mm/year), indicating a general trend of negative velocities that indicate a pervasive subsidence. On top of this general trend, we found two anomalous domains in which the subsidence rate is much larger: a) The northern domain, over a NE elongated surface of more than 500 km^2 (Quilicura, Chicureo, and Colina localities), in which subsidence up to −25.00 mm/year is observed (Figure 4a)the southeastern domain, along a more restricted surface of 250 km^2 (Paine, Huelquen localities), with subsidence in the order of −15.00 mm/year. The two anomalous domains of relatively large subsidence rates are associated with fine deposits of lacustrine origin in the northern domain and gravel deposits in the southern domain (Figure 4b). However, in terms of the deposit densities, both anomalous domains share the presence of low-density deposits in the southern region, probably associated with the distal disposition with respect to the Maipo River. On the other hand, these two anomalous domains show different soil use throughout the last few decades; while the southeastern domain is still dedicated to two agricultural activities, the northern domain has experienced a rapid transition to urban areas. In Section 4.3 we interpret these anomalous domains in terms of the interaction between soil characteristics and the spatial/time evolution of the groundwater processes in the basin.

Finally, the most stable soils are in Santiago's center and they are shown in the PSI of green colors with values between 0.99–5.00 (mm/year). For this area, a slight upward trend can be observed and this is probably caused by the tectonic processes that affect the Santiago Basin regionally. For the area near the San Ramon fault, no indication of fault activity was observed in the analyzed time window; however, an uplift caused by the same regional tectonic effects as those in the central area of Santiago is observed.

4.3. Ground Deformations and Time Series

Multi-temporal satellite radar interferometry is based on the analysis of a series of SAR (Synthetic Aperture Radar) images, which, in our case, were acquired in the period between May 2018–May 2021. By selecting the most stable targets in the anomalous areas (PSI > 0.85) that maintain the electromagnetic scattering signature in each image, it is possible to measure ground displacements by exploiting the phase shift (sensor–target distance) and the amplitude of the signal reflected from the ground surface.

The product generated by this multi-interferometric analysis is a cartographic representation of the time series, with a cumulative displacement value of up to −100 (mm) for isolated PSI targets and values of −74–50 (mm) for homogeneous areas (see Figure 5). The records of interferometric data for both orbits allowed us to know the temporal evolution of the deformation phenomenon in the study area and build the time series in the anomalous areas. For this we have selected 8 pixels of the study area, numbered from 1–8, where each one is assigned the name of the locality where the deformation phenomenon occurs.

Deformation time series represent the most advanced DInSAR product. They provide the history of the deformations during the observed period, which is fundamental to represent and study the ground deformation in the study area and its correlation with the inducing factors. To properly use, interpret, and exploit deformation time series, it is important to consider that they can be affected by geometric and atmospheric distortions. In fact, they contain an estimate of the deformation for each acquisition (SAR image), so they are particularly sensitive to phase noise [40].

Following the results in the deformation area (Pixels 1–8), we focused on a quantitative analysis, comparing each of the ascending and descending time series, based on the parameters of velocity and R2. We present graphs in Figure 6, which show the time series comparisons of the orbits of the interferometric data developed in the processing phase of the images of Sentinel-1, corresponding to LOS, ascending (blue color), and descending orbits (red color). According to the above results, there is very good agreement between the measurements. However, the difference in the SAR-derived time series 7 and 8 (Figure 6) between the ascending and descending orbits is a consequence of the distances between the pixel's PSI, which reach up to 100 m so the pixels can detect different local deformations.

For the time series 1, 2, 3, 4, and 6, the values of R2 (Figure 6) are close to 1.0 with a good fit; therefore, the results of both orbits are very mutually reliable, indicating a negative displacement trend (subsidence). For time series 5, the values of R2 are close to 0, and this is similar for both orbits, indicating a stable deformation. In time series 7 and 8, the R2 values are also close to 1.0; however, there is a difference due to the distance between the PSI pixels, indicating a trend of seasonal subsidence.

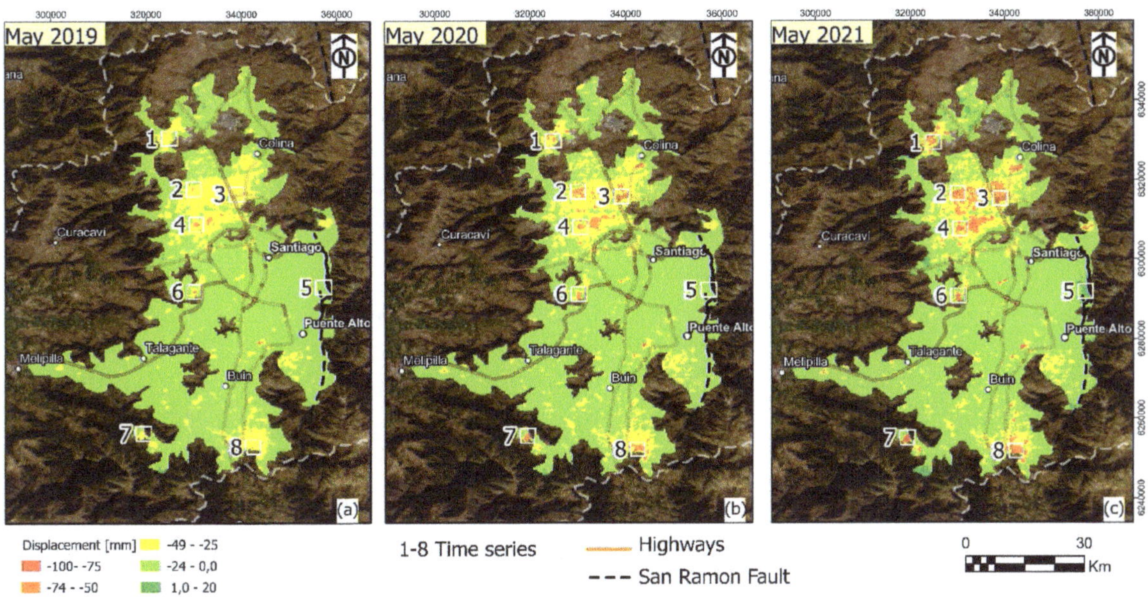

Figure 5. Map of cumulative displacements and temporal evolution (**a**) Recorded in May 2019 (**b**), Recorded in May 2020, and (**c**) Recorded in May 2021, with Maxar image (source Esri) used as background. Numbers in each panel label the localities in which the deformation time series response is described in the text of Section 4.3.

The time series of displacements show values of up to −60 (mm) for the period of May 2018–May 2021 or ~20 mm/year. The greatest deformations are located in the northern area of Santiago, in the towns of Polpaico (Time series 1), Lampa (Time series 1), and Quilicura (Time series 3–4). On the other hand, the area of the San Ramon fault (Time series 5) appears stable and there is no evidence of activity in the fault. In Maipu west (Time series 6), an anthropogenic deformation caused by the increase in urbanization in the area is observed. The wetland "Aculeo" (Time Series 7) is close to the "Aculeo" lake, which has been drying up dramatically [73,74]. Finally, for Paine (Time Series 8), an area of intensive agriculture shows an increase in subsidence in recent years.

4.3.1. Subsidence and Groundwater Spatial and Temporal Evolution

Given the spatial relationship between anomalous deformation, mostly subsidence, soil characteristics, and the known increase of groundwater exploitation, both for agriculture and for human consumption, were explored in this section with the working hypothesis that the main factors that cause the deformation of the Santiago Basin are linked with changes in groundwater flow and static levels.

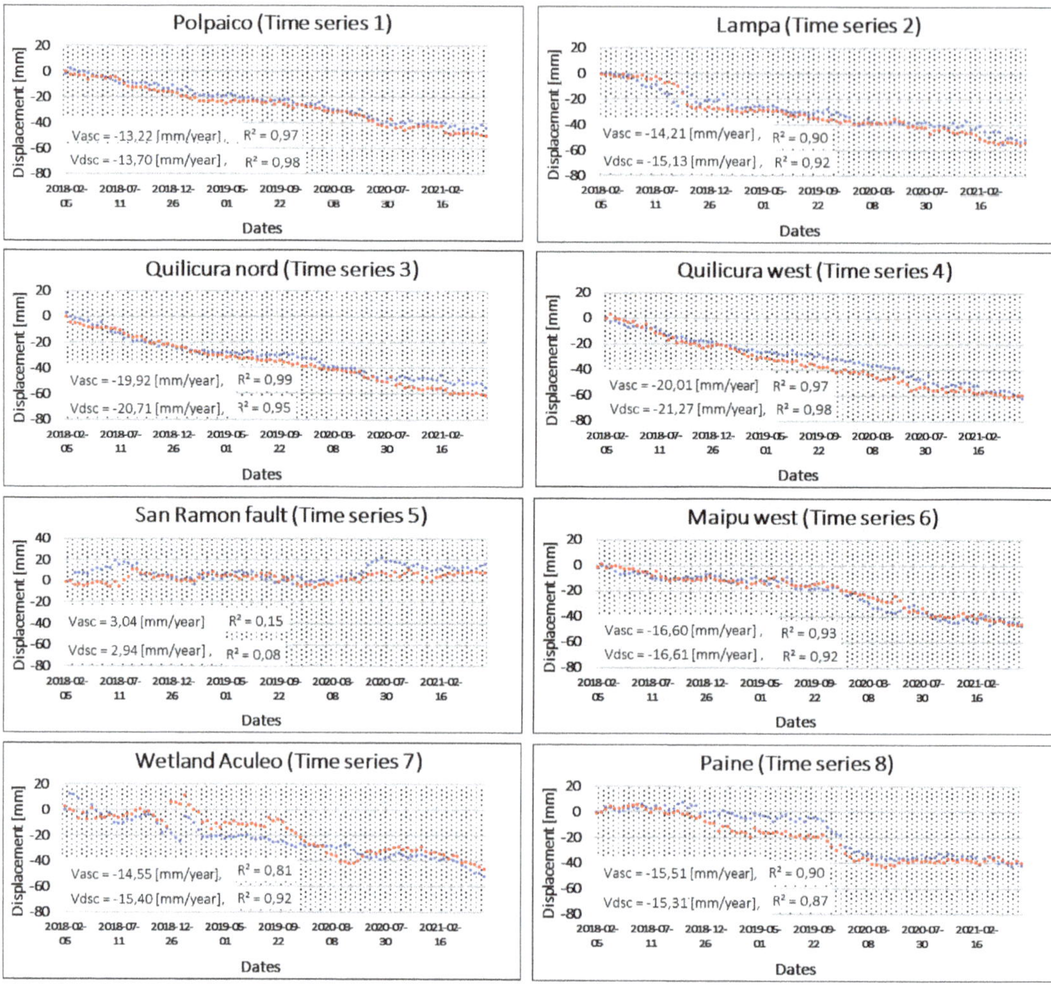

Figure 6. Time series 1–8 in red ascending orbit and blue descending orbit, located at Polpaico, Lampa, Quilicura nord, Quilicura west, San Ramon fault, Maipu west, Wetland Aculeo, and Paine.

In order to determine the water table of the Santiago Basin, we include a water table grid produced by DGA (2000) [75] in Figure 7b, which is a compilation of data up until the year 2000. This map shows that the anomalous DInSAR domains (Figure 7a) are located in shallow areas of the water table. In addition to that, and with the aim to validate this water table map while exploring the time evolution of the water table at the same time, we analyzed 48 wells within the area of interest. From these wells we extract time series evolution of the water table at a rate of 2–4 points per year. From this basic data we calculate a trend to have the gradient per year and an average value for the last 10 years. Comparing Figure 7b,c, we conclude that the DGA 2000 map is basically correct and still valid today, with the deeper water table in the central and eastern part of the basin, and the shallower ones in the northern, southern, and western flanks. The majority of the wells show a decrease in its water table, with some of them in excess of 1–2 m/year; however, in the central part of the basin (Quinta Normal), the well shows that the water table indeed becomes shallower (at rates greater than 1.5 m/year). A qualitative comparison between water table evolution and deformation of the surface according to

DInSAR show an apparent inconsistency, zones with large water table oscillation show no or minor deformation, i.e., in the central part of the basin. However, a closer look shows that, in general, these areas are associated with zones of deeper water table and coarse soil in the area of high energy of the fluvial system. In order to gain a better understanding of the role of groundwater dynamics in surface deformation, we need to consider some basic concepts of pore elasticity.

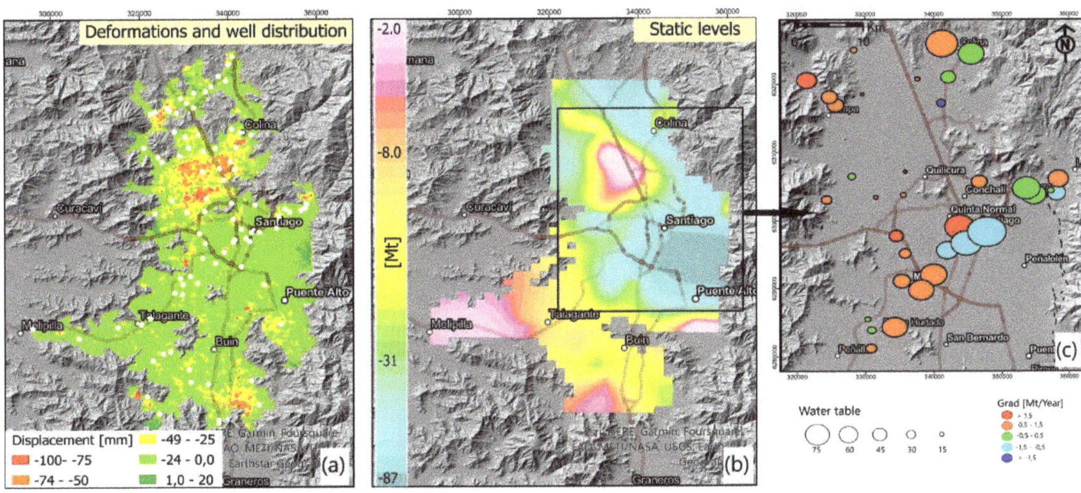

Figure 7. (**a**) Ground deformation map and well distribution in white point color, (**b**) Static level map (modified from DGA 2000), and (**c**) Piezometric variations maps, with ALOS DEM 30 mts used as background.

According to the pore-elastic model developed by Terzaghi, K. [76] and Biot, M. [77], the variation in effective stress (geostatic stress minus pressure head) is related to changes in the hydraulic head (Δh). Then, if such a stress field is applied over a compressible matrix aquifer, they result in a matrix deformation, including surface deformation (Δu). For a simple one-dimensional compaction model, Terzaghi, K. [76] established a linear relationship between (Δu) and (Δh), where the constant of proportion is the skeletal storage (Sk), such that $\Delta u = Sk * \Delta h$. The main factors that control skeletal storage are the compression index (Cc) and the effective stress and the direct and reverse dependence. Effective stress increases with depth, so larger effects are expected in shallow aquifers. Compressional index depends on the granulometry of the soil: fine soil (i.e., clay and silt) show larger Cc, whereas coarse and more rigid soils (i.e., gravels) show Cc values practically zero, regardless of any other factor.

In the Santiago Basin, the DInSAR show a major subsidence effect in the northern anomalous domain (Figure 7a), in good spatial agreement with the region in which the static level is shallower than 10 m (Figure 7b) and where fine sediments of lacustrine origin are emplaced (Figure 4b in Yanez et al. [3]). The southeastern anomalous domain is also located in a shallow water table region, where gravel sediments show low-density, and thus distal, fine sediments. On the other hand, most of the central and southern part of the basin is dominated by the high-energy gravel material, which is highly rigid, having an almost zero compressional index. In addition to that, the water table in this region shows depths deeper than 50 m (Figure 7b,c). Both factors correctly predict a minor or null surface deformation, as observed in the DInSAR data. The $\Delta u/\Delta h$ ratio, which is the skeletal storage in this case, is in the order of 0.01 in the anomalous domains. These values are in good agreement with observations in central Mexico [78], Kumato area, Japan [79], and Tucson Arizona [80].

4.4. Landslide Identification

In recent decades, the metropolitan area of Santiago has experienced sustained growth in urbanization towards the Andes Mountain range, where landslide activity is more frequent. The occurrence of landslides in the front of the mountain and on the slopes of the interior basins and, in particular, debris flows that can reach the alluvial plain are common and represent an increasing risk for populated areas [81,82]. A large debris flow event in 1993 [83,84] is an example of the potential impact of catastrophic landslides in the area. The most common types of landslides in the area are debris flows that occur from the mountain range towards the city, rock falls from steep and fractured slopes, and rock slides [85].

To identify landslides, it is necessary to identify the geomorphological units, because they have different conditions that can define a type of landslide, in addition to the fact that the geomorphological unit is favorable for one type of landslide but not for another. For this reason, we have focused the analysis using the susceptibility map (Figure 8a) to mainly debris and rock flows. We also considered the high coherence of the SAR in highly reflective rocky areas. To compare our PSI interferometric data, we used a landslide susceptibility map [86], which shows areas that have the potential for landslides (see Figure 8a), determined by the correlation of some of the main factors that contribute to landslide generation. The area is affected by the main geometric distortions of shadows, layover and foreshortening, where the differences between the ascending and descending orbits are evident and can be seen in Figure 8b,c.

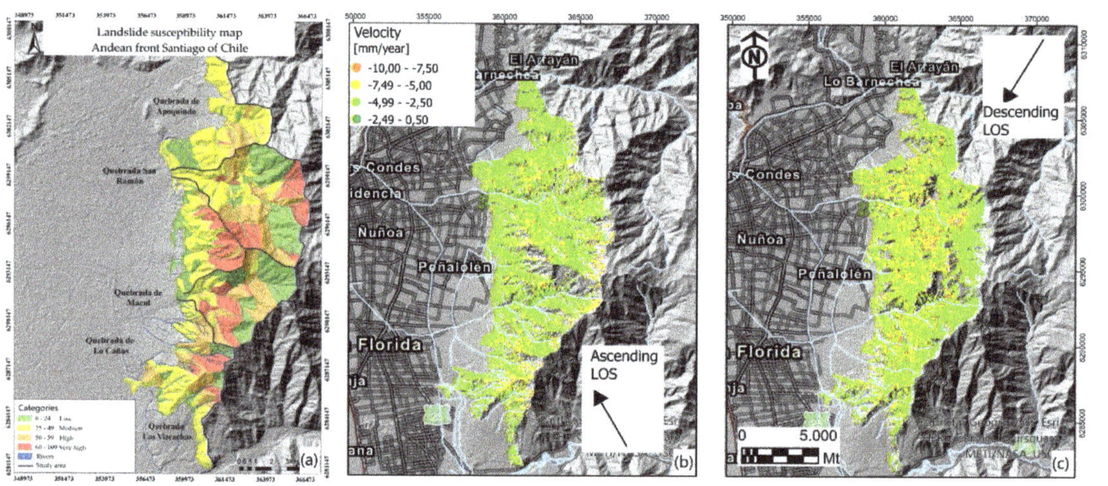

Figure 8. (a) Landslide susceptibility map (modified from Celis, C. [86]), (b) LOS displacement map for ascending orbit, and (c) LOS displacement map for descending orbit, with ALOS DEM 30 mts used as background.

From the geometric relationship, it can be known that in the area of visibility, the closer the slope is to the angle of incidence of the satellite, the more LOS deformation can reflect the real deformation; under this assumption we have selected the descending LOS to continue our analysis. A one-dimensional component along the LOS is suitable for rotational landslides, but not translational landslides. In the study area, the most representative processes are debris flows, which have parallel movements along the direction of the steepest slope. In this case, we project the slope along the steepest slope under the formula proposed by Cascini et al. [71], Colesanti and Wasowski [87], and Plank et al. [88]. The slope and aspect values were calculated in GIS and then used in the calculation of the estimated velocity strain along the steepest slope velocity direction (Vslope) in the post-processing phase, where the velocity targets were recalculated. PSI is presented in Figure 9c. For

Vslope projection, we consider the Slope (β) > 40° and the values of the slope directions aspect (γ) in the northwest, southwest, and northeast, located in the white box (Figure 9b). We notice that Vslope velocity ranges increased along the slope and were mostly recorded on rocky slopes and soils located on higher slopes (Figure 9a). Thesse results confirmed the correspondence of the most susceptible landslides in the highest areas of the ravines (Figure 8a) and the usefulness of the DInSAR approach to identify regions susceptible to landslide processes.

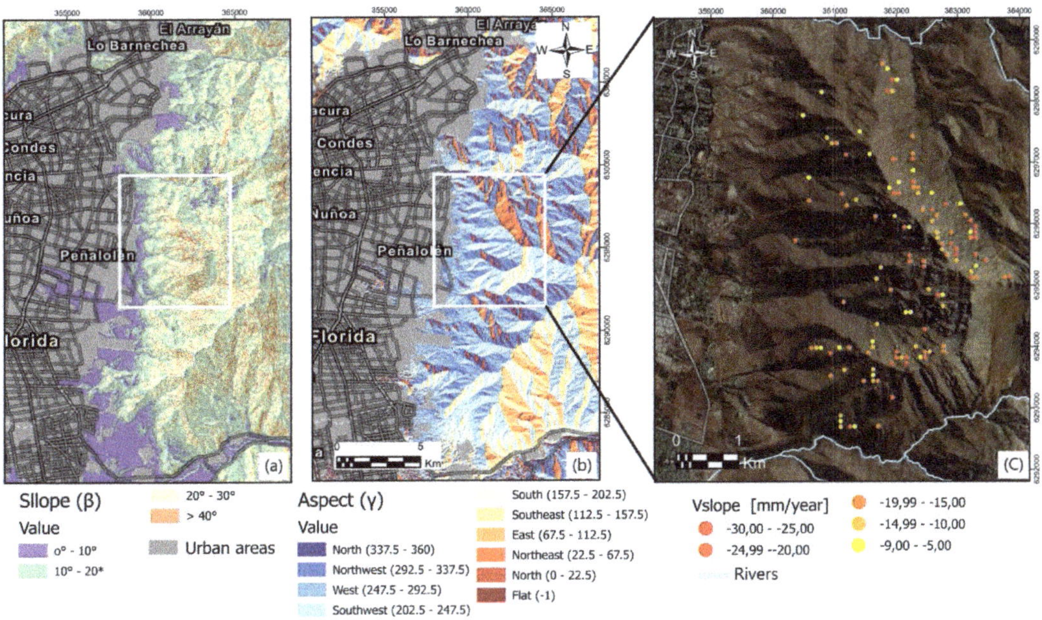

Figure 9. (**a**) Slope map with the highest values in pink color, (**b**) Aspect map indicating the values of the direction of the slopes, with ALOS DEM 30 mts used as background, and (**c**) Velocities along steepest slope (Vslope map), with Maxar image (source Esri) used as background.

5. Conclusions

The metropolitan area of Santiago de Chile is exposed to numerous natural and anthropic processes that induce regional and local deformations that affect the stability of the basin and its surrounding foothills. GNSS-derived velocities show that the Santiago Basin is tectonically stable, showing typical patterns of interseismic deformation, such as slow eastward motion and uplift. DInSAR allowed us to quantify with excellent spatial resolution areas with local subsidence within the Santiago Basin and landslides in mountainous regions. This allowed us to map the Santiago Basin's ground stability and investigate the mechanisms responsible for these surface-level variations. Therefore, our study highlights the use of the SAR interferometry technique for the detection of subsidence caused by groundwater exploitation and geohazards related to landslides. Our estimates of surface level changes in the Santiago basin will also be of great importance for urban planning; these results emphasize the mechanical effect of sediment thickness on surface stability and demonstrate how groundwater extraction induces significant land subsidence.

In general, DInSAR shows that the ground surface level in the Santiago basin is relatively stable, but there are areas showing anomalous subsidence. These areas are located in places where groundwater is exploited at the expense of non-renewable storage in aquifer systems, which causes land subsidence and other environmental impacts such as the degradation of water quality, spring discharge, and flow reduction from the rivers.

In Quilicura, north of Santiago, the exploitation and compaction of the aquifer is more evident, and a criticality is noted. For Paine, the southern area of Santiago, the deformations have been more evident in recent years, which are influenced by intensive agriculture in the area. On the other hand, although the area is located in the Chilean subduction area, there is no evidence of large deformations influenced by tectonic movement, in particular, movement linked to the San Ramon fault activity. However, we do not rule out that the fault is active and that more observation time is needed to estimate potential deformations across the fault.

Our results show a constant evolution in the subsidence of the anomalous zones, indicating that water withdrawals have continued to affect soil stability during the observed period. This work also provides semi-theoretical relationships in Section 4.3.1 to link information on metropolitan-scale groundwater use with compaction and storage loss, which could allow predictions of subsidence rates and volumes for different groundwater management scenarios.

In the reliefs of the Andes mountain range, using SAR interferometry, we were able to record information in places with difficult access, quickly and efficiently, and detect numerous control points obtained with the SAR sensor. Based on the Vslope projection, we were able to estimate the velocities and determine the status of the movement of the slopes, comparing them with the susceptibility catalogs of landslides to obtain reliable results. Our results highlight the potential of interferometry to identify landslides; however, more information, such as materials, type of geomorphological units, and meteorological data, is required in situ in order to confirm and define more precisely the activity of the phenomenon and the types of movements present in the area. Our study demonstrates that InSAR provides a spatial accuracy that allows for the detection of ground instabilities in areas above ~500 m^2. The interferograms have an excellent consistency in the analyzed urban areas of the Santiago basin. In mountainous regions there may be topographic effects that incorporate noise into ground motion estimates, which is why we limited ourselves to using only the descending orbit for landslide identification. Furthermore, the integration of geological information enabled us to interpret the observed ground elevation anomalies in the Santiago Basin. Our results highlight that the thickness of the basin is a key factor for ground stability; in contrast, groundwater extraction induces local instability. These results are fundamental for urban territorial planning in the city of Santiago and demonstrate the importance of geodetic measurements in assessing the impact of climate change on groundwater storage and how this affects the ground surface elevation.

Author Contributions: Conceptualization, F.O., M.M. and G.Y.; methodology, F.O., M.M. and G.Y.; SAR data processing, F.O., validation G.Y. and M.M.; writing F.O., M.M. and G.Y.; supervision, M.M. and G.Y. All authors have read and agreed to the published version of the manuscript.

Funding: This study was supported by the Millennium Nucleus CYCLO (The Seismic Cycle Along Subduction Zones) (ICM) grant NC160025.

Data Availability Statement: The geological maps in Yanez et al. [3]. The catalog of active faults is available at: https://fallasactivas.cl/ (accessed on 10 September 2022, under license https://creativecommons.org/licenses/by/4.0/. The Hydrogeological Data are available on the portal of the DGA 2000, Chilean public water directorate https://dga.mop.gob.cl/servicioshidrometeorologicos/Paginas/default.aspx (accessed on 15 September 2022). Landslide catalogs are available in the master thesis under license https://creativecommons.org/licenses/by-nc-nd/3.0/cl/ (accessed on 10 October 2022).

Acknowledgments: F.O. acknowledges support from the ESA NoR Project ID: 65514. M.M. acknowledges support from the ANID PIA Anillo ACT192169, and National Research Center for Integrated Natural Disaster Management (CIGIDEN).

Conflicts of Interest: The authors declare no conflict of interest.

References

1. National Institute of Statistics of Chile, Census Report 2017. Available online: https://www.ine.cl/censo (accessed on 31 October 2022).
2. Pastén, C.; Sáez, M.; Ruiz, S.; Leyton, F.; Salomón, J.; Poli, P. Deep characterization of the Santiago Basin using HVSR and cross-correlation of ambient seismic noise. *Eng. Geol.* **2016**, *201*, 57–66. [CrossRef]
3. Yáñez, G.; Muñoz, M.; Flores Aqueveque, V.; Bosch, A. Gravity derived depth to basement in Santiago Basin, Chile: Implications for its geological evolution, hydrogeology, low enthalpy geothermal, soil characterization and geo-hazards. *Andean Geol.* **2015**, *42*, 147–172. [CrossRef]
4. González, F.A.; Maksymowicz, A.; Díaz, D.; Villegas, L.; Leiva, M.; Blanco, B.; Bonvalot, S. Characterization of the depocenters and the basement structure, below the central Chile Andean Forearc: A 3D geophysical modelling in Santiago Basin area. *Basin Res.* **2018**, *30*, 799–815. [CrossRef]
5. Leyton, F.; Sepúlveda, S.A.; Astroza, M.; Rebolledo, S.; Acevedo, P.; Ruiz, S.; Foncea, C. Seismic zonation of the Santiago basin, Chile. In Proceedings of the 5th International Conference on Earthquake Geotechnical Engineering, Santiago, Chile, 10–13 January 2011.
6. Leyton, F.; Pérez, A.; Campos, J.; Rauld, R.; Kausel, E. Anomalous seismicity in the lower crust of the Santiago Basin, Chile. *Phys. Earth Planet. Inter.* **2009**, *175*, 17–25. [CrossRef]
7. Armijo, R.; Rauld, R.; Thiele, R.; Vargas, G.; Campos, J.; Lacassin, R.; Kausel, E. The West Andean thrust, the San Ramon fault, and the seismic hazard for Santiago, Chile. *Tectonics* **2010**, *29*. [CrossRef]
8. Vargas, G.; Klinger, Y.; Rockwell, T.K.; Forman, S.L.; Rebolledo, S.; Baize, S.; Armijo, R. Probing large intraplate earthquakes at the west flank of the Andes. *Geology* **2014**, *42*, 1083–1086. [CrossRef]
9. Estay, N.P.; Yanez, G.A.; Maringue, J.I. TEM prospection on quaternary faults: The case of San Ramón Fault (SRF), Central Chile. In *AGU Fall Meeting Abstracts*; American Geophysical Union: Washington, DC, USA, 2016; Volume 2016, p. NH11A-1716.
10. Yáñez, G.; Perez-Estay, N.; Araya-Vargas, J.; Sanhueza, J.; Figueroa, R.; Maringue, J.; Rojas, T. Shallow anatomy of the San Ramón Fault (Chile) constrained by geophysical methods: Implications for its role in the Andean deformation. *Tectonics* **2020**, *39*, e2020TC006294. [CrossRef]
11. van Dinther, Y.; Preiswerk, L.E.; Gerya, T.V. A secondary zone of uplift due to megathrust earthquakes. *Pure Appl. Geophys.* **2019**, *176*, 4043–4068. [CrossRef]
12. Mosca, I.; Console, R.; d'Addezio, G. Renewal models of seismic recurrence applied to paleoseismological and historical observations. *Tectonophysics* **2012**, *564*, 54–67. [CrossRef]
13. Santibáñez, I.; Cembrano, J.; García-Pérez, T.; Costa, C.; Yáñez, G.; Marquardt, C.; González, G. Crustal faults in the Chilean Andes: Geological constraints and seismic potential. *Andean Geol.* **2018**, *46*, 32–65. [CrossRef]
14. Peltier, W.R. Global glacial isostasy and the surface of the ice-age Earth: The ICE-5G (VM2)model and GRACE. *Annu. Rev. Earth Planet. Sci.* **2004**, *32*, 111–149. [CrossRef]
15. Kooi, H.; de Vries, J. Land subsidence and hydrodynamic compaction of sedimentarybasins. *Hydrol. Earth Syst. Sci.* **1998**, *2*, 159–171. [CrossRef]
16. White, A.M.; Gardner, W.P.; Borsa, A.A.; Argus, D.F.; Martens, H.R. A review of GNSS/GPS in hydrogeodesy: Hydrologic loading applications and their implications for water resource research. *Water Resour. Res.* **2022**, *58*, e2022WR032078. [CrossRef] [PubMed]
17. Minderhoud, P.; Middelkoop, H.; Erkens, G.; Stouthamer, E. Groundwater extraction may drown mega-delta: Projections of extraction-induced subsidence and elevation of the Mekong delta for the 21st century. *Environ. Res. Commun.* **2019**, *2*, 011005. [CrossRef]
18. Chen, B.; Gong, H.; Chen, Y.; Li, X.; Zhou, C.; Lei, K.; Zhu, L.; Duan, L.; Zhao, X. Land subsidence and its relation with groundwater aquifers in Beijing Plain of China. *Sci. Total Environ.* **2020**, *735*, 139111. [CrossRef]
19. Turner, R.E.; Mo, Y. Salt Marsh Elevation limit determined after subsidence from hydrologic change and hydro-carbon extraction. *Remote Sens.* **2021**, *13*, 49. [CrossRef]
20. Singh, A.; Rao, G.S. Crustal structure and subsidence history of the Mannar basin through potential field modelling and backstripping analysis: Implications on basin evolution and hydrocarbon exploration. *J. Pet. Sci. Eng.* **2021**, *206*, 109000. [CrossRef]
21. Gahramanov, G.; Babayev, M.; Shpyrko, S.; Mukhtarova, K. Subsidence history and hydrocarbon migration modeling in south caspian basin. Visnyk Taras Shevchenko Natl. *Univ. Kyiv. Geol.* **2020**, *1*, 82–91. [CrossRef]
22. Stramondo, S.; Bozzano, F.; Marra, F.; Wegmuller, U.; Cinti, F.; Moro, M.; Saroli, M. Subsidence induced by urbanisation in the city of Rome detected by advanced InSAR technique and geotechnical investigations. *Remote Sens. Environ.* **2008**, *112*, 3160–3172. [CrossRef]
23. Manunta, M.; Marsella, M.; Zeni, G.; Sciotti, M.; Atzori, S.; Lanari, R. Two-scale surface deformation analysis using the SBAS-DInSAR technique: A case study of the city of Rome, Italy. *Int. J. Remote Sens.* **2008**, *29*, 1665–1684. [CrossRef]
24. Abidin, H.Z.; Andreas, H.; Djaja, R.; Darmawan, D.; Gamal, M. Land subsidence characteristics of Jakarta between 1997 and 2005, as estimated using GPS surveys. *GPS Solut.* **2008**, *12*, 23–32. [CrossRef]
25. Orellana, F.; Blasco, J.D.; Foumelis, M.; D'Aranno, P.; Marsella, M.; Di Mascio, P. DInSAR for Road Infrastructure Monitoring: Case Study Highway Network of Rome Metropolitan (Italy). *Remote Sens.* **2020**, *12*, 3697. [CrossRef]

26. Chang, L.; Dollevoet, R.P.B.J.; Hanssen, R.F. Monitoring Line-Infrastructure with Multisensor SAR Interferometry: Products and Performance Assessment Metrics. *IEEE J. Sel. Top. Appl. Earth Obs. Remote Sens.* **2018**, *11*, 1593–1605. [CrossRef]
27. Li, S.; Xu, W.; Li, Z. Review of the SBAS InSAR Time-series algorithms, applications, and challenges. *Geod. Geodyn.* **2021**, *13*, 114–126. [CrossRef]
28. Orellana, F.; Hormazábal, J.; Montalva, G.; Moreno, M. Measuring Coastal Subsidence after Recent Earthquakes in Chile Central Using SAR Interferometry and GNSS Data. *Remote Sens.* **2022**, *14*, 1611. [CrossRef]
29. Delouis, B.; Nocquet, J.M.; Vallée, M. Slip distribution of the February 27, 2010 Mw = 8.8 Maule earthquake, central Chile, from static and high-rate GPS, InSAR, and broadband teleseismic data. *Geophys. Res. Lett.* **2010**, *37*, 17. [CrossRef]
30. van Natijne, A.L.; Bogaard, T.A.; van Leijen, F.J.; Hanssen, R.F.; Lindenbergh, R.C. World-wide insar sensitivity index for landslide deformation tracking. *Int. J. Appl. Earth Obs. Geoinf.* **2022**, *111*, 102829. [CrossRef]
31. Fobert, M.A.; Singhroy, V.; Spray, J.G. InSAR monitoring of landslide activity in Dominica. *Remote Sens.* **2021**, *13*, 815. [CrossRef]
32. Moretto, S.; Bozzano, F.; Mazzanti, P. The role of satellite InSAR for landslide forecasting: Limitations and openings. *Remote Sens.* **2021**, *13*, 3735. [CrossRef]
33. De Corso, T.; Mignone, L.; Sebastianelli, A.; del Rosso, M.P.; Yost, C.; Ciampa, E.; Ullo, S. Application of DInSAR technique to high coherence satellite images for strategic infrastructure monitoring. In *IGARSS 2020-2020 IEEE International Geoscience and Remote Sensing Symposium*; IEEE: Piscataway, NJ, USA, 2020; pp. 4235–4238. [CrossRef]
34. D'Aranno, P.; Di Benedetto, A.; Fiani, M.; Marsella, M. Remote sensing technologies for linear infrastructure monitoring. *Int. Arch. Photogramm. Remote Sens. Spat. Inf. Sci.* **2019**, *42*, 461–468. [CrossRef]
35. Bonì, R.; Herrera, G.; Meisina, C.; Notti, D.; Béjar-Pizarro, M.; Zucca, F.; González, P.J.; Palano, M.; Tomás, R.; Fernández, J.; et al. Twenty-year advanced DInSAR analysis of severe land subsidence: The Alto Guadalentín Basin (Spain) case study. *Eng. Geol.* **2015**, *198*, 40–52. [CrossRef]
36. Ezquerro, P.; Tomás, R.; Béjar-Pizarro, M.; Fernández-Merodo, J.A.; Guardiola-Albert, C.; Staller, A.; Sánchez-Sobrino, J.A.; Herrera, G. Improving multi—Technique monitoring using Sentinel-1 and Cosmo-SkyMed data and upgrading groundwater model capabilities. *Sci. Total Environ.* **2019**, *703*, 134757. [CrossRef] [PubMed]
37. Taftazani, R.; Kazama, S.; Takizawa, S. Spatial Analysis of Groundwater Abstraction and Land Subsidence for Planning the Piped Water Supply in Jakarta, Indonesia. *Water* **2022**, *14*, 3197. [CrossRef]
38. Ferretti, A.; Prati, C.; Rocca, F. Permanent scatterers in SAR interferometry. *IEEE Trans. Geosci. Remote Sens.* **2001**, *39*, 8–20. [CrossRef]
39. Berardino, P.; Fornaro, G.; Lanari, R.; Sansosti, E. A new algorithm for surface deformation monitoring based on small baseline differential SAR interferograms. *IEEE Trans. Geosci. Remote Sens.* **2002**, *40*, 2375–2383. [CrossRef]
40. Crosetto, M.; Monserrat, O.; Cuevas-González, M.; Devanthéry, N.; Crippa, B. Persistent Scatterer Interferometry: A Review. *ISPRS J. Photogramm. Remote Sens.* **2016**, *115*, 78–89. [CrossRef]
41. Ferretti, A.; Fumagalli, A.; Novali, F.; Prati, C.; Rocca, F.; Rucci, A. A New Algorithm for Processing Interferometric Data-Stacks: SqueeSAR. *IEEE Trans. Geosci. Remote Sens.* **2011**, *49*, 3460–3470. [CrossRef]
42. Hooper, A.J. A Multi-Temporal InSAR Method Incorporating Both Persistent Scatterer and Small Baseline Approaches. *Geophys. Res. Lett.* **2008**, *35*. [CrossRef]
43. Casu, F.; Elefante, E.; Imperatore, P.; Zinno, I.; Manunta, M.; De Luca, C.; Lanari, R. SBAS-DInSAR Parallel Processing for Deformation Time Series Computation. *IEEE JSTARS* **2004**, *7*, 3285–3296. [CrossRef]
44. European Space Agency. Sentinel-1—Missions—Sentinel. Available online: https://sentinel.esa.int/web/sentinel/missions/sentinel-1 (accessed on 13 July 2022).
45. Manunta, M.; De Luca, C.; Zinno, I.; Casu, F.; Manzo, M.; Bonano, M.; Fusco, A.; Pepe, A.; Onorato, G.; Berardino, P.; et al. The Parallel SBAS Approach for Sentinel-1 Interferometric Wide Swath Deformation Time-Series Generation: Algorithm Description and Products Quality Assessment. *IEEE Trans. Geosci. Remote Sens.* **2019**, *57*, 6259–6281. [CrossRef]
46. De Luca, C.; Cuccu, R.; Elefante, S.; Zinno, I.; Manunta, M.; Casola, V.; Rivolta, G.; Lanari, R.; Casu, F. An On-Demand Web Tool for the Unsupervised Retrieval of Earth's Surface Deformation from SAR Data: The P-SBAS Service within the ESA G-POD Environment. *Remote Sens.* **2015**, *7*, 15630–15650. [CrossRef]
47. Manunta, M.; Casu, F.; Zinno, I.; de Luca, C.; Pacini, F.; Brito, F.; Blanco, P.; Iglesias, R.; Lopez, A.; Briole, P.; et al. The Geohazards Exploitation Platform: An advanced cloud-based environment for the Earth Science community. In Proceedings of the 19th EGU General Assembly, EGU2017, Vienna, Austria, 23–28 April 2017; p. 14911.
48. Foumelis, M.; Papadopoulou, T.; Bally, P.; Pacini, F.; Provost, F.; Patruno, J. Monitoring Geohazards Using On-Demand and Systematic Services on Esa's Geohazards Exploitation Platform. In Proceedings of the IGARSS 2019, IEEE International Geoscience and Remote Sensing Symposium, Yokohama, Japan, 28 July–2 August 2019; IEEE: Piscataway, NJ, USA, 2019; pp. 5457–5460. [CrossRef]
49. Galve, J.P.; Pérez-Peña, J.V.; Azañón, J.M.; Closson, D.; Calò, F.; Reyes-Carmona, C.; Jabaloy, A.; Ruano, P.; Mateos, R.M.; Notti, D.; et al. Evaluation of the SBAS InSAR Service of the European Space Agency's Geohazard Exploitation Platform (GEP). *Remote Sens.* **2017**, *9*, 1291. [CrossRef]
50. Reyes-Carmona, C.; Galve, J.P.; Barra, A.; Monserrat, O.; María Mateos, R.; Azañón, J.M.; Perez-Pena, J.V.; Ruano, P. The Sentinel-1 CNR-IREA SBAS service of the European Space Agency's Geohazard Exploitation Platform (GEP) as a powerful tool for landslide activity detection and monitoring. In Proceedings of the EGU General Assembly, Vienna, Austria, 3–8 May 2020; p. 19410.

51. Moreno, M.; Rosenau, M.; Oncken, O. Maule earthquake slip correlates with pre-seismic locking of Andean subduction zone. *Nature* **2010**, *467*, 198–202. [CrossRef] [PubMed]
52. Tilmann, F.; Zhang, Y.; Moreno, M.; Saul, J.; Eckelmann, F.; Palo, M.; Deng, Z.; Babeyko, A.; Chen, K.; Báez, J.C.; et al. The 2015 Illapel earthquake, central Chile: A type case for a characteristic earthquake? *Geophys. Res. Lett.* **2016**, *43*, 574–583. [CrossRef]
53. Carvajal, M.; Cisternas, M.; Catalán, P.A. Source of the 1730 Chilean earthquake from historical records: Implications for the future tsunami hazard on the coast of Metropolitan Chile. *J. Geophys. Res. Solid Earth* **2017**, *122*, 3648–3660. [CrossRef]
54. Sippl, C.; Moreno, M.; Benavente, R. Microseismicity appears to outline highly coupled regions on the Central Chile megathrust. *J. Geophys. Res. Solid Earth* **2021**, *126*, e2021JB022252. [CrossRef]
55. Maldonado, V.; Contreras, M.; Melnick, D. A comprehensive database of active and potentially-active continental faults in Chile at 1:25,000 scale. *Sci. Data* **2021**, *8*, 20. [CrossRef]
56. Donoso, F.; Moreno, M.; Ortega-Culaciati, F.; Bedford, J.; Benavente, R. Automatic Detection of Slow Slip Events Using the PICCA: Application to Chilean GNSS Data. *Front. Earth Sci.* **2021**, *9*, 788054. [CrossRef]
57. Charrier, R.; Farías, M.; Maksaev, V. Evolución tectónica, paleogeográfica y metalogénica durante el Cenozoico en los Andes de Chile norte y central e implicaciones para las regiones adyacentes de Bolivia y Argentina. *Rev. Asoc. Geológica Argent.* **2009**, *65*, 5–35.
58. Pardo-Casas, F.; Molnar, P. Relative motion of the Nazca (Farallon) and South American plates since Late Cretaceous time. *Tectonics* **1987**, *6*, 233–248. [CrossRef]
59. Godoy, E.; Lara, L. Segmentación estructural andina a los 33-34: Nuevos datos en la Cordillera Principal. In *Congreso Geológico Chileno*; Universidad de Concepción, Departamento Ciencias de la Tierra: Concepción, Chile, 1994; pp. 1344–1348.
60. Wyss, A.R.; Flynn, J.J.; Norell, M.; Swisher, C.C.; Novacek, M.J.; McKenna, M.C.; Charrier, R. Paleogene mammals from the Andes of central Chile: A preliminary taxonomic, biostratigraphic, and geochronologic assessment. *Am. Mus. Novit.* **1994**, *5*, 3098.
61. Charrier, R.; Baeza, O.; Elgueta, S.; Flynn, J.J.; Gans, P.; Kay, S.M.; Zurita, E. Evidence for Cenozoic extensional basin development and tectonic inversion south of the flat-slab segment, southern Central Andes, Chile (33–36 SL). *J. South Am. Earth Sci.* **2002**, *15*, 117–139. [CrossRef]
62. Charrier, R.; Bustamante, M.; Comte, D.; Elgueta, S.; Flynn, J.J.; Iturra, N.; Wyss, A.R. The Abanico extensional basin: Regional extension, chronology of tectonic inversion and relation to shallow seismic activity and Andean uplift. *Neues Jahrb. Für Geol. Und Paläontologie Abh.* **2005**, *236*, 43–77. [CrossRef]
63. Fock, A.; Charrier, R.; Farías, M.; Muñoz, M. Fallas de vergencia oeste en la Cordillera Principal de Chile Central: Inversión de la cuenca de Abanico (33-34 S). *Rev. Asoc. Geológica Argent. Publicación Espec.* **2006**, *6*, 48–55.
64. Moreno, T.; Gibbons, W.; Cembrano, J.; Lavenue, A.; Yáñez, G. (Eds.) *The Geology of Chile, Chapter 9*; Geological Society of London: London, UK, 2007.
65. Farías, M.; Vargas, G.; Tassara, A.; Carretier, S.; Baize, S.; Melnick, D.; Bataille, K. Land-level changes produced by the M w 8.8 2010 Chilean earthquake. *Science* **2010**, *329*, 916. [CrossRef] [PubMed]
66. Garreaud, R.D.; Alvarez-Garreton, C.; Barichivich, J.; Boisier, J.P.; Christie, D.; Galleguillos, M.; LeQuesne, C.; McPhee, J.; Zambrano-Bigiarini, M. The 2010–2015 megadrought in central Chile: Impacts on regional hydroclimate and vegetation. *Hydrol. Earth Syst. Sci.* **2017**, *21*, 6307–6327. [CrossRef]
67. Torres, R.; Snoeij, P.; Geudtner, D.; Bibby, D.; Davidson, M.; Attema, E.; Rostan, F. GMES Sentinel-1 mission. *Remote Sens. Environ.* **2012**, *120*, 9–24. [CrossRef]
68. Cigna, F.; Tapete, D. Sentinel-1 big data processing with P-SBAS InSAR in the geohazards exploitation platform: An experiment on coastal land subsidence and landslides in Italy. *Remote Sens.* **2021**, *13*, 885. [CrossRef]
69. Farr, T.G.; Rosen, P.A.; Caro, E.; Crippen, R.; Duren, R.; Hensley, S.; Kobrick, M.; Paller, M.; Rodriguez, E.; Roth, L.; et al. The Shuttle Radar Topography Mission. *Rev. Geophys.* **2007**, *45*, RG2004. [CrossRef]
70. Yague-Martinez, N.; De Zan, F.; Prats-Iraola, P. Coregistration of Interferometric Stacks of Sentinel-1 TOPS Data. *IEEE Geosci. Remote Sens. Lett.* **2017**, *14*, 1002–1006. [CrossRef]
71. Cascini, L.; Fornaro, G.; Peduto, D. Advanced low- and full-resolution DInSAR map generation for slow-moving landslide analysis at different scales. *Eng. Geol.* **2010**, *112*, 29–42. [CrossRef]
72. Cigna, F.; Bianchini, S.; Casagli, N. How to assess landslide activity and intensity with Persistent Scatterer Interferometry (PSI): The PSI-based matrix approach. *Landslides* **2013**, *10*, 267–283. [CrossRef]
73. Barría, P.; Chadwick, C.; Ocampo-Melgar, A.; Galleguillos, M.; Garreaud, R.; Díaz-Vasconcellos, R.; Poblete-Caballero, D. Water management or megadrought: What caused the Chilean Aculeo Lake drying? *Reg. Environ. Chang.* **2021**, *21*, 19. [CrossRef]
74. Ocampo-Melgar, A.; Barria, P.; Chadwick, C.; Villoch, P. Restoration perceptions and collaboration challenges under severe water scarcity: The Aculeo Lake process. *Restor. Ecol.* **2021**, *29*, e13337. [CrossRef]
75. DGA. *Modelo de Simulación Hidrológico Operacional: Cuencas de los ríos Maipo y Mapocho*; Ayala, Cabrera y Asociados; Series SIT; Dirección General de Aguas, Ministerio de Obras Públicas: Santiago, Chile, 2000; p. 87.
76. Terzaghi, K. Principles of soil mechanics: IV. Settlement and consolidation of clay. *Erdbaummechanic* **1925**, *95*, 874–878.
77. Biot, M. General theory of three-dimensional consolidation. *J. Appl. Phys.* **1941**, *12*, 155–164. [CrossRef]
78. Castellazzi, P.; Martel, R.; Rivera, A.; Huang, J.; Pavlic, G.; Calderhead, A.I.; Chaussard, E.; Garfias, J.; Salas, J. Groundwater depletion in Central Mexico: Use of GRACE and InSAR to support water resources management. *Water Resour. Res.* **2016**, *52*, 5985–6003. [CrossRef]

79. Mourad, M.; Tsuji, T.; Ikeda, T.; Ishitsuka, K.; Senna, S.; Ide, K. Mapping Aquifer Storage Properties Using S-Wave Velocity and InSAR-Derived Surface Displacement in the Kumamoto Area, Southwest Japan. *Remote Sens.* **2021**, *13*, 4391. [CrossRef]
80. Miller, M.M.; Shirzaei, M.; Argus, D. Aquifer mechanical properties and decelerated compaction in Tucson, Arizona. *J. Geophys. Res. Solid Earth* **2017**, *122*, 8402–8416. [CrossRef]
81. Antinao, J.; Fernández, J.C.; Naranjo, J.A.; Villarroel, P. Peligro de Remociones en Masa e Inundaciones de la Cuenca de Santiago, Región Metropolitana. Servicio Nacional de Geología y Minería, Carta Geológica de Chile, Serie Geología Ambiental 2, 1 Mapa Escala 1:100,000. Santiago 2003. Available online: https://snia.mop.gob.cl/repositoriodga/handle/20.500.13000/3929 (accessed on 15 September 2022).
82. Sepúlveda, S.A.; Padilla, C. Rain-induced debris and mudflow triggering factors assessment in the Santiago cordilleran foothills, Central Chile. *Nat. Hazards* **2008**, *47*, 201–215. [CrossRef]
83. Naranjo, J.A.; Varela, J. Flows of debris and mud that affected the eastern sector of Santiago on May 3, 1993. *Natl. Serv. Geol. Min. Bull.* **1996**, *47*, 42.
84. Lara, M.; Sepúlveda, S.A.; Celis, C.; Rebolledo, S.; Ceballos, P. Landslide susceptibility maps of Santiago city Andean foothills, Chile. *Andean Geol.* **2018**, *45*, 433–442. [CrossRef]
85. Sepúlveda, S.A.; Moreiras, S.M.; Lara, M.; Alfaro, A. Debris flows in the Andean ranges of central Chile and Argentina triggered by 2013 summer storms: Characteristics and consequences. *Landslides* **2015**, *12*, 115–133. [CrossRef]
86. Celis Saez, C. Susceptibility of Mass Removals and Danger of Flows in the Andean Front of Santiago, Metropolitan Region. 2018. Available online: https://repositorio.uchile.cl/handle/2250/168618 (accessed on 10 October 2022).
87. Colesanti, C.; Wasowski, J. Investigating landslides with space-borne Synthetic Aperture Radar (SAR) interferometry. *Eng. Geol.* **2006**, *88*, 173–199. [CrossRef]
88. Plank, S.; Singer, J.; Minet, C.; Thuro, K. Pre-survey suitability evaluation of the differential synthetic aperture radar interferometry method for landslide monitoring. *Int. J. Remote Sens.* **2012**, *33*, 6623–6637. [CrossRef]

remote sensing MDPI

Article

Characterizing Spatiotemporal Patterns of Land Subsidence after the South-to-North Water Diversion Project Based on Sentinel-1 InSAR Observations in the Eastern Beijing Plain

Yuanyuan Liu [1,2,3], Xia Yan [1], Yuanping Xia [1,2], Bo Liu [1,2,*], Zhong Lu [4] and Mei Yu [1,2]

1 Faculty of Geomatics, East China University of Technology, Nanchang 330013, China
2 Key Laboratory of Mine Environmental Monitoring and Improving around Poyang Lake, Ministry of Natural Resources, Nanchang 330013, China
3 Key Laboratory for Digital Land and Resources of Jiangxi Province, East China University of Technology, Nanchang 330013, China
4 Roy M. Huffington Department of Earth Sciences, Southern Methodist University, Dallas, TX 75275, USA
* Correspondence: liubo@ecut.edu.cn

Abstract: The eastern Beijing plain has been suffering severe subsidence for the last decades, mainly associated with the long-term excessive extraction of groundwater resource. Since the end of 2014, the annual water supply in Beijing plain has reached several hundred million cubic meters because of the South-to-North Water Diversion (SNWD) Project, which has reduced the groundwater exploitation and changed the status of land subsidence. In this work, we first obtain the current spatiotemporal variations of land subsidence in the eastern Beijing plain by using progressive small baseline subsets (SBAS) InSAR time series analysis method with Sentinel-1 SAR data acquired from July 2015 to December 2021. Then, we analyze the correlations between InSAR-derived subsidence and groundwater level change by applying the cross wavelet method. The results show that two major subsidence zones are successfully detected with the maximum deformation rate of −150 mm/yr and maximum cumulative deformation of −950 mm. Besides, the ground deformation at different stages from 2016 to 2021 reveal that the area and magnitude of major deformation significantly slow down, even in the regions with severe subsidence, especially in the year of 2017, which is about two years later than the start time of SNWD Project in Beijing. Further, we find the InSAR-derived subsidence lags groundwater level change with about 1–2-month lagging time, indicating that the dynamic variation of groundwater level fluctuation may be the main factor affecting the uneven subsidence in the severe subsiding zones. Last, differential subsidence rates are identified at both sides of geological faults, such as Nankou-Sunhe fault and Nanyuan-Tongxian fault, from the observed deformation map, which could be explained that the groundwater flow is blocked when a fault is encountered. These findings can provide significant information to reveal the deformation mechanisms of land subsidence, establish the hydrogeological models and assist decision-making, early warning and hazard relief in Beijing, China.

Keywords: South-to-North Water Diversion (SNWD) Project; InSAR; land subsidence; cross wavelet method; Sentinel-1; Beijing plain

Citation: Liu, Y.; Yan, X.; Xia, Y.; Liu, B.; Lu, Z.; Yu, M. Characterizing Spatiotemporal Patterns of Land Subsidence after the South-to-North Water Diversion Project Based on Sentinel-1 InSAR Observations in the Eastern Beijing Plain. *Remote Sens.* **2022**, *14*, 5810. https://doi.org/10.3390/rs14225810

Academic Editors: Ioannis Papoutsis, Konstantinos G. Nikolakopoulos and Constantinos Loupasakis

Received: 13 October 2022
Accepted: 15 November 2022
Published: 17 November 2022

1. Introduction

As a common environmental geological phenomenon, land subsidence refers to the decline of regional ground elevation owing to the influence of natural factors or human engineering activities [1]. With long-term over-exploitation of groundwater and rapid urban expansion, land subsidence has become a worldwide widespread geological hazard, such as Mexico [2–4], Houston [5], Willcox Basin, Arizona [6], Northern Italy [7], Semarang, Indonesia [8]. In China, many cities, including Beijing [9–16], Shanghai [17], Xi'an [18], Wuhan [19] and Taiyuan [20], are suffering from the potential land subsidence.

In order to alleviate and mitigate the disasters related to land subsidence, it is essential to characterize the spatiotemporal patterns of land subsidence with high precision. Several ground-based geodetic surveying techniques (e.g., GNSS, leveling) have been applied to obtain the ground displacements in the typical regions. Because these techniques are time-consuming, labor-intensive and only have a few sparsely distributed monitoring stations, the applications are limited in long-term and large-scale deformation monitoring. Since the end of the 20th century, Synthetic Aperture Radar Interferometry (InSAR) technique has been an effective geodetic remote sensing tool for measuring large-scale deformation from space, with the advantages of large space coverage, high spatial resolution and millimeter-scale accuracy [21] and has been successfully used in many areas [2–20].

However, there are also several following challenges for retrieving the high-precision ground deformation with conventional InSAR technique: spatial or geometrical decorrelation [22], temporal decorrelation [22] and atmospheric disturbances [23]. Subsequently, a series of advanced InSAR time series analysis methods based on multi-interferograms have been proposed and developed by many research institutions to obtain the deformation with high precision [24–30]. Regarding the deformation monitoring of land subsidence in Beijing with InSAR observations, several related results with ERS-1/2, ENVISAT ASAR, ALOS PALSAR-1/2, RADARSAT-2, TerraSAR-X and Sentienl-1 SAR datasets have been issued in recent years [9–16,31–44], which has shown that land subsidence has been developed in the eastern Beijing plain for the last decades (e.g., 2003–2015, 2015–2017) with large deformation rate (>100 mm/yr) and is mainly caused by the excessive extraction of groundwater resource and the construction of urban buildings and infrastructures [31]. However, the middle route of the South-to-North Water Diversion (SNWD) Project, which starts from Danjiangkou Reservoir and ends in Beijing, has reduced the groundwater exploitation and altered the water consumption structure of Beijing plain since December 2014, changing the status of land subsidence [43]. Unfortunately, few studies are published concerning the correlations between SNWD Project and land subsidence in Beijing plain [40,45]. Therefore, the present development, more detailed spatiotemporal changes of land subsidence after the SNWD Project should be answered in the eastern Beijing plain.

The main objective of this work is to reveal the latest spatiotemporal changes of land subsidence over the eastern Beijing plain and assess the roles that groundwater exploitation, urban construction and geological faults played in the deformation. Firstly, a progressive SBAS-InSAR approach was adopted to obtain the current deformation rate and deformation time series in the eastern Beijing plain (2015–2021) by using 72 C-band Sentinel-1 SAR images. Then, we combined the InSAR-derived subsidence and groundwater level measurements to reveal the potential relationship and lagging time between the InSAR-derived subsidence and groundwater level changes. Finally, the possible links between the InSAR-derived subsidence, urban expansion and geological faults are discussed.

2. Study Area and Dataset

2.1. Study Area

Beijing, the capital of China, covers an area of about 16,412 km^2. It is surrounded by mountains in the west, north and northeast and the average ground elevation is approximately 43.5 m. As shown in Figure 1, the study area of this paper is situated in the eastern part of Beijing plain, which presents a typical semi-humid and semi-arid continental monsoon climate. The average annual temperature is about 11–12 °C [15]. The seasonal rainfall is unevenly distributed, and approximately 80% of the total annual precipitation concentrates between mid-June and September [16]. Furthermore, this region is a typical piedmont alluvial diluvial plain composed of sediments from three major river systems, namely Yongding, Wenyu and Chaobai Rivers. Based on the stratigraphic age, genesis type and burial depth, the Quaternary aquifers are characterized by a multi-layer aquifer system in our study area (Table 1). At present, the groundwater withdrawal is mainly from the Quaternary aquifer within the burial depth of 300 m [9]. For more detailed information on the spatial extension of the aquifer systems in our study area, readers can refer to [40,44].

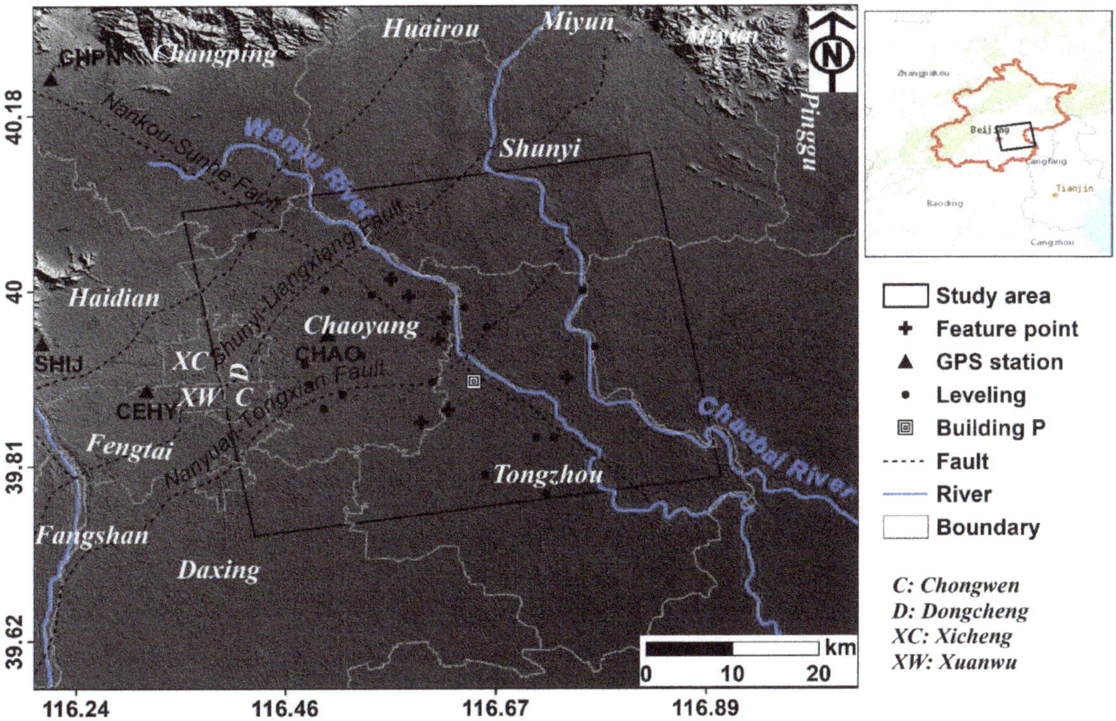

Figure 1. Shaded relief map of the eastern Beijing plain, where major rivers (blue solid lines), main active faults (black dashed lines), GPS stations (black triangles), leveling benchmarks (black dots) and the county boundary of Beijing administrative district (gray dashed lines) are all represented. The red polygon shown in the inset on the top-right corner indicates the location of Beijing and the black box illustrates the study area in this paper.

Table 1. The detailed information of aquifer groups in the study area.

Aquifer Group	Main Lithological Features	Burial Depth (m)
The unconfined aquifer (I)	silt, silty sand and sandy clay	0~50
The first confined aquifer (II)	multiple types of gravel, sand and clay soil	80~100
The second confined aquifer (III)	multiple types of gravel, sand and clay soil	100~180
The third confined aquifer (IV)	mainly sand	180~300

There are several main active geological faults cross throughout the study area, including Nankou-Sunhe fault, Nanyuan-Tongxian fault and Shunyi-Liangxiang fault. Nankou-sunhe fault is a NW-trending left-lateral normal fault, with the dip angle of 60°~68°, while Nanyuan-tongxian fault is a NNE-trending normal fault, with the dip angel of 40°~60°. According to the results from the Beijing earthquake station, the latest activity era of Nankou-Sunhe fault and Nanyuan-Tongxian fault are late Pleistocene and Holocene, whose mean slip rates are about 0.3 mm/a and 0.75 mm/a, respectively [46]. For more detailed information of the active geological faults, readers can refer to [46].

2.2. Dataset

In this work, the data set includes SAR images, GPS observations and leveling measurements (for the validation of InSAR-derived results). 72 C-band IW (Interferometric Wide) Sentinel-1 SAR images acquired from 30 July 2015 to 15 December 2021 are collected to reveal the latest land subsidence after the South-to-North Water Diversion (SNWD) Project in the eastern part of Beijing plain. For the progressive SBAS-InSAR processing, we divide the SAR data into two different groups. Among which, the first group refers to the archived SAR images (i.e., the first 56 SAR images acquired from 30 July 2015 to 24 December 2019) to generate possible differential interferograms (shown in black solid lines in Figure 2) with traditional SBAS-InSAR technique for deformation parameter initialization. The second group is the newly acquired SAR data to connect images with relatively new acquisition dates in the archived SAR data (i.e., the 55th and 56th SAR images in the first group) and generate new differential interferograms under the same spatial and temporal thresholds (shown in black dashed lines in Figure 2) to update new deformation parameters. In this study, the spatial and temporal baseline thresholds are set as 150 m and 96 days, respectively. GPS data were obtained from the GPS monitoring stations, which were measured by the China Earthquake Administration within the period of August 2015–December 2017. The measurements of 18 leveling benchmarks were measured by the Beijing Institute of Geological & Prospecting Engineering in China within the year of 2017. Meanwhile, the corresponding precise orbit ephemeris are obtained to correct the orbit information. Furthermore, the digital elevation model (DEM) data with a resolution of 30 m obtained from Shuttle Radar Topography Mission-1 (SRTM) is used to remove the topographic phases and geocode the deformation results.

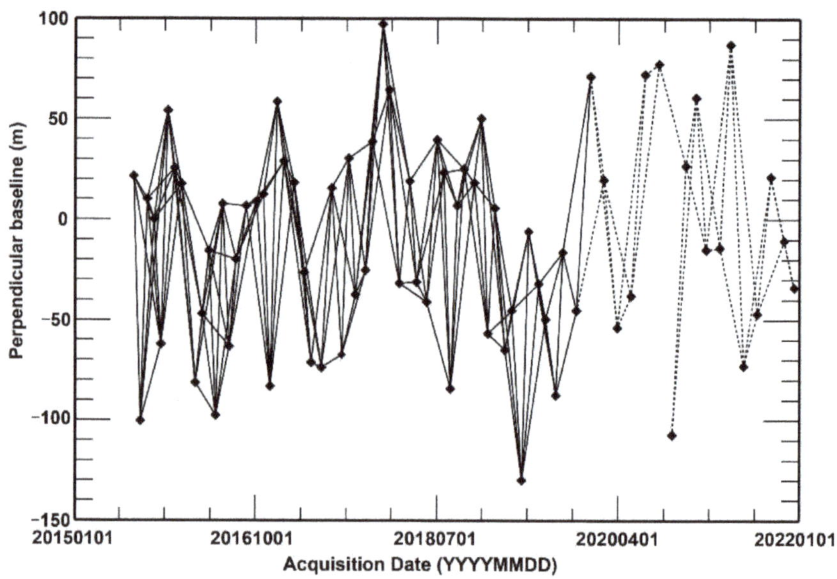

Figure 2. The spatiotemporal baseline distribution of differential interferograms used in this study. The black dots denote the acquisition dates of SAR data. The black solid lines represent the interferometric combinations of the archived SAR data in the first group for parameter initialization. The black dashed lines represent the new generated interferometric combinations between the newly acquired SAR data and relatively new acquisition dates in the archived SAR data for the sequential estimation.

3. Methodology

3.1. Progressive SBAS-InSAR Technique

In this paper, the deformation time series is calculated by using the progressive small baseline subsets (SBAS) InSAR approach. As shown in Figure 3, the data processing of this approach includes two major parts: (I) calculating preliminary time series of archived SAR images using traditional SBAS-InSAR technique and (II) updating deformation time series corresponding to the new SAR acquisition date with sequential estimation.

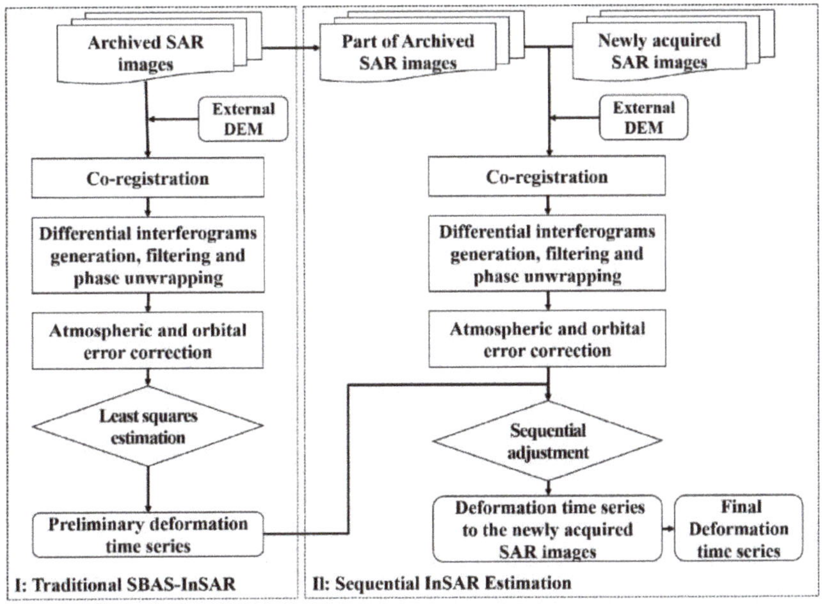

Figure 3. Flowchart of progressive SBAS-InSAR processing.

3.1.1. Traditional SBAS-InSAR

Preliminary deformation time series of archived SAR images is calculated with the small baseline subsets (SBAS) InSAR technique [27,47] under the frame of GAMMA software. Considering $N+1$ archived SAR acquisitions in chronological order $(t_0, t_1, \cdots, t_N)$, thus M interferometric combinations can be obtained by setting the appropriate spatiotemporal baseline thresholds (t_c and b_c). After removing topography-related phase and other phases induced by atmospheric disturbances and inaccurate orbital information from the interferograms, the *j-th* unwrapped differential interferogram $\delta\varphi_j$ generated between t_A and t_B can be expressed:

$$\sum_{k=t_A}^{t_{B-1}} v_{k+1}(t_{k+1} - t_k) = \delta\varphi_j \tag{1}$$

where k is the acquisition time indexes, which ranges from t_A to t_{B-1}; v_k denotes the unknown deformation phase rate between the time-adjacent acquisitions.

Based on Equation (1), we can successfully obtain the deformation phase rates with the least squares (LS) method. Subsequently, under the assumption that the deformation at the first SAR acquisition is 0, we retrieve the cumulative deformation phases through an additional integration operation. Finally, we convert the deformation phases to ground displacements.

3.1.2. Sequential InSAR Estimation

Unlike traditional SBAS-InSAR, in which the cumulative deformation is re-estimated by using all SAR acquisitions when one or more new SAR images are added, we adopt the sequential adjustment to dynamically update the deformation parameters by just using the unwrapped interferograms obtained from those newly acquired SAR images. Assuming N_1 newly SAR acquisitions in chronological order $(t_{N+1}, t_{N+2}, \cdots, t_{N+N1})$, M_1 interferograms are generated under the same temporal and spatial baseline constraints, part of which are obtained by combining the newly acquired SAR data and n images with relatively new acquisition dates (after t_{n1}) in the archived SAR data. Accordingly, the function model can be expressed as:

$$\underbrace{\begin{bmatrix} t_1-t_0 & 0 & \cdots & 0 \\ t_1-t_0 & t_2-t_1 & \cdots & 0 \\ \vdots & \vdots & \vdots & \vdots \\ 0 & 0 & \cdots & t_{n1-1}-t_{n1-2} \end{bmatrix}}_{A} \underbrace{\begin{bmatrix} v_1 \\ v_2 \\ \vdots \\ v_{t_{n1-1}} \end{bmatrix}}_{X_1} + \underbrace{\begin{bmatrix} t_{n1}-t_{n1-1} & 0 & \cdots & 0 \\ t_{n1}-t_{n1-1} & t_{n1+1}-t_{n1} & \cdots & 0 \\ \vdots & \vdots & \vdots & \vdots \\ 0 & 0 & \cdots & t_N-t_{N-1} \end{bmatrix}}_{B_1} \underbrace{\begin{bmatrix} v_{t_{n1}} \\ v_{t_{n1+1}} \\ \vdots \\ v_{t_N} \end{bmatrix}}_{X_2} = \underbrace{\begin{bmatrix} \delta\varphi_1 \\ \delta\varphi_2 \\ \vdots \\ \delta\varphi_M \end{bmatrix}}_{L_1} \tag{2}$$

$$\underbrace{\begin{bmatrix} 0 & 0 & \cdots & 0 \\ 0 & t_{n1+1}-t_{n1} & \cdots & 0 \\ \vdots & \vdots & \vdots & \vdots \\ 0 & 0 & \cdots & t_N-t_{N-1} \end{bmatrix}}_{B_2} \underbrace{\begin{bmatrix} v_{t_{n1}} \\ v_{t_{n1+1}} \\ \vdots \\ v_{t_N} \end{bmatrix}}_{X_2} + \underbrace{\begin{bmatrix} t_{N+1}-t_N & 0 & \cdots & 0 \\ t_{N+1}-t_N & t_{N+2}-t_{N+1} & \cdots & 0 \\ \vdots & \vdots & \vdots & \vdots \\ 0 & 0 & \cdots & t_{N+N_1}-t_{N+N_1-1} \end{bmatrix}}_{C} \underbrace{\begin{bmatrix} v_{t_{N+1}} \\ v_{t_{N+2}} \\ \vdots \\ v_{t_{N+N_1}} \end{bmatrix}}_{X_3} = \underbrace{\begin{bmatrix} \delta\varphi_{M+1} \\ \delta\varphi_{M+2} \\ \vdots \\ \delta\varphi_{M+M_1} \end{bmatrix}}_{L_2} \tag{3}$$

where A, B_1, B_2, C denotes the design matrix with the element of 0 or the interval between the time-adjacent acquisitions; X_1, X_2, X_3 denotes the unknown parameters associated with the deformation phase rates. It's worth noting that the initial solutions (X_1' and X_2') of X_1 and X_2 have been obtained from traditional SBAS-InSAR technique. Thus, X_1 and X_2 can be rewritten as:

$$\begin{aligned} X_1 &= X_1' + \delta X_1 \\ X_2 &= X_2' + \delta X_2 \end{aligned} \tag{4}$$

where δX_1 and δX_2 represents the corrections of X_1 and X_2.

Assuming that all SAR images are mutually independent and the weight matrix equals to 1, the unknown parameters can be inversed by using sequential adjustment [48]:

$$\begin{aligned} &\begin{bmatrix} \delta X_2 \\ X_3 \end{bmatrix} = \begin{bmatrix} Q + B_2^T B_2 & B_2^T C \\ C^T B_2 & C^T C \end{bmatrix}^{-1} \begin{bmatrix} B_2^T L_2' \\ C^T L_2' \end{bmatrix} \\ &Q = B_1^T B_1 - B_1^T A (A^T A)^{-1} A^T B_1 \\ &\delta X_1 = -(A^T A)^{-1} A^T B_1 \delta X_2 \\ &L_2' = L_2 - B_2 X_2' \end{aligned} \tag{5}$$

where Q represents the cofactor matrix derived from traditional SBAS-InSAR technique.

According to Equations (4) and (5), the updated deformation phase rates are obtained and the updated cumulative deformation phases are then solved through an additional integration operation. Therefore, the deformation parameters can be updated as quickly as possible once the new SAR images are involved.

3.2. *Wavelet Transform for Time Series Analysis*

Wavelet transform, one of the common time-frequency analysis methods, has been widely applied in various SAR/InSAR applications, especially in the analysis and interpretation of SAR data [15,49–51]. Cross wavelet transform (XWT) [52], which is a development of continuous wavelet transform (CWT), is used to explore the correlation between two different time series (i.e., the InSAR-derived time series deformation and hydraulic level) across time and scales. In this work, the complex Morlet wavelet is selected to serve as mother wavelet in the processing. Assuming W^X and W^Y are the CWT results of two different time series $a(t)$ and $b(t)$, the cross wavelet transform W^{XY} can be expressed as:

$$W^{XY} = W^X \cdot W^{Y*} \tag{6}$$

where * denotes the complex conjugation.

Generally speaking, the product of the XWT is a complex number; therefore, the cross wavelet power spectrum ($|W^{XY}|$) is defined to reveal the correlation between $a(t)$ and $b(t)$, exposing the regions where the common power is high in the time-frequency domain [52].

Furthermore, the wavelet coherence (WTC) is introduced, which reveals the correlation between W^X and W^Y in the time-frequency domain. The wavelet coherence can be calculated as:

$$R^2(s) = \frac{\left|S(s^{-1}W^{XY}(s))\right|^2}{S\left(s^{-1}|W^X(s)|^2\right)S\left((s^{-1}|W^Y(s)|^2\right)} \tag{7}$$

where $R_n^2(s)$ denotes the wavelet coherence; s denotes the scale of wavelet transform; S is a smoothing operator. The value of $R_n^2(s)$ ranges from 0 to 1, with 1 indicating highest wavelet coherence in both CWT results.

Additionally, the phase angle of the cross wavelet spectrum is introduced to represent the phase difference between two different time series (i.e., $a(t)$ and $b(t)$). It can be calculated from the real and imaginary parts of W^{XY}, whose graphical result is shown as an arrow. In the graph, the arrow pointing east (west) indicates the time series is positively (negatively) correlated in that point. Meanwhile, the arrow pointing south (north) indicates that the former time series (i.e., $a(t)$) precedes the latter one (i.e., $b(t)$) by a ratio of $\pi/2$, and vice versa. In general, the arrow is shown as a combination of horizontal and vertical directions. For example, an arrow pointing south-west direction indicates that the two different time series (i.e., $a(t)$ and $b(t)$) are negatively correlated with the former time series (i.e., $a(t)$) preceding the latter one (i.e., $b(t)$).

4. Results

4.1. Deformation Rate of Eastern Beijing Plain

Previous studies have shown that the deformation of our study area mainly occurred in the vertical direction (e.g., [14,15]). Therefore, in this work, it is reasonable to convert the line-of-sight (LOS) deformation into the vertical direction. The vertical deformation rate of the eastern Beijing plain was retrieved from C-band Sentinel-1 observations by using the abovementioned progressive small baseline subsets (SBAS) InSAR approach. It can be seen from Figure 4 that the land subsidence of the eastern Beijing plain was relatively severe and the annual average deformation velocity was −150 mm/yr~20 mm/yr from July 2015 to December 2021. Significant land subsidence regions were mainly concentrated in the eastern part of Chaoyang District (e.g., Jinzhan, Guanzhuang, Shuangqiao and Heizhuanghu) and the northwestern part of Tongzhou District (e.g., Songzhuang, Yongshun, Liyuan, Taihu and Lucheng). During the monitoring period, the maximum annual deformation velocity was about −150 mm/yr, which was located in the south of Jinzhan in Chaoyang District. Another noteworthy phenomenon was that the deformed regions of Chaoyang District and Tongzhou District developed to connect into a whole piece with wide spatial coverage.

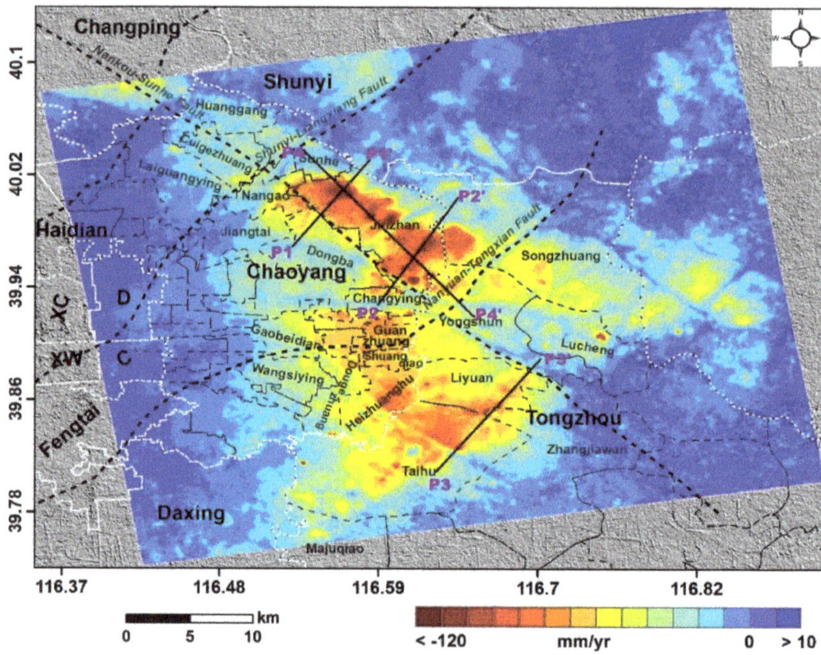

Figure 4. Annual average deformation rate map of eastern Beijing plain from July 2015 to December 2021 in the vertical direction. The positive values mean uplift, while the negative values mean subsidence. The thick black dashed line represents the main active faults, while the thick black solid line represents the location of the profiles across the faults.

4.2. Time Series Deformation of Eastern Beijing Plain

The cumulative time series deformation from 30 July 2015 to 13 December 2021 was also calculated in this study (Figure 5). It can be seen that obvious patterns were revealed in the spatial distribution, where the cumulative deformation magnitude and extent gradually increased with time in the eastern part of Chaoyang District and the northwestern part of Tongzhou District. From 30 July 2015 to 13 December 2021, the area with the cumulative deformation greater than −100 mm was about 651.1 km^2 and the maximum cumulative subsidence reached around −950 mm during the observation period, which was located in the south of Jinzhan in Chaoyang District.

In addition, the ground deformation at different stages from 2016 to 2021 were illustrated in Figure 6. From 2016 to 2018, the maximum deformation reduced from −192 mm to −173 mm, while the areas with the deformation greater than −50 mm and −100 mm decreased from 332.52 km^2 to 249.66 km^2, and from 46.6 km^2 to 8.47 km^2, respectively. From 2019 to 2021, the maximum deformation reduced from −133 mm to −102 mm, while the areas with the deformation greater than −50 mm and −100 mm decreased from 63.06 km^2 to 14.64 km^2, and from 0.58 km^2 to 0.01 km^2, respectively. Then, the deformation evolution of the feature points (labeled as FP1~FP6 in Figure 6a) at the typical deformation centers were obtained and analyzed (Figure 7). From Figure 7, FP1 showed a maximum deformation of −148.6 mm in 2016 and then suddenly decreased to −79.1 mm in 2019. Subsequently, the deformation continued to slow down, with a minimum of −50.85 mm in 2021. FP2 also showed a maximum deformation of −170.94 mm in 2016, and then fell to −75.49 mm in 2020. Subsequently, the deformation increased slightly to −86.37 mm in 2021. FP3 reached the maximum deformation of −173.21 mm in 2017, and then gradually declined to −30.55 mm in 2021. For FP4 and FP5, the deformation trend was similar to that of FP3. For FP6, the deformation was relatively steady and maintained at −60~−80 mm, except for a slight increase in 2019 and 2020. Generally speaking, the time series deformation of typical deformation centers showed that the land subsidence slowed down in 2017, which was about two years later than the start time of SNWD Project. Moreover, the corresponding complete time series of the feature points was shown in Figure 8. From Figures 7 and 8, it could be concluded that the area and magnitude of major deformation showed a significant decay trend from 2016 to 2021, even in the severe subsiding regions.

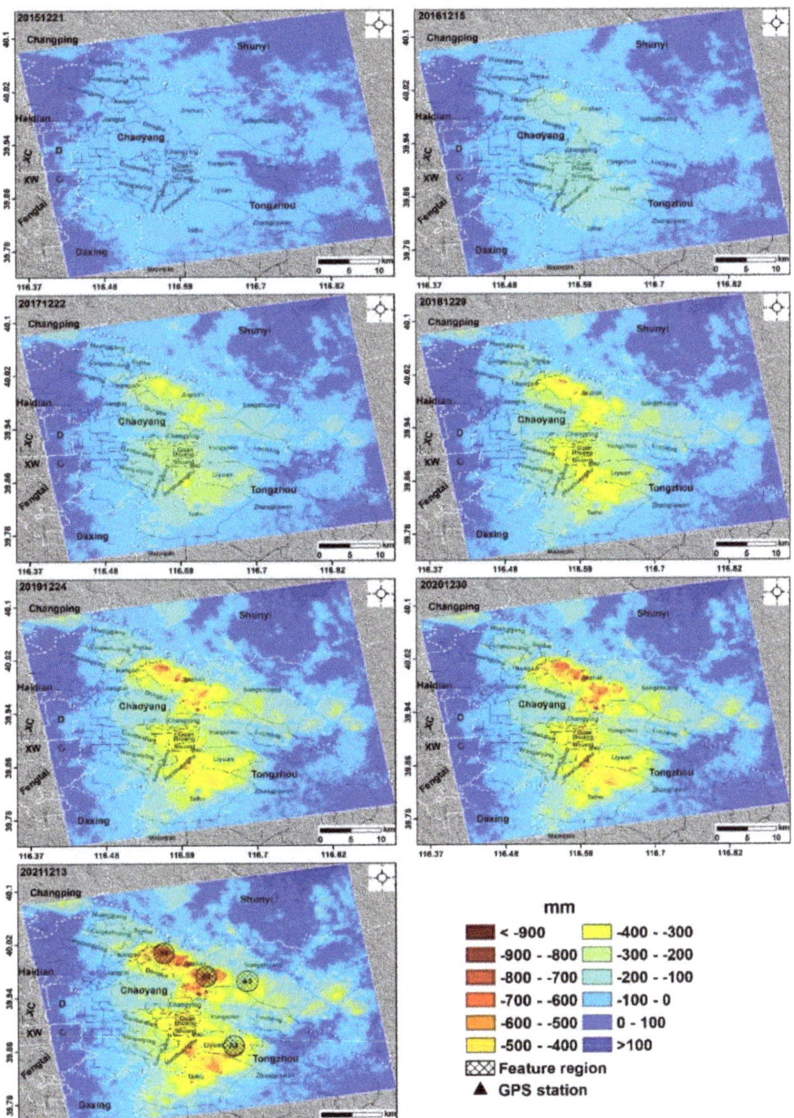

Figure 5. Cumulative vertical deformation from 30 July 2015 to 13 December 2021. The color bar indicates the vertical deformation in millimeters. The dashed white line is the county boundary of Beijing administrative district.

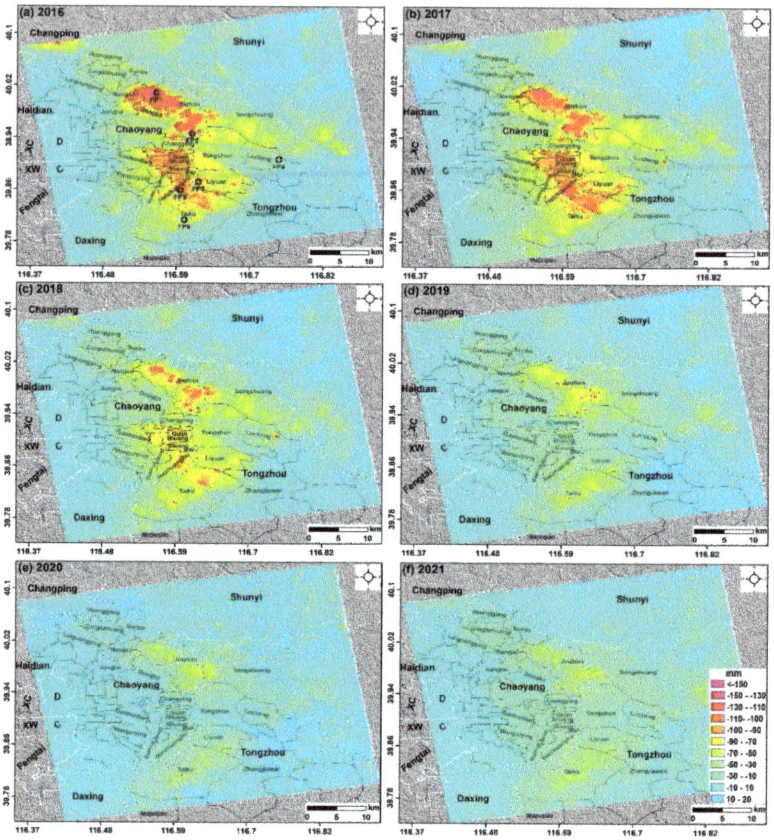

Figure 6. The distribution characteristics of land subsidence at different stages ((**a**) 2016, (**b**) 2017, (**c**) 2018, (**d**) 2019, (**e**) 2020, and (**f**) 2021) during the observation period in the eastern Beijing plain.

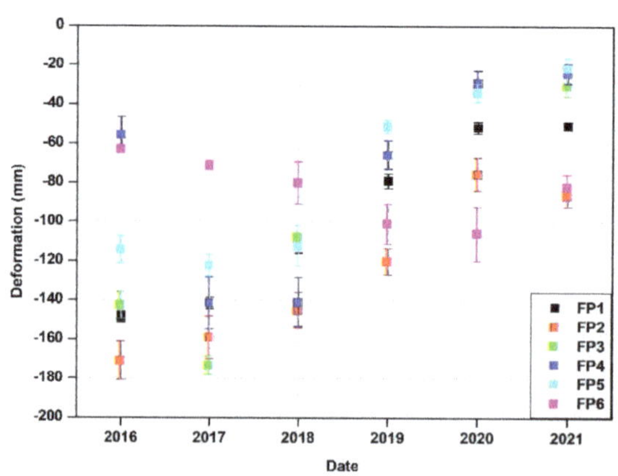

Figure 7. Annual Deformation of feature points located at typical deformed regions of Jinzhan (FP1~FP2), Heizhuanghu (FP3), Lucheng (FP4), Liyuan (FP5) and Taihu (FP6). Here the point targets within a 100-m radius buffer zone are selected to obtain the corresponding uncertainty of the InSAR-derived deformation. The error bar in each point represents the uncertainty of 1σ.

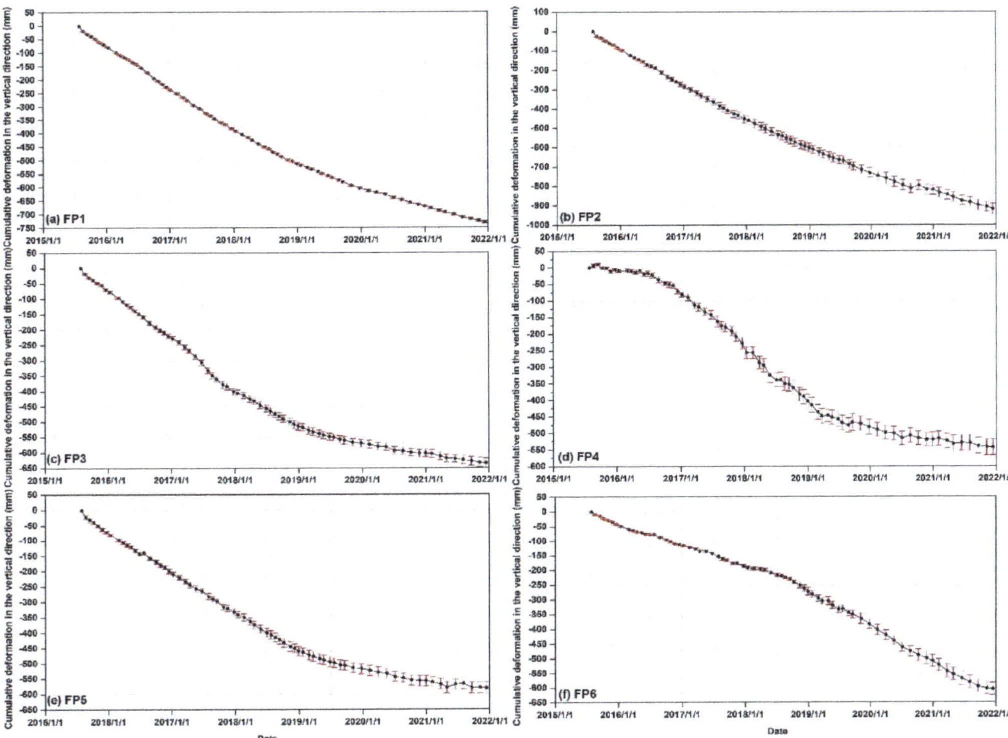

Figure 8. Time series deformation of feature points located at typical deformed regions of Jinzhan ((**a**) FP1~(**b**) FP2), Heizhuanghu ((**c**) FP3), Lucheng ((**d**) FP4), Liyuan ((**e**) FP5) and Taihu ((**f**) FP6).

4.3. Validation of InSAR Measurements

In order to reasonably validate the InSAR-derived deformation, we adopted two different validation strategies in this study. The first strategy referred to the full validation, which assessed the accuracy of InSAR-derived time series deformation with the continuous GPS observations. The second strategy referred to the trend validation, which conducted the comparison of deformation rates between InSAR and leveling measurements.

In this work, one GPS station and eighteen second-order leveling observations are collected, whose positions were shown in Figure 1. To reasonably compare the results from different techniques, we calibrated them into the common reference point. Then, we selected the coherent pixels within a 50-m radius buffer zone of each GPS station or leveling benchmark and averaged the corresponding InSAR-derived deformations around them. The cumulative time series deformation of the GPS station (shown as triangle in Figure 1) was presented in Figure 9, indicating that the InSAR-derived deformation agreed well with the GPS observations. The root mean square error (RMSE) between them was about 6.4 mm. A good agreement was also presented from the comparison result between InSAR observations and leveling measurements (Figure 10). The root mean square error (RMSE) and R^2 value (the confidence level at 95%) between them was about 4.3 mm and 0.99, respectively. The above comparison results suggested that the InSAR-derived deformation had a high accuracy and could provide reliable information for the following analysis on the potential influences of land subsidence in our study area.

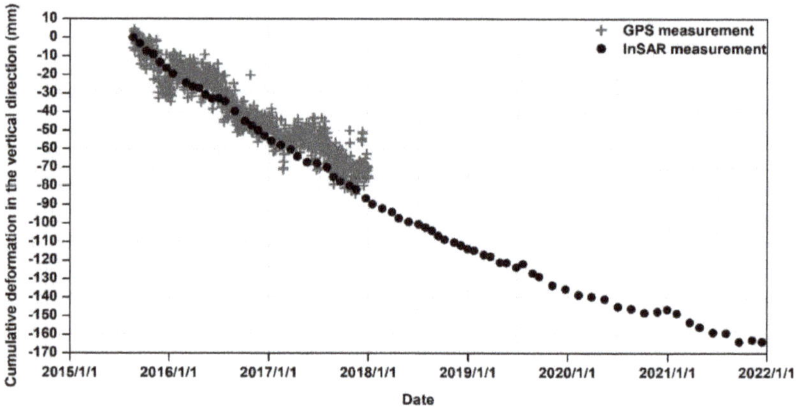

Figure 9. Cumulative time series deformation obtained from InSAR-derived and GPS observations.

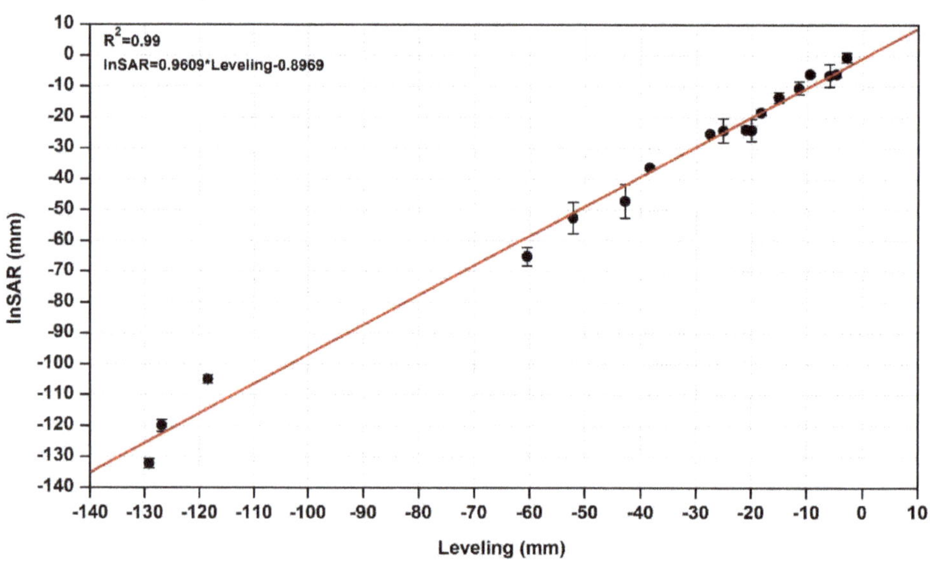

Figure 10. Comparison of InSAR-derived land subsidence and leveling measurements in the year of 2017. Here the point targets within a 50-m radius buffer zone are selected to obtain the corresponding uncertainty of the InSAR-derived deformation. The error bar in each point represents the uncertainty of 1σ.

5. Discussion

5.1. Cross Wavelet Transform (XWT) and Wavelet Transform Coherence (WTC) on Groundwater Level Change and InSAR-Derived Deformation

To reveal the correlations between the groundwater level change and ground deformation in the eastern Beijing plain, we conducted XWT and WTC on the groundwater level change and InSAR-derived deformation. Due to the lack of groundwater level information before June 2019, the correlation analysis was just conducted from June 2019 to December 2021 in this study. In addition, only the monthly groundwater level data could be collected, so XWT and WTC were applied based on the monthly groundwater level change and InSAR deformation change. Furthermore, considering the variation of groundwater exploitation in space, a buffer zone within the radius of 1.5 km was adopted to define the study area (A1~A4 shown in Figure 5). The mean value of each buffer zone was adopted to represent the groundwater level change and InSAR deformation change of each defined study area. Figure 11 showed the time series of monthly groundwater level change and InSAR-derived deformation change of the selected regions, while Figure 12 represented the corresponding cross wavelet power spectrum and wavelet coherence spectrum. The XWT power spectrum showed the distribution of frequency component with time, while the WTC power described the coherence of the different time series in the time-frequency space. Arrows indicated the relative phase difference between groundwater level change and InSAR-derived deformation change. The thick contour designated the area that has passed the red noise test with confidence level at 95%, while the cone of influence (COI) was displayed as the thinner black line and the regions outside the COI were displayed as lighter shadows. From Figures 11a and 12a,e, at A1, the groundwater level change and InSAR-derived deformation change were basically in-phase correlation between June 2019 and December 2021 with the 10-month resonance period. If we converted the cross wavelet coherence into a lagging time, the InSAR-derived deformation change lagged behind the groundwater level change by 0~1 month. At A2, the groundwater level change and InSAR-derived deformation change were anti-phase correlation between May 2020 and September 2020, indicating that groundwater level change was not the main control factor of land subsidence. From July 2020 to October 2021, in-phase correlation was observed and the lagging time between InSAR-derived deformation change and groundwater level change was about 0~2 months (Figures 11b and 12b,f). The similar in-phase correlation and lagging time were observed at A3 (Figures 11c and 12c,g). However, the InSAR-derived deformation change preceded the groundwater level change from June 2019 to December 2019, indicating that the groundwater exploitation may not be the key influencing factor. At A4, in-phase correlation was clearly observed during the whole period with resonance period of 4~10 months and the lagging time of 0~3 months (Figures 11d and 12d,h)), revealing that the land subsidence was mainly controlled by the groundwater level change. Our results may provide guiding information on the lagging characteristic between potential influencing factors and the observed deformation [49], which is of great significance to reveal the deformation mechanism of regional subsidence and establish the hydrogeological models.

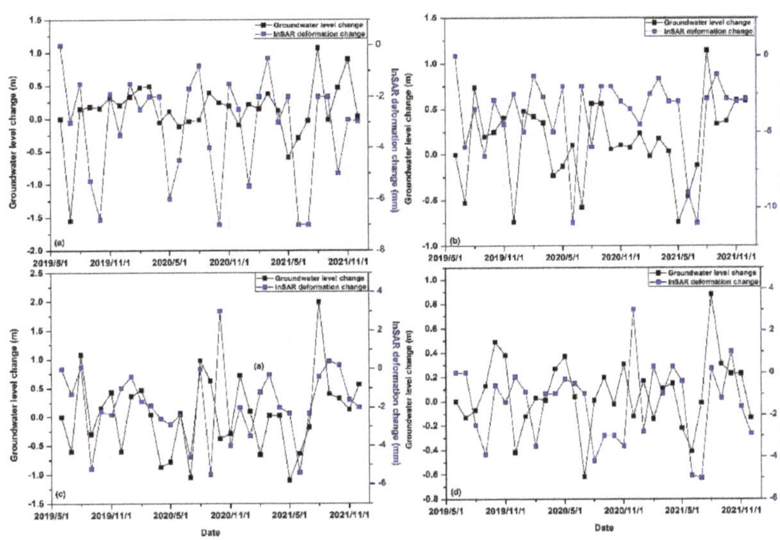

Figure 11. Time series of the monthly groundwater level change and InSAR-derived deformation change of the selected regions ((**a**) A1, (**b**) A2, (**c**) A3 and (**d**) A4).

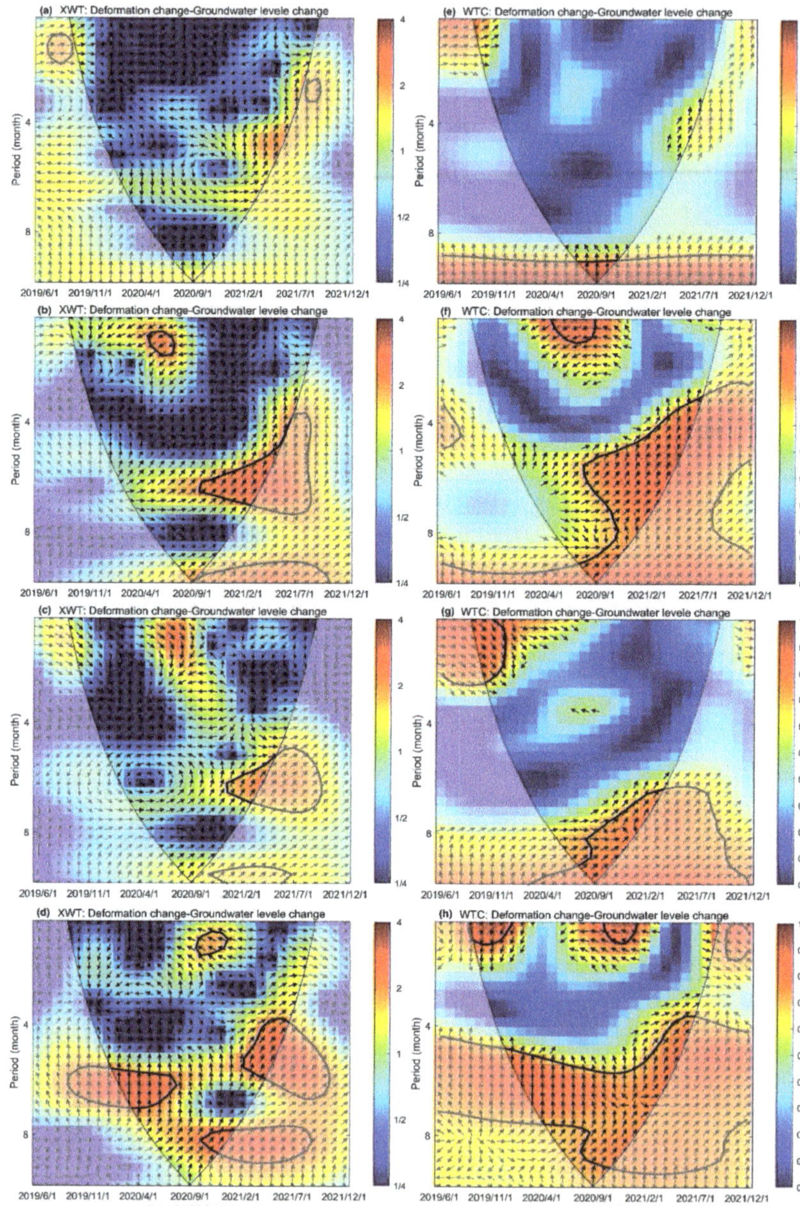

Figure 12. The cross wavelet power spectrum and the wavelet coherence between groundwater level change and InSAR-derived deformation change at four selected regions (A1: (**a**,**e**); A2: (**b**,**f**); A3: (**c**,**g**); A4: (**d**,**h**)). Pointing-right arrows: in-phase correlations, Pointing-left arrows: anti-phase correlations; Straight-down arrows: groundwater level change preceding InSAR-derived deformation change by 90°. The color bar of XWT and WTC shows the common cross-wavelet power of groundwater level change and InSAR-derived deformation change. The blue represents the low power while red represents the high power.

5.2. Correlation between Land Subsidence and Faults

Bai et al. [43] recorded that the compressible layer related to ground subsidence in the Beijing plain mainly occurred in the Quaternary sedimentary layer, and the sedimentary process was affected by the active geological faults. Therefore, the thickness of Quaternary sediments varied in spatial distribution, which provided the geological conditions to induce uneven ground subsidence. To reveal the correlation between ground subsidence and active geological faults, the annual average deformation rates along four profiles were analyzed (Figure 13), whose positions were marked in Figure 4. As shown in Figures 4 and 13, the boundaries of some subsiding regions corresponded to the traces of active geological faults, such as Nankou-Sunhe fault in the NW direction and Nanyuan-Tongxian fault in the NNE direction, which indicated that the spatial distribution of land subsidence was restricted by the active geological faults. Our results agreed well with previous studies [53]. In fact, many studies have reported similar phenomena in other regions, such as Mexico [4], Taiyuan [20], San Bernardino [54]. According to their opinions, the geological fault acts as a barrier to groundwater flow field, changing the horizontal hydraulic conductivity of aquifer systems and resulting in the differential groundwater level on both sides of the fault. This is the reason why the observed deformation rates are different on both sides of the Nankou-Sunhe fault and Nanyuan-Tongxian fault. The observed structural control of ground subsidence caused differential deformation rates on both sides of the fault, which may have an impact on urban infrastructures, for example, causing the damage to roads and railways. Furthermore, InSAR has proven to be an effective technique for mapping the active geological faults, which should be taken into consideration in future urban planning to avoid the damage of urban infrastructures owing to different deformations. At last, it is essential to conduct the continuous long-term deformation monitoring in these regions in order to update the kinematic data of geological faults and identify serious situations that could lead to the damage of urban infrastructures.

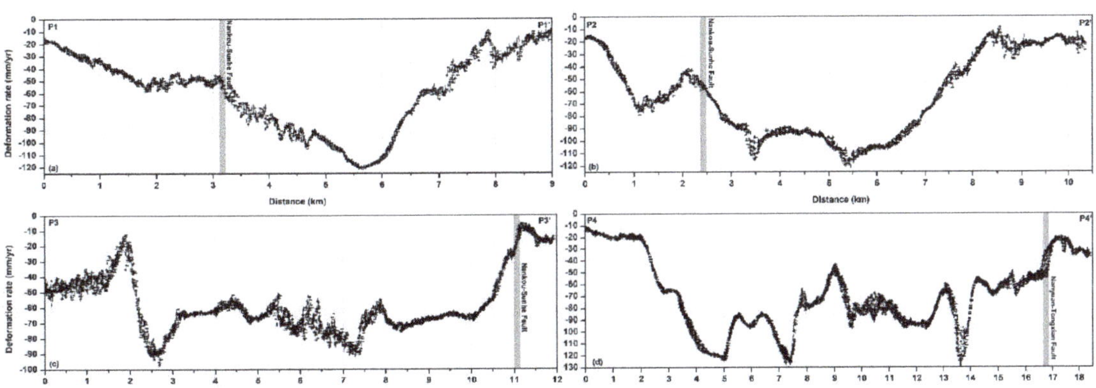

Figure 13. Vertical deformation rates along the profiles of P1P1′, P2P2′, P3P3′ and P4P4′. The locations of these profiles are shown in Figure 4. The gray bar shows the location of the active geological fault shown in Figure 4.

5.3. Correlation between Land Subsidence and Urban Expansion in Beijing's Sub-Administrative Center

Beijing's sub-administrative center (BSAC), located in the Tongzhou District, is of great significance for the construction and development of the Beijing-Tianjin-Hebei Metropolitan Area [55]. In general, the rapid urban expansion could aggravate the ground subsidence and threaten the safe operation of cities. Because the urban expansion was likely highly heterogeneous both in time and space, the potential relationship between ground subsidence and engineering construction was only revealed at the single- building scale in this study. Thus, a single building (labeled P) was selected for the deformation time series analysis, which belonged to a newly built Central Business District in the BSAC (Figure 14). The location of building P was shown in Figure 1. The cumulative deformation from 2015 to 2021 was shown in Figure 14a, which revealed that the time-varying deformation rates were clearly observed at point P during the whole monitoring period. The deformation of the building P was mainly characterized by two different stages: rapid settlement stage (T1) with the average deformation rate of −66.9 mm/yr and relatively stable stage (T2) with the deformation rate of −13.38 mm/yr after the completion of engineering construction. In order to further explain the deformation characteristics of the building P during different stages, the geological drilling information was described in Figure 14b, which showed that there was an aquifer filled with gravel and silt-fine sand below the building P. The total thickness of the aquifer was 13.5 m, where the effective stress on soil could increase rapidly and then the gradual compression of sediments occurred accordingly. During the process of foundation pit dewatering of building P, the pore pressure disappeared quickly. After the completion of foundation pit drainage, the groundwater level gradually rose. With the increase of pore water pressure, the effective stress on soil slowed down, leading to the decrease of deformation rate and even rebound. Previous studies showed that the compressive modulus of silty clay was approximately 8.64~11.41 and that of silt was about 9.71~17.22 in this area [55], indicating that the compressibility of silty clay and silt was high. When the effective stress increased gradually, the deformation characteristic of building P was established during the stage of T2. Based on the effective stress principle proposed by Terzaghi [56], once the pore water pressure disappears completely, the effective stress is equal to the total stress. Subsequently, the main consolidation of the soil will be finished, but it may last for many years. According to the collected Google Earth images shown in Figure 14c, it could be concluded that urban construction may be one of the influencing factor affecting the land subsidence in this area to some extent.

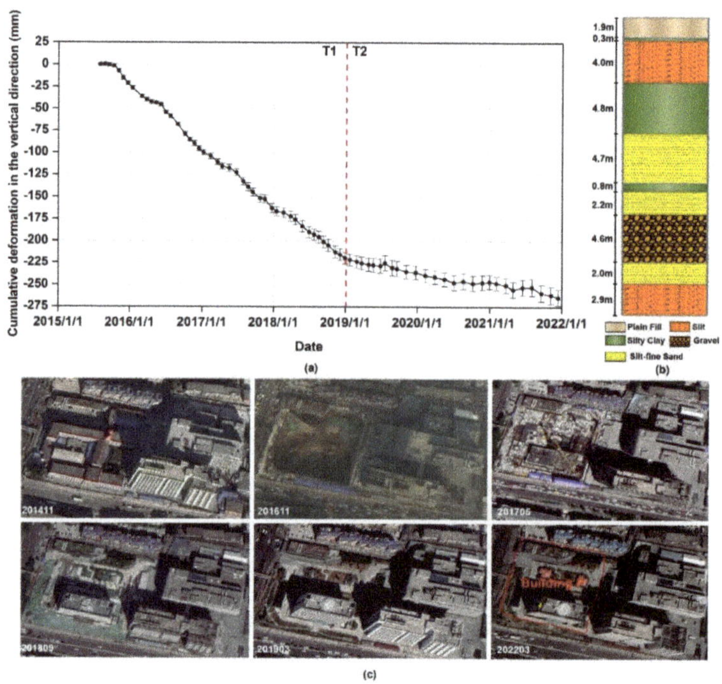

Figure 14. Cumulative deformation (a) and the geological drilling information of Building P (b), and (c) several typical enlarged optical remote sensing images obtained from Google earth of Building P.

Based on the above analysis, the InSAR-derived ground subsidence in the eastern Beijing plain is mainly induced by the excessive exploitation of groundwater, while its spatial distribution and extension is restricted by the local geological faults. In general, under the implementation of the SNWD Project, the excessive exploitation of groundwater in the eastern Beijing plain has been restricted and alleviated to a certain degree. However, it is a long-term and challenging task for the prevention and control of land subsidence caused by excessive exploitation of groundwater. It is necessary to wait for a long time to assess the role of the SNWD Project in replacing partial water demand and slowing down the excessive exploitation of groundwater in the capital. In addition, the urban construction (e.g., new buildings) may be one of the influencing factors affecting the ground subsidence to some extent. Therefore, in order to better study the spatiotemporal evolutions and influencing factors of land subsidence before and after the SNWD Project, we will adopt multi-source (e.g., Sentinel-1A/B, TerraSAR-X, ALOS PALSAR-2) or multi-track (e.g., ascending-track, descending-track) SAR data to retrieve the multi-dimensional deformations, combine optical remote sensing images to explore more influencing factors (e.g., floor areas of building, precipitation, land use type, urban expansion at the regional scale) on land subsidence and predict the future deformation (e.g., the deformation in 2050) with machine learning algorithms in the future research.

6. Conclusions

In this study, the deformation after the South-to-North Water Diversion (SNWD) Project in the eastern Beijing plain was retrieved by using the progressive small baseline subsets (SBAS) InSAR approach using Sentinel-1 SAR observations obtained from July 2015 to December 2021. The results showed that the eastern Beijing plain was still suffering severe ground subsidence and the maximum deformation rate was about −150 mm/yr in the vertical direction, which agreed well with the results published in the previous studies (e.g., −146 mm/yr [11]). Two main subsiding bowls with the deformation rate more than −100 mm/yr were identified in the eastern part of Chaoyang District (namely Jinzhan and Guanzhuan-Heizhuanghu) and the northwestern part of Tongzhou District (namely Songzhuang-Yongshun-Liyuan-Taihu). The InSAR-derived time series deformation agreed well with that from GPS observations and the RMSE was about 6.4 mm. Additionally, the validation between InSAR-derived deformation and leveling measurements indicated a high agreement, with the RMSE of 4.3 mm and the R^2 value up to 0.99 (the confidence level at 95%). Then, the deformation of six typical feature points at different stages revealed that the ground subsidence slowed down in 2017, which was about two years later than the start time of SNWD Project. Furthermore, the results obtained from cross wavelet transforms and wavelet transform coherence on the groundwater level change and InSAR-derived deformation change showed that the dynamic variation of groundwater fluctuation was the main influencing factor in the severe subsiding zones. The groundwater level change preceded land subsidence by 0~2 months in the eastern Beijing plain. In addition, the urban construction (e.g., new buildings) may be considered as one of the influencing factor on the ground subsidence to some extent.

This research provides important information about the latest land subsidence in the eastern Beijing plain, which may serve the decision-making in reducing the subsidence-related disasters and the management of water resources in Beijing.

Author Contributions: Conceptualization, Y.L. and Z.L.; methodology, Y.L.; software, X.Y.; validation, Y.L. and X.Y.; formal analysis, Y.L.; investigation, X.Y.; resources, Y.L.; writing—original draft preparation, Y.L. and X.Y.; writing—review and editing, M.Y. and B.L.; visualization, Y.X. and B.L.; funding acquisition, Y.L., Y.X. and B.L. All authors have read and agreed to the published version of the manuscript.

Funding: This work was supported by the National Natural Science Foundation of China (grant numbers 42104030, 42174055, 41962018), the Jiangxi Natural Science Foundation (grant numbers 20202BABL212016, 20212BAB204003), Key Laboratory for Digital Land and Resources of Jiangxi Province, East China University of Technology (grant number DLLJ202013) and the Doctoral Scientific Research Foundation of East China University of Technology (grant number DHBK2018004).

Data Availability Statement: Not applicable.

Conflicts of Interest: The authors declare no conflict of interest.

References

1. Herrera-García, G.; Ezquerro, P.; Tomás, R.; Béjar-Pizarro, M.; López-Vinielles, J.; Rossi, M.; Mateos, R.M.; Carreón-Freyre, D.; Lambert, J.; Teatini, P.; et al. Mapping the global threat of land subsidence. *Science* **2021**, *371*, 34–36. [CrossRef] [PubMed]
2. Cigna, F.; Tapete, D. Present-day land subsidence rates, surface faulting hazard and risk in Mexico City with 2014–2020 Sentinel-1 IW InSAR. *Remote Sens. Environ.* **2021**, *253*, 112161. [CrossRef]
3. Castellazzi, P.; Arroyo-Domínguez, N.; Martel, R.; Calderhead, A.I.; Normand, J.C.L.; Gárfias, J.; Rivera, A. Land subsidence in major cities of Central Mexico: Interpreting InSAR-derived land subsidence mapping with hydrogeological data. *Int. J. Appl. Earth Obs. Geoinf.* **2016**, *47*, 102–111. [CrossRef]
4. Cigna, F.; Osmanoglu, B.; Cabral-Cano, E.; Dixon, T.H.; Avila-Olivera, J.A.; Garduno-Monroy, V.H.; DeMets, C.; Wdowinski, S. Monitoring land subsidence and its induced geological hazard with Synthetic Aperture Radar Interferometry: A case study in Morelia, Mexico. *Remote Sens. Environ.* **2012**, *117*, 146–161. [CrossRef]
5. Qu, F.; Lu, Z.; Zhang, Q.; Bawden, G.W.; Kim, J.W.; Zhao, C.; Qu, W. Mapping ground deformation over Houston-Galveston, Texas using multi-temporal InSAR. *Remote Sens. Environ.* **2015**, *169*, 290–306. [CrossRef]
6. Peng, M.; Lu, Z.; Zhao, C.; Motagh, M.; Bai, L.; Conway, B.D.; Chen, H. Mapping land subsidence and aquifer system properties of the Willcox Basin, Arizona, from InSAR observations and independent component analysis. *Remote Sens. Environ.* **2022**, *271*, 112894. [CrossRef]
7. Fiaschi, S.; Fabris, M.; Floris, M.; Achilli, V. Estimation of land subsidence in deltaic areas through differential SAR interferometry: The Po River Delta case study (Northeast Italy). *Int. J. Remote Sens.* **2018**, *39*, 8724–8745. [CrossRef]
8. Yastika, P.E.; Shimizu, N.; Abidin, H.Z. Monitoring of long-term land subsidence from 2003 to 2017 in coastal area of Semarang, Indonesia by SBAS DInSAR analyses using Envisat-ASAR, ALOS-PALSAR, and Sentinel-1A SAR data. *Adv. Space Res.* **2019**, *63*, 1719–1736. [CrossRef]
9. Guo, L.; Gong, H.; Li, J.; Zhu, L.; Xue, A.; Liao, L.; Sun, Y.; Li, Y.; Zhang, Z.; Hu, L.; et al. Understanding uneven land subsidence in Beijing, China, using a novel combination of geophysical prospecting and InSAR. *Geophys. Res. Lett.* **2020**, *47*, e2020GL088676. [CrossRef]
10. Zhou, C.; Gong, H.; Chen, B.; Gao, M.; Cao, Q.; Cao, J.; Duan, L.; Zuo, J.; Shi, M. Land subsidence response to different land use types and water resource utilization in Beijing-Tianjin-Hebei, China. *Remote Sens.* **2020**, *12*, 457. [CrossRef]
11. Zhou, C.; Gong, H.; Chen, B.; Li, J.; Gao, M.; Zhu, F.; Chen, W.; Liang, Y. InSAR time-series analysis of land subsidence under different land use types in the eastern Beijing Plain, China. *Remote Sens.* **2017**, *9*, 380. [CrossRef]
12. Zhou, C.; Gong, H.; Zhang, Y.; Warner, T.; Wang, C. Spatiotemporal evolution of land subsidence in the Beijing Plain 2003–2015 using Persistent Scatterer Interferometry (PSI) with multi-source SAR data. *Remote Sens.* **2018**, *10*, 552. [CrossRef]
13. Chen, M.; Tomás, R.; Li, Z.; Motagh, M.; Li, T.; Hu, L.; Gong, H.; Li, X.; Yu, J.; Gong, X. Imaging land subsidence induced by groundwater extraction in Beijing (China) using satellite radar interferometry. *Remote Sens.* **2016**, *8*, 468. [CrossRef]
14. Du, Z.; Ge, L.; Ng, A.H.M.; Lian, X.; Zhu, Q.; Horgan, F.G.; Zhang, Q. Analysis of the impact of the South-to-North water diversion project on water balance and land subsidence in Beijing, China between 2007 and 2020. *J. Hydrol.* **2021**, *603*, 126990. [CrossRef]
15. Gao, M.; Gong, H.; Chen, B.; Li, X.; Zhou, C.; Shi, M.; Si, Y.; Chen, Z.; Duan, G. Regional land subsidence analysis in eastern Beijing plain by InSAR time series and wavelet transforms. *Remote Sens.* **2018**, *10*, 365. [CrossRef]
16. Zhu, L.; Gong, H.; Li, X.; Wang, R.; Chen, B.; Dai, Z.; Teatini, P. Land subsidence due to groundwater withdrawal in the northern Beijing plain, China. *Eng. Geol.* **2015**, *193*, 243–255. [CrossRef]
17. Yang, M.; Yang, T.; Zhang, L.; Lin, J.; Qin, X.; Liao, M. Spatio-temporal characterization of a reclamation settlement in the Shanghai coastal area with time series analyses of X-, C-, and L-band SAR datasets. *Remote Sens.* **2018**, *10*, 329. [CrossRef]
18. Qu, F.; Zhang, Q.; Lu, Z.; Zhao, C.; Yang, C.; Zhang, J. Land subsidence and ground fissures in Xi'an, China 2005–2012 revealed by multi-band InSAR time-series analysis. *Remote Sens. Environ.* **2014**, *155*, 366–376. [CrossRef]
19. Zhao, Y.; Zhou, L.; Wang, C.; Li, J.; Qin, J.; Sheng, H.; Huang, L.; Li, X. Analysis of the spatial and temporal evolution of land subsidence in Wuhan, China from 2017 to 2021. *Remote Sens.* **2022**, *14*, 3142. [CrossRef]
20. Liu, Y.; Zhao, C.; Zhang, Q.; Yang, C.; Zhang, J. Land subsidence in Taiyuan, China, monitored by InSAR technique with multi-sensor SAR datasets from 1992 to 2015. *IEEE J. Sel. Top. Appl. Earth Obs. Remote Sens.* **2018**, *11*, 1509–1519. [CrossRef]
21. Hanssen, R.F. *Radar Interferometry: Data Interpretation and Error Analysis*; Kluwer Academic Publishers: Dordrecht, The Netherlands, 2001.

22. Zebker, H.A.; Villasenor, J. Decorrelation in interferometric radar echoes. *IEEE Trans. Geosci. Remote Sens.* **1992**, *30*, 950–959. [CrossRef]
23. Zebker, H.A.; Rosen, P.A.; Hensley, S. Atmospheric effects in interferometric synthetic aperture radar surface deformation and topographic maps. *J. Geophys. Res. Solid Earth* **1997**, *102*, 7547–7563. [CrossRef]
24. Ferretti, A.; Prati, C.; Rocca, F. Nonlinear subsidence rate estimation using permanent scatterers in differential SAR interferometry. *IEEE Trans. Geosci. Remote Sens.* **2000**, *38*, 2202–2212. [CrossRef]
25. Hooper, A.; Segall, P.; Zebker, H. Persistent scatterer interferometric synthetic aperture radar for crustal deformation analysis, with application to Volcán Alcedo, Galápagos. *J. Geophys. Res.* **2007**, *112*. [CrossRef]
26. Zhang, L.; Ding, X.L.; Lu, Z. Modeling the PSInSAR time-series without phase unwrapping. *IEEE Trans. Geosci. Remote Sens.* **2011**, *49*, 547–556. [CrossRef]
27. Berardino, P.; Fornaro, G.; Lanari, R.; Sansosti, E. A new algorithm for surface deformation monitoring based on small baseline differential SAR interferograms. *IEEE Trans. Geosci. Remote Sens.* **2002**, *40*, 2375–2383. [CrossRef]
28. Hetland, E.A.; Musé, P.; Simons, M.; Lin, Y.N.; Agram, P.S.; DiCaprio, C.J. Multiscale InSAR time series (MInTS) analysis of surface deformation. *J. Geophys. Res. Solid Earth* **2012**, *117*. [CrossRef]
29. Samsonov, S.V.; d'Oreye, N.; González, P.J.; Tiampo, K.F.; Ertolahti, L.; Clague, J.J. Rapidly accelerating subsidence in the Greater Vancouver region from two decades of ERS-ENVISAT-RADARSAT-2 DInSAR measurements. *Remote Sens. Environ.* **2014**, *143*, 180–191. [CrossRef]
30. Ferretti, A.; Fumagalli, A.; Novali, F.; Prati, C.; Rocca, F.; Rucci, A. A New algorithm for processing interferometric data-stacks: SqueeSAR. *IEEE Trans. Geosci. Remote Sens.* **2011**, *49*, 3460–3470. [CrossRef]
31. Gong, H.; Pan, Y.; Zheng, L.; Li, X.; Zhu, L.; Zhang, C.; Huang, Z.; Li, Z.; Wang, H.; Zhou, C. Long-term groundwater storage changes and land subsidence development in the North China Plain (1971–2015). *Appl. Hydrogeol.* **2018**, *26*, 1417–1427. [CrossRef]
32. Chen, B.; Gong, H.; Li, X.; Lei, K.; Zhang, Y.; Li, J.; Gu, Z.; Dang, Y. Spatial-temporal characteristics of land subsidence corresponding to dynamic groundwater funnel in Beijing Municipality, China. *Chin. Geogr. Sci.* **2011**, *21*, 753–764. [CrossRef]
33. Ng, A.H.-M.; Ge, L.; Li, X.; Zhang, K. Monitoring ground deformation in Beijing, China with persistent scatterer SAR interferometry. *J. Geod.* **2012**, *86*, 375–392. [CrossRef]
34. Zhang, Y.; Wu, H.; Kang, Y.; Zhu, C. Ground subsidence in the Beijing-Tianjin-Hebei region from 1992 to 2014 revealed by multiple SAR stacks. *Remote Sens.* **2016**, *8*, 675. [CrossRef]
35. Chen, B.; Gong, H.; Li, X.; Lei, K.; Zhu, L.; Gao, M.; Zhou, C. Characterization and causes of land subsidence in Beijing, China. *Int. J. Remote Sens.* **2017**, *38*, 808–826. [CrossRef]
36. Yang, Q.; Ke, Y.; Zhang, D.; Chen, B.; Gong, H.; Lv, M.; Zhu, L.; Li, X. Multi-scale analysis of the relationship between land subsidence and buildings: A case study in an eastern Beijing urban area using the PS-InSAR technique. *Remote Sens.* **2018**, *10*, 1006. [CrossRef]
37. Zhou, C.; Lan, H.; Gong, H.; Zhang, Y.; Warner, T.A.; Clague, J.J.; Wu, Y. Reduced rate of land subsidence since 2016 in Beijing, China: Evidence from Tomo-PSInSAR using RadarSAT-2 and Sentinel-1 datasets. *Int. J. Remote Sens.* **2020**, *41*, 1259–1285. [CrossRef]
38. Duan, L.; Gong, H.; Chen, B.; Zhou, C.; Lei, K.; Gao, M.; Yu, H.; Cao, Q.; Cao, J. An improved multi-sensor MTI time-series fusion method to monitor the subsidence of Beijing Subway Network during the past 15 years. *Remote Sens.* **2020**, *12*, 2125. [CrossRef]
39. Lyu, M.; Ke, Y.; Li, X.; Zhu, L.; Guo, L.; Gong, H. Detection of seasonal deformation of highway overpasses using the PS-InSAR technique: A case study in Beijing urban area. *Remote Sens.* **2020**, *12*, 3071. [CrossRef]
40. Chen, B.; Gong, H.; Chen, Y.; Li, X.; Zhou, C.; Lei, K.; Zhu, L.; Duan, L.; Zhao, X. Land subsidence and its relation with groundwater aquifers in Beijing Plain of China. *Sci. Total Environ.* **2020**, *735*, 139111. [CrossRef]
41. Chen, B.; Gong, H.; Chen, Y.; Lei, K.; Zhou, C.; Si, Y.; Li, X.; Pan, Y.; Gao, M. Investigating land subsidence and its causes along Beijing high-speed railway using multi-platform InSAR and a maximum entropy model. *Int. J. Appl. Earth Obs. Geoinf.* **2021**, *96*, 102284. [CrossRef]
42. Wang, Y.; Bai, Z.; Zhang, Y.; Qin, Y.; Lin, Y.; Li, Y.; Shen, W. Using TerraSAR X-band and Sentinel-1 C-band SAR interferometry for deformation along Beijing-Tianjin Intercity Railway analysis. *IEEE J. Sel. Top. Appl. Earth Obs. Remote Sens.* **2021**, *14*, 4832–4841. [CrossRef]
43. Bai, Z.; Wang, Y.; Balz, T. Beijing land subsidence revealed using PS-InSAR with long time series TerraSAR-X SAR data. *Remote Sens.* **2022**, *14*, 2529. [CrossRef]
44. Zhang, X.; Chen, B.; Gong, H.; Lei, K.; Zhou, C.; Lu, Z.; Zhao, D. Inversion of groundwater storage variations considering lag effect in Beijing plain, from Radarsat-2 with SBAS-InSAR Technology. *Remote Sens.* **2022**, *14*, 991. [CrossRef]
45. Mu, C. Analysis of the influence of South to North Water Transfer on land subsidence in Beijing. *Hebei Dizhidaxue Xuebao* **2017**, *40*, 22–27.
46. Xu, Y.; Wu, W.; Zhang, X.; Ma, S.; Ma, W.; Yu, G.; Gu, M.; Jiang, W. *Crust Newly Tectonic Deformation and Earthquake in Capital Zone*; Science Press: Beijing, China, 2002.
47. Lanari, R.; Mora, O.; Manunta, M.; Mallorquí, J.J.; Berardino, P.; Sansosti, E. A small-baseline approach for investigating deformations on full-resolution differential SAR interferograms. *IEEE Trans. Geosci. Remote Sens.* **2004**, *42*, 1377–1386. [CrossRef]
48. Sui, L.; Liu, Y.; Wang, W. Adaptive sequential adjustment and its application. *Geomat. Inf. Sci. Wuhan Univ.* **2007**, *32*, 51–54.

49. Liu, Y.; Qiu, H.; Yang, D.; Liu, Z.; Ma, S.; Pei, Y.; Zhang, J.; Tang, B. Deformation responses of landslides to seasonal rainfall based on InSAR and wavelet analysis. *Landslides* **2022**, *19*, 199–210. [CrossRef]
50. Shirzaei, M. A wavelet-based multitemporal DInSAR algorithm for monitoring ground surface motion. *IEEE Geosci. Remote Sens. Lett.* **2013**, *10*, 456–460. [CrossRef]
51. Tomás, R.; Li, Z.; Lopez-Sanchez, J.M.; Liu, P.; Singleton, A. Using wavelet tools to analyse seasonal variations from InSAR time-series data: A case study of the Huangtupo landslide. *Landslides* **2016**, *13*, 437–450. [CrossRef]
52. Grinsted, A.; Moore, J.C.; Jevrejeva, S. Application of the cross wavelet transform and wavelet coherence to geophysical time series. *Nonlinear Proc. Geoph.* **2004**, *11*, 561–566. [CrossRef]
53. Lei, K.; Ma, F.; Chen, B.; Luo, Y.; Cui, W.; Zhou, Y.; Liu, H.; Sha, T. Three-dimensional surface deformation characteristics based on time series InSAR and GPS technologies in Beijing, China. *Remote Sens.* **2021**, *13*, 3964. [CrossRef]
54. Lu, Z.; Danskin, W.R. InSAR analysis of natural recharge to define structure of a ground-water basin, San Bernardino, California. *Geophys. Res. Lett.* **2001**, *28*, 2661–2664. [CrossRef]
55. Cao, J.; Gong, H.; Chen, B.; Shi, M.; Zhou, C.; Lei, K.; Yu, H.; Sun, Y. Land subsidence in Beijing's Sub-Administrative Center and its relationship with urban expansion inferred from Sentinel-1/2 observations. *Can. J. Remote Sens.* **2021**, *47*, 802–817. [CrossRef]
56. Terzaghi, K.V. *Theoretical Soil Mechanics*; John Wiley and Sons: New York, NY, USA, 1943.

remote sensing

MDPI

Article

Prediction of Mine Subsidence Based on InSAR Technology and the LSTM Algorithm: A Case Study of the Shigouyi Coalfield, Ningxia (China)

Fei Ma [1,2,*], Lichun Sui [2] and Wei Lian [1]

[1] Department of Computer Science, Changzhi University, Changzhi 046011, China; lianwei@czc.edu.cn
[2] College of Geological Engineering and Geomatics, Chang'an University, Xi'an 710054, China; sui1011@chd.edu.cn
* Correspondence: mafei@czc.edu.cn; Tel.: +86-1553-458-7722

Abstract: The accurate prediction of surface subsidence induced by coal mining is critical to safeguarding the environment and resources. However, the precision of current prediction models is often restricted by the lack of pertinent data or imprecise model parameters. To overcome these limitations, this study proposes an approach to predicting mine subsidence that leverages Interferometric Synthetic Aperture Radar (InSAR) technology and the long short-term memory network (LSTM). The proposed approach utilizes small baseline multiple-master high-coherent target (SBMHCT) interferometric synthetic aperture radar technology to monitor the mine surface and applies the long short-term memory (LSTM) algorithm to construct the prediction model. The Shigouyi coalfield in Ningxia Province, China was chosen as a study area, and time series ground subsidence data were obtained based on Sentinel-1A data from 9 March 2015 to 7 June 2016. To evaluate the proposed approach, the prediction accuracies of LSTM and Support Vector Regression (SVR) were compared. The results show that the proposed approach could accurately predict mine subsidence, with maximum absolute errors of less than 2 cm and maximum relative errors of less than 6%. The findings demonstrate that combining InSAR technology with the LSTM algorithm is an effective and robust approach for predicting mine subsidence.

Keywords: InSAR technology; mines; time series subsidence monitoring; deformation prediction; deep learning; LSTM algorithm

Citation: Ma, F.; Sui, L.; Lian, W. Prediction of Mine Subsidence Based on InSAR Technology and the LSTM Algorithm: A Case Study of the Shigouyi Coalfield, Ningxia (China). *Remote Sens.* **2023**, *15*, 2755. https://doi.org/10.3390/rs15112755

Academic Editors: Constantinos Loupasakis, Ioannis Papoutsis and Konstantinos G. Nikolakopoulos

Received: 26 March 2023
Revised: 21 May 2023
Accepted: 22 May 2023
Published: 25 May 2023

1. Introduction

In recent years, satellite remote sensing images have become increasingly utilized for monitoring land surface changes due to the continuous development of computer and space satellite technology. In particular, InSAR has emerged as a popular technology for monitoring surface subsidence changes, especially goaf deformation, as it is not impacted by weather and has extensive coverage [1]. Goaf is formed after coal extraction from underground, and the continuous and efficient monitoring of surface subsidence above goaf can facilitate the understanding of surface subsidence damage to surface structures, explore mining subsidence mechanisms, and provide a decision-making basis for geological disaster prevention and ecological restoration in mining areas [2]. The authors of [3] estimated that the total economic losses due to subsidence from coal mining in China were approximately 32 billion CNY (about 4.9 billion USD) from 2001 to 2010. These losses were primarily due to damage to buildings, roads, and other infrastructure caused by surface subsidence above goaf. Surface subsidence caused by goaf formation can lead to accidents such as landslides, rockfalls, and collapses, which can result in injuries and fatalities; [4] and [5] reported that coal mining-related subsidence caused accidents, resulting in deaths or injuries every year in the world. Efficient monitoring of surface subsidence above goaf can help in preventing accidents and reducing economic losses [6–8]. The traditional form of goaf surface subsidence monitoring involves point-like monitoring stations, which are characterized by high consumption, low efficiency, limited coverage, and insufficient monitoring capability. Thus, it is of immense theoretical value and practical significance to use the new monitoring method of goaf surface subsidence (InSAR) to explore the formation mechanism behind subsidence in key mining areas and predict the evolution law and development trend of subsidence based on new monitoring methods and technology. Recent advances in deep learning theory have had a significant impact on time series prediction. As a result, an increasing number of deep-learning algorithms are being utilized to research long time series prediction, thereby making it possible to obtain mine subsidence characteristic information and dynamic forecasting in mining areas [9,10]. Deep-learning algorithms such as artificial neural networks (ANNs) [11] and recurrent neural networks (RNNs) [12] can be used to analyze long time series data and predict mine subsidence characteristics and dynamics. However, these methods need further improvement to improve the accuracy of prediction.

InSAR, an active remote-sensing technology, has been widely used for monitoring subsidence and surface deformation [13]. Initially, this technology was employed for ground elevation mapping [14] and subsequently extended to surface deformation monitoring [15]. However, the atmospheric phase delay and temporal and spatial decorrelation associated with two-pass InSAR technology can lead to phase unwrapping failure. Therefore, researchers proposed time series InSAR monitoring technology [16], including PS-InSAR [17] and SBAS-InSAR [18], with the latter being more suitable for deformation monitoring in mining areas. SBAS-InSAR technology involves registering interferences in pairs of SAR data sets covering the same area and selecting interferograms whose temporal and spatial baselines meet the threshold [19]. The highly coherent points in the images are then reconstructed based on the interferograms' phase. SBAS-InSAR has shown high-quality monitoring results with sub-centimeter monitoring accuracy [20,21]. G. Herrera et al. [22] demonstrated the monitoring capacity of InSAR technology using multi-sensor and multi-temporal SAR data in very slow landslides. Dario Peduto et al. [23] used DInSAR technology to analyze building deformation and presented a multi-scale procedure tailored to analyze the settlement-induced building damage; it could forecast building damage in urban areas. M. P. Sanabria et al. [24] proposed a methodology to produce subsidence activity maps based on PSInSAR data; these displacement map measurements are interpolated based on conditional Sequential Gaussian Simulation complement, and they are helpful for the identification of wide subsiding areas.

In the context of mining subsidence, InSAR technology has been increasingly recognized as a valuable tool for monitoring surface deformation. However, predicting subsidence movement remains a challenge and requires a prediction model that integrates InSAR data. To date, two broad categories of prediction models have been employed: traditional and late models. Traditional models use various technical methods to obtain surface deformation data post-mining and predict the maximum deformation value using mathematical functions or numerical models. Examples include numerical simulation, similar material simulation, probability integration, and other static prediction models [25].

The mining subsidence process is a complex spatio-temporal phenomenon, posing challenges for applying static prediction models that cannot account for dynamic changes [26]. Alternatively, continuous multi-period surface deformation data obtained by various technical means can be analyzed to predict the location and timing of maximum surface movement deformation by incorporating time functions such as Knothe [27], Weibull [28], and Logistic [29]. However, these models can only capture the linear relationship between two vectors and are limited in their ability to predict nonlinear deformation in mining areas. Moreover, due to the dynamic changes in mining practices, such as mode, speed, and roof management, the accuracy of dynamic time function simulations is often compromised [30]. Due to the complexity of the mining subsidence process in both time and space, static prediction models have limited practical application as they cannot simulate the dynamic changes in the subsidence process. In addition, the actual surface subsidence is different under different geological and mining conditions, but the prediction result is the same if the same time function is used, which is contradictory to the actual situation. Therefore, researchers have focused on "late models", such as the grey model [31], regression analysis [32,33], support vector machine regression [9], Bayesian network [10], wavelet analysis [34], and artificial neural network [35,36]. These models rely on modern and efficient monitoring means such as GNSS, InSAR, and LIDAR to obtain long-term series monitoring data and analyze internal statistical laws and trends. However, these methods are sensitive to model parameters, and adding mining geological parameters for goaf prediction can be challenging.

Compared to traditional mining subsidence prediction methods and late models, deep learning offers a novel approach to address this challenge. By formulating the relationship between response variables in a regression equation, deep-learning algorithms can accurately capture the impact of independent variables influenced by one or more dependent variables. While previous deep-learning techniques such as BP neural networks [37] and recurrent neural networks (RNN) [12] have been developed, they have not been ideal for long-term series prediction [38]. However, recent studies have shown that LSTM models, which combine RNN and attention mechanisms, are better suited for long-term prediction [39]. These models use a cellular structure, with the forgetting gate discarding unnecessary information while the memory gate retains important information.

Homa Ansari et al. [40] conducted an experiment on the Lazufre Volcanic Complex, situated in the central volcanic region, concluding that signal error associated with InSAR technology is a crucial factor contributing to inaccurate predictions when combined with LSTM. Hill et al. [41] focused on the influence of seasonal perturbations on forecasting outcomes. The LSTM prediction methodology proved efficient for short-term projections (less than three months). Qinghao Liu et al. [42] proposed a heterogeneous LSTM network model, which integrates spatial heterogeneity into predicting ground subsidence, successfully achieving accurate and efficient large-scale subsidence forecast. Yi Chen et al. [43] demonstrated the effectiveness of an unimproved LSTM neural network approach for time-series InSAR land subsidence prediction. Despite using InSAR technology and LSTM network models to forecast deformation, these studies produce contradictory findings due to the insufficient recognition of the significance of the InSAR training data, particularly in areas affected by error signals, such as volcanic and mining regions. Above all, the traditional InSAR needs to be improved to monitor mine deformation; inaccurate training data do not help improve the prediction accuracy of deep learning [44,45]. The purpose of this study is to obtain fine subsidence characteristics and accurate data on mining surfaces by improving InSAR technology and realize the accurate prediction of mining surface subsidence combined with the LSTM algorithm. This study presents an integrated monitoring and prediction model for the goaf surface that combines SBMHCT-InSAR and LSTM algorithms. The utilization of SBMHCT-InSAR technology enhanced multiple aspects of image processing and interpretation, such as the image registration algorithm, interferogram filtering method, and high coherence point extraction method. The objective of these advancements is to mitigate the disruptive influence of noise signal and optimize the training data of the prediction model, ultimately resulting in an enhanced level of accuracy. In practice, this technology is leveraged in the monitoring of goaf surface deformation, facilitating the retrieval of settlement values from equal interval time series training data. Additionally, the LSTM algorithm is employed to establish a deformation prediction model for coal mining areas by drawing the global dependence relationship between input and output and learning the nonlinear patterns and features of the training data. The main aim of this research is to forecast geological hazards while also addressing practical problems associated with the prediction of goaf subsidence using InSAR technology with deep learning theory. The study findings not only offer methodological support for mining subsidence management but also promote the quantitative application and development of InSAR technology. Therefore, the proposed model carries significant scientific and practical value.

2. Materials and Methods

2.1. Site Selection

The Shigouyi (SGY) mine area, one of the Ningdong coalfields situated in the eastern region of the Ningxia province, is depicted in Figure 1, with its geographical coordinates falling between 37°39′15″–37°45′17″N and 106°27′49″–106°30′44″E. It stretches westward to the Liupan Mount tectonic zone and eastward to the Erdos coal seam; it comprises a series of folds and faults. However, the Shigouyi mine experiences a significant, concentrated distribution of surface damage and subsidence because of its location on the Locss Plateau. The geological structure above the coal beds susceptible to mining is highly fragile.

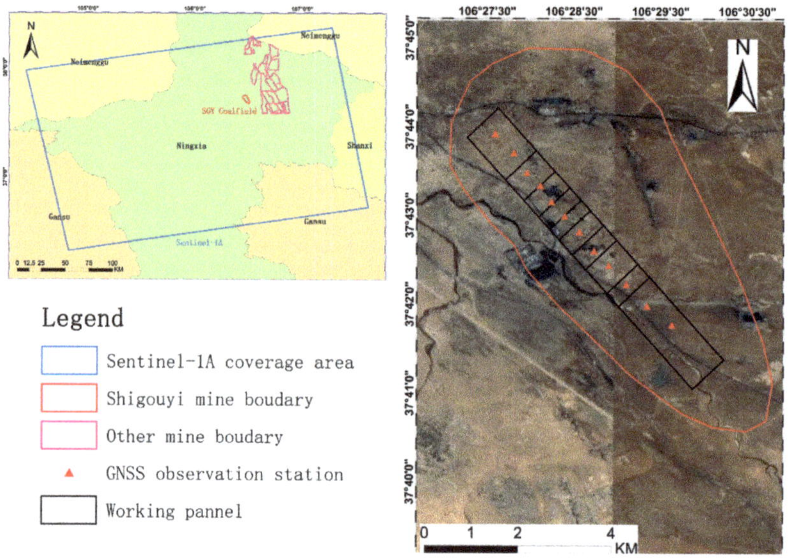

Figure 1. Geographical location of Shigouyi coalfield. The background is optical imagery of SGY coalfield.

Considering space constraints, this study focuses primarily on monitoring and predicting surface subsidence in the SGY coal mine. In this regard, a total of 13 GNSS observation stations have been installed above the working face of the mine, and the corresponding settlement data have been collected for verification of the experimental findings. The Chinese Southsurvey GNSS receivers were utilized to collect GNSS data in the real-time kinematic mode. One receiver was situated at the base station, which was positioned on a stable surface, while the others monitored displacements at the GNSS stations. The GNSS receiver exhibited horizontal and vertical accuracy of $\pm(10 + 1 \times 10^{-6} \times D)$ mm and $\pm(20 + 1 \times 10^{-6} \times D)$ mm (where D is the distance), respectively. Over the period of 9 March 2015 to 1 July 2016, GNSS-RTK measurements were taken at intervals of 24 days. Initially, a GNSS receiver with a tripod was installed on the reference station situated on the stable surface, where the antenna height was measured; receivers were opened; and the reference station height, antenna height, and WGS84 coordinate were inputted. The radio channel was then turned on and checked. The roving station GNSS receiver was subsequently opened with a centering rod, exact parameters were inputted, the radio channel and number of satellites were checked, and simultaneous observation with the reference station GNSS receiver was completed. Using the roving station GNSS receiver, the 12 GNSS stations' coordinates and heights were measured. Data were obtained at a sampling rate of 20 Hz, with the observation time being more than 180 s. These 12 GNSS stations were measured again in the same manner at intervals of 24 days in the ensuing months, with the deformation value calculated by the difference value of these times, while quality control measures were taken each time. Verification of the reliability of the RTK results was conducted using the method of comparison with quick static measurement, where at least three points were selected as checkpoints, and the observation time of quick static measurement exceeded 600 s. After data processing, the maximum error between quick static measurement and RTK was less than 2 cm in height.

2.2. Data Selection

This study processed 19 SAR images (C-band) obtained from the Sentinel-1 satellite over a 15-month period from 9 March 2015 to 7 June 2016. Table 1 shows the parameters of the SAR data used in this study; their temporal resolution is 24 days, and their spatial resolutions are 20 and 5 m in azimuth and range, respectively.

Table 1. Sentinel-1A data covering SGY mine area used in this work.

No.	Acquisition Date (ddmmyyyy)	Track	Product Type	Mode	Polarization	Orbit Direction
1	9 March 2015	157	SLC	IW	VV	ascending
2	2 April 2015	157	SLC	IW	VV	ascending
3	26 April 2015	157	SLC	IW	VV	ascending
4	20 May 2015	157	SLC	IW	VV	ascending
5	13 June 2015	157	SLC	IW	VV	ascending
6	7 July 2015	157	SLC	IW	VV	ascending
7	31 July 2015	157	SLC	IW	VV	ascending
8	24 August 2015	157	SLC	IW	VV	ascending
9	17 September 2015	157	SLC	IW	VV	ascending
10	11 October 2015	157	SLC	IW	VV	ascending
11	4 November 2015	157	SLC	IW	VV	ascending
12	22 December 2015	157	SLC	IW	VV	ascending
13	15 January 2016	157	SLC	IW	VV	ascending
14	8 February 2016	157	SLC	IW	VV	ascending
15	3 March 2016	157	SLC	IW	VV	ascending
16	27 March 2016	157	SLC	IW	VV	ascending
17	20 April 2016	157	SLC	IW	VV	ascending
18	14 May 2016	157	SLC	IW	VV	ascending
19	7 June 2016	157	SLC	IW	VV	ascending

The SAR data cover a large area, including the SGY mine area, and due to computational efficiency, SAR data are clipped in pre-processed procedure. The European Space Agency (ESA) released precise orbit ephemerides (POD) data for all the Sentinel-1 SAR data. POD data are important for reducing registration errors. These data are used for phase re-flattening and orbital refinement. To eliminate the impact of topography on the measured surface deformation, the authors employed the three-arc-second Shuttle Radar Topography Mission (SRTM) Digital Elevation Model (DEM) obtained from the National Aeronautics and Space Administration (NASA).

2.3. Fundamental Principle of SBMHCT-InSAR Technique

The present study puts forward SBMHCT-InSAR technology for precise inversion of surface deformation. The proposed approach integrates the Permanent Scattering (PS) and high-coherence target methods. Linear and nonlinear deformation inversion methods are employed using the coherent target and singular value decomposition methods, respectively. SBMHCT-InSAR technology comprises key steps such as interference pair combination of multi-principal images, high-precision image registration, interferometric phase noise filtering, high-coherence target extraction, and the deformation inversion method. The SBMHCT-InSAR processing steps are as follows:

2.3.1. Group the SAR Pairs

There are 19 scenes of SAR images ordered at times ($t_0, \ldots, t_N$) over the SGY mine area. M interferograms are constructed using installed multiple thresholds. The quantity M is such that it adheres to the following inequality:

$$\frac{N+1}{2} \leq M \leq N\left(\frac{N+1}{2}\right) \tag{1}$$

Figure 2 illustrates the experimental setup of the SAR pairs' connection diagram. The experiment set thresholds of spatial and temporal baselines as 300 m and 200 days, respectively. The SAR data acquired on 2 April 2015 were selected as the super master image; others were co-registered and resampled. Other images that meet the threshold condition also generate interferometric pairs, resulting in 78 differential interferometric images.

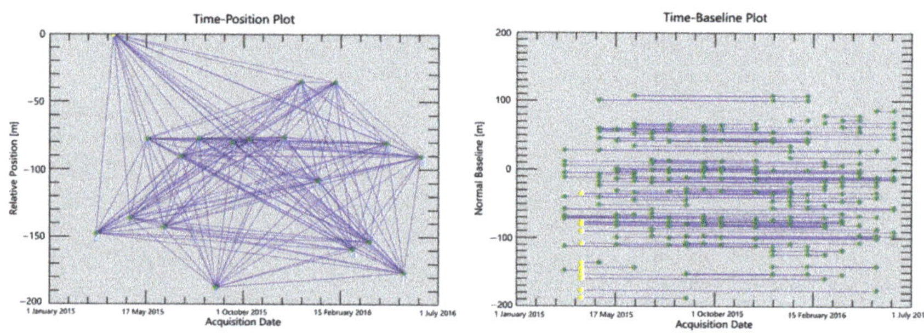

(a) **(b)**

Figure 2. (**a**) Timeposition plot of interferometric pairs; (**b**) timebaseline of interferometric pairs. The yellow diamond denotes the super master image.

2.3.2. Highly Accurate Image Registration

This paper presents an optimal matching point-based InSAR image registration method. Initially, an external DEM is emulated as a synthetic SAR image, and matching features are extracted from the SAR image to be registered in the simulated image. Then, the vector field consistent point set matching algorithm is employed to eliminate the homonymous feature points between the primary and secondary SAR images, remove the external points, and compute the polynomial transformation parameters for accurate registration. Ultimately, high-precision registration of the InSAR image is achieved.

2.3.3. Noise Filtering of Interferometric Phase

This study proposes an interferogram filtering method based on binary decomposition, which has the potential to effectively address the issue of noise in SAR images. The proposed approach decomposes the interferogram using a binary empirical mode algorithm into image and noise information. Filtering is then performed using a local window signal-to-noise ratio as the filtering factor, with strong filtering applied in regions of high noise and weak filtering in regions of low noise. Specifically, the method decomposes the original interferogram into fourth-order intrinsic mode function (IMF) signals and uses the signal-to-noise ratio of local windows as the filtering factor of the Goldstein filter to filter the first third-order IMF signals, which contain most of the noise information. The method demonstrates a strong noise-filtering ability while also preserving the edge details of interference fringes. As a result, the coherence of the interferogram is improved significantly after filtering.

2.3.4. High-Coherence Target Extraction

In 2004, Hooper [46] introduced the StaMPS method, which identifies highly coherent points based on the stability of their phase values and high coherence and signal-to-noise ratio. Another method used to identify high-coherence points involves selecting a radius of a circle around known high-coherence points and then applying the amplitude dispersion threshold method to find candidate high-coherence points. Next, iterative analysis is carried out on the phase stability of the candidate points, and the high-coherence points are determined. This method reduces the computational workload and improves efficiency [47].

2.3.5. Deformation Inversion

The proposed method aims to generate a Delaunay triangulation network for the highly coherent points after differential interferogram generation and identification of the highly coherent points. A linear model of velocity and elevation errors is then established based on the phase difference between two adjacent highly coherent points on the interferogram. By solving the coherence coefficient equation of the model, the incremental values of deformation velocity and elevation error are determined. The absolute value is obtained by incremental integration of the velocity and elevation error of several points. Next, the residual phase is unwrapped and calibrated by the discrete point phase after removing the linear model phase. Subsequently, the residual phase of a single SAR image is inverted using an interference combination matrix. Finally, nonlinear deformation and atmospheric influence phases are separated through time and space filtering, and a time series deformation sequence is obtained by calculating the linear deformation rate and nonlinear deformation phase.

Above all, the SBMHCT-InSAR technology introduced above is improved on the basis of SBAS-InSAR technology. Interactive Data Language (IDL) used for programming to improve the key steps of SBAS-InSAR technology, including image registration, filtering, and high coherence point extraction, in order to improve the adaptability and accuracy of SBAS-InSAR technology in mining deformation application.

2.4. Principles of LSTM

The LSTM network introduces a gate mechanism in the hidden layer to regulate information loss and dynamically adjusts the backpropagation process, enabling the network to learn long-distance time series data. This mechanism is crucial for the successful application of the LSTM model in large-scale surface subsidence prediction over extended time periods.

2.4.1. The Framework of LSTM

Figure 3 illustrates the prediction framework for time series mine subsidence based on LSTM. The original settlement data are pre-processed to meet the network input requirements in the first step, while the hidden layer uses the cell structure to construct the circulating neural network. Then, predicted values are exported by the output layer. The network training calculates the loss value between the predicted and true values and uses the ADAM algorithm to optimize the model. By dynamically adjusting the long and short-term memory network, the network can fully learn the nonlinear correlation of different subsidence time series and thus capture the complex subsidence mechanism in the study area. This approach not only reduces the requirement for high-quality diachronic data but also improves the accuracy and interpretability of subsidence prediction.

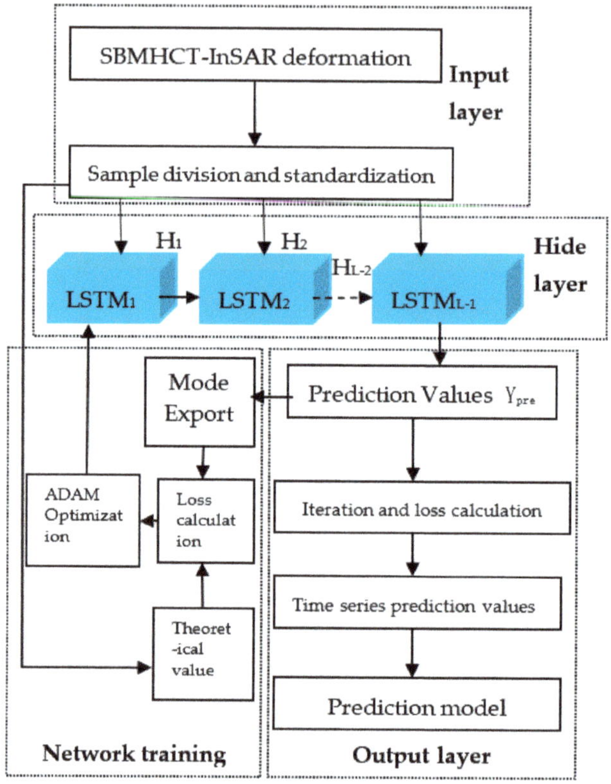

Figure 3. The prediction framework of LSTM.

2.4.2. Cell Structure of LSTM

The LSTM network comprises a set of cell units that serve as the central structure in the hidden layer. Figure 4 illustrates that the hidden layer contains three cell units. In the LSTM model, the input data at time t in the sample time series are represented by x_t, while the corresponding output data of the cell unit in the implicit state are represented by h_t. The flow of data in each cell unit is executed sequentially for input, information forgetting, cell state update, and implicit state output. The forward calculation method can be expressed as follows:

$$i_t = \sigma(W_{xi}x_t + W_{hi}h_{t-1} + W_{ci}c_{t-1} + b_i) \tag{2}$$

$$f_t = \sigma\left(W_{xf}x_t + W_{hf}h_{t-1} + W_{cf}c_{t-1} + b_f\right) \tag{3}$$

$$c_t = f_t c_{t-1} + i_t \tanh(w_{xc}x_t + w_{hc}h_{t-1} + b_c) \tag{4}$$

$$o_t = \sigma(W_{xo}x_t + W_{ho}h_{t-1} + W_{co}c_t + b_o) \tag{5}$$

$$h_t = o_t \tanh(c_t) \tag{6}$$

where i, f, c, and o represent the input gate, forgetting gate, cell state, and output gate, respectively; W and b represent the corresponding weight coefficient matrix and bias, respectively; σ and tanh refer to the sigmoid and the hyperbolic tangent activation function, respectively.

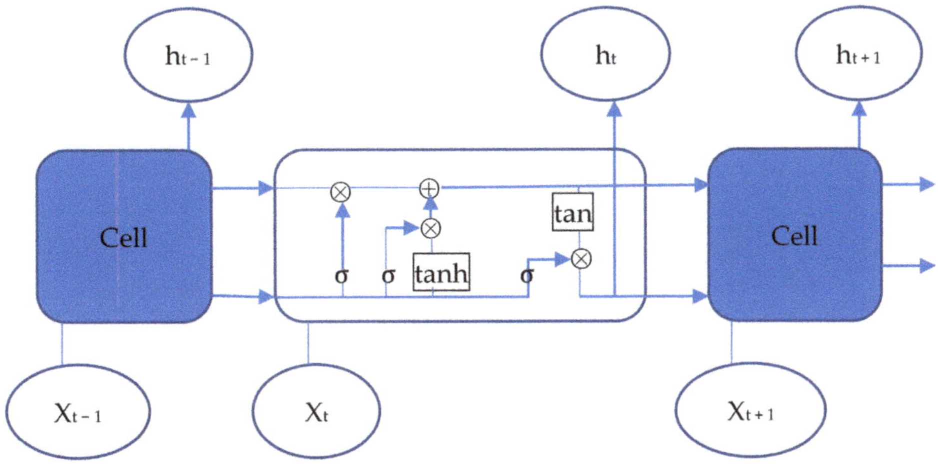

Figure 4. The cell structures of LSTM.

The training process of the LSTM network adopts time backpropagation (BPTT), which is similar to the traditional backpropagation algorithm [48]. The algorithm involves four steps: First, the output of the cells is calculated based on the forward-computation method specified in Equation (5). The error term for each cell is then calculated in reverse, including time and network level backpropagation. Then, the gradient of each weight is determined according to the corresponding error term. Finally, the weights are updated using a gradient optimization algorithm.

2.5. Time Series Prediction Model Combining SBMHCT-InSAR Results and LSTM

Drawing on the fundamental tenets of the LSTM algorithm, the time series of coal mine subsidence obtained via InSAR technology are leveraged as training samples. Notably, these data exhibit nonlinear relationships, taking the form of $\{H_t\} = \{H_1, H_2, \cdots, H_n\}$. As such, the values of these data serve as the training samples for the LSTM algorithm, whereby a predictive model is established, the model parameters are solved, and the corresponding predicted values are obtained. The accuracy of the predictive model is evaluated by comparing the expected value with the corresponding truth value. Ultimately, the mine forecasting method is implemented in Python language.

Step 1: Data pretreatments. The settlement time sequence data are processed by extracting a training sample of length L, denoted as H_s, from which the last Y values are designated as sample labels, and the first (L−Y) values are used as sample inputs, subject to the constraints of 2 ≤ L < m and 1 ≤ Y < L. Figure 5 depicts the form of sample division. By implementing this segmentation method, all highly coherent target points are pre-processed, and n training samples can be extracted. The paper adopts a single-step prediction method to construct the network model, whereby the length of the output sequence Y is set to y, and the settlement at the L moment is predicted based on the settlement information at the first (L−Y) moment.

Step 2: Network training. In Figure 3, the hidden layer output value Y_{pre} is the final output through all LSTM hidden layer cell units. The input sample $\{x_1, x_2, \cdots x_{L-Y}\}$ of the hidden layer is a two-dimensional array; output Y_{pre} of the hidden layer and sample label Y are both one-dimensional arrays (n,1), where n represents the number of highly coherent points. In this paper, the statistical error index is the mean square error, and the following formula is defined as the loss function of the training process:

$$loss = \sum_{t=1}^{n} \left(Y_{pre} - Y\right)^2 / n \tag{7}$$

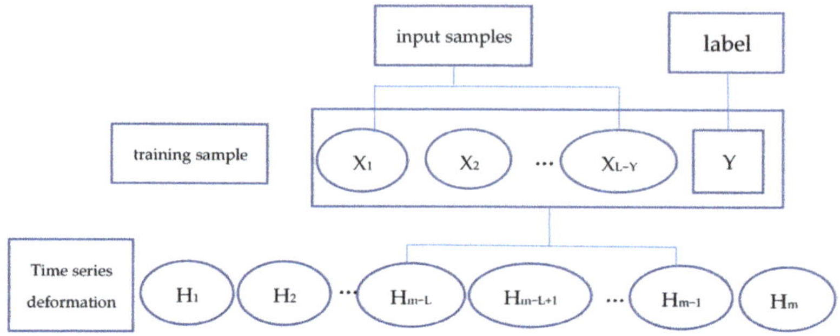

Figure 5. The structures of data pretreatments.

Step 3: Parameter optimization. To construct an accurate LSTM prediction model, several parameters need to be considered, including the sample partition length (L), network layer number (K), and feature number (S) of each LSTM hidden layer [49,50]. This paper employs a multi-layer grid search method to explore these parameters and selects the parameter combination with the highest average prediction accuracy as the optimal choice. The accuracy is determined by minimizing the prediction error (ε) between the predicted sample (Y_{pre}) and the actual sample (Y). The objective function is expressed as follows:

$$\min\varepsilon(Y, Y_{pre}) = |Y_{pre} - Y| \tag{8}$$

$$s.t. \begin{cases} 2 \le L < M \; STEP_L \parallel L \\ 2 \le K < i \; STEP_K \parallel K \\ 10 \le S \le S_{max} \; STEP_S \parallel S \\ L, K, S \in N \end{cases} \tag{9}$$

where $STEP_L$, $STEP_K$, and $STEP_S$ are the grid search $STEP_S$ of corresponding parameters, respectively.

Step 4: Output and accuracy assessment. The LSTM model can adapt the parameters during the training and validation process simultaneously, leading to the attainment of the optimal model $LSTM^*_{net}$. This model is then used to predict the future settlement amounts by inputting all standardized prediction samples in a sequential manner. The output of the model is represented as $Y_{pre} = \{y_1, y_2, y_3 \cdots y_n\}$, where Y_{pre} denotes the set of prediction results of different highly coherent points. Finally, the discrepancy between the output Y_{pre} and the actual measured data Y_{survey} in the course of deep learning prediction is computed, thereby providing a quantitative assessment of the training and prediction accuracy of the model.

Above all, the LSTM neural network was built based on Python 3.9 language and the Pytorch 1.10 deep learning framework [43]. The input dataset includes all highly coherent point feature vectors obtained by SBMHCT-InSAR technology, containing longitude, latitude, coherence value, cumulative time, deformation rate, and cumulative subsidence value, among which cumulative subsidence value is the label data predicted in the model. The grid search algorithm was applied to select the hyperparameters in the LSTM ground deformation prediction model.

The absolute errors (AE) and the relative error (RE) are defined as follows:

$$AE = |m_i - m_i'| \tag{10}$$

$$RE = \left|\frac{m_i - m_i'}{m_i}\right| \times 100\% \tag{11}$$

where m_i represents the truth value and m_i' represents the predicted value obtained by the LSTM model, and the absolute value is taken to avoid negative errors. The absolute error reflects the magnitude of the errors between the predicted and truth values, while the relative error indicates the proportion of the error relative to the truth value.

The relative error of the predicted results was evaluated using the Mean Absolute Percentage Error (MAPE). The generalization performance and degree of error of the prediction model were evaluated using the Wilmot Consistency Index (WIA), with values ranging from 0 to 1. Specifically, the MAPE was defined as the average of the absolute difference between the predicted and truth values, normalized by the observed value, expressed as a percentage. On the other hand, WIA was defined as the ratio of the observed variance to the sum of the observed variance and the variance of the prediction residuals, which were used to measure the degree of deviation of the model from the true values. They are defined as follows:

$$\text{MAPE} = \frac{\sum_{i=1}^{n}|m_i - m_i'|/m_i}{n} \times 100\% \tag{12}$$

$$WIA = 1 - \frac{\sum_{i=1}^{n}(m_i - m_i')^2}{\sum_{i=1}^{n}\left(|m_i - \overline{m_i}| + |\text{m}_i' - \overline{m_i}|\right)^2} \tag{13}$$

where m_i represents the truth value, m_i' represents the predicted value obtained by the LSTM model, n is the number of samples, and $\overline{m_i}$ is the average of m_i'.

3. Results and Discussion

3.1. Analysis of InSAR Results

In order to confirm the precision and dependability of our experimental results, we conducted an analysis and comparison of InSAR data and the GNSS values in the SGY mining area. As discussed in [51], the study area's InSAR monitoring and precision verification outcomes have been documented and will not be reiterated here. Figure 6a shows the cumulative deformation and coherence maps in the Ningdong coalfield; the graphs show a broader area to exhibit the mesoscale results, and the SGY mine is located in an oval area in the southwest. The coherence diagram shows that the coherence of the interferogram is greater than 0.41, and the coherence is good. Figure 6b shows the cumulative deformation maps in the SGY mining area from March 2015 to June 2016, where the red triangle represents the GNSS observation stations and the black rectangles represent the location of the underground coal seam. Here, a total of 4795 pixels obtained deformation values from InSAR data; these values were calculated by dividing by the cosine of the incident angle to obtain the vertical deformations.

Based on this research, a significant deformation basin has emerged in the study area since March 2015, which is caused by the continuous mining activities in the study area. The maximum observed deformation in the mining area was as high as −94 cm, which is a typical example of the subsidence basin commonly observed during the excavation and mining activities in this area.

The InSAR monitoring of accumulated deformation value error within the study area satisfies standard specifications, validating the applicability of this research outcome for settlement prediction. In contrast to GNSS scattered point deformation monitoring with time series values, InSAR technology acquires continuous and planar deformation information, which can be utilized to predict surface subsidence and provide an effective representation of both the evolving surface trends and the regional distribution of subsidence.

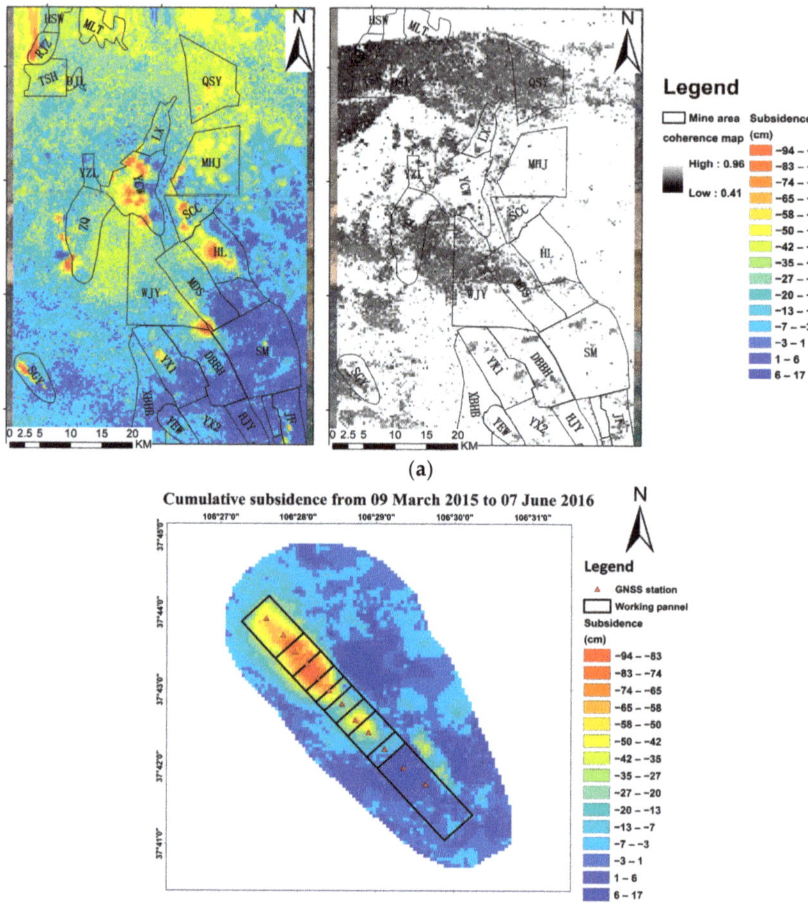

Figure 6. (**a**) Cumulative deformation and coherence maps in the Ningdong coalfield. (**b**) Cumulative deformation map in the SGY from March 2015 to June 2016.

3.2. LSTM Prediction Results

In the present study, 15 sets of InSAR monitoring data obtained between 9 March 2015 and 27 March 2016 were utilized as training samples, while 3 sets of InSAR monitoring data obtained between 27 March 2016 and 7 June 2016 were employed as test samples. Following the deformation results, the settlement sequence of each observation point was pre-processed, and 12 observation points were selected for further analysis.

In this study, the settlement time sequence of each observation point was pre-processed based on the inversion results, and a total of 4795 points were selected as training samples. To determine the optimal network parameters, a grid search method was utilized to investigate the impact of the number of network layers K and the number of hidden layer nodes S on the prediction accuracy. The resulting heatmap of prediction errors, shown in Figure 7, reveals that the number of network layers and hidden layer nodes are not solely responsible for prediction accuracy. While increasing the number of network layers generally enhances the prediction accuracy, and increasing the number of hidden nodes generally does the same, these correlations are not absolute. The prediction error reaches a trough when the number of network layers K reaches 5, and the number of hidden nodes S

reaches 55. Further increasing the number of network layers and hidden nodes results in decreased efficiency of the prediction model without improving the prediction accuracy. Therefore, the optimal configuration for this study's LSTM network consists of 5 layers and 55 hidden layer nodes.

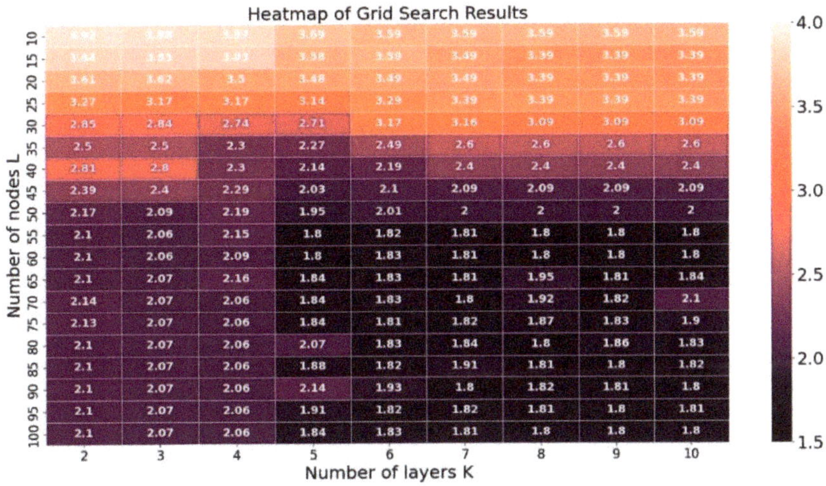

Figure 7. Heatmap of grid search results.

The focus of this study is on the 12 observation stations located in the center of the subsidence area. The objective is to investigate the relationship between the length of training samples and the prediction error. Additionally, the performance of each observation point in both single-step prediction and multi-step prediction is evaluated.

Figure 8 presents a line chart depicting the relationship between the length of training samples and the average prediction error. The predicted step is set as a single-step prediction. The 12 observation stations located in the center of the subsidence area are analyzed in this study to understand the error performance of each observation point. The results indicate that when the training sample length is 5, the prediction errors for all points are considerably higher compared to the prediction results of others. However, when the training sample length is set to 10, some observation points, such as G1, G2, and G9, exhibit better prediction results, while others show larger or similar errors with the prediction results of the blue line. The reason for this could be the fluctuation in subsidence values at these points. On the other hand, when the training sample length is set as 15, the prediction errors for all observation points are significantly lower. The line chart highlights that increasing the number of training samples results in lower prediction errors. Moreover, the local minimum prediction error of grid search gradually decreases with the increase in training sample length. Overall, the LSTM model performs better in predicting settlement values that exhibit stable and orderly changes over a longer period.

Figure 9 presents a line chart depicting the relationship between the predicted step size and the average prediction error. Based on the analysis, it can be observed that the prediction error increases with the increase in the predicted step size. This is because the longer the prediction sequence, the greater the cumulative error, resulting in a larger error in the predicted values. The error value of the light blue curve, which represents the predicted value error when the predicted step size is 3, is the largest among the three curves, indicating that the error accumulates with the increase in the predicted step size. The red curve and the dark blue curve have similar error values, indicating that the error accumulation is relatively small for multi-step prediction with a predicted step size of 2, and the prediction accuracy is still relatively high. However, it is worth noting that for some observation points with small cumulative form variables, the single-step prediction error is slightly smaller than that of multi-step prediction, which may be related to the characteristics of the subsidence process at these points. Overall, single-step prediction is a more accurate and reliable prediction method for SBMHCT-InSAR deformation monitoring values based on the LSTM model.

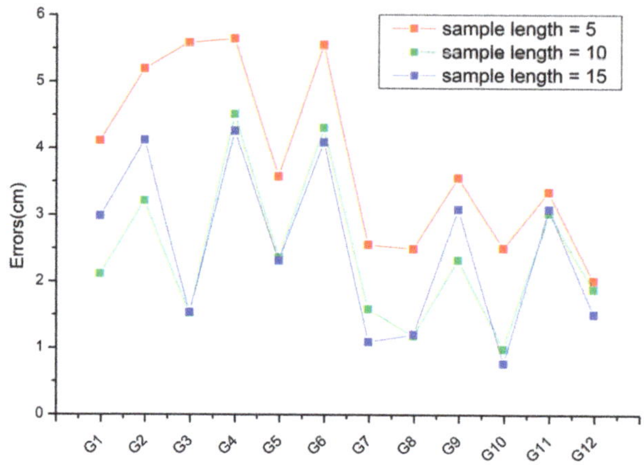

Figure 8. Relationship between training sample length and errors.

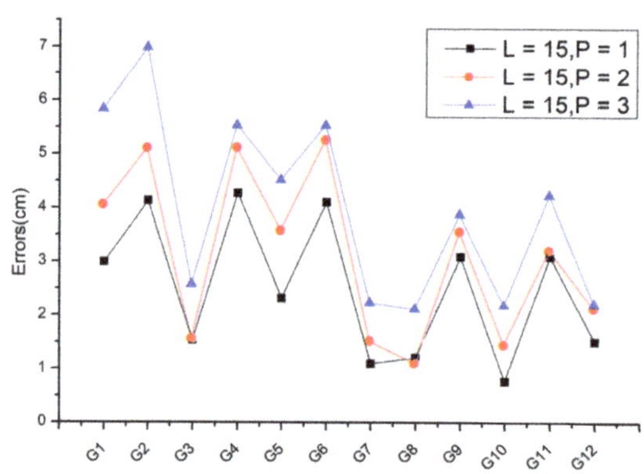

Figure 9. Relationship between prediction steps and errors.

In multi-step prediction, the model needs to predict several time steps ahead, which increases the complexity of the problem. As a result, the prediction error may accumulate over time, leading to a less-accurate prediction. On the other hand, in single-step prediction, the model only needs to predict the next time step, which is a simpler problem, and therefore the average error is better than that of multi-step prediction.

The present study adopts a strategic approach for selecting the sample segmentation length by leveraging the outcomes of the experiments conducted. Specifically, given the small ordinal number and equal time interval of the data, and taking into consideration the LSTM model's ability to learn long-distance time-series data, longer sample lengths were preferred to achieve better prediction results. In this regard, 15 sets of InSAR monitoring data obtained from 9 March 2015 to 27 March 2016 were selected as the training samples, whereas 3 sets of InSAR monitoring data from 27 March 2016 to 7 June 2016 were chosen as the test samples. The table below presents the final combination of LSTM prediction model parameters.

Table 2 shows the parameters of LSTM. Under the aforementioned parameter configuration, the experiment predicting surface subsidence in the SGY coal mine has yielded favorable results. Specifically, the average absolute difference (cumulative) form variable error has been effectively constrained to within 3 cm, thereby satisfying the accuracy standards prescribed within current InSAR data processing protocols.

Table 2. Optimal search results of LSTM parameters.

Parameters	Training Sample Length	Test Sample Length	LSTM Layers	Hidden Layer Nodes	Optimizer	Loss Function	Iterations
Values	15	1	4	55	ADAM	MSE	16

Figure 10 presents the settlement prediction results of the SGY mining area at the overall experimental scale. To obtain these results, the last 15 groups of deformation data from 20 May 2015 to 14 May 2016 were used as training samples to predict the deformation results of 7 June 2016. The predicted results were compared with the measured deformation results of InSAR, and the analysis showed a high degree of consistency between the predicted and real shape variables, indicating that the predicted settlement center area was clear and accurate. Moreover, the prediction errors of each monitoring point in the subsidence area were within 3 cm. Figure 11 presents the prediction errors of LSTM. Out of 4795 observation points, the maximum difference (cumulative) shape variable had a prediction error of 2.6 cm, and the average prediction accuracy reached 93.6%. The calculation method of relevant indicators is given in Equations (9)–(12). Therefore, the model proposed in this paper has a small deviation when compared with the actual settlement amount, effectively reflecting the basic law of land surface settlement changing with time, and the predicted results are reliable.

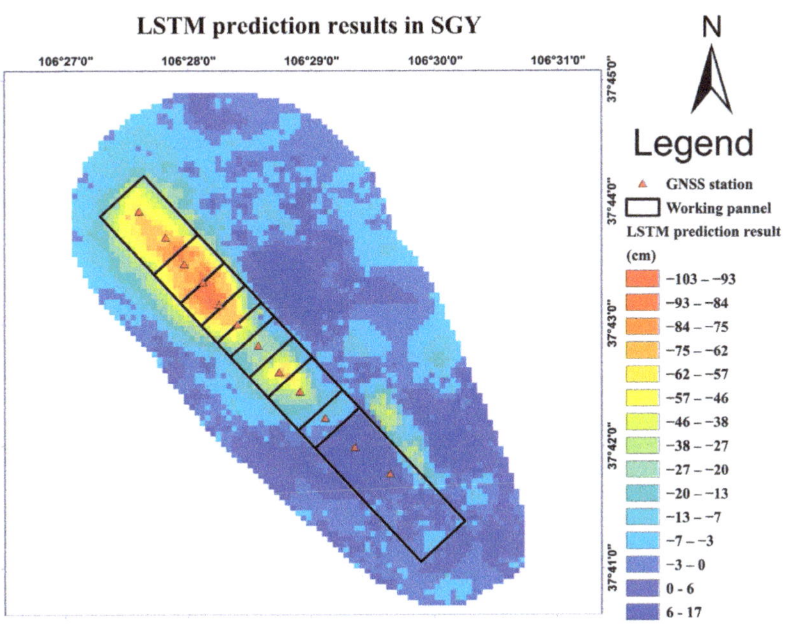

Figure 10. Prediction results of LSTM in SGY mine area.

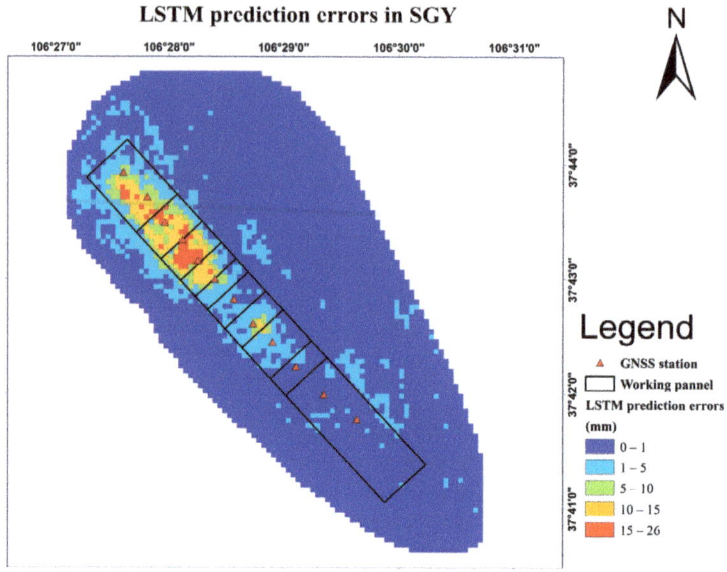

Figure 11. Prediction errors of LSTM in SGY mine area.

4. Discussion

To assess the validity of the prediction model introduced in this study, GNSS observation stations situated on the surface of the mining area were selected for analysis. Real-time kinematic (RTK) technology utilizing the double-difference mode was employed across all stations, with a precision of up to 1 mm for horizontal displacement monitoring. As previously reported, the measurement error of elevation direction is approximately twice that of the horizontal displacement error [52]. The monitoring values derived from the continuous time series deformation of the GNSS observation stations were extracted to verify the dependability of the prediction results obtained from the fusion of InSAR technology with LSTM. The table below exhibits the deformation values over time for each GNSS observation station based on their respective coordinates.

To demonstrate the superiority of the proposed prediction model, a comparison is made between the LSTM model and the traditional machine-learning model, using model establishment time and prediction error as evaluation metrics. Specifically, the SVR model is chosen as the representative of the traditional machine-learning model, which utilizes a nonlinear kernel function to map multidimensional inputs to higher dimensional feature space for regression analysis. In this study, the C penalty parameter and g kernel function parameter are optimized through the cross-verification method and grid search technique to identify the optimal parameter settings. Further details on the parameter searching process and experimental methodology can be found in [53].

Table 3 presents the InSAR monitoring deformation cumulants of the SGY coal mine surface GNSS observation points over 432 consecutive days in the second column, while the third column displays the GNSS monitoring deformation cumulants over 456 consecutive days. Additionally, the fourth and fifth columns exhibit the cumulative shape variables predicted for 456 consecutive days by the LSTM algorithm and SVR algorithm, respectively. Figure 12 corresponds to the data presented in Table 3, showcasing the close proximity of the predicted values of the LSTM algorithm and the SVR algorithm to the GNSS monitoring results. Furthermore, the LSTM algorithm's predicted values align with the GNSS monitoring values at multiple monitoring points, suggesting a high level of prediction accuracy. Thus, qualitatively, the results indicate the efficacy of the LSTM method in predicting cumulative shape variables.

Table 3. SGY coal mine observation points' deformation values.

Observation Stations	InSAR Results during 13 June 2015–7 June 2016 (cm)	GNSS Results during 13 June 2015–1 July 2016 (cm)	LSTM Prediction Results (cm)	SVR Prediction Results (cm)
G1	−59	−66	−66	−62
G2	−63	−72	−74	−78
G3	−75	−84	−84	−86
G4	−88	−105	−103	−99
G5	−81	−91	−90	−88
G6	−58	−70	−68	−64
G7	−23	−26	−26	−26
G8	−49	−56	−56	−55
G9	−37	−43	−45	−47
G10	−8	−10	−10	−9
G11	−6	−12	−12	−8
G12	−12	−12	−12	−10

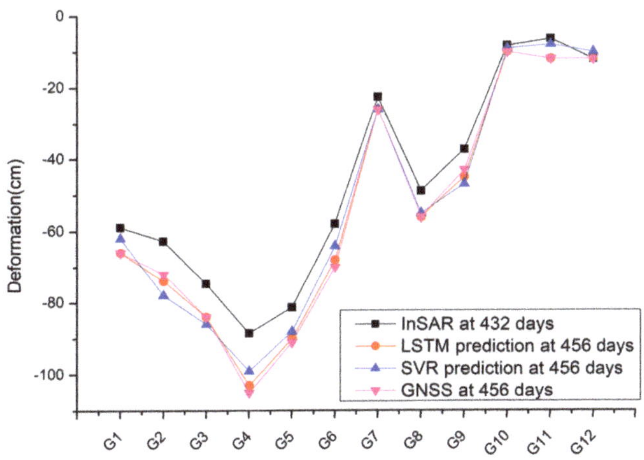

Figure 12. Monitoring and forecast results in SGY mine area.

The bar charts displayed in Figures 13 and 14 demonstrate the error distribution of prediction results for 12 GNSS monitoring points, revealing both the absolute and relative prediction errors of the LSTM and SVR prediction methods. These error metrics serve as quantitative indicators to assess the prediction accuracy of the two methods. The absolute and relative errors of the LSTM prediction at the 12 monitoring stations are smaller than those of the SVR prediction results. Specifically, the LSTM prediction method reports a zero error at G1, G3, G7, G8, G10, G11, and G12, whereas the SVR prediction method only has a zero error at the G7 observation station. The highest prediction error for both methods was observed at G2, G4, and G4 observation stations, with the LSTM method reporting an absolute error of 2 cm and the SVR method reporting an absolute error of 6 cm. This may be attributed to their central location within the subsidence basin, where mining activities were intensive over the 456-day inland. The deformation of these measuring points is influenced by multiple factors, such as geological structure, mining speed, and coal pillar, resulting in a more complex settlement pattern. The LSTM algorithm, with its superior learning ability, demonstrated a higher prediction accuracy compared to the SVR algorithm. Notably, the relative error between the predicted value and the truth value of the SVR method at the G12 station is at most 14%, with an absolute error of 2 cm. This is because G12 is situated at the edge of the subsidence basin with a relatively small subsidence value. However, the prediction accuracy of observation stations such as G1, G11, and G12 located in other subsidence basins is relatively high. These findings suggest that the LSTM method is better equipped to learn the intricate details of the settlement pattern.

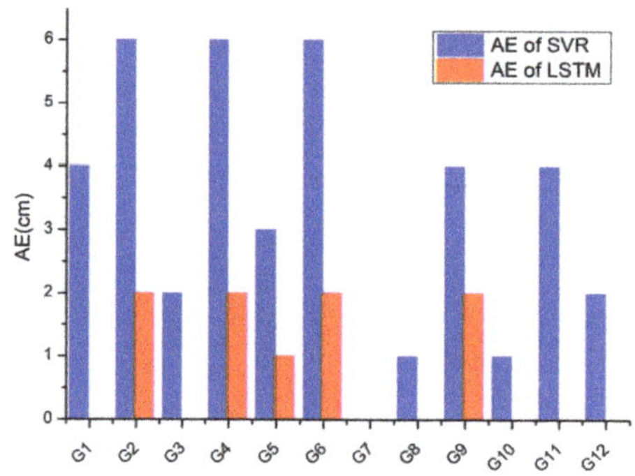

Figure 13. Comparison of absolute error between LSTM and SVR.

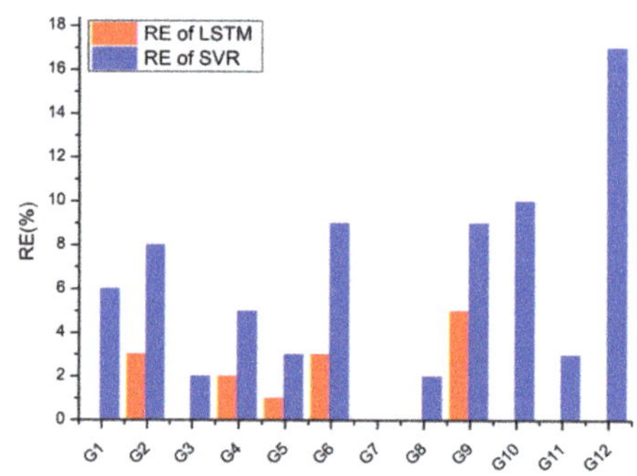

Figure 14. Comparison of relative error between LSTM and SVR.

Table 4 compares the prediction accuracy of the LSTM and SVR methods for the SGY coal mine. The table contains important error metrics such as the maximum absolute error, maximum relative error, average absolute error, and Wilmot consistency index, which provide a quantitative assessment of the accuracy of the prediction results.

Table 4. Comparison of prediction accuracy between LSTM and SVR.

Prediction Method	MAX.AE (cm)	MAX.RE (%)	MAPE (%)	WIA (0–1)
LSTM	2	6%	1.1%	0.999
SVR	6	33%	8.8%	0.996

The results in Table 4 demonstrate that the LSTM model outperforms the SVR time series prediction method in terms of prediction accuracy. Specifically, the maximum prediction error of cumulative deposition using the LSTM model is less than 2 cm, while the maximum relative error and average relative error are lower compared to the SVR method. These findings indicate that the deep-learning-based prediction model proposed in this study is highly accurate and robust. The WIA and MAPE values for the LSTM model are 0.999 and 1.1%, respectively, and the Wilmot consistency index is close to 1, demonstrating the effectiveness of the prediction function established by Equation (13). The mining settlement prediction model based on InSAR monitoring data and LSTM is robust and outperforms the representative machine-learning model, SVR, in various evaluation indicators, including the Wilmot consistency index. Based on the above analysis, it is evident that the LSTM-based prediction method used for large-scale surface subsidence is highly accurate, efficient, and a better option to ensure production safety.

Comparative with the findings in references [44,45], it shows that the SBMHCT-InSAR technology improved from SBAS-InSAR is more suitable for deformation monitoring in mining areas. It can obtain the accurate deformation value of mining area surface, which is conducive to the training of the LSTM prediction model. The prediction results show that the prediction of mining area deformation by combining SBMHCT-InSAR technology and LSTM model is highly reliable and robust.

5. Conclusions

The current study focuses on the surface settlement monitoring of the SGY coal mine in the SGY mining area, utilizing time series InSAR technology with an optimization algorithm. Through this technique, the study acquired series settlement values with equal time intervals. Subsequently, a deep-learning mining settlement prediction model was built, which utilized InSAR monitoring data to predict the future settlement values of the next time series. The results of this study suggest that the proposed deep-learning-based method exhibits higher accuracy, lower time cost, and better performance in various evaluation indicators than the representative machine-learning-based SVR method. Based on these findings, it can be concluded that the proposed approach is a superior method for fulfilling production safety requirements. These conclusions are drawn as follows:

(1) A novel approach, referred to as the SBMHCT-InSAR, is proposed in this study to address the issue of limited high-coherence points obtained using traditional small baseline set time series technology. This is achieved by utilizing an optimal matching-point-based strategy to improve the registration accuracy of interferometric synthetic aperture radar (SAR) images. Furthermore, the adaptive Goldstein filtering method, based on bivariate empirical mode decomposition (BEMD), is employed to enhance the mean coherence of high-coherence points in the interferogram. As a result, this method provides highly accurate training data for the deep-learning prediction model.

(2) According to the experimental results, the proposed SBMHCT-InSAR technology, which utilizes an optimization algorithm, has proved to be an effective method for monitoring surface settlement in mining areas. Specifically, 19 Sentinel-1A satellite images from 2015 to 2016 were utilized with the SBAS small baseline set technique to retrieve the cumulative deformation sequence values of the SGY coal mine. As a result, the settlement time series of 4795 high-coherence points were obtained. The findings reveal that the surface subsidence of the SGY coal mine gradually sank from northwest to southeast, forming a long and narrow subsidence basin. The maximum subsidence occurred near the G4 monitoring site, reaching −94 cm during the monitoring period.

(3) A deep-learning prediction model for mining subsidence based on InSAR monitoring data is developed, wherein the time series settlement values obtained through InSAR technology at equal time intervals are utilized as the training data. The LSTM deep-learning algorithm is employed to establish a nonlinear function relationship between the InSAR monitoring data and unknown predicted values, thus enabling dynamic prediction of mining surface and monitoring the change trend of surface settlement. The resultant deep-learning mining settlement prediction model based on InSAR monitoring data exhibits accuracy and robustness.

(4) The accuracy of the LSTM-based prediction method for large-scale surface subsidence is higher, and the evaluation indicators are better than those of the SVR-based method. The comparative analysis between the LSTM and SVR prediction methods reveals that the LSTM deep-learning algorithm is more effective in learning surface subsidence patterns in mining areas. The algorithm can accurately learn the fine deformation rules of the subsidence basin edge and achieve high-precision prediction for the central area of the subsidence basin with large subsidence values. The proposed method's reliability is verified by comparing it with ground GNSS observation station data,

thereby providing theoretical support for the expansion and application of InSAR monitoring technology in mining areas.

Author Contributions: Conceptualization, F.M. and L.S.; methodology, F.M.; software, W.L.; validation, F.M., L.S. and W.L.; formal analysis, F.M.; investigation, F.M.; resources, W.L.; data curation, L.S.; writing—original draft preparation, F.M.; writing—review and editing, W.L.; visualization, F.M.; supervision, L.S.; project administration, L.S.; funding acquisition, F.M. All authors have read and agreed to the published version of the manuscript.

Funding: This research was funded by the Fundamental Research Program of Shanxi Province (grant number 202103021223381).

Data Availability Statement: The data presented in this study are available on request from the corresponding author. The data are not publicly available due to confidentiality.

Acknowledgments: The authors acknowledge the provision of the Sentinel-1 data by ESA and DEM data by NASA.

Conflicts of Interest: The authors declare no conflict of interest.

References

1. Carnec, C.; Delacourt, C. Three years of mining subsidence monitored by SAR interferometry, near Gardanne, France. *J. Appl. Geophys.* **2000**, *43*, 43–54. [CrossRef]
2. Worlanyo, A.S.; Jiangfeng, L. Evaluating the environmental and economic impact of mining for post-mined land restoration and land-use: A review. *J. Environ. Manag.* **2021**, *279*, 111623. [CrossRef] [PubMed]
3. Zhou, L.; Li, Z.; Liu, W.; Liu, S.; Zhang, L.; Zhong, L.; Luo, X.; Liang, H. Restoration of rare earth mine areas: Organic amendments and phytoremediation. *Environ. Sci. Pollut. Res.* **2015**, *22*, 17151–17160. [CrossRef] [PubMed]
4. Cordy, P.; Veiga, M.M.; Salih, I.; Al-Saadi, S.; Console, S.; Garcia, O.; Mesa, L.A.; Velásquez-López, P.C.; Roeser, M. Mercury contamination from artisanal gold mining in Antioquia, Colombia: The world's highest per capita mercury pollution. *Sci. Total Environ.* **2011**, *410*, 154–160. [CrossRef] [PubMed]
5. Mensah, A.K.; Mahiri, I.O.; Owusu, O.; Mireku, O.D.; Wireko, I.; Kissi, E.A. Environmental impacts of mining: A study of mining communities in Ghana. *Appl. Ecol. Environ. Sci.* **2015**, *3*, 81–94.
6. Miao, Z.; Marrs, R. Ecological restoration and land reclamation in open-cast mines in Shanxi Province, China. *J. Environ. Manag.* **2000**, *59*, 205–215. [CrossRef]
7. Palekhov, D.; Palekhova, L. Responsible Mining: Challenges, Perspectives and Approaches. In *Sustainable Global Value Chains*; Springer: Cham, Switzerland, 2019; pp. 521–544.
8. Wang, J.; Zhao, F.; Yang, J.; Li, X. Mining site reclamation planning based on land suitability analysis and ecosystem services evaluation: A case study in Liaoning province, China. *Sustainability* **2017**, *9*, 890. [CrossRef]
9. Hui, G.; Qi-Chao, S.; Jun, H. Subgrade settlement prediction based on least square support vector regession and real-coded quantum evolutionary algorithm. *Int. J. Grid Distrib. Comput.* **2016**, *9*, 83–90. [CrossRef]
10. Shahin, M.; Maier, H.; Jaksa, M. Settlement prediction of shallow foundations on granular soils using B-spline neurofuzzy models. *Comput. Geotech.* **2003**, *30*, 637–647. [CrossRef]
11. Kim, Y.; Lee, S.S. Application of artificial neural networks in assessing mining subsidence risk. *Appl. Sci.* **2020**, *10*, 1302. [CrossRef]
12. Kumar, S.; Kumar, D.; Donta, P.K.; Amgoth, T. Land subsidence prediction using recurrent neural networks. *Stoch. Environ. Res. Risk Assess.* **2022**, *36*, 373–388. [CrossRef]
13. Havivi, S.; Rotman, S.R.; Blumberg, D.G.; Maman, S. Damage assessment in rural environments following natural disasters using multi-sensor remote sensing data. *Sensors* **2022**, *22*, 9998. [CrossRef] [PubMed]
14. Graham, L.C. Synthetic interferometer radar for topographic mapping. *Proc. IEEE* **1974**, *62*, 763–768. [CrossRef]
15. Gabriel, A.K.; Goldstein, R.M.; Zebker, H.A. Mapping small elevation changes over large areas: Differential radar interferometry. *J. Geophys. Res. Solid Earth* **1989**, *94*, 9183–9191. [CrossRef]
16. Ng, A.H.-M.; Ge, L.; Yan, Y.; Li, X.; Chang, H.-C.; Zhang, K.; Rizos, C. Mapping accumulated mine subsidence using small stack of SAR differential interferograms in the Southern coalfield of New South Wales, Australia. *Eng. Geol.* **2010**, *115*, 1–15. [CrossRef]
17. Deng, Z.; Ke, Y.; Gong, H.; Li, X.; Li, Z. Land subsidence prediction in Beijing based on PS-InSAR technique and improved Grey-Markov model. *GISci. Remote Sens.* **2017**, *54*, 797–818. [CrossRef]
18. Yin, H.; Zhu, J.; Li, Z.; Ding, X.; Wang, C. Ground Subsidence Monitoring in Mining Area Using DInSAR SBAS Algorithm. *Acta Geod. Cartogr. Sin.* **2011**, *40*, 52–58.
19. Wasowski, J.; Bovenga, F. Investigating landslides and unstable slopes with satellite Multi Temporal Interferometry: Current issues and future perspectives. *Eng. Geol.* **2014**, *174*, 103–138. [CrossRef]
20. Berardino, P.; Fornaro, G.; Lanari, R.; Sansosti, E. A new algorithm for surface deformation monitoring based on small baseline differential SAR interferograms. *IEEE Trans. Geosci. Remote Sens.* **2002**, *40*, 2375–2383.

21. Samsonov, S.; d'Oreye, N.; Smets, B. Ground deformation associated with post-mining activity at the French–German border revealed by novel InSAR time series method. *Int. J. Appl. Earth Obs. Geoinf.* **2013**, *23*, 142–154. [CrossRef]
22. Herrera, G.; Gutiérrez, F.; García-Davalillo, J.C.; Guerrero, J.; Notti, D.; Galve, J.P.; Fernández-Merodo, J.; Cooksley, G. Multi-sensor advanced DInSAR monitoring of very slow landslides: The Tena Valley case study (Central Spanish Pyrenees). *Remote Sens. Environ.* **2013**, *128*, 31–43. [CrossRef]
23. Peduto, D.; Maccabiani, J.; Ferlisi, S.; Nicodemo, G. Multi-scale analysis of settlement-induced building damage using damage surveys and DInSAR data: A case study in The Netherlands. *Eng. Geol.* **2017**, *218*, 117–133. [CrossRef]
24. Sanabria, M.P.; Guardiola-Albert, C.; Tomás, R.; Herrera, G.; Prieto, A.; Sánchez, H.; Tessitore, S. Subsidence activity maps derived from DInSAR data: Orihuela case study. *Nat. Hazards Earth Syst. Sci.* **2014**, *14*, 1341–1360. [CrossRef]
25. Hao, Y.; Wu, L.; Dai, H. Establishing a ground settlement prediction model with elastic slab theory. *Yanshilixue Yu Gongcheng Xuebao/Chin. J. Rock Mech. Eng.* **2006**, *25*, 2958–2962.
26. Yahong, Z.; Weina, W.; Peihua, J.; Jiahui, L. An improved MMF subsidence prediction model based on the Markov chain and its application. *Bull. Surv. Mapp.* **2022**, *1*, 79–83.
27. Jarosz, A.; Karmis, M.; Sroka, A. Subsidence development with time—Experiences from longwall operations in the Appalachian coalfield. *Int. J. Min. Geol. Eng.* **1990**, *8*, 261–273. [CrossRef]
28. Kowalski, A. Surface subsidence and rate of its increaments based on measurements and theory. *Arch. Min. Sci.* **2001**, *46*, 391–406.
29. Gonzalez-Nicieza, C.; Alvarez-Fernandez, M.; Menendez-Diaz, A.; Alvarez-Vigil, A. The influence of time on subsidence in the Central Asturian Coalfield. *Bull. Eng. Geol. Environ.* **2007**, *66*, 319–329. [CrossRef]
30. Cui, X.; Wang, J.; Liu, Y. Prediction of progressive surface subsidence above longwall coal mining using a time function. *Int. J. Rock Mech. Min. Sci.* **2001**, *38*, 1057–1063. [CrossRef]
31. Wang, Y.; Yang, G. Prediction of composite foundation settlement based on multi-variable gray model. *Proc. Appl. Mech. Mater.* **2014**, *580–583*, 669–673.
32. Shearer, T. A numerical model to calculate land subsidence, applied at Hangu in China. *Eng. Geol.* **1998**, *49*, 85–93. [CrossRef]
33. Hussain, M.A.; Chen, Z.; Zheng, Y.; Shoaib, M.; Shah, S.U.; Ali, N.; Afzal, Z. Landslide susceptibility mapping using machine learning algorithm validated by persistent scatterer In-SAR technique. *Sensors* **2022**, *22*, 3119. [CrossRef]
34. Li, L.; Huang, G. Prediction of goaf settlement with time sequence of wavelet neural network. *Procedia Eng.* **2011**, *15*, 4723–4727. [CrossRef]
35. Yinghua, W.; Chang, X. Using genetic artificial neural network to model dam monitoring data. In Proceedings of the 2010 Second International Conference on Computer Modeling and Simulation, Sanya, China, 22–24 January 2010; pp. 3–7.
36. Yang, Z.; Xi, W.; Yang, Z.; Shi, Z.; Qian, T. Monitoring and Prediction of Glacier Deformation in the Meili Snow Mountain Based on InSAR Technology and GA-BP Neural Network Algorithm. *Sensors* **2022**, *22*, 8350. [CrossRef] [PubMed]
37. Li, Z.W.; Yang, Z.F.; Zhu, J.J.; Hu, J.; Wang, Y.J.; Li, P.X.; Chen, G.L. Retrieving three-dimensional displacement fields of mining areas from a single InSAR pair. *J. Geod.* **2015**, *89*, 17–32. [CrossRef]
38. Zhou, D.; Zuo, X.; Zhao, Z. Constructing a Large-Scale Urban Land Subsidence Prediction Method Based on Neural Network Algorithm from the Perspective of Multiple Factors. *Remote Sens.* **2022**, *14*, 1803. [CrossRef]
39. Qinghao, L.; Yonghong, Z.; Min, D.; Hongan, W.; Yonghui, K.; Jujie, W. Time series prediction method of large-scale surface subsidence based on deep learning. *Acta Geod. Cartogr. Sin.* **2021**, *50*, 396.
40. Ansari, H.; Rußwurm, M.; Ali, M.; Montazeri, S.; Parizzi, A.; Zhu, X.X. InSAR displacement time series mining: A machine learning approach. In Proceedings of the 2021 IEEE International Geoscience and Remote Sensing Symposium IGARSS, Brussels, Belgium, 11–16 July 2021; pp. 3301–3304.
41. Hill, P.; Biggs, J.; Ponce-López, V.; Bull, D. Time-Series Prediction Approaches to Forecasting Deformation in Sentinel-1 InSAR Data. *J. Geophys. Res. Solid Earth* **2021**, *126*, e2020JB020176. [CrossRef]
42. Liu, Q.; Zhang, Y.; Wei, J.; Wu, H.; Deng, M. HLSTM: Heterogeneous long short-term memory network for large-scale InSAR ground subsidence prediction. *IEEE J. Sel. Top. Appl. Earth Obs. Remote Sens.* **2021**, *14*, 8679–8688. [CrossRef]
43. Chen, Y.; He, Y.; Zhang, L.; Chen, Y.; Pu, H.; Chen, B.; Gao, L. Prediction of InSAR deformation time-series using a long short-term memory neural network. *Int. J. Remote Sens.* **2021**, *42*, 6919–6942. [CrossRef]
44. Yao, S.; He, Y.; Zhang, L.F.; Yang, W.; Chen, Y.; Sun, Q.; Zhao, Z.A.; Cao, S.P. A ConvLSTM Neural Network Model for Spatiotemporal Prediction of Mining Area Surface Deformation Based on SBAS-InSAR Monitoring Data. *IEEE Trans. Geosci. Remote Sens.* **2023**, *61*, 22. [CrossRef]
45. Chen, B.; Yu, H.; Zhang, X.; Li, Z.; Kang, J.; Yu, Y.; Yang, J.; Qin, L. Time-Varying Surface Deformation Retrieval and Prediction in Closed Mines through Integration of SBAS InSAR Measurements and LSTM Algorithm. *Remote Sens.* **2022**, *14*, 788. [CrossRef]
46. Hooper, A.; Spaans, K.; Bekaert, D.; Cuenca, M.C.; Arıkan, M.; Oyen, A. *StaMPS/MTI Manual*; Delft Institute of Earth Observation and Space Systems Delft University of Technology, Kluyverweg: Delft, The Netherlands, 2010; Volume 1, p. 2629.
47. Vadivel, S.K.P.; Kim, D.-j.; Kim, Y.C. Time-series InSAR Analysis and Post-processing Using ISCE-StaMPS Package for Measuring Bridge Displacements. *Korean J. Remote Sens.* **2020**, *36*, 527–534.
48. Graves, A.; Schmidhuber, J. Framewise phoneme classification with bidirectional LSTM and other neural network architectures. *Neural Netw.* **2005**, *18*, 602–610. [CrossRef]
49. Gers, F.A.; Schmidhuber, J.; Cummins, F. Learning to forget: Continual prediction with LSTM. *Neural Comput.* **2000**, *12*, 2451–2471. [CrossRef]

50. Graves, A.; Graves, A. Long short-term memory. In *Supervised Sequence Labelling with Recurrent Neural Networks*; Springer: Cham, Switzerland, 2012; pp. 37–45.
51. Ma, F.; Sui, L. Investigation on mining subsidence based on sentinel-1A Data by SBAS-InSAR technology-Case Study of Ningdong Coalfield, China. *Earth Sci. Res. J.* **2020**, *24*, 373–386. [CrossRef]
52. Xiaohong, Z.; Jiahuan, H.; Xiaodong, R. New progress of PPP/PPP-RTK and positioning performance comparison of BDS/GNSS PPP. *Acta Geod. Cartogr. Sin.* **2020**, *49*, 1084.
53. Sui, L.; Ma, F.; Chen, N. Mining Subsidence Prediction by Combining Support Vector Machine Regression and Interferometric Synthetic Aperture Radar Data. *ISPRS Int. J. Geo-Inf.* **2020**, *9*, 390. [CrossRef]

remote sensing

MDPI

Article

Integration of DInSAR-PS-Stacking and SBAS-PS-InSAR Methods to Monitor Mining-Related Surface Subsidence

Yuejuan Chen [1,2], Xu Dong [1,2], Yaolong Qi [1,2,3], Pingping Huang [1,2,*], Wenqing Sun [4], Wei Xu [1,2], Weixian Tan [1,2], Xiujuan Li [1,2] and Xiaolong Liu [1,2]

1 College of Information Engineering, Inner Mongolia University of Technology, Hohhot 010051, China; chen_yj@imut.edu.cn (Y.C.); 20201800109@imut.edu.cn (X.D.); qiyaolong@imut.edu.cn (Y.Q.); xuwei1983@imut.edu.cn (W.X.); wxtan@imut.edu.cn (W.T.)
2 Inner Mongolia Key Laboratory of Radar Technology and Application, Hohhot 010051, China
3 Inner Mongolia Mypattern Technology Co., Ltd., Hohhot 010000, China
4 China Coal Pingshuo Group Co., Ltd., Shuozhou 036006, China
* Correspondence: hwangpp@imut.edu.cn

Abstract: Over-exploitation of coal mines leads to surface subsidence, surface cracks, collapses, landslides, and other geological disasters. Taking a mining area in Nalintaohai Town, Ejin Horo Banner, Ordos City, Inner Mongolia Autonomous Region, as an example, Sentinel-1A data from January 2018 to October 2019 were used as the data source in this study. Based on the high interference coherence of the permanent scatterer (PS) over a long period of time, the problem of the manual selection of ground control points (GCPs) affecting the monitoring results during refinement and re-flattening is solved. A DInSAR-PS-Stacking method combining the PS three-threshold method (the coherence coefficient threshold, amplitude dispersion index threshold, and deformation velocity interval) is proposed as a means to select ground control points for refinement and re-flattening, as well as a means to obtain time-series deformation by weighted stacking processing. A SBAS-PS-InSAR method combining the PS three-threshold method to select PS points as GCPs for refinement and re-flattening is also proposed. The surface deformation results monitored by the DInSAR-PS-Stacking and SBAS-PS-InSAR methods are analyzed and verified. The results show that the subsidence location, range, distribution, and space–time subsidence law of surface deformation results obtained by DInSAR-PS-Stacking, SBAS-PS-InSAR, and GPS methods are basically the same. The deformation results obtained by these two InSAR methods have a good correlation with the GPS monitoring results, and the MAE and RMSE are within the acceptable range. The error showed that the edge of the subsidence basin was small and that the center was large. Both methods were found to be able to effectively monitor the coal mine, but there were also shortcomings. DInSAR-PS-Stacking has a strong ability to monitor the settlement center. SBAS-PS-InSAR performed well in monitoring slow and small deformations, but its monitoring of the settlement center was insufficient. Considering the advantages of these two InSAR methods, we proposed fusing the time-series deformation results obtained using these two InSAR methods to allow for more reliable deformation results and to carry out settlement analysis. The results showed that the automatic two-threshold (deformation threshold and average coherence threshold) fusion was effective for monitoring and analysis, and the deformation monitoring results are in good agreement with the actual situation. The deformation information obtained by the comparison, and fusion of multiple methods can allow for better monitoring and analysis of the mining area surface deformation, and can also provide a scientific reference for mining subsidence control and early disaster warning.

Keywords: InSAR; mining area; surface subsidence monitoring; DInSAR-PS-Stacking; SBAS-PS-InSAR; ground control point; fusion

Citation: Chen, Y.; Dong, X.; Qi, Y.; Huang, P.; Sun, W.; Xu, W.; Tan, W.; Li, X.; Liu, X. Integration of DInSAR-PS-Stacking and SBAS-PS-InSAR Methods to Monitor Mining-Related Surface Subsidence. *Remote Sens.* **2023**, *15*, 2691. https://doi.org/10.3390/rs15102691

Academic Editors: Constantinos Loupasakis, Ioannis Papoutsis and Konstantinos G. Nikolakopoulos

Received: 20 April 2023
Revised: 19 May 2023
Accepted: 19 May 2023
Published: 22 May 2023

1. Introduction

Coal mine resources play an important role in China's energy resources. The large-scale exploitation of coal mine resources has promoted the development of China's economy, but it has also caused some ecological environment and surface subsidence problems [1–4]. Land subsidence caused by coal mining is a destructive disaster that often occurs in mining areas and is one of the most severe geological disasters in China [1,5]. Large-scale underground coal mining can cause the formation of underground cavities, which can lead to the loss of support for rocks and soils, resulting in ground subsidence and sinking. During the subsidence process, substances such as groundwater, sediment, and coal seams are squeezed and displaced, causing the appearance of depressions and funnel-shaped pits on the surface, known as ground subsidence funnels [6]. Coal mining can also cause changes in geological stress, resulting in the formation of cracks and faults in rock layers, which may trigger ground fissures. These cracks may expand and lead to geological disasters such as ground collapse [7,8]. The large-scale mining of underground coal mines can lead to the occurrence of surface subsidence funnels, ground cracks, and collapses in mining areas, which cause certain safety hazards and affect the local ecological environment and the safety of the surrounding residents. Identifying the causes and risks of surface subsidence by means of monitoring and analyzing the surface deformation of mining areas is of great significance for protecting residents and the safety of their properties, as well as mining subsidence disaster warnings, control, and management [9,10].

Interferometric synthetic aperture radar technology (InSAR) is a new all-weather, all-time Earth observation method [11–16]. With the rapid development of differential interferometric synthetic aperture radar (DInSAR) technology, radar line-of-sight deformation can be obtained up to the millimeter level [17,18]. DInSAR technology has been used to monitor mining subsidence and related research [19]. Berardino et al. proposed small baseline subset InSAR (SBAS) technology, which uses small baseline combinations for measurement, and uses the singular value decomposition (SVD) method to calculate multiple small baseline combinations to effectively obtain deformation information on time series [20–23]. Stacking-InSAR technology is one of the relatively simple time-series InSAR technologies. Specifically, it refers to the linear superposition and weighted average of multiple unwrapped differential interferometry pairs during the study period, which obtains more accurate deformation information [24–28]. Ferretti et al. proposed persistent scatterer InSAR (PS-InSAR) technology. In the long time series of SAR images, the points with high coherence and stability are selected as PS point targets, and then the phase characteristics of these target points are analyzed, and the corresponding atmospheric phases are separated to obtain relatively accurate surface subsidence information [29–32]. Guo Shanchuan et al. used DInSAR technology to effectively monitor and verify the mining area of the Loess Plateau with complex and dangerous terrain [33]. Xia Yuanping et al. combined DInSAR and GIS technology to identify illegal underground mining in the Shanxi Yangquan mining area and provided technical support for monitoring underground mining [34]. Li Da et al. used SBAS-InSAR technology to monitor and analyze the time-series deformation of the Yulin mining area [35]. Ma Fei et al. introduced SBAS-InSAR technology to the Ningdong mining area for monitoring, and compared the subsidence value of ground monitoring points in the Shigouyi coal mine with GPS monitoring values, revealing the effectiveness of SBAS-InSAR technology for mining area subsidence monitoring [36].

The InSAR method has achieved remarkable results in monitoring mining deformation, and different methods show their own advantages. Researchers in China and abroad have carried out relevant research on comparisons between and combinations of various InSAR methods for mining area deformation monitoring and analysis [28,37–39]. For example, Wei Jicheng et al. [40] combined DInSAR and PS-InSAR technology to effectively monitor mining subsidence in the Ordos area in the north of the Shendong mining area, revealing the space–time evolution process of surface deformation in this area. Depin Ou et al. [41] combined DInSAR and pixel offset tracking methods to monitor coal mine deformation. The selection of ground control points (GCP) used for refinement and re-flattening in

InSAR processing is crucial for deformation inversion. Some scholars have used the PS points obtained by PS-InSAR as ground control points in the SBAS-InSAR method to monitor surface deformation [42,43] and compared this with the traditional time-series InSAR method to verify its feasibility and improve the deformation accuracy. As mentioned above, combining the advantages of various InSAR methods can allow for more effective monitoring of the deformation of mining areas.

The limitations of some InSAR technologies should be considered, for example, SBAS-InSAR is insufficient in monitoring the subsidence center of large subsidence and areas with poor coherence, but it is better for monitoring small deformations. In this study, a mining area in Nalintaohai Town, Ejin Horo Banner, Ordos City, Inner Mongolia Autonomous Region, was selected as the study area, and Sentinel-1A data were used as the data source. The improved DInSAR-PS-Stacking and SBAS-PS-InSAR methods are proposed as a means to improve the inversion accuracy by combining the PS three-threshold method (coherence coefficient threshold, amplitude deviation index threshold, and deformation velocity interval) to select ground control points. The time-series deformation results obtained from the DInSAR-PS-Stacking and SBAS-PS-InSAR methods were compared, verified, and analyzed. The ground displacement time series were fused using the OTSU (Otsu 2007, named after the author) [44] automatic extraction method with two thresholds (the deformation and the average coherence threshold). The deformation analysis and subsidence rule of the mining area were studied in relation to the time-series deformation results after fusion. The fusion of multiple InSAR methods can allow us to overcome the shortcomings of individual deformation monitoring methods and obtain more complete and accurate deformation results. This study provides a scientific basis and technical support for mining subsidence prevention and sustainable development.

2. Materials and Methods

2.1. Study Area

The object of this study is a coal mine in Nalintaohai Town, Ejin Horo Banner, Ordos, Inner Mongolia Autonomous Region. Located at the north end of the Loess Plateau, the landform has been severely cut after long-term rain erosion, forming a standard beam and hilly platform, and its surface vegetation mostly consists of semi-barren areas. The coal mine is in the local village, the terrain is high in the north and low in the south, where it is relatively flat. The coal mine is close to the traffic line, and the traffic is relatively convenient. The location belongs to the temperate continental monsoon climate, and the temperature changes significantly in the four seasons of the year, with a large temperature difference. The annual average precipitation is low, and the precipitation is mostly concentrated between July and August, and the climate is dry. Its geographical location and scope are shown in Figure 1.

The geological structure of the working face is simple: the elevation of the coal seam floor is 1060~1075 m, the ground elevation is 1193.8~1260 m, the thickness of the overlying bedrock of the 4-2 coal seam is 120~140 m, and the thickness of the loose layer is 20~45 m. The mining period was from 1 January 2018 to 30 September 2019.

2.2. Data

The experimental data involved in this study are as follows: (1) Sentinel-1A provided C-band synthetic aperture radar data [45], and the satellite revisit period was 12 days/time. The data obtained in this study were 52 scenes of Sentinel-1 A satellite image data from 6 January November 2018 to 4 October 2019. Based on these data, the target area was analyzed, and these data types were SLC. The data was downloaded from the ASF Data Search (URL: https://search.asf.alaska.edu/, accessed on 8 February 2023). (2) SRTM DEM, which are released by NASA of the United States, with a ground resolution of 30 m [46]. POD precise orbit ephemerides were provided by ESA. The main parameters are shown in Tables 1 and 2.

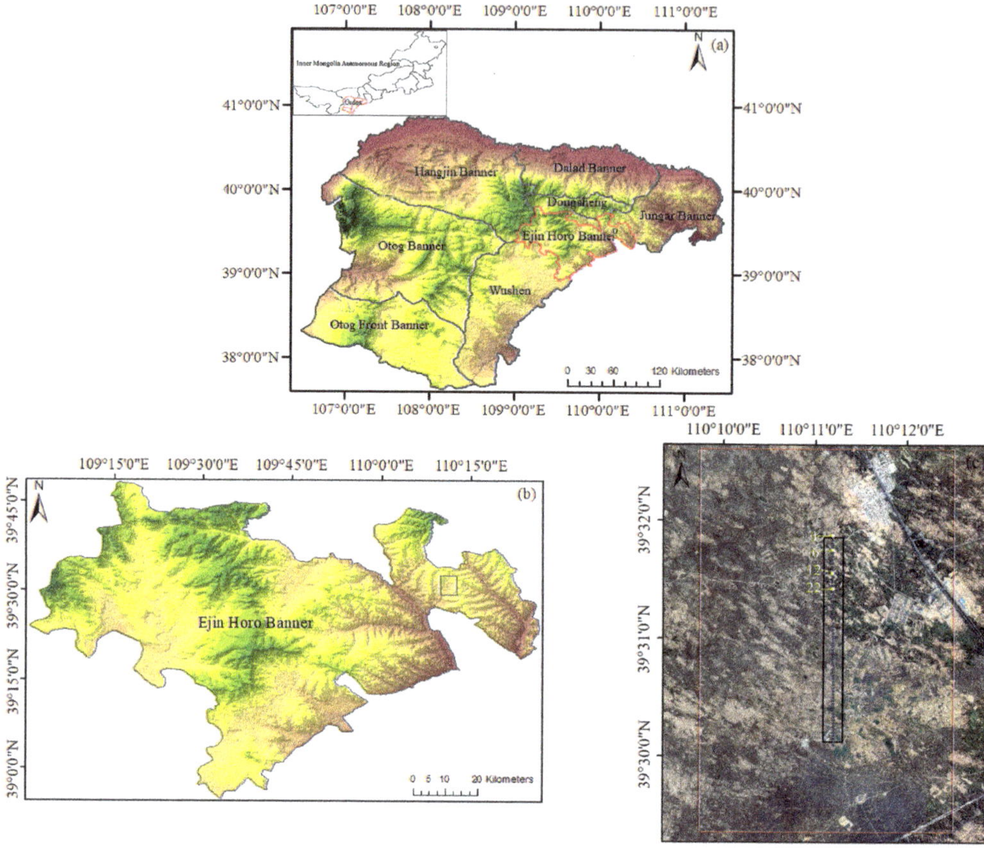

Figure 1. Geographical location and scope of the mining area: (**a**) location of the mining area, (**b**) regional location of mining area, (**c**) and location of the working face and monitoring points.

Table 1. The parameters of the Sentinel-1A images.

No	Image Data	Orbit	No	Image Data	Orbit	No	Image Data	Orbit
1	6 January 2018	020033	19	22 August 2018	023358	37	26 March 2019	026508
2	30 January 2018	020383	20	3 September 2018	023533	38	7 April 2019	026683
3	11 February 2018	020558	21	15 September 2018	023708	39	19 April 2019	026858
4	23 February 2018	020733	22	27 September 2018	023883	40	1 May 2019	027033
5	7 March 2018	020908	23	9 October 2018	024058	41	13 May 2019	027208
6	19 March 2018	021083	24	21 October 2018	024233	42	6 June 2019	027558
7	31 March 2018	021258	25	2 November 2018	024408	43	18 June 2019	027733
8	12 April 2018	021433	26	14 November 2018	024583	44	30 June 2019	027908
9	24 April 2018	021608	27	26 November 2018	024758	45	12 July 2019	028083
10	6 May 2018	021783	28	8 December 2018	024933	46	24 July 2019	028258
11	18 May 2018	021958	29	20 December 2018	025108	47	5 August 2019	028433
12	30 May 2018	022133	30	1 January 2019	025283	48	17 August 2019	028608
13	11 June 2018	022308	31	13 January 2019	025458	49	29 August 2019	028783
14	23 June 2018	022483	32	25 January 2019	025633	50	10 September 2019	028958
15	5 July 2018	022658	33	6 February 2019	025808	51	22 September 2019	029133
16	17 July 2018	022833	34	18 February 2019	025983	52	4 October 2019	029308
17	29 July 2018	023008	35	2 March 2019	026158			
18	10 August 2018	023183	36	14 March 2019	026333			

Table 2. Main parameters of the Sentinel-1A data used in this study.

Parameter	Value
Pass direction	Ascending
Beam mode	IW
Polarization	VV
Wave band	C
Wavelength/cm	5.6
Number of images	52
Monitored period	6 January 2018–4 October 2019

2.3. Methods

The deformation monitoring and analysis of the mining areas based on InSAR proposed in this study are divided into seven main steps:

(1) Data preprocessing. Select the required 52 Sentinel-1A image data, DEM data, and orbit data. Data clipping and baseline estimations are performed according to the scope of the study area.
(2) Ground control points (GCPs) screening. The obtained 52 Sentinel-1A SAR image data are processed using the PS-InSAR three-threshold method (the coherence coefficient threshold, amplitude dispersion index threshold, and deformation velocity interval) to obtain stable and qualified PS points as ground control points.
(3) DInSAR-PS-Stacking processing. The GCPs selected using the PS-InSAR three-threshold method are used for the refinement and re-flattening step of DInSAR-PS-Stacking processing, and then the cumulative time-series deformation phase information is obtained by weighted stacking; finally, the time-series cumulative deformation results are obtained.
(4) SBAS-PS-InSAR processing. The GCPs selected using the PS-InSAR three-threshold method are used for the orbit refinement and re-leveling steps of SBAS-InSAR processing, and then the time-series deformation information is obtained by deformation inversion.
(5) Comparative verification and analysis. The deformation information monitored by the DInSAR-PS-Stacking and SBAS-PS-InSAR methods are compared and analyzed according to comparative validation and the deformation results.
(6) DInSAR-PS-Stacking and SBAS-PS-InSAR fusion. The time-series deformation information of these two methods is fused using the two-threshold method (OTSU method sets the average coherence threshold and deformation threshold), and the time-series deformation information after fusion is obtained for complementary advantages.
(7) Deformation analysis after fusion. The settlement analysis of the fused deformation results obtained by the DInSAR-PS-Stacking and SBAS-PS-InSAR methods provides a scientific reference for coal mining subsidence control and disaster warning. The technical flow of the data processing is shown in Figure 2.

2.3.1. Ground Control Point Screening

The ground control points (GCPs) were used for refinement and re-flattening to estimate and remove the residual constant phase and residual phase ramp after unwrapping to improve the accuracy of deformation monitoring. The GCPs should be located in a flat terrain, with no phase jump, and far away from the deformation zone. Manual selection of the GCPs causes large errors. Therefore, we propose a method based on permanent scatterers (PS) to determine the GCPs. Firstly, the PS point targets with high coherence coefficients were identified using the coherence coefficient threshold method. Then, the amplitude deviation index threshold was set to further screen the PS points with strong, stable scattering as the target. Finally, a deformation velocity interval was set to select the final PS points [47].

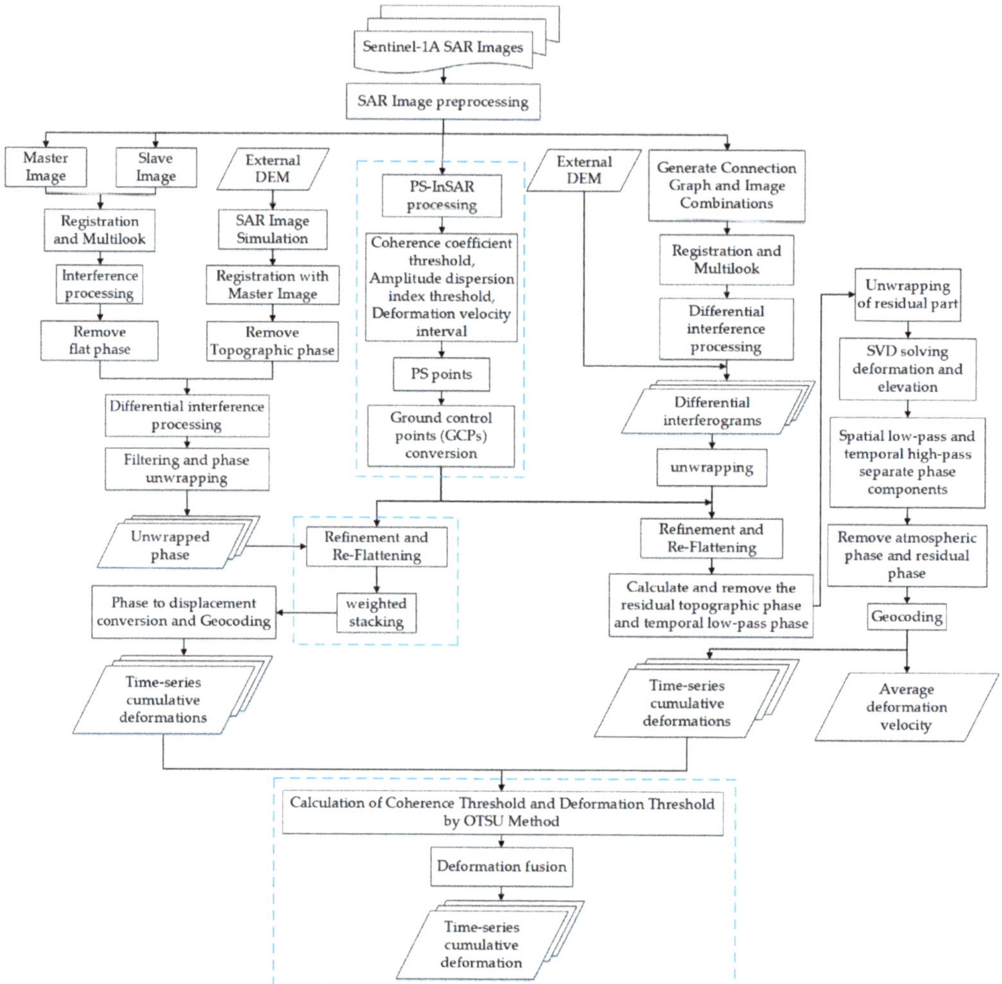

Figure 2. Technical flow chart.

(1) Coherence coefficient method

The coherence coefficient is an important index used to measure the interference quality of interference image pairs. It is mainly used to describe the similarity between the master and slave images in the same area [48]. The expression of the coherence coefficient is:

$$\gamma = \frac{\left| \sum_{i=1}^{m} \sum_{j=1}^{n} M(i,j) \cdot S^*(i,j) \right|}{\sqrt{\sum_{i=1}^{m} \sum_{j=1}^{n} |M(i,j)|^2 \cdot \sum_{i=1}^{m} \sum_{j=1}^{n} |S^*(i,j)|^2}} \tag{1}$$

where M is the master image, S is the slave image, and $*$ is the conjugate multiplication. After calculating the value of each pixel, the average value $\overline{\gamma}$ of each pixel in the time series is obtained:

$$\overline{\gamma} = \frac{1}{N} \sum_{i=1}^{N} \gamma_i \tag{2}$$

In general cases, the larger the γ value, the more stable the pixel, the lower the noise, and the better the interference phase quality. The coherence threshold is set, and when the $\overline{\gamma}$ value of a pixel point is greater than that of the threshold, it is determined to be an effective PS point [49].

(2) Amplitude dispersion index method

Ferretti et al. proposed that the stability of the interference phase can be measured by the time series of the amplitude information in the pixel [50]. R and I represent the real and imaginary parts of the image, respectively. If there is Gaussian noise with the standard deviation of σ_n, the amplitude value A obeys the Rice distribution [51]:

$$f_A(a) = \frac{a}{\sigma_n^2} I_0\left(\frac{ag}{\sigma_n^2}\right) e^{-\frac{a^2+g^2}{2\sigma_n^2}}, a > 0 \tag{3}$$

In the above formula, g is the target reflection, and $g > 0$, I_0 is a Bessel function. When the signal-to-noise ratio g/σ_n is small, the Rice distribution evolves into the Rayleigh distribution. In the high-SNR target ($g/\sigma_n > 4$), the distribution tends towards a Gaussian distribution [42]. Therefore, when $\sigma_n \ll g$, the phase dispersion index can be estimated by the amplitude dispersion index:

$$\sigma_v \cong \frac{\sigma_{nI}}{g} \cong \frac{\sigma_A}{m_A} \triangleq D_A = \frac{\sigma_A}{\mu_A} \tag{4}$$

In the above formula, σ_v is the phase dispersion index; σ_{nI} is the standard deviation of the imaginary part; μ_A and σ_A are the mean and standard deviation of the time series amplitude, respectively; and D_A is the amplitude dispersion index.

This method selects stable PS points by comprehensively considering the coherence coefficient, amplitude deviation index, and deformation velocity via PS-InSAR processing. Firstly, the pixels with high coherence were selected as PS points by the coherence coefficient method. The coherence threshold was set to 0.95, and the pixels with a coherence higher than that of the threshold were selected as PS points. Secondly, the PS points with a stable phase were further selected by the amplitude dispersion method, and the amplitude dispersion index threshold was set to 0.40. The PS points were selected if the amplitude dispersion index was lower than that of the threshold. Finally, the PS points were determined by the deformation velocity interval, and the deformation velocity interval was set to be [−1 mm/a, 1 mm/a]. Based on the three-threshold method described above, a total of 21 eligible PS points were selected as ground control points (GCPs) for refinement and re-flattening, so as to obtain more accurate surface deformation monitoring results.

2.3.2. DInSAR-PS-Stacking Processing

We selected multiple SAR images at different times in the same area from 6 January 2018 to 4 October 2019. Using the DInSAR-PS-Stacking method, we first preprocessed them by registration and multi-look processing (range/azimuth = 5:1). Then, the two–two interference processing was performed to generate the time-series interferogram, which was processed using the external DEM data to obtain the time-series differential interferogram, and then processed by filtering (Goldstein filtering), phase unwrapping (minimum cost flow method (MCF)), and other data processing steps. The GCPs obtained using the PS-InSAR three-threshold method was used for refinement and re-flattening, and then weighted stacking, phase-to-displacement conversion, and geocoding were performed [52,53] to obtain the accumulated time-series surface deformation information.

The interference phase φ is expressed as:

$$\varphi = \varphi_{flat} + \varphi_{topo} + \varphi_{def} + \varphi_{atm} + \varphi_{noise} + \varphi_{orbit} \tag{5}$$

In the above formula: φ_{flat} is the flat phase, which can be removed by accurate calculation of the baseline length; φ_{topo} is the topographic phase, which can be removed by the DEM data simulation; φ_{atm} and φ_{noise} are the atmospheric delay and noise phase, respectively, which can be reduced by filtering; φ_{orbit} is the phase caused by the orbit error, which can be eliminated using precise orbit data of the image pair; and φ_{def} is the deformation phase.

The deformation variable calculated by the deformation phase is expressed as:

$$\Delta r = -\frac{\lambda}{4\pi}\varphi_{def} \tag{6}$$

In the above formula, Δr is the deformation variable of the ground target in the direction of the radar line-of-sight (LOS); λ is the working wavelength of the radar sensor, 5.6 cm; and φ_{def} is the phase of the LOS surface deformation.

In this study, the PS points were selected using the PS-InSAR three-threshold method for conversion into GCPs for refinement and re-flattening so as to remove orbit errors and residual phases and to improve inversion accuracy. The unwrapping phases after refinement and re-flattening were stacked to reduce errors. The cumulative time-series subsidence maps of the study area were obtained by phase-to-displacement conversion, geocoding, and clipping.

2.3.3. SBAS-PS-InSAR Processing

SBAS-InSAR technology is one of the branches of time-series InSAR technology [54]. The SBAS-PS-InSAR method is an improvement of the SBAS-InSAR method. According to the spatio-temporal baseline threshold, all the SAR images from the same area are divided into several small baseline sets. The least squares method solves the deformation phase of each set. All the small baseline sets are connected, and the least squares solution, in the sense of the minimum norm of the deformation phase, is obtained by singular value decomposition (SVD) [55].

Basic principle: $N+1$ SAR images are selected according to the time sequence, and one scene is selected as the super master image, which is registered and resampled with other N SAR images [56]. According to the space–time baseline threshold, M differential interference pairs are obtained, where M is:

$$\frac{N+1}{2} \leq M \leq \left(\frac{N+1}{2}\right)N \tag{7}$$

Assuming that the $j-th(j=1,2,\ldots M)$ interferogram is at time t_A and t_B $(t_B > t_A)$, the phase value of the pixel (x,y) can be expressed as Equation (8).

$$\begin{aligned}\delta\varphi_j(i,j) &= \varphi(t_B,x,y) - \varphi(t_A,x,y)\\ &\approx \tfrac{4\pi}{\lambda}[d(t_B,x,y) - d(t_A,x,y)] + \varphi_{\text{topo}}(x,y) + \varphi_{orb} + \varphi_{res}(x,y)\end{aligned} \tag{8}$$

In the above formula, $d(t_A,x,y)$ and $d(t_B,x,y)$ are the LOS directional variables of the pixel at time t_A and t_B relative to the initial time; $\varphi_{\text{topo}}(x,y)$ is the topographic phase error caused by elevation data; φ_{orb} is the orbit error phase; and $\varphi_{res}(x,y)$ is the residual phase.

The topographic phase and flat phase were removed using external DEM data and geometric imaging relationships. Then, the required deformation phase information was obtained by filtering and phase unwrapping; the stable PS points obtained using the PS-InSAR three-threshold method were used as GCPs for refinement and re-flattening to further correct the deformation phase and improve the subsequent in-version accuracy. If $d(t_0,x,y)=0$, the corresponding time-series phase is:

$$\varphi(t_i,x,y) \approx \frac{4\pi}{\lambda}d(t_i,x,y) \tag{9}$$

The temporal deformation phase sequence of the pixel to be solved can be expressed as:

$$\varphi = [\varphi(t_1),\ldots\varphi(t_N)]^T \tag{10}$$

$$\delta\varphi = [\delta\varphi(t_1), \ldots \delta\varphi(t_N)]^T \quad (11)$$

The primary image and the secondary image sequences arranged in chronological order are represented by sets $IE = [IE_1 \cdots IE_M]$ and $IS = [IS_1 \cdots IS_M]$, respectively, and satisfy $IE_j > IS_j (j = 1, 2, \ldots, M)$. Then, all the differential phases can be expressed as:

$$\delta\varphi_j = \varphi(t_{IE_j}) - \varphi(t_{IS_j}), (j = 1, 2, \ldots, M) \quad (12)$$

Converting it to a matrix form, it can be expressed as:

$$A\varphi = \delta\varphi \quad (13)$$

In the above formula, A is the coefficient matrix. When all the interference pairs belong to a subset, the rank of A is $N(M \geq N)$, which can be obtained by the least square method:

$$\hat{\varphi} = A^{+}\delta\varphi, A^{+} = \left(A^T A\right)^{-1} A^T \quad (14)$$

In actual processing, A is mostly a non-full rank matrix and $A^T A$ is a singular matrix, so there are numerous solutions, which are solved by SVD.

$$A = USV^T \quad (15)$$

Finally, the phase-change velocity V is solved. After the phase-change velocity is solved, the cumulative deformation can be calculated by integrating the phase-change rate in the corresponding time period [57], so as to obtain the time-series deformation.

The SBAS-PS-InSAR method selected one of the images as the super primary image, and the remaining images were secondary images. The connection graph was determined using the spatio-temporal baseline threshold, and the deformation monitoring accuracy was improved with the reduction in the spatio-temporal baseline [58]. The spatio-temporal baseline thresholds were set to 2% and 70 days of the maximum critical baseline, respectively, and the image interference pairs were generated according to the principle of a small baseline set. The interference processing included the removal of the flat phase and topographic phase, the generation of a differential interferogram, Goldstein filtering, and minimum cost flow method (MCF) phase un-wrapping. The PS-InSAR three-threshold method was used to automatically select the PS points for the GCPs for refinement and re-flattening, to estimate and remove the residual constant phase and phase ramp after unwrapping, and to reduce orbit errors and improve the inversion accuracy. Then, we estimated and removed the residual terrain phase and time low-pass phase and performed secondary phase unwrapping on the residual part. The average deformation velocity and elevation topography were inverted. Then, the phase time series was filtered by atmospheric space–time filtering to filter out the atmospheric delay phase [59]. Finally, phase-to-displacement conversion, geocoding, and clipping were used to obtain the cumulative time-series deformation sequence of the study area.

2.3.4. DInSAR-PS-Stacking and SBAS-PS-InSAR Fusion

The cumulative time-series deformations monitored by the DInSAR-PS-Stacking and SBAS-PS-InSAR methods were compared and analyzed, and the monitoring ability and the advantages and disadvantages of these two methods for surface deformation in the goaf of the working face were studied. The SBAS-InSAR method performs poorly in areas with large subsidence gradients and poor coherence, but it can achieve higher accuracy in subsidence edge regions. On the contrary, the DInSAR method performs well in subsidence central areas and can achieve higher accuracy, but its accuracy is relatively low in subsidence edge regions [60,61]. In this study, the DInSAR-PS-Stacking method can obtain the cumulative subsidence in the center area with a large motion gradient, and SBAS-PS-InSAR can more accurately monitor slow and small deformations. Therefore, OTSU threshold segmentation is used to automatically extract double thresholds (coherence coefficient threshold and deformation threshold) to fuse the simultaneous cumulative time-series deformations obtained by the DInSAR-PS-Stacking and SBAS-PS-InSAR methods.

Firstly, the coherence maps were obtained using Formula (1) in the coherence coefficient method, and then the multi-temporal average coherence maps were obtained using Formula (2).

Secondly, OTSU was used to automatically obtain the threshold of the time series average coherence map. If the coherence is less than the threshold, the cumulative deformation result of DInSAR-PS-Stacking is fused, and the cumulative deformation result of fused SBAS-PS-InSAR with coherence greater than or equal to the threshold is fused.

Thirdly, on this basis, the OTSU method is used to extract the deformation threshold of the cumulative deformation results monitored by DInSAR-PS-Stacking. When the cumulative deformation is less than the threshold, it is the settlement center deformation with a large settlement gradient, and it is fused with the cumulative deformation map after the fusion in the previous step; finally, the cumulative time series deformation results after the fusion is obtained.

The problems of SBAS-PS-InSAR monitoring having an insufficient subsidence center and DInSAR-PS-Stacking monitoring having poor subsidence edge accuracy are solved, and the subsidence of the working face goaf is further studied and analyzed.

3. Results and Analysis

3.1. Analysis of Refinement and Re-Flattening Results

In this study, the above DInSAR-PS-Stacking and SBAS-PS-InSAR methods were used to process 52 scenes of SAR image data covering a coal mine in Nalintaohai Town, Yijinhuoluo Banner, Ordos City, Inner Mongolia Autonomous Region, from 6 January 2018 to 4 October 2019. Refinement and re-flattening are important steps in InSAR technology, which play an important role in improving the accuracy and reliability of measurement. It is necessary to select ground control points for refinement and re-flattening to determine the position and elevation information of satellite orbit and terrain for improving measurement accuracy and eliminating the influence of non-surface deformation factors. The ground control points are selected in areas with high coherence, flat terrain, no phase jump and deformation stripes. The stable PS points selected by the PS-InSAR three-threshold method basically meet the selection requirements of ground control points. The PS points obtained by the PS-InSAR three-threshold method are used as ground control points for refinement and re-flattening. We selected the representative results of differential interference and unwrapping. Figure 3 shows the differential interferogram after refinement and re-flattening and the phase diagram after unwrapping. It can be seen that the effect after processing is very good, and the root mean square error is within 15 or even 10. The results of other differential interference and unwrapping after refinement and re-flattening are also better, and the root mean square errors basically satisfy the requirements. The differential interferogram and phase unwrapping diagram after refinement and re-flattening are relatively smooth, with less noise, no obvious slope or step-like phase deviation, and the deformation area can be clearly seen and is more consistent. The differential interferogram and phase unwrapping diagram after data processing at different times have a more obvious deformation area, the positions are relatively close, and there is a certain repeatability and space–time continuity. Through the PS-InSAR three-threshold method, the stable PS points are selected as ground control points for refinement and re-flattening, which can correct the satellite orbit and phase offset and effectively eliminate the influence of orbit phase. This method basically meets the selection requirements of ground control points and meets the research needs of deformation monitoring.

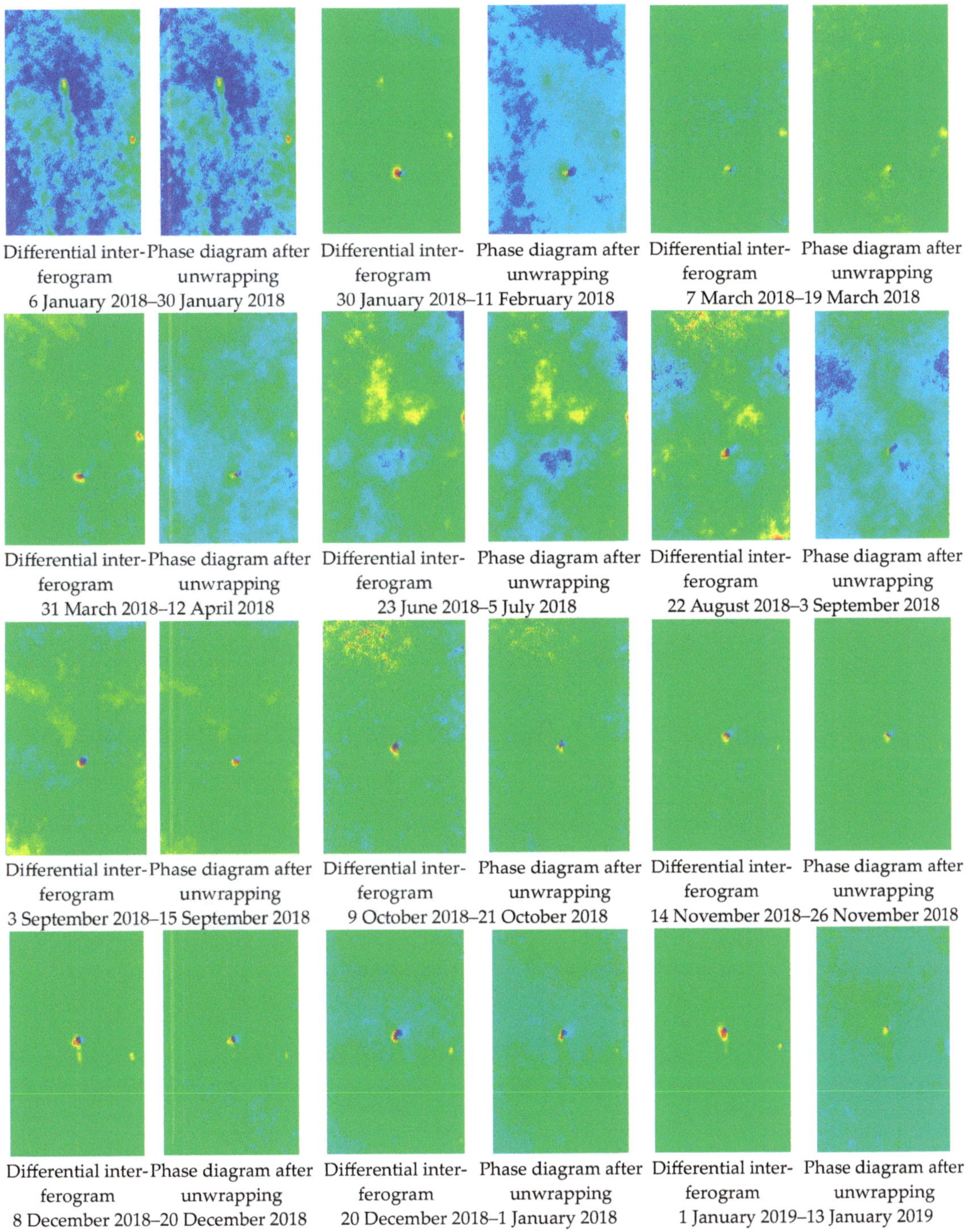

Figure 3. *Cont.*

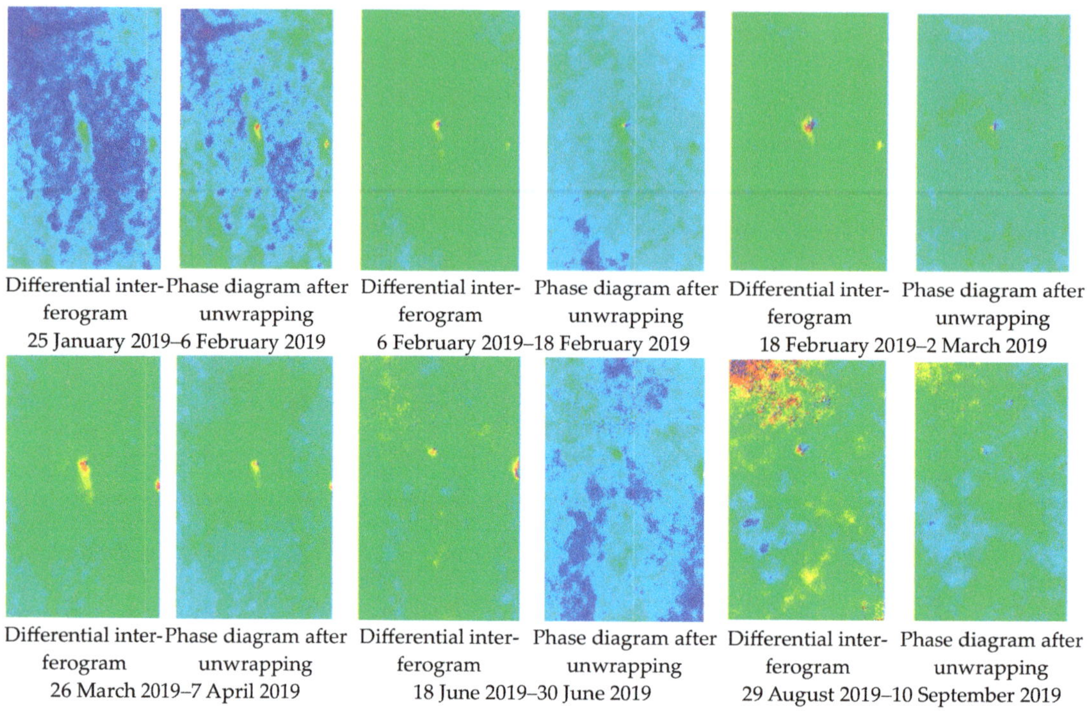

Figure 3. Differential interferogram and unwrapped phase diagram after refinement and re-flattening.

3.2. Monitoring and Analysis of DInSAR-PS-Stacking and SBAS-PS-InSAR

In this study, the DInSAR-PS-Stacking and SBAS-PS-InSAR methods were combined and compared to monitor the surface subsidence of a mining area in Nalintaohai Town, Yijinhuoluo Banner, Ordos City, Inner Mongolia Autonomous Region, from 6 January 2018 to 4 October 2019. After that, the monitoring deformation results were verified and analyzed. The proposed method improved the inaccuracy of mining area monitoring using the single-InSAR method, improved the accuracy and integrity of deformation monitoring in the mining area, reduced the atmospheric error, improved the monitoring effect, and was conducive to the efficient identification of goaf in the mining area. Figure 4 shows the average annual displacement velocity of the coal mines monitored using SBAS-PS-InSAR. Figure 5 shows the comparison of the monitoring results of the DInSAR-PS-Stacking and SBAS-PS-InSAR methods in the same period.

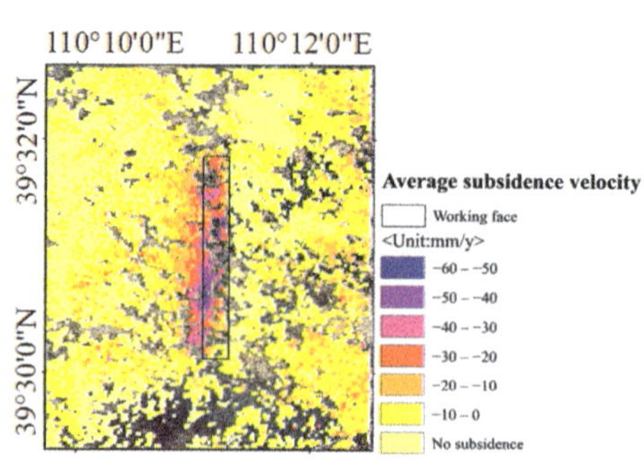

Figure 4. The average subsidence velocity of the mining area monitored using SBAS-PS-InSAR.

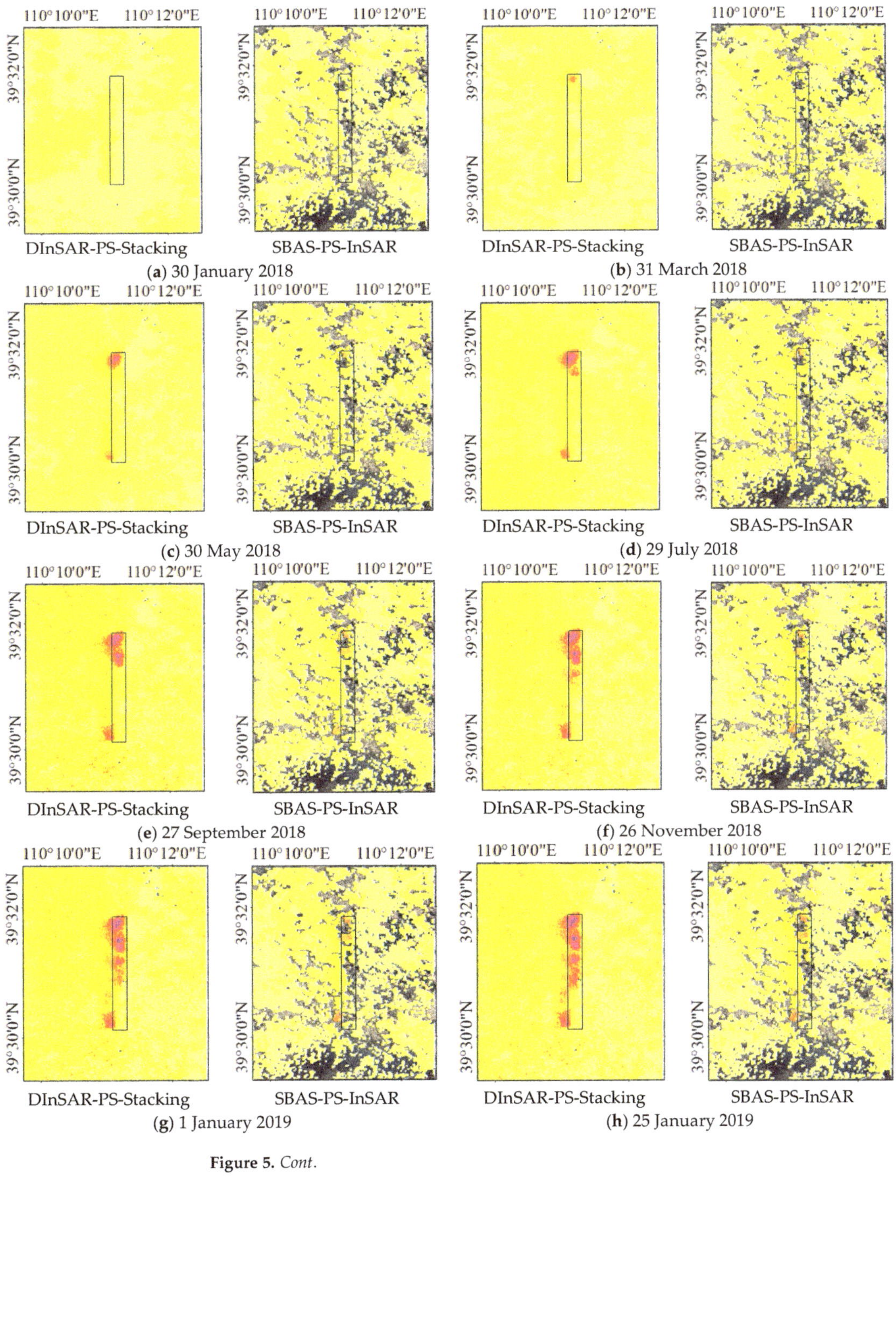

Figure 5. *Cont.*

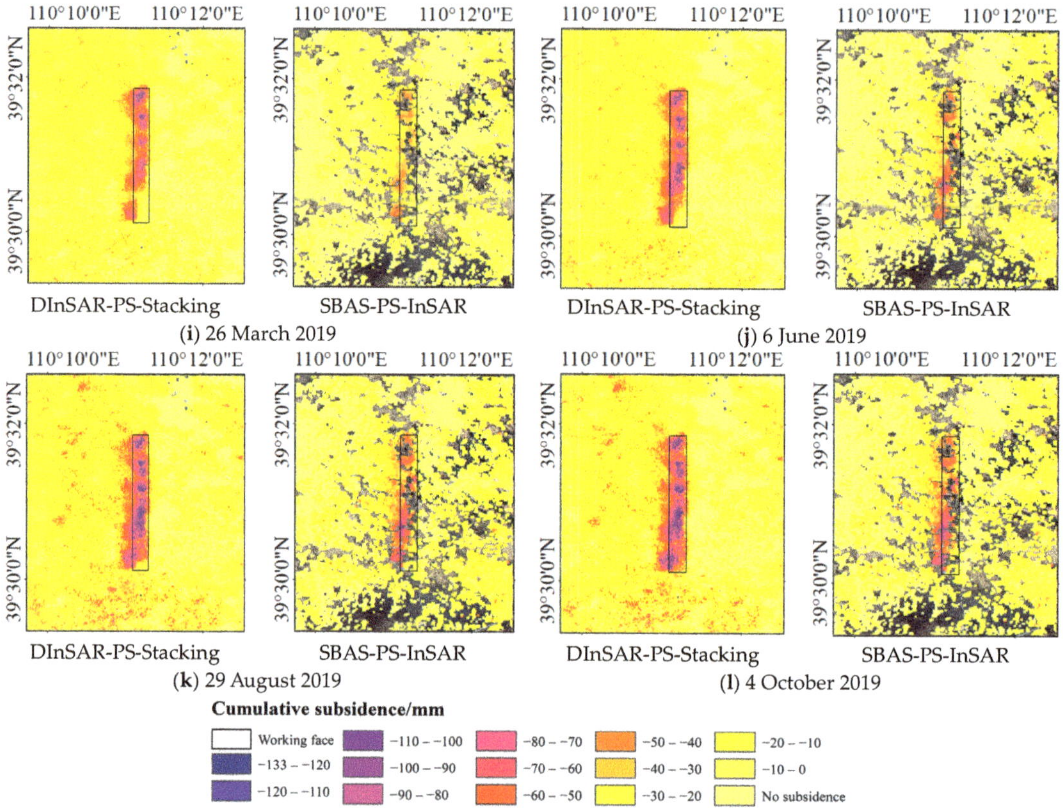

Figure 5. Comparison of the cumulative subsidence monitored using DInSAR-PS-Stacking and SBAS-PS-InSAR in the same period.

Coal mining often produces large subsidence in a short time, which can lead to loss of coherence. SBAS-PS-InSAR cannot select high-coherence points in the mining subsidence center, which leads to loss of information.

It can be noticed from the surface subsidence results shown in Figures 4 and 5 that from 6 January 2018 to 4 October 2019, both the DInSAR-PS-Stacking and SBAS-PS-InSAR methods monitored obvious subsidence funnels, which extended around the center of the subsidence funnel and were gradually distributed as strip. The subsidence range gradually extends from north to south along the working face, which is consistent with the advancement of mining progress. In the mining face area, there was basically no settlement at the beginning. With the advancement of mining progress, the settlement gradually increased, and then gradually stabilized. The subsidence change in the mining face is closely related to the advancement of mining progress. The position and spatio-temporal changes in the monitored mining subsidence basin were basically the same, which were highly consistent with the actual mining area. These two methods reveal that the ground subsidence in the mining area of the coal mining working face is gradually increasing, and the coal mine has been mined in large quantities, resulting in large-scale ground subsidence. As the mining intensity increased, the subsidence range spread around the center of the subsidence funnel. For the mining area, the maximum cumulative subsidence monitored by the DInSAR-PS-Stacking and SBAS-PS-InSAR methods was −131.4 mm and −96.2 mm, respectively. The annual average maximum subsidence velocity obtained by SBAS-PS-InSAR method was −59.3 mm/year. Overall, the two methods used to monitor coal mine surface subsidence locations, scope, distribution, and temporal and spatial subsidence laws showed a high degree of agreement. However, the specific analysis showed that: (1) in the same time period, the monitored cumulative subsidence using DInSAR-PS-Stacking is larger than that of SBAS-PS-InSAR, and that (2) in the area with a large subsidence gradient (such as the subsidence center) and poor coherence, DInSAR-PS-Stacking was able to monitor the subsidence information, while SBAS-PS-InSAR technology failed to effectively monitor the subsidence information, causing the obtained cumulative time-series subsidence map to be lacking.

Through the cumulative time-series subsidence map obtained using the DInSAR-PS-Stacking and SBAS-PS-InSAR monitoring methods, the time-series analysis and comparative verification of the selected four subsidence points were carried out. The location of the deformation monitoring point is shown in Figure 1c. Point 1 is at the starting position of the mining face and also at the edge of the settlement. Point 6, point 12, and point 22 are located in the larger settlement area. The large settlement area of the mining face in the mining area is more significant, and the settlement in most areas is larger considering the SBAS-PS-InSAR monitoring results are partially missing and the location of the GPS monitoring points. These four points can be used as a representation of the subsidence of the mining face, which can better reflect the deformation of the mining face and the surrounding ground. Choosing these points can make people better understand the subsidence near the mining face.

As shown in Figure 6, in the goaf of the coal mine's working face, it can be seen that the four characteristic subsidence points monitored using these two methods were basically the same for the overall subsidence trend. The DInSAR-PS-Stacking and SBAS-PS-InSAR methods both accurately monitor the temporal subsidence trends of the ground surface, which are consistent with GPS measurements. The settlement curves of the three show similar trends. However, the DInSAR-PS-Stacking method detects a greater amount of subsidence compared with the SBAS-PS-InSAR method. Taking Figure 6d as an example, the time-series settlement plot of this point shows the following information: from 6 January 2018 to 29 July 2018, the settlement was relatively small and changed slowly, with a slight downward trend. A significant settlement change occurred between 29 July 2018 and 27 September 2018. After 27 September 2018, although settlement still occurred, the settlement trend became relatively flat. In summary, this point experienced two different settlement stages during the observation period, with relatively slow settlement changes in the early and late stages and a significant change in the middle stage. The time series subsidence trend of Figure 6d is related to the mining progress. In Figure 6, it can be seen that the time series change basically shows a sinking trend. The time series settlement trend is basically that the settlement in the early and late stages is small and gentle, and the settlement in the middle stage is obviously larger, but the large settlement inflection point in the middle stage of each figure is different. The time of this inflection point is related to the degree of mining advancement.

The settlement point in Figure 6a is at the edge of the settlement, and the settlement points in Figure 6b–d are close to the large settlement center. In Figure 6a, the time series subsidence monitored by the SBAS-PS-InSAR method is closer to the GPS-measured time series subsidence. The time-series subsidence monitored by the DInSAR-PS-InSAR method in Figure 6b–d is closer to the GPS-measured time-series subsidence. In contrast, the SBAS-PS-InSAR method is more suitable for monitoring the slow and small deformation of the edge of the mining area, and its monitoring settlement is closer to the GPS-measured results. The DInSAR-PS-Stacking method is more suitable for monitoring the deformation of the large subsidence area in the mining area, and its monitoring subsidence and subsidence trend are closer to the GPS-measured results. Combining or fusing the DInSAR-PS-Stacking method and the SBAS-PS-InSAR method can better monitor the surface subsidence of the mining area and improve the monitoring accuracy and ability.

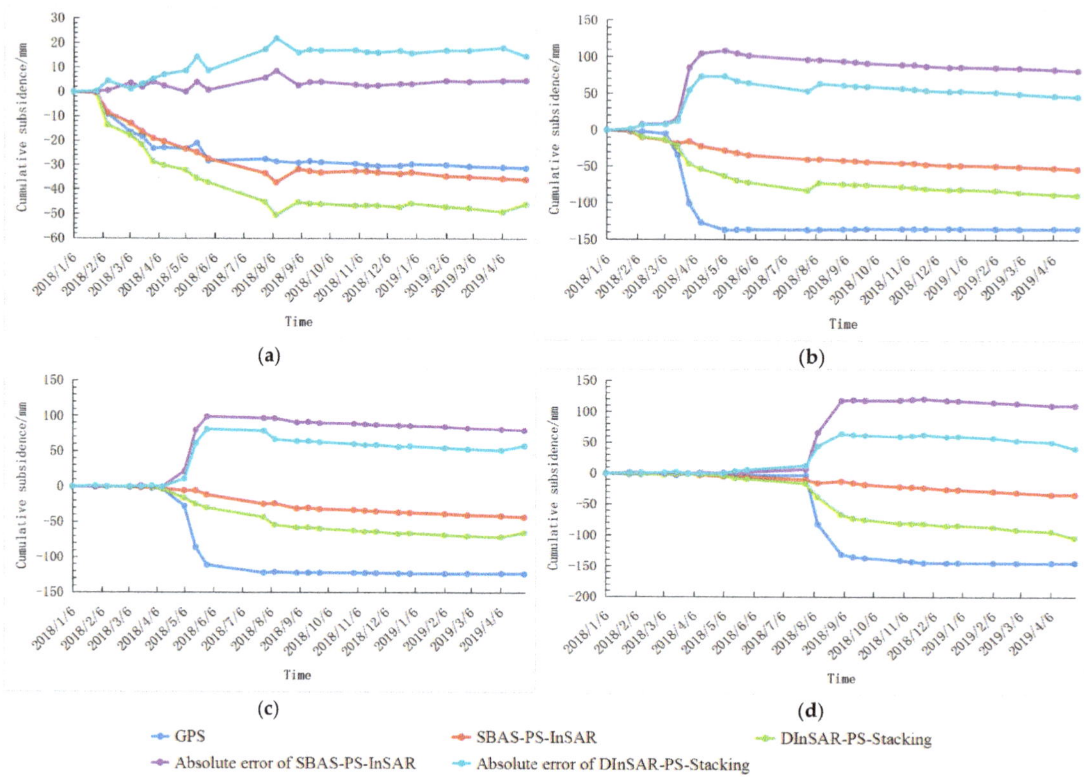

Figure 6. Results and comparison of time-series subsidence points monitored using DInSAR-PS-Stacking and SBAS-PS-InSAR: (**a**) point 1, (**b**) point 6, (**c**) point 12, and (**d**) point 22.

This paper compares and analyzes the correlation between the monitoring results of the DInSAR-PS-Stacking and SBAS-PS-InSAR methods and GPS monitoring results at four subsidence feature points. Pearson correlation coefficient is used to measure the strength of the linear relationship between the two, with a higher coefficient (closer to 1) indicating a more consistent subsidence trend. Figure 7 shows a comparison with the correlation diagrams of settlement curves measured by DInSAR-PS-Stacking, SBAS-PS-InSAR, and GPS. The time-series subsidence results monitored by the DInSAR-PS-Stacking and SBAS-PS-InSAR methods were compared with GPS monitoring results, and indicators such as Pearson correlation coefficient, mean absolute error (MAE), and root-mean-square error (RMSE) were calculated, verified, and analyzed. Table 3 shows the comparison and verification of the DInSAR-PS-Stacking time series settlement results and the GPS monitoring results. Table 4 shows the comparison and verification of the SBAS-PS-InSAR time series settlement results and the GPS monitoring results.

Table 3. Comparison and verification of DInSAR-PS-Stacking time series settlement results and GPS monitoring results.

Monitoring Points	Correlation Coefficient (Pearson)	Mean Absolute Error (MAE)/mm	Root Mean Square Error (RMSE)/mm
1	0.9735	12.2	13.9
6	0.9637	47.2	52.2
12	0.9568	41.9	51.0
22	0.9858	32.1	42.3

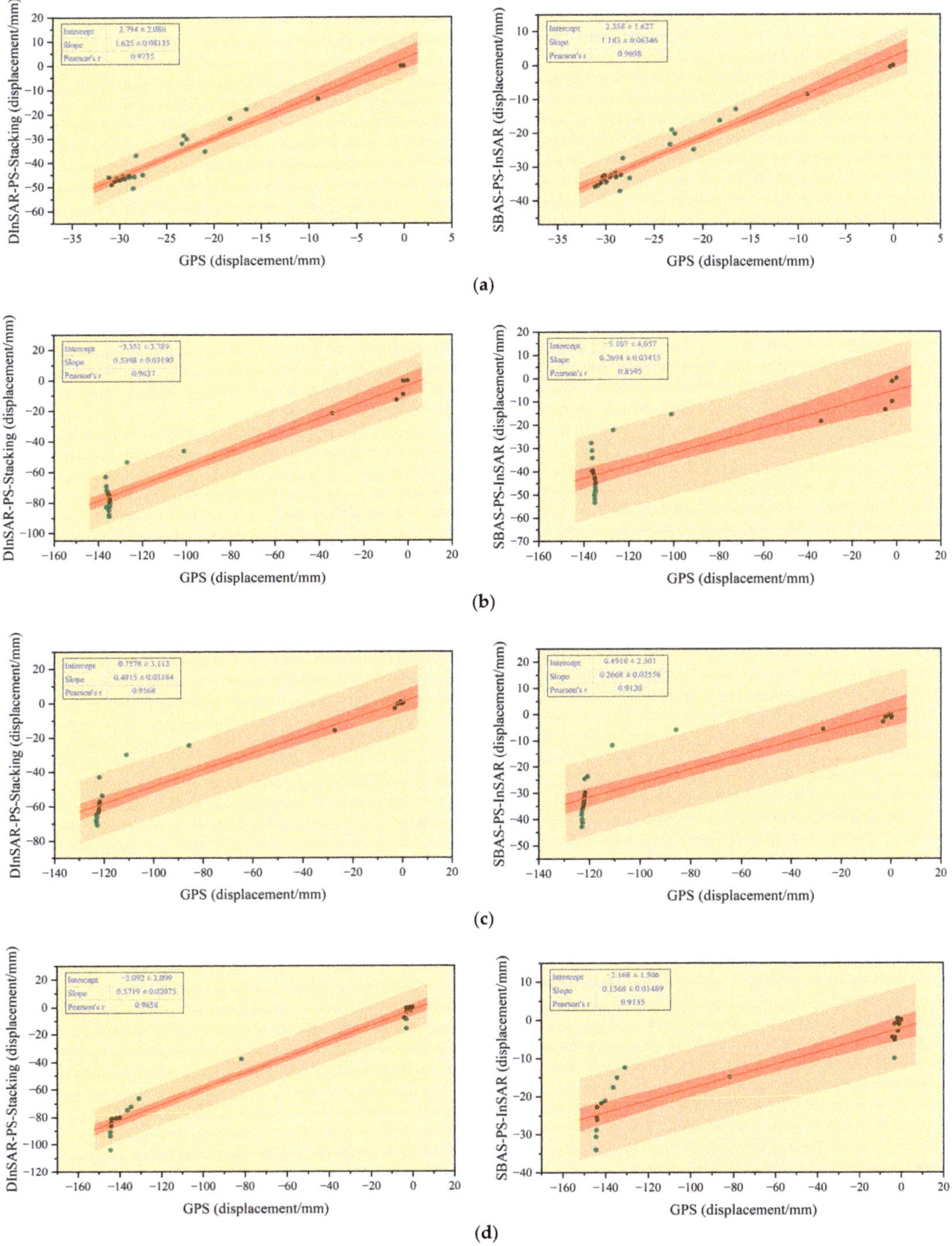

Figure 7. Comparing the correlation diagrams of settlement curves measured by DInSAR-PS-Stacking, SBAS-PS-InSAR, and GPS: (**a**) point 1, (**b**) point 6, (**c**) point 11, and (**d**) point 22.

Table 4. Comparison and verification of SBAS-PS-InSAR time-series settlement results and GPS monitoring results.

Monitoring Points	Correlation Coefficient (Pearson)	Mean Absolute Error (MAE)/mm	Root Mean Square Error (RMSE)/mm
1	0.9698	3.2	3.8
6	0.8595	74.5	82.6
12	0.9120	59.9	72.4
22	0.9135	62.0	83.9

There are differences between the monitoring results of the DInSAR-PS-Stacking and SBAS-PS-InSAR methods. Considering Figure 7 and Tables 3 and 4, an analysis and verification were carried out. The color bands in the correlation map in Figure 7 indicate better correlation with narrower bands. Based on Figure 7 and Pearson correlation coefficients, both InSAR methods show good correlation with GPS monitoring results, indicating that the subsidence trend detected by these two InSAR methods is consistent with GPS measurements. By comparing Tables 3 and 4, the Pearson correlation coefficient between the time-series subsidence results monitored by the DInSAR-PS-Stacking method and GPS monitoring results is higher than that of the SBAS-PS-InSAR method, approaching 1. This suggests that the monitoring results of the DInSAR-PS-Stacking method are more consistent with the GPS monitoring results, indicating higher reliability and accuracy. Combining Tables 3 and 4 with Figure 6, at point 1, the absolute error, mean absolute error (MAE), and root mean square error (RMSE) between the time-series subsidence results monitored by the SBAS-PS-InSAR method and GPS monitoring results are smaller than those of the DInSAR-PS-Stacking method. At points 6, 12, and 22, the absolute error, MAE, and RMSE between the time-series subsidence results monitored by the DInSAR-PS-Stacking method and GPS monitoring results are smaller than those of the SBAS-PS-InSAR method. In comparison, the monitoring results of the SBAS-PS-InSAR method have smaller errors in the subsidence marginal area, while the monitoring results of the DInSAR-PS-Stacking method show smaller errors in the large subsidence area. The errors of the surface deformation monitoring results of these two InSAR methods are within an acceptable range, indicating that the monitoring results of these two InSAR methods represent the same deformation field of the mining area. These results indicate that the surface subsidence monitoring results of the DInSAR-PS-Stacking and SBAS-PS-InSAR methods are effective and reliable.

In summary:

(1) The DInSAR-PS-Stacking and SBAS-PS-InSAR methods can accurately locate and detect the change trend of mining subsidence, which is in good agreement with the mining process of the coal mining face. The surface subsidence was found to gradually increase with the mining of the working face. The location, range, distribution, and space–time subsidence laws of the surface subsidence of the coal mine monitored by the two InSAR methods had good consistency.

(2) The point subsidence results obtained by these two InSAR techniques are well correlated with GPS monitoring results. The settlement trend of each point is basically the same, and the monitoring results are effective and reliable. The time series settlement errors monitored by these two InSAR methods show that the settlement edge is small and the large settlement area is large.

(3) Underground coal mining leads to surface subsidence, which makes radar images partially incoherent. Especially in the area with large settlement, the settlement gradient was large and the decoherence was substantial. For the large subsidence area of the goaf, DInSAR-PS-Stacking was found to be more effective than SBAS-PS-InSAR according to the monitoring results. The SBAS-PS-InSAR method is more effective for monitoring slow and small deformations.

3.3. Deformation Fusion Monitoring and Analysis of DInSAR-PS-Stacking and SBAS-PS-InSAR

Comparison analysis of monitoring mining area subsidence using the DInSAR-PS-Stacking and SBAS-PS-InSAR methods. The DInSAR-PS-Stacking method performs better in areas with large subsidence, while the SBAS-PS-InSAR method is more effective in monitoring slow and small subsidence in the outer regions. Therefore, the cumulative subsidence maps obtained from the DInSAR-PS-Stacking and SBAS-PS-InSAR technologies are fused. The fused cumulative subsidence map combines the advantages of monitoring large subsidence areas using the DInSAR-PS-Stacking method and slow subsidence areas using the SBAS-PS-InSAR method. The fused cumulative subsidence map, as shown in Figure 8, is combined with the working face and goaf zone shown in Figure 1c. Overall, the fusion of these two InSAR methods is effective, especially in the subsidence area of the working face and goaf zone where the obvious subsidence trend can be observed, and the missing phenomenon is not serious. The fusion of these two InSAR methods compensates for the deformation loss of the large subsidence area in the mining area monitored by the SBAS-PS-InSAR method, and the slow subsidence edge can be effectively monitored using the SBAS-PS-InSAR method. The mining area is less affected by factors such as climate and roads, and the subsidence basin is mainly caused by underground coal mining. As the mining operation progresses, subsidence gradually occurs in the area of the working face, which spreads to the surrounding areas and deepens gradually, eventually becoming stable. This subsidence gradually forms a subsidence basin that eventually matches the actual location of the mining face. By fusing the DInSAR-PS-Stacking and SBAS-PS-InSAR methods to monitor mining area subsidence, their respective advantages can be fully utilized to complement each other, thereby improving the ability and effectiveness of monitoring mining area surface deformation.

The fusion monitoring results of DInSAR-PS-Stacking method and SBAS-PS-InSAR method are compared with GPS subsidence monitoring results. The distribution of GPS subsidence monitoring points is shown in Figure 1c, and there are 66 monitoring points from north to south. Figure 9 compares the fused cumulative subsidence results from the two InSAR methods with the GPS measurement monitoring results. In Figure 9, it is possible to observe the settlement of monitoring points oriented from north to south on the mining working face very well.

From Figure 9, it can be observed that there are obvious subsidence basins in the mining face from north to south, and the subsidence trend is basically the same. The settlement changes in the middle and front sections of Figure 9a,b are consistent. The settlement of the rear section in Figure 9b is larger than that in Figure 9a, and the position of the last inflection point in Figure 9b is later than that in Figure 9a. This indicates that most areas in the northern part of the mining working face had stabilized subsidence trends by 29 August 2019, while the southernmost part continued to subside until 4 October 2019, before stabilizing. The subsidence of the goaf area of the working face is mainly caused by large-scale coal mining. As the coal mining progresses, the subsidence basin becomes increasingly apparent, and the subsidence volume and subsidence area gradually increase, with steep and uneven subsidence edges. The average absolute errors of the cumulative subsidence at each point on the two subsidence curves in Figure 9a,b are 55.8 mm and 56.1 mm, respectively, and the root mean square errors are 60.8 mm and 59.3 mm, respectively. This method can effectively monitor subsidence and is consistent with the actual situation.

The mining of underground coal mines can cause movement in the surrounding and overlying strata of the goaf, resulting in deformation and destruction of the surface. The surface deformation information obtained by combining the two methods was analyzed, and the distribution of subsidence in the goaf of the mining face and its surroundings can be clearly seen. Combined with the above analysis, the mining face subsidence range and the coincidence level with the working face became more obvious, and the surface continued to sink; according to this trend, in the future, the area will continue to settle. The ground cracks were generally parallel to the boundary of the mined-out area. In the edge area of the subsidence funnel, the ground surface was greatly affected by tensile deformation, resulting in cracks. As shown in Figure 9, the edge of the subsidence funnel is slightly steep, and there is a high possibility of surface cracking. Therefore, preventive measures should be taken before these conditions deteriorate. During the coal mining process, more preventive measures should be taken to prevent large-scale collapse.

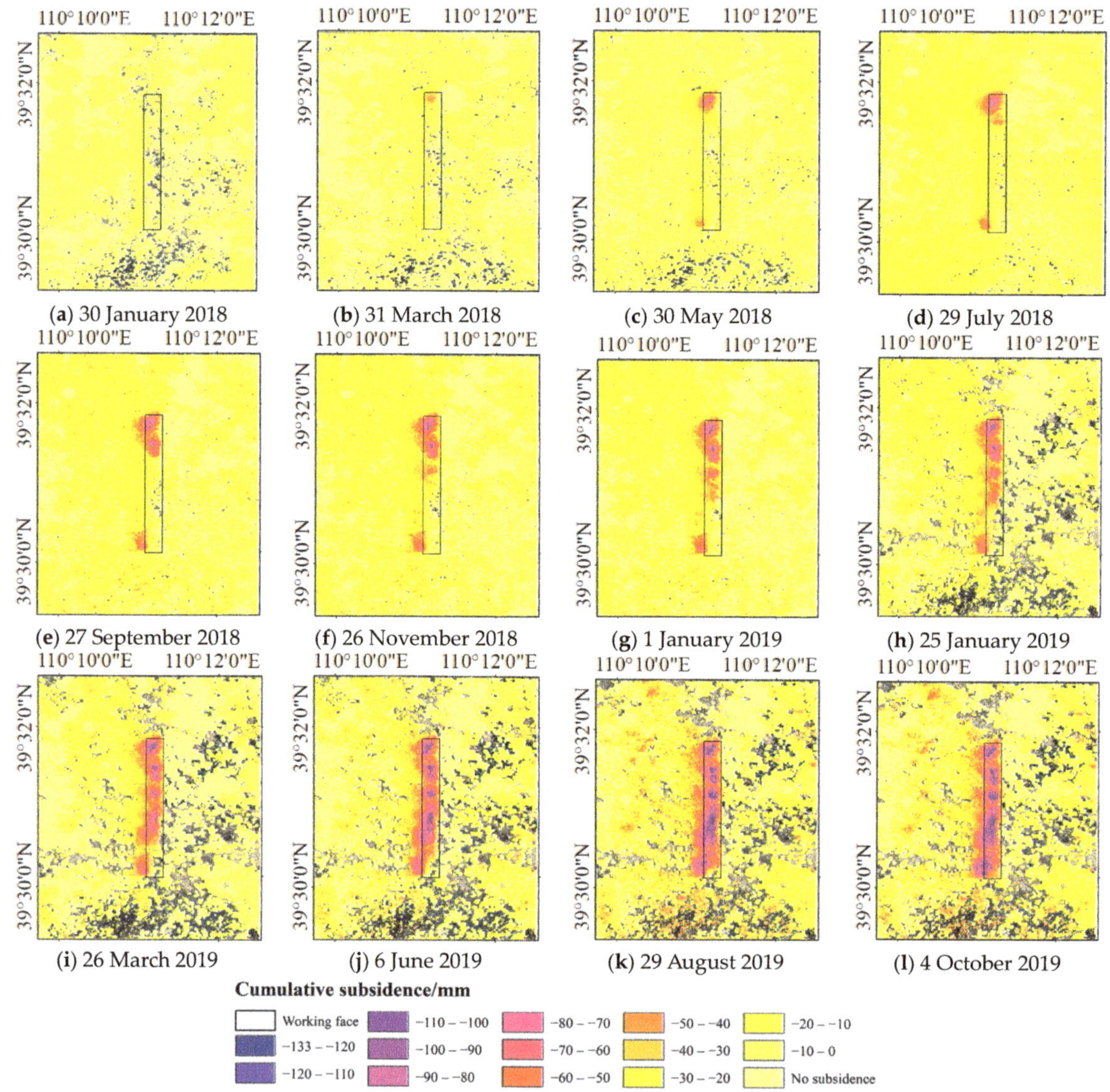

Figure 8. The cumulative subsidence maps after the fusion of DInSAR-PS-Stacking and SBAS-PS-InSAR.

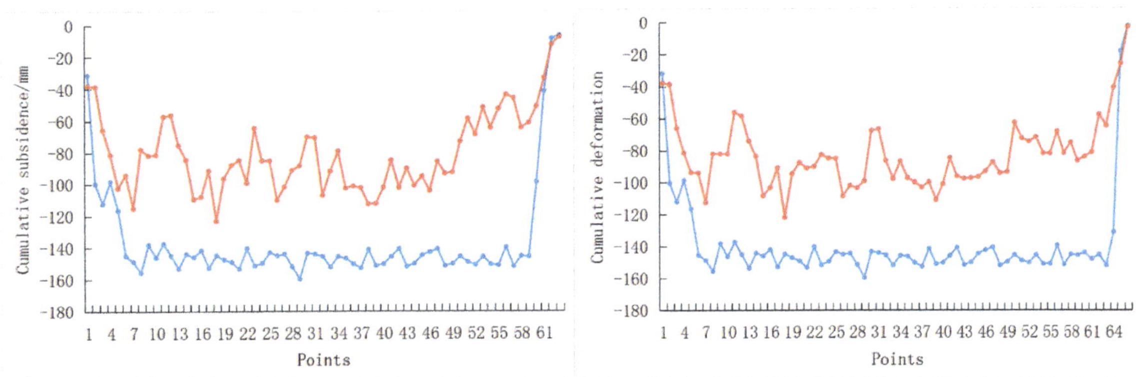

(**a**) (**b**)

Figure 9. Comparison between the fusion results of the two InSAR methods and GPS monitoring results: (**a**) 29 August 2019 and (**b**) 4 October 2019.

4. Discussion

The experimental results show that both the DInSAR-PS-Stacking and SBAS-PS-InSAR methods combine the PS-InSAR three-threshold selection method (the coherence coefficient threshold, amplitude dispersion index threshold, and deformation velocity interval) to select ground control points for orbit refinement and reinterferometric processing, solving the problem of manually selecting ground control points (GCPs) that affect monitoring results during orbit refinement and reinterferometric processing. They can effectively correct satellite orbits and phase offsets, reduce the influence of orbit errors, and improve deformation monitoring effectiveness.

Both the DInSAR-PS-Stacking and SBAS-PS-InSAR methods can be used as effective methods for real-time monitoring of subsidence caused by coal mining. Both methods can accurately monitor the location, extent, and spatio-temporal distribution of coal mining subsidence, with good correlation and consistency. The spatio-temporal subsidence trends monitored by both methods are consistent with the mining progress. It can be seen from Figure 6, Tables 3 and 4, that the deformation error of the DInSAR-PS-Stacking and SBAS-PS-InSAR methods is smaller in the edge region of the subsidence basin and larger in the area with significant subsidence. In the edge region with less subsidence, both methods can reflect the spatio-temporal subsidence trends well, with SBAS-PS-InSAR having higher monitoring accuracy. In areas with greater subsidence, the DInSAR-PS-Stacking method can better monitor the spatio-temporal subsidence trends, with a subsidence monitoring error smaller than that of SBAS-PS-InSAR. However, the deformation results monitored by the SBAS-PS-InSAR method have some missing information, with slightly inferior monitoring capabilities in the subsidence center and the area with significant subsidence. Nevertheless, the SBAS-PS-InSAR method has higher monitoring accuracy in areas with smaller subsidence and is suitable for slow and small deformation monitoring.

In order to better monitor the surface deformation of the mining area, the above characteristics of DInSAR-PS-InSAR and SBAS-PS-InSAR for monitoring the surface deformation of the mining area are considered. The double thresholds (deformation threshold and average coherence threshold and deformation threshold) are extracted by OTSU threshold segmentation to fuse the simultaneous time-series deformation results monitored by the two InSAR methods. The fused deformation results combine the advantages of DInSAR-PS-InSAR's effective monitoring of significant subsidence in large deformation areas and SBAS-PS-InSAR's effective monitoring of slow and small subsidence in the subsidence edge area, thereby improving the accuracy and completeness of mining area surface deformation monitoring. By comparing and fusing the DInSAR-PS-Stacking and SBAS-PS-InSAR methods, mining area surface deformation can be monitored more effectively, and the problem of incomplete monitoring in some areas can be avoided. The comparison and fusion of these methods can quickly and effectively obtain the deformation distribution of the mining area and more accurate mining area deformation information, achieving more effective monitoring of mining area surface deformation. The results of the space–time analysis show that subsidence will continue to occur in this area, which needs to be further studied to form an integrated research system for subsidence monitoring and prediction in mining areas. This will help to provide early warnings before disasters occur and will also provide auxiliary decision support for safe production in mining areas.

The deformation results of the mining area monitored by this research method are good, but the monitoring accuracy is still affected by some error factors. In a follow-up study, DEM data with higher precision and sourced from closer to the mining period can be considered, and the atmospheric delay error can be corrected by using external meteorological data (such as GACOS data [62]) to improve the monitoring accuracy. Subsequently, the phase filtering and phase unwrapping algorithms can be optimized for the study area to improve the monitoring accuracy. This study has scientific guiding significance for rational mining planning, accident prevention and control, and disaster prediction.

5. Conclusions

Taking a mining area in the Inner Mongolia Autonomous Region of China as the research area, this study used the DInSAR-PS-Stacking and SBAS-PS-InSAR methods to monitor the surface deformation of the mining area and compared and analyzed the deformation results monitored by the DInSAR-PS-Stacking and SBAS-PS-InSAR methods. The deformation results monitored by the two InSAR methods were fused and the fused deformation results were analyzed. The following conclusions are drawn:

(1) Both the DInSAR-PS-Stacking and SBAS-PS-InSAR methods can monitor the surface deformation of the mining area in real time and effectively, and can accurately monitor the location, range, and spatial and temporal distribution of coal mine subsidence.

(2) There is a subsidence basin gradually expanding from north to south in the coal mining face, and the edge is steep, which can easily produce surface cracks. The change trend of the subsidence basin is obvious and consistent with the mining situation of the working face. Large-scale mining of underground coal mines is the main factor causing surface subsidence.

(3) The DInSAR-PS-Stacking and SBAS-PS-InSAR methods are compared to better monitor the surface deformation of the mining area, effectively obtain the subsidence distribution of the mining area, and analyze the deformation of the coal mine goaf. The two InSAR methods have small deformation errors at the edge of the subsidence basin and large deformation errors in the large subsidence area. In contrast, the SBAS-PS-InSAR method performs better in monitoring slow and small deformation in the edge area of the subsidence basin. The DInSAR-PS-Stacking method is more effective in monitoring large deformation in large subsidence areas.

(4) The surface deformation results monitored by the DInSAR-PS-Stacking and SBAS-PS-InSAR methods are fused, and the fused deformation monitoring results are better. The fused method improves the inaccuracy of the traditional single InSAR method used to monitor the mining area, reduces the error, and improves the accuracy and integrity of the mining area deformation monitoring. Furthermore, the fused method provides more comprehensive deformation information for the comprehensive management of mining subsidence and realizes the effective monitoring of surface deformation in mining areas.

Author Contributions: Conceptualization, Y.C. and X.D.; methodology, Y.C. and X.D.; data curation, Y.C., X.D., P.H. and W.S.; formal analysis, Y.C. and X.D.; investigation, Y.C., Y.Q. and X.L. (Xiujuan Li); visualization, Y.C. and X.D.; writing—original draft, Y.C. and X.D.; writing—review and editing, P.H., W.X. and W.T.; project administration, Y.C., P.H. and X.L. (Xiaolong Liu). All authors have read and agreed to the published version of the manuscript.

Funding: This research was funded by the National Natural Science Foundation of China, grant number 52064039; the Natural Science Foundation of Inner Mongolia Autonomous Region, grant numbers 2020ZD18 and 2021MS06004; the Science and technology planning project of Inner Mongolia Autonomous Region, grant numbers 2020GG0073 and 2019GG139; the Major Project of Science and Technology of Inner Mongolia Autonomous Region, grant number 2019ZD022; the Basic scientific research business expenses of colleges and universities directly under Inner Mongolia Autonomous Region, grant number JY20220134; the Scientific research project of Inner Mongolia University of Technology, grant numbers BS201945 and DC2000001249; and the Equipment Engineering Special Project of natural disaster prevention and control, grant number TC210H00L-49.

Data Availability Statement: Not applicable.

Acknowledgments: Many thanks to Inner Mongolia Key Laboratory of Radar Technology and Application for the great support. Thanks to anonymous reviewers for their constructive comments to improve the quality of this paper.

Conflicts of Interest: The authors declare no conflict of interest.

References

1. Song, S.; Zhao, M.; Zhu, C.; Wang, F.; Cao, C.; Li, H.; Ma, M. Identification of the potential critical slip surface for fractured rock slope using the FLOYD algorithm. *Remote Sens.* **2022**, *14*, 1284. [CrossRef]
2. Meng, X. Application of InSAR Technology in Deformation Monitoring of Hancheng Mining Area. Master's Thesis, Xi'An University of Science and Technology, Xi'an, China, 2019.
3. Dong, M.; Zhang, F.; Lv, J.; Hu, M.; Li, Z. Study on deformation and failure law of soft-hard rock interbedding toppling slope base on similar test. *Bull. Eng. Geol. Environ.* **2020**, *79*, 4625–4637. [CrossRef]
4. Tao, Z.; Geng, Q.; Zhu, C.; He, M.; Cai, H.; Pang, S.; Meng, X. The mechanical mechanisms of large-scale toppling failure for counter-inclined rock slopes. *J. Geophys. Eng.* **2019**, *16*, 541–558. [CrossRef]
5. Dong, M.; Zhang, F.; Hu, M.; Liu, C. Study on the influence of anchorage angle on the anchorage effect of soft-hard interbedded toppling deformed rock mass. *KSCE J. Civ. Eng.* **2020**, *24*, 2382–2392. [CrossRef]
6. Ma, S.; Qiu, H.; Yang, D.; Wang, J.; Zhu, Y.; Tang, B.; Sun, K.; Cao, M. Surface multi-hazard effect of underground coal mining. *Landslides* **2023**, *20*, 39–52. [CrossRef]
7. Yang, D.; Qiu, H.; Ma, S.; Liu, Z.; Du, C.; Zhu, Y.; Cao, M. Slow surface subsidence and its impact on shallow loess landslides in a coal mining area. *Catena* **2022**, *209*, 105830. [CrossRef]
8. Ma, S.; Qiu, H.; Zhu, Y.; Yang, D.; Tang, B.; Wang, D.; Wang, L.; Cao, M. Topographic Changes, Surface Deformation and Movement Process before, during and after a Rotational Landslide. *Remote Sens.* **2023**, *15*, 662. [CrossRef]
9. Deng, Y. Monitoring and Forecasting of Surface subsidence in Mining Area Based on InSAR Technology. Master's Thesis, East China University of Technology, Nanchang, China, 2017.
10. Zhang, X.; Zhu, C.; He, M.; Dong, M.; Zhang, G.; Zhang, F. Failure mechanism and long short-term memory neural network model for landslide risk prediction. *Remote Sens.* **2022**, *14*, 166. [CrossRef]
11. Bamler, R.; Hartl, P. Synthetic aperture radar interferometry. *Inverse Probl.* **1998**, *14*, R1. [CrossRef]
12. Massonnet, D.; Feigl, K.L. Radar interferometry and its application to changes in the Earth's surface. *Rev. Geophys.* **1998**, *36*, 441–500. [CrossRef]
13. Hooper, A. A multi-temporal InSAR method incorporating both persistent scatterer and small baseline approaches. *Geophys. Res. Lett.* **2008**, *35*, L16302. [CrossRef]
14. Osmanoğlu, B.; Sunar, F.; Wdowinski, S.; Cabral-Cano, E. Time series analysis of InSAR data: Methods and trends. *ISPRS J. Photogramm. Remote Sens.* **2016**, *115*, 90–102. [CrossRef]
15. Ma, C.; Cheng, X.; Yang, Y.; Zhang, X.; Guo, Z.; Zou, Y. Investigation on Mining Subsidence Based on Multi-Temporal InSAR and Time-Series Analysis of the Small Baseline Subset—Case Study of Working Faces 22201-1/2 in Bu'ertai Mine, Shendong Coalfield, China. *Remote Sens.* **2016**, *8*, 951. [CrossRef]
16. Zhu, J.; Li, Z.; Hu, J. Research progress and methods of InSAR for deformation monitoring. *Acta Geod. Cartogr. Sin.* **2017**, *46*, 1717.
17. Gabriel, A.K.; Goldstein, R.M.; Zebker, H.A. Mapping small elevation changes over large areas: Differential radar interferometry. *J. Geophys. Res. Solid Earth* **1989**, *94*, 9183–9191. [CrossRef]
18. Xiao, Y.; Zhao, P.; Fan, Z.; Chen, G. Applicantion of TerraSAR-X Data for Subsidence Monitoring in Huainan Mining Area. *Remote Sens. Technol. Appl.* **2017**, *32*, 95–103.
19. Pawluszek-Filipiak, K.; Borkowski, A. Mining-induced tremors in the light of deformations estimated by satellite SAR interferometry in the Upper Silesian Coal Basin, Poland. *Procedia Comput. Sci.* **2021**, *181*, 685–692. [CrossRef]
20. Berardino, P.; Fornaro, G.; Lanari, R.; Sansosti, E. A new algorithm for surface deformation monitoring based on small baseline differential SAR interferograms. *IEEE Trans. Geosci. Remote Sens.* **2002**, *40*, 2375–2383. [CrossRef]
21. Lanari, R.; Mora, O.; Manunta, M.; Mallorquí, J.J.; Berardino, P.; Sansosti, E. A small-baseline approach for investigating deformations on full-resolution differential SAR interferograms. *IEEE Trans. Geosci. Remote Sens.* **2004**, *42*, 1377–1386. [CrossRef]

22. Darvishi, M.; Destouni, G.; Aminjafari, S.; Jaramillo, F. Multi-Sensor InSAR Assessment of Ground Deformations around Lake Mead and Its Relation to Water Level Changes. *Remote Sens.* **2021**, *13*, 406. [CrossRef]
23. Li, S.; Xu, W.; Li, Z. Review of the SBAS InSAR Time-series algorithms, applications, and challenges. *Geod. Geodyn.* **2022**, *13*, 114–126. [CrossRef]
24. Sandwell, D.T.; Price, E.J. Phase gradient approach to stacking interferograms. *J. Geophys. Res. Solid Earth* **1998**, *103*, 30183–30204. [CrossRef]
25. Dai, K.; Liu, G.; Li, Z.; Ma, D.; Wang, X.; Zhang, B.; Tang, J.; Li, G. Monitoring Highway Stability in Permafrost Regions with X-band Temporary Scatterers Stacking InSAR. *Sensors* **2018**, *18*, 1876. [CrossRef] [PubMed]
26. Liu, B.; Ge, D.; Wang, S.; Li, M.; Zhang, L.; Wang, Y.; Wu, Q. Combining Application of TOPS and ScanSAR InSAR In Large-Scale Geohazards Identification. *Geomat. Inf. Sci. Wuhan Univ.* **2020**, *45*, 1756–1762.
27. Dong, J.; Ma, Z.; Liang, J.; Liu, B.; Zhao, C.; Zeng, S.; Yan, S.; Ma, X. A comparative study of the identification of hidden landslide hazards based on time series InSAR techniques. *Remote Sens. Nat. Resour.* **2022**, *34*, 73–81.
28. Xu, Y.; Li, T.; Tang, X.; Zhang, X.; Fan, H.; Wang, Y. Research on the Applicability of DInSAR, Stacking-InSAR and SBAS-InSAR for Mining Region Subsidence Detection in the Datong Coalfield. *Remote Sens.* **2022**, *14*, 3314. [CrossRef]
29. Ferretti, A.; Prati, C.; Rocca, F. Permanent scatterers in SAR interferometry. *IEEE Trans. Geosci. Remote Sens.* **2001**, *39*, 8–20. [CrossRef]
30. Li, D.; Liao, M.; Wang, Y. Progress of Permanent Scatterer Interferometry. *Geomat. Inf. Sci. Wuhan Univ.* **2004**, *29*, 664–668.
31. Hooper, A.; Zebker, H.; Segall, P.; Kampes, B. A new method for measuring deformation on volcanoes and other natural terrains using InSAR persistent scatterers. *Geophys. Res. Lett.* **2004**, *31*, L23611. [CrossRef]
32. Hooper, A.; Segall, P.; Zebker, H. Persistent scatterer interferometric synthetic aperture radar for crustal deformation analysis, with application to Volcán Alcedo, Galápagos. *J. Geophys. Res. Solid Earth* **2007**, *112*, B07407. [CrossRef]
33. Guo, S.; Hou, H.; Zhang, S.; Xu, Y.; Yang, Y. Surface deformation monitoring of the mining area in Loess Plateau based on D-InSAR. *Sci. Surv. Mapp.* **2017**, *42*, 207–212.
34. Xia, Y.; Wang, Y.; Du, S.; Liu, X.; Zhou, H. Integration of D-InSAR and GIS technology for identifying illegal underground mining in Yangquan District, Shanxi Province, China. *Environ. Earth Sci.* **2018**, *77*, 319. [CrossRef]
35. Li, D.; Deng, K.; Gao, X.; Niu, H. Monitoring and Analysis of Surface in Mining Area Based on SBAS-InSAR. *Geomat. Inf. Sci. Wuhan Univ.* **2018**, *43*, 1531–1537.
36. Ma, F. Research on InSAR Monitoring and Prediction Method of Mining Subsidence. Ph.D. Thesis, Chang'an University, Xi'an, China, 2020.
37. Pawluszek-Filipiak, K.; Borkowski, A. Integration of DInSAR and SBAS Techniques to Determine Mining-Related Deformations Using Sentinel-1 Data: The Case Study of Rydułtowy Mine in Poland. *Remote Sens.* **2020**, *12*, 242. [CrossRef]
38. Zhang, P.; Guo, Z.; Guo, S.; Xia, J. Land Subsidence Monitoring Method in Regions of Variable Radar Reflection Characteristics by Integrating PS-InSAR and SBAS-InSAR Techniques. *Remote Sens.* **2022**, *14*, 3265. [CrossRef]
39. Zhang, L.; Dai, K.; Deng, J.; Ge, D.; Liang, R.; Li, W.; Xu, Q. Identifying Potential Landslides by Stacking-InSAR in Southwestern China and Its Performance Comparison with SBAS-InSAR. *Remote Sens.* **2021**, *13*, 3662. [CrossRef]
40. Wei, J.; Zhang, H.; Bai, Z.; Wang, S. Integrated Monitoring Method of the Mining subsidence of Shendong Mining Area based on D-InSAR and PS-InSAR Technology. *Met. Mine* **2019**, *10*, 55–60.
41. Ou, D.; Tan, K.; Du, Q.; Chen, Y.; Ding, J. Decision Fusion of D-InSAR and Pixel Offset Tracking for Coal Mining Deformation Monitoring. *Remote Sens.* **2018**, *10*, 1055. [CrossRef]
42. Luo, X.; Xiang, X.; Lyu, Y. PS corrrection of InSAR time series deformation monitoring for a certain collapse in Longli County. *Remote Sens. Nat. Resour.* **2022**, *34*, 82–87.
43. Wang, S.; Lu, X.; Liu, X.; Fu, S. A SBAS InSAR time series ground deformation extraction approach considering permanent scatterers. *Bull. Surv. Mapp.* **2019**, *2*, 58–62+70.
44. Otsu, N. A threshold selection method from gray-level histograms. *IEEE Trans. Syst. Man Cybern.* **1979**, *9*, 62–66. [CrossRef]
45. Snoeij, P.; Attema, E.; Davidson, M.; Duesmann, B.; Floury, N.; Levrini, G.; Rommen, B.; Rosich, B. The Sentinel-1 radar mission: Status and performance. In Proceedings of the 2009 International Radar Conference Surveillance for a Safer World (RADAR 2009), Bordeaux, France, 12–16 October 2009.
46. Li, P.; Li, Z.; Shi, C.; Liu, J. Quality evaluation of 1 arc second version SRTM DEM in China. *Bull. Surv. Mapp.* **2016**, *9*, 24–28.
47. Nie, Y.; Xiong, J.; Cheng, P.; Luo, Y. SBAS surface deformation monitoring with PS feature points. *Bull. Surv. Mapp.* **2022**, *4*, 91–95.
48. Pan, J.; Deng, F.; Xu, Z.; Xiang, Q.; Tu, W.; Fu, Z. Time series InSAR surface deformation monitoring in extremely difficult area based on track refining control points selection. *Chin. J. Geol. Hazard Control* **2021**, *32*, 98–104.
49. Li, N.; Wu, J. Research on methods of high coherent target extraction in urban area based on PSINSAR technology. *Int. Arch. Photogramm. Remote Sens. Spat. Inf. Sci.* **2018**, *42*, 901–908. [CrossRef]
50. Ferretti, A.; Prati, C.; Rocca, F. Analysis of Permanent Scatterers in SAR interferometry. In Proceedings of the (IGARSS 2000) IEEE 2000 International Geoscience and Remote Sensing Symposium—Taking the Pulse of the Planet: The Role of Remote Sensing in Managing the Environment, Honolulu, HI, USA, 24–28 July 2000; IEEE: Piscataway, NJ, USA, 2000; Volume 2, pp. 761–763.
51. Mao, Q. Research on Ground Deformation Detection Based on SBAS-InSAR Technology. Master's Thesis, Xidian University, Xi'an, China, 2019.
52. Li, D.; Hou, X. A case study of how to measure surface deformation using Sentinel-1A data. *Bull. Surv. Mapp.* **2020**, *3*, 118–122.

53. Tong, Y.; Huang, Y.; Chen, Y.; Tan, K.; Ou, D.; Han, F. Surface subsidence monitoring and spatio-temporal analysis in mining area based on D-InSAR. *Sci. Surv. Mapp.* **2020**, *45*, 67–73.
54. Cao, F.; Liao, W. Application of SBAS technology in ground subsidence monitoring of mining area. *Bull. Surv. Mapp.* **2020**, *3*, 156–158+163.
55. Wang, J. Study on the Key Technology and Applications of the Small Baseline Subset Time Series in SAR Technology for Surface Deformation Inversion. Ph.D. Thesis, University of Chinese Academy of Sciences, Beijing, China, 2019.
56. Wang, Z.; Ren, J.; You, H.; Liu, J.; Wang, L. Application of InSAR technology combined with probability integral method in deformation monitoring of mining area periphery. *Sci. Surv. Mapp.* **2022**, *47*, 84–94.
57. An, B.; Luo, H.; Ding, H.; Zhang, Z.; Wang, W.; Shi, X.; Ke, F.; Wang, M. Monitoring of Surface Deformation in Xining based on SBAS-InSAR. *Remote Sens. Technol. Appl.* **2021**, *36*, 838–846.
58. Du, Q.; Li, G.; Zhou, Y.; Chai, M.; Chen, D.; Qi, S.; Wu, G. Deformation Monitoring in an Alpine Mining Area in the Tianshan Mountains Based on SBAS-InSAR Technology. *Adv. Mater. Sci. Eng.* **2021**, *2021*, 9988017. [CrossRef]
59. Huang, J.; Guo, Y.; Luo, W. Monitoring land subsidence in Wuhan city based on Sentinel-1. *Bull. Surv. Mapp.* **2021**, *9*, 53–58.
60. Chen, Y.; Yu, S.; Tao, Q.; Liu, G.; Wang, L.; Wang, F. Accuracy Verification and Correction of D-InSAR and SBAS-InSAR in Monitoring Mining Surface Subsidence. *Remote Sens.* **2021**, *13*, 4365. [CrossRef]
61. Liu, X.; Tao, Q.; Niu, C.; Li, B.; Zhang, Y.; Ren, Y. Comparative analysis and verification of DInSAR and SBAS InSAR in mining subsidence monitoring. *Prog. Geophys.* **2022**, *37*, 1825–1833.
62. Li, Z.; Cao, Y.; Wei, J.; Duan, M.; Wu, L.; Hou, J.; Zhu, J. Time-series InSAR ground deformation monitoring: Atmospheric delay modeling and estimating. *Earth-Sci. Rev.* **2019**, *192*, 258–284. [CrossRef]

remote sensing

MDPI

Article

Seasonal Ground Movement Due to Swelling/Shrinkage of Nicosia Marl

Ploutarchos Tzampoglou [1,*], Dimitrios Loukidis [1] and Niki Koulermou [2]

[1] Department of Civil and Environmental Engineering, University of Cyprus, Nicosia 1678, Cyprus; loukidis@ucy.ac.cy
[2] Cyprus Geological Survey Department, Ministry of Agriculture, Rural Development and Environment, Nicosia 1301, Cyprus; nkoulermou@gsd.moa.gov.cy
* Correspondence: tzampoglou.ploutarchos@ucy.ac.cy

Abstract: This research investigates the seasonal ground heave/settlement of an area covered by an expansive soil of Cyprus called Nicosia marl, highlighting the degree of influence of the main causal factors. For this purpose, existing geotechnical data from the archives of the Cyprus Geological Survey were first collected and processed to compile maps of the key geotechnical parameters in the study area. In order to estimate the ground movements in the area, Earth Observation (EO) techniques for the period between 16 November 2002–30 December 2006 were processed. The correlation of these movements with the existing geotechnical data indicates that there is a statistically significant correlation between plasticity index and the ground movements. Multivariate linear regression analysis using Lasso revealed that the plasticity index ranks first in importance among the examined variables.

Keywords: ground heave/settlement; expansive clay; seasonal motion; rainfall

Citation: Tzampoglou, P.; Loukidis, D.; Koulermou, N. Seasonal Ground Movement Due to Swelling/Shrinkage of Nicosia Marl. *Remote Sens.* **2022**, *14*, 1440. https://doi.org/10.3390/rs14061440

Academic Editor: Andrea Ciampalini

Received: 7 January 2022
Accepted: 9 March 2022
Published: 16 March 2022

1. Introduction

Expansive soils significantly affect many countries worldwide, damaging buildings, road networks and other infrastructure. The mechanisms involved in this phenomenon are complex, and the factors that play an active role in its development can be divided into two categories: the preparatory and the triggering ones. In the first category are the geotechnical and the geological conditions, while in the second one, the moisture variation of the ground caused by rainfall infiltration and evapotranspiration, the leakage of water supply and sewage pipes, the fluctuations of the water table, etc. The economic losses caused by this hazard is in the scale of billions of dollars each year [1,2]. In fact, the annual cost of damages due to expansive soils surpasses that of other natural geohazards, namely earthquakes and landslides [3]. Examples of regions with expansive clays include: China [4]; Sudan [5,6]; Australia [7,8]; Saudi Arabia [9,10]; United Kingdom [1,11]; Canada [12], and Sweden [13]. Several studies (numerical or experimental) have been carried out in order to estimate the connection between the degree of damage to buildings and the distribution of ground moisture underneath their foundation, e.g., [14–16].

To date, Earth Observation (EO) technologies have been widely used to carry out studies investigating the failure mechanism of various natural hazard, such as landslides [17–19], land subsidence [20–22], flood [23,24], etc., along with field surveys and numerical modeling. However, the use of these technologies for the investigation of vertical ground movements due to the seasonal swelling and shrinkage of clays is at a relatively early stage [25–27]. Nonetheless, InSAR has been successfully used in numerous studies in the last decades for studying a phenomenon also pertaining to predominantly vertical ground movements, i.e., the land subsidence due to changes in the groundwater level. These studies estimate the degree of influence of the factors that play an active role on the future extent and magnitude of land subsidence, such as the geological conditions [28,29], tectonic structure [30], hydrological conditions [31–39], the thickness of the compressible formations [36,40], the urbanization [20,41] and the land use [40,42].

The present study employs Interferometric Synthetic Aperture Radar (InSAR) data to investigate the phenomenon of ground swelling/shrinkage due to seasonal moisture changes in an expansive clay formation and assess the degree of influence of the causal factors. The methodology developed herein relies on the integration of EO technologies, pre-existing local data and experience, conventional geotechnical research and the use of statistical analysis within a GIS environment. All the factors which affect this phenomenon, namely plasticity index [43–45], clay and montmorillonite content [4,46,47], expansive soil layer thickness and depth [6,15], were compiled and thematic maps were constructed using spatial analysis tools. It has long been established that the larger are the values of plasticity index, clay content and montmorillonite content, the larger the tendencies for swelling/shrinkage are expected to be [1,2]. Subsequently, the ground movements in the area as inferred by InSAR for the period between 16 November 2002–30 December 2006 [48] were processed. Finally, multivariate linear regression analysis using Lasso was carried out, elucidating the degree of influence of the main causal factors [49].

The study area is the central and eastern part of the city of Nicosia, which is the capital of the Republic of Cyprus (Figure 1). Expansive clays cover extensive areas of the island of Cyprus. The majority of these geomaterials are bentonitic clays deposited during the late Cretaceous period as deep-water sediments produced by the hydrothermal weathering of the basaltic rocks of the Tethyan oceanic crust. Nowadays, the outcrops of these extremely expansive materials are rather limited and can be found on the ground surface in a relatively small number of areas along the southern coast and in the mountainous region of Paphos District. However, the weathering of these old formations produced sediments that were redeposited in basins north and south of the axis of the Troodos mountain range, eventually resulting in the creation of the marls of the Nicosia geological formation, which are outcroppings particularly in areas of important urban development in the island of Cyprus, such as the cities of Nicosia, Larnaca and Paphos. These outcrops in combination with the climate of the region, which is characterized by rainy winters and very dry summers, are responsible for various types of damages to buildings and other infrastructure. More specifically, the seasonal variations of the degree of ground saturation that happen inside the so-called active zone, the depth of which is of the order of 3m to 8m, cause soil expansion (during the rainy winter months) or soil shrinkage (during the dry summer months). This phenomenon puts strain on buildings founded on such soils, which more often results in cracks in in-fill walls. In more extreme cases, these cracks may propagate to the structural frame. According to a previous study [50], which surveyed a part of the city of Nicosia where the marl is extremely expansive, 58 out of 96 buildings showed various degrees of damage due to ground swelling/shrinkage, characterized as moderate/severe for 7 of them. It is worth mentioning that in areas where the slope of the ground is >3° damages are substantially more frequent. This phenomenon may be due to the ratcheting effect on sloping terrain [51,52]. The annual cost of repairs in the surveyed area of 1.75 km^2 was estimated to be more than 2.4 million euros (based on 2002 prices), while the decrease in real estate value was estimated at 2 million euro per year.

Based on the above, it is obvious that the social and economic impact of expansive soils has been historically significant and is becoming more important in recent years as urban centers expand into areas covering such soils. The results of the present study are meant to help in understanding the relationship between the spatial distribution of the causal factors and the magnitude of this phenomenon. Its findings can be of benefit to state authorities and other policy makers, as well as the construction industry, in planning future urban development and the maintenance of existing infrastructure.

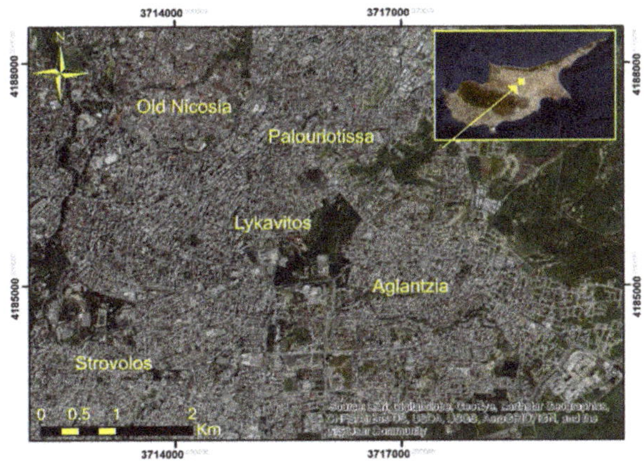

Figure 1. Satellite image of the study area.

2. Geomorphological, Geological and Climatic Setting

The study's location is mostly urban with the exception of the northeastern part where it consists of an industrial-agricultural zone. In the Old Nicosia area, there are mainly older types of buildings, while in the area west of Lykavitos the modern center of the city is developed, with a significant concentration of high-rise buildings (applying high vertical stress to the ground). The areas southeast of Aglantzia and Strovolos are considered predominantly residential with relatively newly built modern houses. The elevation of the study area ranges from 110 to 190 m. The lowest parts are located in the northeast (110 to 140 m) while the elevation increases southwest, where it reaches 170–190 in the suburb of the Strovolos. The city center presents gentle slopes, with elevation varying from 150 to 170 m.

The geological maps of the Cyprus Geological Survey Department (GSD) pertaining to the wider region of Nicosia, namely the Bedrock Geologic Map of Nicosia and the Surficial Geologic Map of Nicosia (scale 1:25000) are exploited. After spatial analysis and digitization processes, the geological formations with similar characteristics were merged and the surficial geological map of the study area was constructed (Figure 2A). The stratigraphy of the study area can be summarized as follows (starting with the younger formations and moving to the older ones):

Manmade fills: It consists of earthfills and manmade materials that have been placed irregularly on the natural ground.

Alluvial deposits: Depending on their deposition age, they can be divided into two groups:

The first group contains two distinct series. The more modern and the older one. The modern, which was deposited in rivers that were eroded into the older alluvial deposits, consist mainly of gravel, sand, silt and clay, as well as organic material. Its thickness reaches 1–2 m. The older consists of fine sand and silty clay with some pebbles.

The second group overlays the Apalos Formation. It can be mainly distinguished into two categories. The upper one underlies the alluvial terrace and it mainly consists of fine sand, silt, clay, pebbles, cobbles, and sporadic amount of manmade and organic materials. Its thickness varies from 1 to 4 m, while in some areas it reaches 10 m. The lower one consists of sands and gravels, the thickness of which vary from 1 to 5m. In some areas, fluvial stream deposits can be found, which are composed of sand and gravel with varying size of cobbles. These deposits do not exceed the 6 m in thickness.

Colluvial deposits: They consist of sand, calcareous sand, silt, clay, gravel and sporadic boulders. The total thickness reaches 4 to 6 m.

Apalos Formation (Pleistocene): Fluvial deposits, consisting of alternating layers of sand and gravel, sand, silt and clay. Its total thickness is approximately 22 m.

Nicosia Formation: The Nicosia Formation constitutes the bedrock of the entire study region and consists of several members:

Athalassa member of Nicosia Formation (Pliocene-Pleistocene): It mainly consists of calcarenite, calcareous sandstone and sandy marls.

Kephales member of Nicosia Formation (Pliocene): It consists of alternations of fine to coarse sands with varying sizes of pebble and cobble gravels. It also contains marine fossils. The thickness of this formation reaches 8 m.

Marl member of Nicosia Formation (Pliocene): This member is the dominant one in the region and is overlain unconformably by the Athalassa member and it is made up of marl and silty marl alternations with small amounts of sandy marl. It appears in two distinct horizons, the grey marl and the overlying light brown (khaki) marl. Its total thickness is the order of hundreds of meters. The grey marl is not outcropping in the Nicosia region and, thus, focus has traditionally been placed on the overlying khaki marl, which in certain locations has a montmorillonite content that can be up to 25% [50,53], rendering the khaki marl of the Nicosia formation medium to very highly expansive. The marl member outcrops in almost half of the study region, as seen in Figure 2B.

Lapatza Formation: It mainly consists of siltstone, marl, khaki marl, and limestone with thin intercalations of gypsum and silicified zones.

Among the geomaterials described above, only the Marl Member of the Nicosia Formation (and possibly the sandy marl of the Athalassa member, but to a much lesser degree) consists of expansive soils. All other materials are coarse to very coarse grained, and wherever they are encountered overlying the Nicosia marl they act as a buffer zone that lessens the negative effects of the marl swelling/shrinkage.

Figure 2B shows the bedrock geology, i.e., the geology excluding alluvial-colluvial deposits and manmade fills. These geomaterials are still largely inside the so-called active zone (the zone in which the ground moisture varies seasonally). This map is a useful tool as it presents the prevailing geological conditions pertaining to building foundations. Most building foundations in Nicosia are constructer either inside the Nicosia marl or are affected by its presence at some larger but proximal depth.

According the Köppen–Geiger climate classification, Cyprus has a subtropical-mediterranean and semi-arid climate [54], while it is characterized as arid based on the Thornthwaite climate classification [55]. In the area of Nicosia, the winter is mild, while during the summer season (from June to August) the temperature often exceeds 40 °C. Most of the rainfalls take place during the period between November and March. The average annual rainfall is of the order of 300–350 mm.

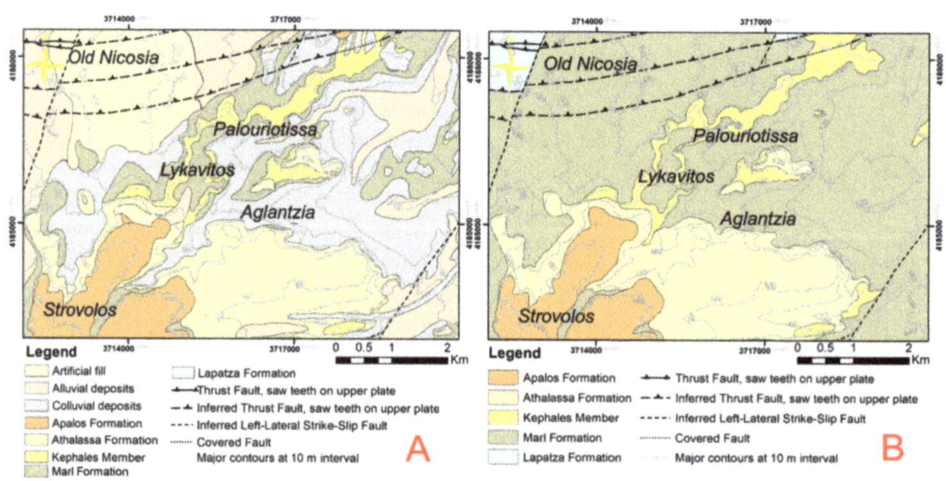

Figure 2. Maps of: (**A**) surficial and (**B**) bedrock geology of the study area.

3. Geotechnical Properties of the Nicosia Marl

In order to investigate the spatial distribution of the physical and mechanical characteristics of the Nicosia marl, over 200 drillings were collected from the numerous geotechnical investigation reports contained in the archives of GSD. The retrieved soil information was mainly about clay content (and occasionally montmorillonite content estimated using the "methylene blue" method), silt content, calcium carbonate content, the Atterberg limits, as well as the free swelling strain and/or swelling pressure measured in conventional oedometers. After data processing in the ArcGIS software (interpolation of geotechnical borehole data by using spatial analysis tools such as Kriging and topo to raster), thematic layers were constructed for the factors controlling ground swelling/shrinkage due to seasonal moisture changes, namely plasticity index (PI), clay and montmorillonite content, expansive soil layer thickness inside the active zone and depth. It should be stressed that the grey marl was not investigated due to the fact that it lies deeper than the zone of ground moisture change (active zone). This database was enriched with geotechnical data from the two campuses of the University of Cyprus, which also lie in the study area, as well as data from the reports of pertinent research projects [25,29].

The physical and mechanical properties of the Nicosia khaki marl are listed in Table 1. The khaki marl consists of horizons that can be classified either as silty clay or silty clay with sand intercalations. The highest clay amounts inside the study region were observed in Lykavitos, the northern part of Aglantzia and at southern part of Palouriotissa. The clay content decrease as we move towards Old Nicosia and Strovolos (Figure 3A). The montmorillonite content on average is 15% (reaching values up to 25%) for the silty clay horizons and the 12% (with maximum 26%) for the silty clay with sandy intercalations. The largest montmorillonite content values are observed at the area of Old Nicosia, Lykavitos and at the northeast of Aglantzia. The lower values are registered in the area around Strovolos (Figure 3B). It should be noted that the montmorillonite content over clay fraction ratio reaches 0.7 in the area around Old Nicosia. This rate, although quite high, is confirmed by an earlier study in which 25 mineralogical tests were performed and the montmorillonite content to clay fraction ratio was found to range from 0.4 to 0.75 [53].

Table 1. Physical and mechanical parameters of the khaki marl.

	Khaki Marl of Nicosia Formation					
	Silty Clay			**Silty Clay with Sand Intercalations**		
	Min	Max	Av.	Min	Max	Av.
Gravel content (%)	0	2	0.01	0	2	0.07
Sand content (%)	0	38	10.56	38	71	51.29
Silt content (%)	27	78	53.87	21	55	33.86
Clay content (%)	5	65	35.90	4	33	15.25
$CaCO_3$ content %	14	59	33.75	23.12	67.5	48.25
Montmorillonite content %	4.5	25	16.58	5	26.2	12.05
Specific Gravity	2.51	2.87	2.72	2.81	2.81	2.81
Bulk density (gr/cm^3)	1.96	2.2	2.02	-	-	-
Liquid Limit LL	41	128	78.63	37	104	55.52
Plastic Limit PL	14	66	31.85	18	67	28.33
Plastic Index PI	17	86	46.79	13	48	27.19
Free swelling strain (%)	1.04	43.87	9.22	-	-	-

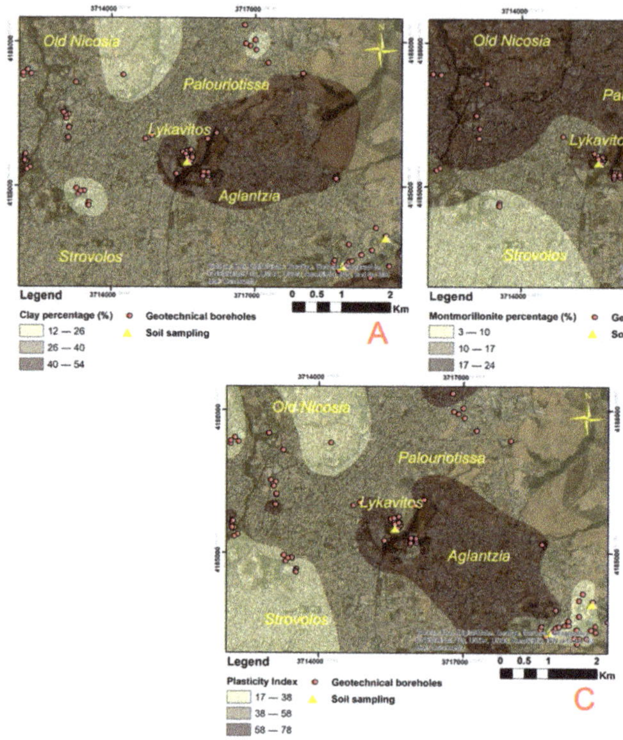

Figure 3. Thematic layers of Nicosia khaki marl geotechnical properties: (**A**) clay content (%), (**B**) montmorillonite content (%) and (**C**) plasticity index.

As for the Plasticity Index (PI), the highest values are observed in the silty clay horizons, reaching up to 86 with an average of 46. For the sandier horizons, the corresponding values are maximum 48 and average 27. These results place the khaki Nicosia marl in the high plasticity group according to the USCS classification. The highest values were observed at Lykavitos and Aglantzia, while the lower ones in Old Nicosia and Strovolos (Figure 3C).

Finally, taking into account the oedometer tests results, it seems that the average free swelling strain is 9%, while in some cases this rate reaches 43%. As such, the khaki marl can be characterized in general as very highly expansive.

In order to create maps with the variation in the thickness of the Nicosia marl inside the presumed active zone (upper 8 m of the ground profile) (Figure 4A) and of the depth of its upper boundary (Figure 4B), more than 230 boreholes that contained stratigraphic information (not only geotechnical investigation boreholes) were processed. The lowest values of the marl thickness in the assumed 8 m thick active zone were observed in Old Nicosia and at the south part of Strovolos, while the highest ones were mainly in the area around Aglantzia and the Eastern part of Palouriotissa (Figure 4A). The formation is practically outcropping in the areas around Lykavitos, Palouriotissa and Aglantzia, while the depth of the upper boundary increases substantially towards Old Nicosia (Figure 4B).

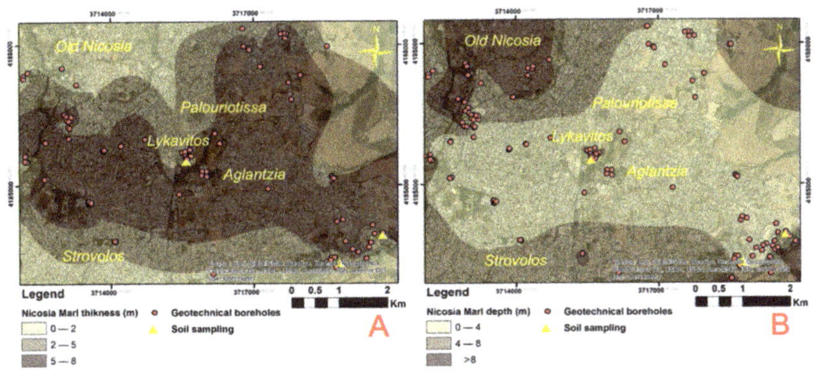

Figure 4. Thematic layers of: (**A**) marl thickness inside active zone and (**B**) upper boundary depth of Nicosia marl in the ground profile.

4. InSAR Data and Processing

For the purpose of this study, the results of the European Union's research project titled 'PanGeo' are used. The project's objective was to enable free and open access to geohazard information in all EU countries, based on the collection of data via satellites [56]. At the moment, 52 European cities provide open access maps, which consist of vector polygon Ground Stability Layer (GSL) and their GeoHazard Description (GHD).

The Geological Survey Department of Cyprus took part in this EU project. In the context of Cyprus's participation, an urban area of 489 km^2 of the city of Nicosia was chosen for studying ground movements. Data by the Persistent Scattering Interferometry (PSI) InSAR technique (Interferometric of Synthetic Aperture Radar) from 20 radar scenes of Envisat [Environmental satellite (Table 2) operated by the European Space Agency (ESA)] were taken and analyzed in the GAMMA IPTA software (version V1.2 June 2007).

The date range of the analysis was between 16 November 2002 and 30 December 2006. The scene of 14 May 2005 was taken as a master scene. The Georeference (X,Y) accuracy was 15 m and the reference data used for georeference was Quickbird (0.6 spatial resolution). A total number of 23818 PS points were identified with density of about 49 PS points per km^2. The average annual motion rate of the entire processed area is about −0.173 mm per year while the standard deviation of the average annual motion rate reaches the 1.523.

Table 2. Radar data used.

Number of Scenes	Slave Date	Perpendicular Baseline (m)	Temporal Separation (Days)
1	16 November 2002	253.8828	−910
2	27 September 2003	−705.023	−595
3	6 December 2003	−729.462	−525
4	29 May 2004	12.3725	−350
5	3 July 2004	−1038.95	−315
6	7 August 2004	−507.004	−280
7	11 September 2004	290.7173	−245
8	16 October 2004	100.2115	−210
9	25 December 2004	−469.766	−140
10	29 January 2005	−503.047	−105
11	9 April 2005	−522.133	−35

Table 2. *Cont.*

Number of Scenes	Slave Date	Perpendicular Baseline (m)	Temporal Separation (Days)
12	14 May 2005	0	0
13	23 July 2005	100.5969	70
14	27 August 2005	106.5261	105
15	5 November 2005	176.0639	175
16	18 February 2006	−694.437	280
17	25 March 2006	−234.957	315
18	3 June 2006	−1172.88	385
19	8 July 2006	428.7025	420
20	30 December 2006	−102.028	595

5. Results

The investigation of this phenomenon through the SAR data processing presents some inherent weaknesses. Initially due to the nature of the phenomenon (i.e., repeated small amount of swelling followed by shrinkage without a characteristic monotonic trend), the creation of time series of the velocity or the average vertical displacements per year could not provide us with meaningful conclusions. In addition, it has to be recognized that the ground heave decreases for larger applied vertical stress and, thus, depends strongly on the pressures applied by the structures. Indeed, beyond a threshold stress point (the "swelling pressure"), the addition of moisture in the Nicosia marl does not produce swelling, but generates a collapse strain and settlement. Therefore, it is clear that, in an urban environment where high rise buildings (applying high vertical stress) coexist with smaller ones (lower vertical stress) and building with shallow foundations coexist with building founded on piles (in which case the vertical movement of the superstructure is drastically limited), the SAR data should be treated with great caution.

In order to accommodate the non-monotonic evolution of the ground movement, the amplitude of the oscillations of the vertical movement, in the form of the difference between the maximum and minimum displacement observed in the given time period, was calculated for all the PS points. This way, the intensity of the vertical deformation fluctuations (and not their tendency) becomes evident. Secondly, to mitigate the fact that ground swelling depends on the applied vertical stress, the PS points that did not belong to buildings were selected and treated separately as Free Field points and assumed to not be influenced by the loads or the rigidity of the neighboring structures. This process was done manually. More specifically, reflectors such as lighting poles, sidewalks, as well as objects and very small structures in parks and unbuilt areas, were searched and separated from PS points that obviously belonged to buildings.

Figure 5A shows all the PS points in the study area for the period between 16 November 2002 and 30 December 2006. All the PS points were divided into five categories according to Natural Breaks method. It can be seen that the largest vertical displacements amplitudes, which can be as high as 40 mm, are observed in the areas around Lykavitos and Aglantzia. This observation is in good agreement with the Plasticity Index Thematic Layer (Figure 6A), and both the clay content and Nicosia Marl depth, which are also very unfavorable in these areas (Figure 6C,E). The smallest vertical displacement amplitudes appear in the Strovolos wider area, in the neighborhoods bordering the Pedieos river and in Old Nicosia and its vicinity.

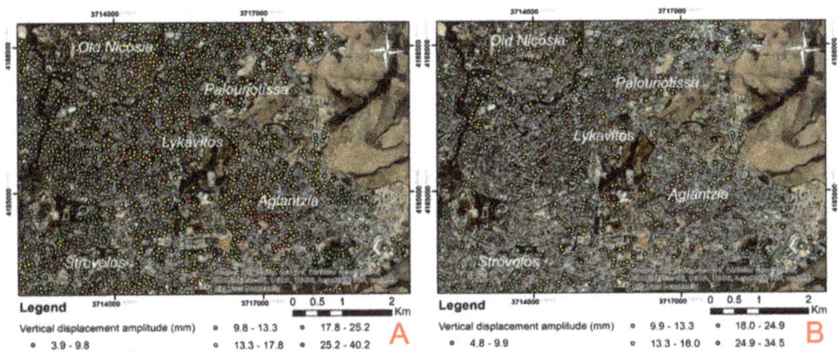

Figure 5. Spatial distribution of the vertical displacement amplitude for the period 16 November 2002–30 December 2006: (**A**) All PS points, (**B**) free field PS points.

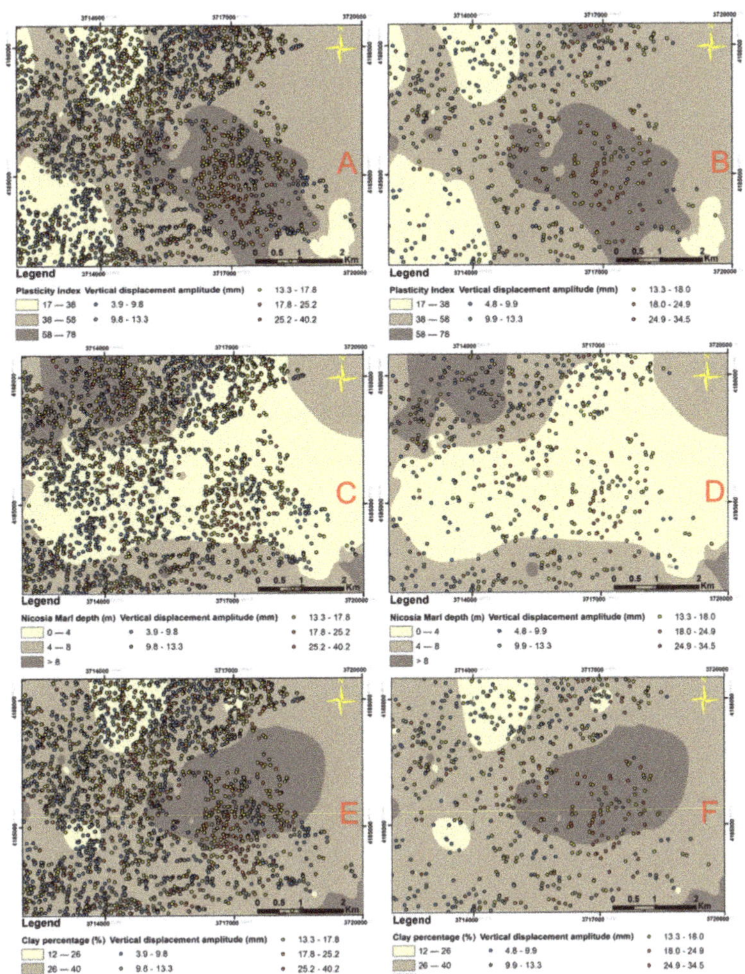

Figure 6. Spatial distribution of the vertical displacement amplitude along with the factors which strongly affect the phenomenon: (**A**,**C**,**E**) PS points, (**B**,**D**,**F**) free field PS points.

A similar but more clear image is observed in Figure 5B, where the PS Free Field points are isolated and plotted. This map pinpoints the highest values of ground movement in the areas of Lykavitos and Aglantzia. These areas present marl of high plasticity index and clay content (Figure 6B,F). In addition, an increase in the variation in the vertical displacement amplitude is marked when moving to the north part of Palouriotissa. This can be attributed to the effect of the relative shallowness of the Nicosia Marl (Figure 6D).

In order to investigate the vertical displacements amplitudes for the three hydrological years, six new maps were created. It should be stressed that the first hydrological year is not presented due to incomplete remote sensing data. Figure 7A,C,E, in which all the PS points are included, shows large vertical displacement amplitudes, especially in the area around Aglantzia and Lykavitos, reaching values as high as 33 mm. Similar conclusions can be drawn for the Free Field PS points (Figure 7B,D,F). More specifically, the highest values are observed in the area around Aglantzia and Lykavitos, while the lowest ones at the Old Nicosia and Strovolos. It is worth stressing the fact that in the western part of study area, the PS points with the lower amplitude values in Figure 7A,C,E disappeared from Figure 7B,D,F, while Free Field PS points with relatively large movement amplitudes remained. This observation can lead to the conclusion that in this area, high vertical displacements amplitudes can occur, but due to the PS points reflectors locations, i.e., on high-rise buildings applying high vertical stresses to the ground, the observed displacements are small.

It should be stressed that the vertical displacement amplitudes are similar both throughout the 4-year examination period (40 mm for the PS points and 35 mm for the Free Field PS points) and for each hydrological year separately (33 mm for the general PS points and 32 mm for the Free Field PS points) (Figure 7). This observation confirms the periodicity of the phenomenon and leads in the conclusion that the main cause of ground volume changes is the seasonal fluctuation in the climatic factors (precipitation, temperature, sunshine).

In order to investigate the periodicity of this phenomenon, the rainfall data from 2 meteorological stations, during the time period were obtained. As Figure 8 shows, the driest period was between June 2004 and September 2004, as the cumulative amount of rainfall did not exceed 1 mm. On the contrary, the periods November 2002–May 2003 and October 2003–May 2004 presented the highest amount of precipitation. A correlation between three free field PS points (19322, 19126, 17779) and rainfall for the period November 2002–November 2006 can be discerned in Figure 8, despite gaps in the time series. It can be seen that rainy periods are followed shortly by an upward movement of the ground, while a steady downward movement develops during the course of the dry periods. The seasonal changes in the displacement of the PS points lead to the conclusion that the repeated changes in the amount of rainfall in conjunction with the high seasonal temperature variations cause significant changes in the ground moisture inside the active zone, resulting in noticeable volumetric changes in the clay-rich horizons of the Nicosia marl.

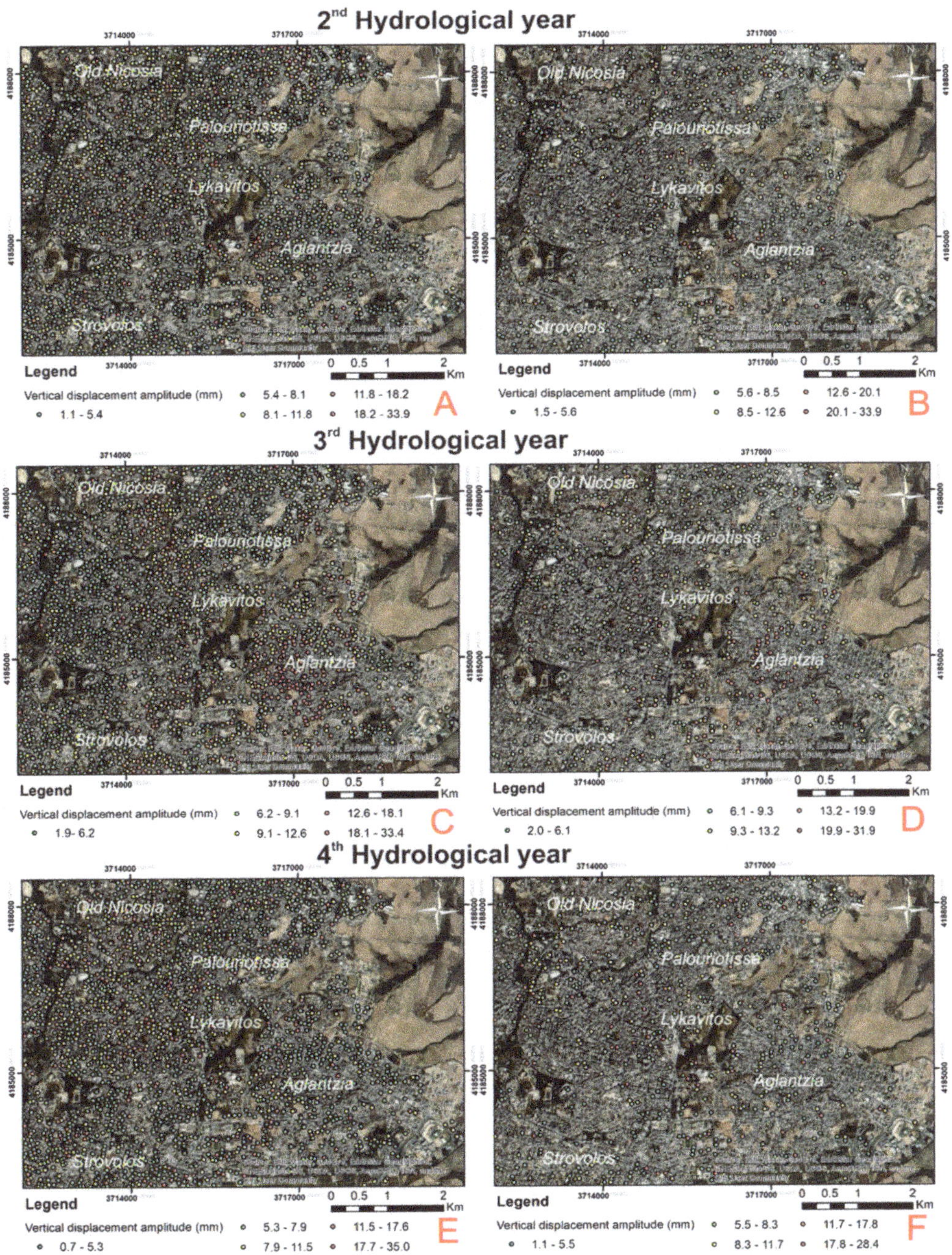

Figure 7. Spatial distribution of the vertical displacement amplitude for the 3 hydrological years: (**A**,**C**,**E**) PS points, (**B**,**D**,**F**) free field PS points.

Hydrological year	Period	Total Rainfall ammount (mm)	
		Nicosia station	Athalassa station
1	November 2002 - May 2003	481	422
1	June 2003 - September 2003	24	21
2	October 2003 - May 2004	366	401
2	June 2004 - September 2004	<1	<1
3	October 2004 - May 2005	327	353
3	June 2005 - September 2005	31	27
4	October 2005 - May 2006	303	283
4	June 2006 - September 2006	67	84

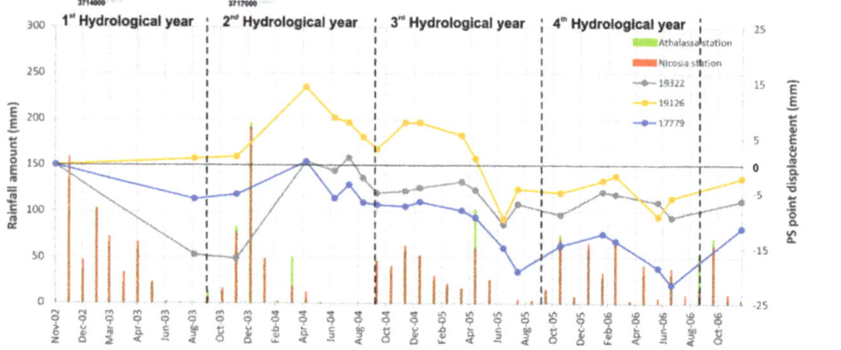

Figure 8. Times series of displacement (right axis) and rainfall (left axis) for three characteristic PS points.

5.1. *Correlation of the Factors Controlling Ground Swelling/Shrinkage with the Remote Sensing Data*

The results of the spatial analysis carried out in the GIS environment are presented below and concern the calculation of the frequency of classes of vertical ground displacement amplitude in correlation with the main causal factors. For this purpose, the PS points were divided into five classes according to the movement amplitude. Figures 9–13 show the results of the statistical analysis in the form of diagrams, accounting both for all PS points and for only the Free Field PS points. It is noted that the relative density is estimated by considering the number of PS points in each individual class depending on the area which occupied.

Regarding the clay content, it was observed that the study area is covered mainly by khaki marl that contains clay percentage between 26–40%. Moreover, the area in which the khaki marl contains more than 40% of clay occupies 29% of the map. The highest frequency of small movement amplitude (blue bar) PS points (824) and large movement amplitude (red bar) PS points (64) occurs in the areas where the clay content is between 26–40% (Figure 9a). A similar picture is observed for the Free Field Ps points (Figure 9b). After the calculation of the relative density, it was observed that the highest rate of low displacement amplitude PS points was noticed in the clay content area ≤26% for both Free Field PS Point and the general PS Points (Figure 9c,d).

Significant variation in the high vertical displacement amplitude class between the results of the Free Field PS points and general PS points was observed. More specifically, the relative density of Free Field red PS points (42.62%) is higher in the area with clay content >40%, while the relative density of the general red PS Points is the lowest (14.29%) in same area.

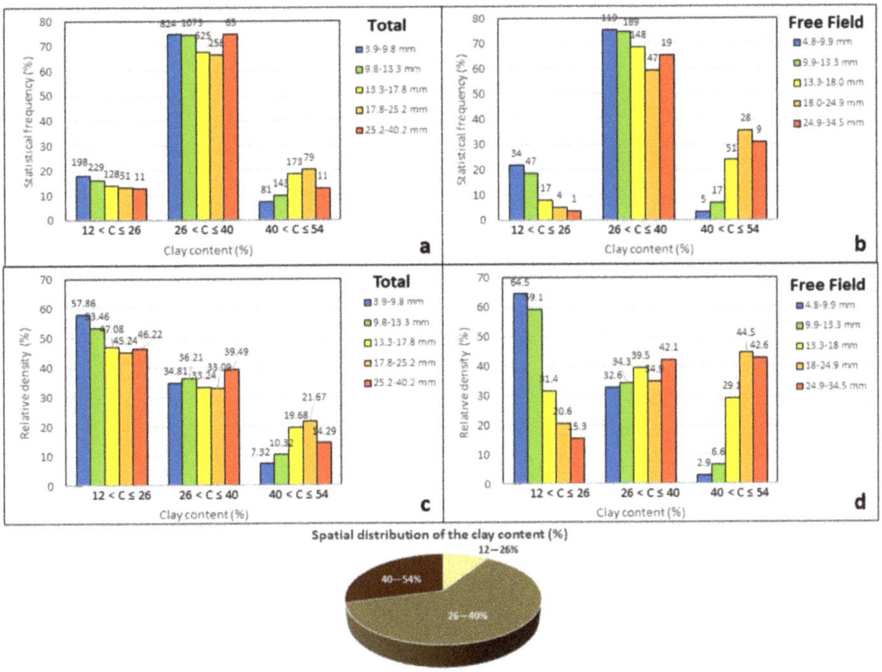

Figure 9. Statistical frequency (**a**,**b**) and relative density (**c**,**d**) of the vertical displacement amplitude points as a function of the khaki marl clay content for total and free field PS points.

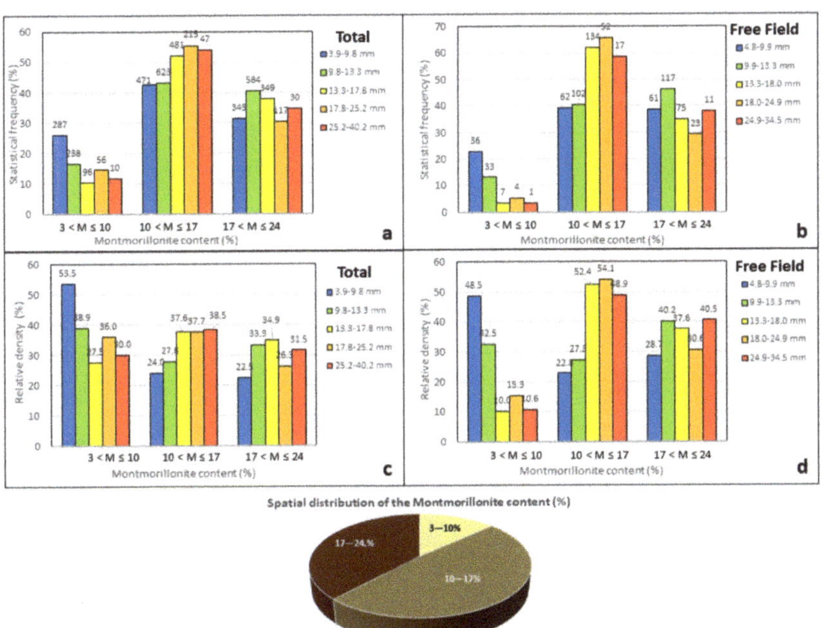

Figure 10. Statistical frequency (**a**,**b**) and relative density (**c**,**d**) of the vertical displacement amplitude points as a function of the khaki marl montmorillonite content for total and free field PS points.

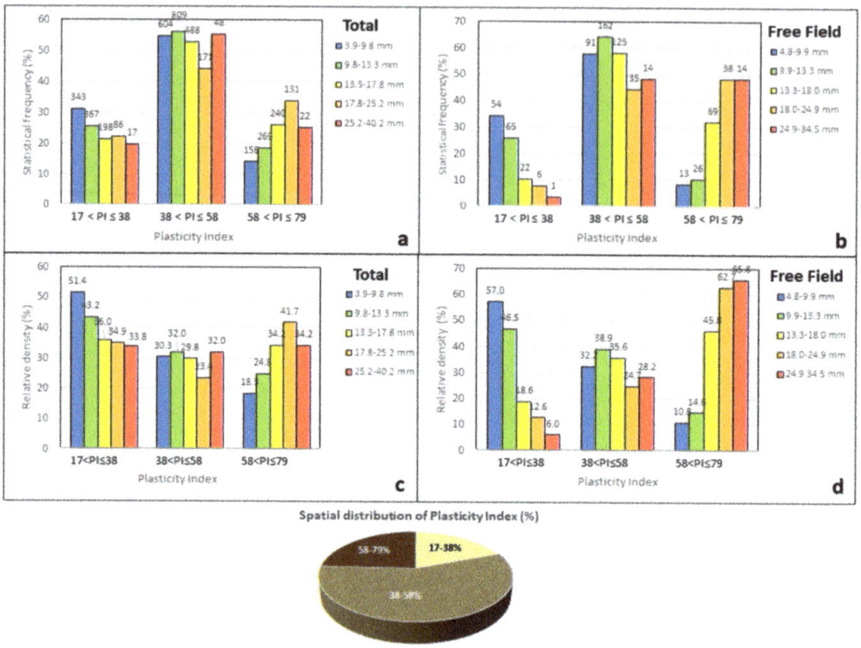

Figure 11. Statistical frequency (**a**,**b**) and relative density (**c**,**d**) of the vertical displacement amplitude points as a function of the khaki marl Plasticity Index for total and free field PS points.

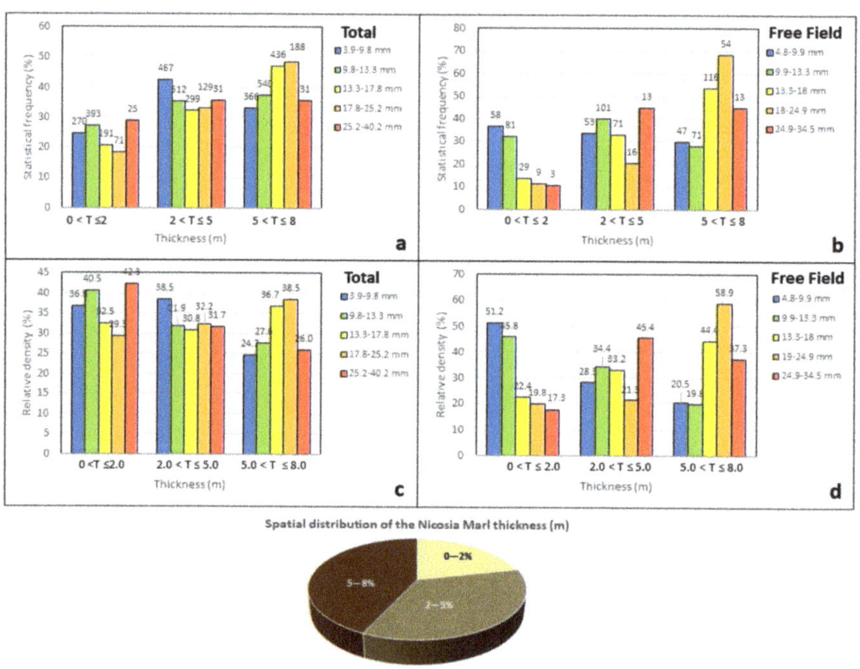

Figure 12. Statistical frequency (**a**,**b**) and relative density (**c**,**d**) of the vertical displacement amplitude points as a function of the khaki marl thickness for total and free field PS points.

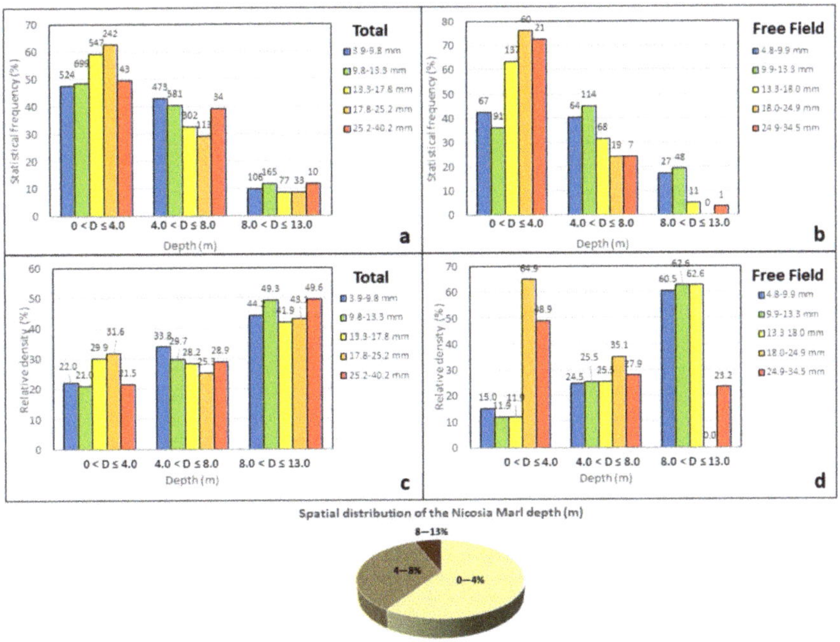

Figure 13. Statistical frequency (**a**,**b**) and relative density (**c**,**d**) of the vertical displacement amplitude points as a function of the khaki marl upper boundary depth for total and free field PS points.

From Figure 9d, it can be seen that there is strong correlation between clay content and ground movement amplitude. For example, the relative density of highest amplitude free field PS points (red bar) increase as the clay content increases, while the lowest amplitude points (blue bar) decreases as the clay content increases. This is not observed when all PS points are considered (Figure 9c). This fact demonstrates clearly the necessity of removing the PS points that do not belong in the free field from the statistical dataset. Contrarily to the clay content, a clear trend is not observed for the montmorillonite content (Figure 10d). For example, the relative density for the high amplitude and the low amplitude movement varying non-monotonically with increasing montmorillonite content. The same is true for the marl thickness inside the active zone (Figure 12).

An even stronger trend than clay content gives the plasticity index (Figure 11d) if only the free-field PS points are considered. The relative density of the highest amplitude displacement rises from 6% to 66% as PI rises to the range 58 to 79. As in the previous cases, if non free-field points are included in the statistical analysis, the correlation with PI is much weaker.

In addition, the depth of the upper boundary of the Nicosia marl (Figure 13d) gives an equally strong correlation with the clay content. Generally, the relative density of the highest amplitude displacement decreases from 48.9% to 23% as the Nicosia marl depth increase up to 8 m. On the contrary, the relative density of the lower amplitude displacement increases proportional to the Nicosia marl depth. Comparing the figures of the total points (Figure 13a,c) with free field points (Figure 13b,d), it is obvious that in latter case the correlation with the Nicosia marl depth is much clearer.

5.2. *Lasso Regression Analysis*

In previous section, the factors controlling ground swelling/shrinkage were investigated by considering each time a single causal variable without taking into account the possible interplay between them. In this paragraph, we fit a linear equation in which all the factors that play an important role in the activation of this phenomenon (Plasticity Index, Clay content, Montmorillonite content, depth of the Nicosia marl) were combined. The Nicosia marl thickness inside the active zone was not considered in this equation since it is firmly correlated with the upper boundary depth (i.e., the thickness of the non-expansive soil cover). After the normalization of each causal variable based on its minimum to maximum range, the linear equation was formulated as follows:

$$AVM_{total,free} = C_0 + C_1\frac{PI - PI_{min}}{PI_{max} - PI_{min}} + C_2\frac{CL - CL_{min}}{CL_{max} - CL_{min}} + C_3\frac{MN - MN_{min}}{MN_{max} - MN_{min}} + C_4\frac{DP - DP_{min}}{DP_{max} - DP_{min}} \quad (1)$$

where *AVM* is the amplitude of vertical movement for either total or free field PS points (in mm), *PI* is the Plasticity Index (%), *CL* is the Clay content (%), *MN* is the Montmorillonite content (%) and *DP* is the upper boundary depth of the Nicosia marl layer (in m).

The main goal of the linear equation was to investigate the degree of influence of the causal factors by ranking them in accordance to their importance. To achieve this, the Lasso regression analysis was performed using the pertinent built-in function in MATLAB. In this method, a penalty term, which is the sum of the absolute values of the linear equation coefficients C_k (k = 1,2 ... n) times a user specified multiplier λ (i.e., $\lambda\Sigma|C_k|$), is inserted in the objective function [31]. As λ increased, the equations coefficients became seriatim zero and causal factors are effectively removed from the equation. The larger the λ for which C_k becomes zero (threshold value λ_T), the greater is the importance of the respective variable.

Figure 14 presents the evolution of the coefficients Ck with increasing λ. It can be seen that the plasticity index drops last out of the linear equation and is thus judged as the most important parameter controlling the phenomenon of ground heave/settlement. Figure 15 shows the λ_T values (values at which the C_k = 0) normalized with respect to the λ_T for the plasticity index, thus quantifying the relative importance of each variable. The plasticity index turns out to be by far the most important variable, followed by the marl depth and the clay content.

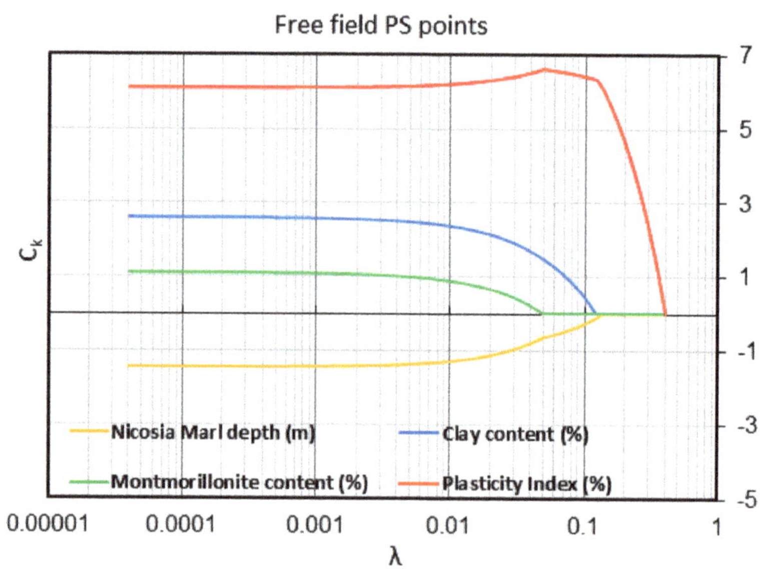

Figure 14. Trace plots of coefficients from Lasso regression for free field PS points.

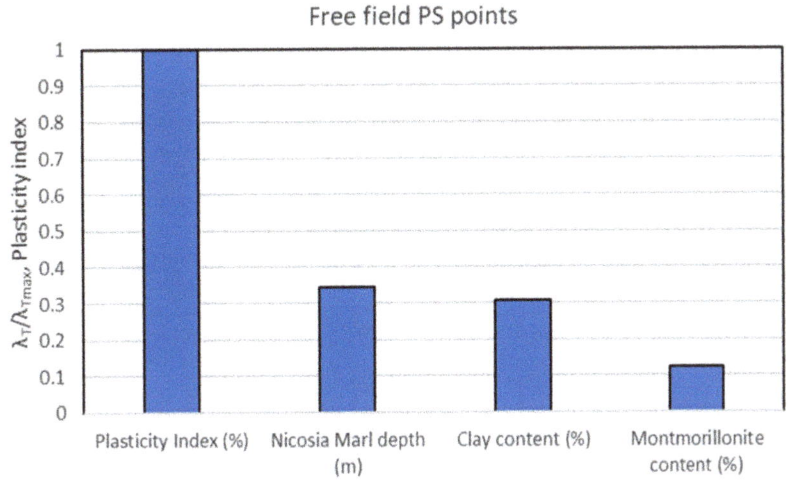

Figure 15. Rank of relative importance of causal variables based on Lasso regression for free field PS points.

6. Discussion

The investigation of this phenomenon through the EO technologies presents some inherent difficulties. Firstly, the creation of time series with the average vertical displacements per year could not provide us with meaningful conclusions due to the repeated small amount of swelling followed by shrinkage on a yearly basis. Secondly, the magnitude of the vertical displacement is affected the pressures applied by the structures. Therefore, in an urban environment where high rise buildings coexist with smaller ones and buildings with shallow foundations coexist with buildings founded on piles (in which case the vertical movement of the superstructure is drastically limited), the SAR data should be treated with great caution. In order to overcome the above weaknesses, this research relied on estimates of the vertical movement amplitude, in the form of the difference between the maximum and minimum displacement observed in a given time period, calculated for each PS point. This way, the intensity of the vertical deformation fluctuations (and not their tendency) becomes evident. Furthermore, the PS points that did not belong to buildings were selected and treated separately as free field points aiming to mitigate the fact that ground movements depend on the applied vertical stress.

Based on the Interferometric Synthetic Aperture Radar (InSAR) data from the PanGeo project between 16 November 2002 and 30 December 2006, it was found that the seasonal fluctuations of the ground surface are of the order of a few tens of millimeters, with a maximum value of the order of 30–35 mm. This amplitude is clearly larger than the 5 mm to 10 mm observed in regions of France [25,26], signifying the highly expansive nature of the Nicosia marl in conjunction with the arid climate of Cyprus. Nonetheless, the value of 30–35 mm is not far from the 20–25 mm amplitude observed in eastern parts of Paris where severe damages to buildings due to soil swelling/shrinkage frequently occur [27]. It should be stressed that the vertical displacement amplitudes are similar both throughout the examined 4-year period and for each hydrological year separately. This observation confirms the periodicity of the phenomenon and indicates that the main cause of ground volume changes is the seasonal fluctuation of ground moisture inside the active zone of Nicosia marl due to the climatic factors (precipitation, temperature, sunlight, wind). Furthermore, as presented in Figure 8, the ground heave follows the periods of heavy precipitation with a delay of two to four months. This observation suggests that ground wetting in Nicosia marl happens slowly and abrupt activations are not expected (in the absence of other triggering factors). Nonetheless, a monotonic trend in the vertical displacement (subsidence) can also be noted in Figure 8, which extends from the middle of the 2nd hydrological year to the end of the examined time period. Given that the water table in the study area is not exploited and lies very deep in the largest part (and, thus, is largely unaffected by the atmospheric factors), this non-periodic subsidence can be attributed to the differences between each year in the total amount of precipitation that occurs during the months in which the evapotranspiration is small (i.e., the winter months).

From the geotechnical point of view, the vertical displacements amplitudes appear to correlate well with the thematic layers of plasticity index and clay content. A similarly strong correlation is observed between the average vertical displacement rate and the clay content in the upper 1m of the soil profile in large regions of Australia [57]. These observations are in good agreement with numerous laboratory studies, which corelate the swelling pressure (the pressure required to hold the soil, or restore the soil, to its initial void ratio when given access to water) to soil plasticity and clay content [58,59]. Nonetheless, future research could also examine other factors that are known to affect this phenomenon, such as liquid limit and dry unit weight.

Statistical analysis shows that strong and meaningful correlations can be established between ground movement amplitude and the causal factors (plasticity index, clay content, montmorillonite content, marl thickness inside the active zone and depth of its upper boundary), provided that PS points that are located on buildings are excluded and only "free-field" PS points are retained in the dataset. Lasso regression reveals that the most important variable controlling the amplitude of ground heave and settlement is the marl's plasticity index, followed by the depth of the upper boundary of the marl layer (i.e., the thickness of the non-expansive soil cover) and the clay content. Yet, it must be pointed out that this study relied on a geotechnical data from boreholes that are sparsely and unevenly distributed. The present findings could be reinforced by applying the proposed approach to a region with a denser and more evenly spaced cloud of points of geotechnical data.

7. Conclusions

The research work presented herein had a goal to develop a methodological approach for investigating the phenomenon of ground swelling/shrinkage due to seasonal moisture changes in an expansive clay formation. This methodology was based on the integration of EO technologies (InSAR techniques), pre-existing local data and experience, conventional geotechnical research and the use of statistical analysis within a GIS environment. The study area is the city of Nicosia, in which an expansive soil formation, called Nicosia marl, dominates the surficial geology. The spatial correlation of seasonal heave/settlement of Nicosia marl with its physical characteristics was investigated using available geological/geotechnical data and InSAR ground motion measurements. It was found that the seasonal fluctuations of the ground surface are of the order of a few tens of millimeters, with maximum value of the order of 30–35 mm. Statistical analysis shows that there is significant correlation between all causal factors examined, namely plasticity index, clay content, montmorillonite conte, marl thickness and thickness of non-expansive cover, provided that PS points that are located on buildings are excluded and only "free-field" PS points are retained in the dataset. Lasso regression reveals that the most important variable controlling the amplitude of ground heave settlement is the marl's plasticity index, followed by the thickness of the non-expansive cover.

This study presented a new methodological approach for the investigation of ground movement without a characteristic trend (such as land subsidence due to the overexploitation of the aquifer), highlighting the important role of EO technologies as a useful tool for the assessment of ground deformation. The use of the vertical deformation amplitude instead of the velocity can limit the weaknesses of PS data and, provided that the quality of the geotechnical thematic layers is high, the results can be rewarding. The methods and findings of the present study could be applied in other regions around the world that suffer from the adverse effects of expansive soils, helping the state authorities and other policy makers in planning future urban development and the maintenance of existing infrastructure.

Author Contributions: Conceptualization, P.T. and D.L.; methodology, P.T. and D.L.; software, N.K.; validation, P.T.; formal analysis, P.T. and N.K.; investigation, P.T. and D.L.; data curation, P.T.; writing—original draft preparation, P.T.; writing—review and editing, D.L and N.K.; supervision, D.L.; project administration, D.L.; funding acquisition, D.L. All authors have read and agreed to the published version of the manuscript.

Funding: This study was part of the research project EXPASOL funded by the Research & Innovation Foundation of Cyprus under contract INTEGRATED/0916/049.

Data Availability Statement: The datasets used and/or analyzed during the current study are available from the corresponding author on reasonable request.

Conflicts of Interest: The authors declare no conflict of interest.

References

1. Jones, L.D.; Jefferson, I. *Expansive Soils*; ICE Publishing: Birmingham, UK, 2012.
2. Loukidis, D.; Bardanis, M.; Lazarou, G. Classification, Soil-Water Characteristic Curve and Swelling/Collapse Behaviour of the Nicosia Marl, Cyprus. In Proceedings of the E3S Web of Conferences, Paris, France, 12–14 September 2016; p. 11009.
3. Chen, F.H. *Foundations on Expansive Soils*; Elsevier: Amsterdam, The Netherlands, 2012; Volume 12.
4. Bao, S.; Shangguan, Y.; Wang, Z. Environmental Geology of Pore-distribution Characteristics of Expansive Soil in the Yanbian Area, Jilin, China. *Ekoloji* **2018**, *27*, 1247–1258.
5. Charlie, W.A.; Osman, M.A.; Ali, E.M. Construction on expansive soils in Sudan. *J. Constr. Eng. Manag.* **1984**, *110*, 359–374. [CrossRef]
6. Zumrawi, M.M. Construction problems of light structures founded on expansive soils in Sudan. *Int. J. Sci. Res.* **2015**, *4*, 896–902.
7. Mitchell, P.W. Footing design for residential type structures in arid climates. *Aust. Geomech. J.* **2008**, *43*, 51–68.
8. Fityus, S.; Smith, D.; Allman, M. Expansive soil test site near Newcastle. *J. Geotech. Geoenviron. Eng.* **2004**, *130*, 686–695. [CrossRef]
9. Abduljauwad, S.; Al-Sulaimani, G.; Basunbul, I.; Al-Buraim, I. Laboratory and field studies of response of structures to heave of expansive clay. *Géotechnique* **1998**, *48*, 103–122. [CrossRef]
10. Dhowian, A.W. Heave prediction techniques and design consideration on expansive soils. *J. King Saud Univ.-Eng. Sci.* **1990**, *2*, 355–376.
11. Freeborough, K.; Kirkham, M.; Jones, L. *Determination of the Shrinking and Swelling Properties of the London Clay Formation: Laboratory Report*; British Geological Survey: Nottingham, UK, 2006.
12. Yoshida, R.; Fredlund, D.G.; Hamilton, J. The prediction of total heave of a slab-on-grade floor on Regina clay. *Can. Geotech. J.* **1983**, *20*, 69–81. [CrossRef]
13. Messing, I.; Jarvis, N. Seasonal variation in field-saturated hydraulic conductivity in two swelling clay soils in Sweden. *J. Soil Sci.* **1990**, *41*, 229–237. [CrossRef]
14. Poulos, H. Parametric Solutions for Strip Footings on Swelling and Shrinking Soils. In Proceedings of the Fifth International Conference on Expansive Soils 1984: Preprints of Papers, Adelaide, Australia, 21–23 May 1984; p. 149.
15. Lazarou, G.; Loukidis, D.; Bardanis, M. Moisture migration under mat foundations in Nicosia Marl. *Geotech. Geol. Eng.* **2019**, *37*, 1585–1608. [CrossRef]
16. Lytton, R.L. Prediction of Movement in Expansive Clays. In Proceedings of the Vertical and Horizontal Deformations of Foundations and Embankments, College Station, TX, USA, 16 June 1994; pp. 1827–1845.
17. Zhao, X.; Chen, W. Optimization of computational intelligence models for landslide susceptibility evaluation. *Remote Sens.* **2020**, *12*, 2180. [CrossRef]
18. Chen, Y.; Wei, Y.; Wang, Q.; Chen, F.; Lu, C.; Lei, S. Mapping Post-Earthquake Landslide Susceptibility: A U-Net Like Approach. *Remote Sens.* **2020**, *12*, 2767. [CrossRef]
19. Hong, H.; Tsangaratos, P.; Ilia, I.; Loupasakis, C.; Wang, Y. Introducing a novel multi-layer perceptron network based on stochastic gradient descent optimized by a meta-heuristic algorithm for landslide susceptibility mapping. *Sci. Total Environ.* **2020**, *742*, 140549. [CrossRef] [PubMed]
20. Kaitantzian, A.; Loupasakis, C.; Tzampoglou, P.; Parcharidis, I. Ground Subsidence Triggered by the Overexploitation of Aquifers Affecting Urban Sites: The Case of Athens Coastal Zone along Faliro Bay (Greece). *Geofluids* **2020**, *2020*, 8896907. [CrossRef]
21. Tzampoglou, P.; Loupasakis, C. Mining geohazards susceptibility and risk mapping: The case of the Amyntaio open-pit coal mine, West Macedonia, Greece. *Environ. Earth Sci.* **2017**, *76*, 542. [CrossRef]
22. Tzampoglou, P.; Loupasakis, C. Numerical simulation of the factors causing land subsidence due to overexploitation of the aquifer in the Amyntaio open coal mine, Greece. *HydroResearch* **2019**, *1*, 8–24. [CrossRef]
23. Hong, H.; Tsangaratos, P.; Ilia, I.; Liu, J.; Zhu, A.-X.; Chen, W. Application of fuzzy weight of evidence and data mining techniques in construction of flood susceptibility map of Poyang County, China. *Sci. Total Environ.* **2018**, *625*, 575–588. [CrossRef]
24. Nhu, V.-H.; Thi Ngo, P.-T.; Pham, T.D.; Dou, J.; Song, X.; Hoang, N.-D.; Tran, D.A.; Cao, D.P.; Aydilek, İ.B.; Amiri, M. A new hybrid firefly–PSO optimized random subspace tree intelligence for torrential rainfall-induced flash flood susceptible mapping. *Remote Sens.* **2020**, *12*, 2688. [CrossRef]

25. Burnol, A.; Aochi, H.; Raucoules, D.; Veloso, F.M.; Koudogbo, F.N.; Fumagalli, A.; Chiquet, P.; Maisons, C. Wavelet-based analysis of ground deformation coupling satellite acquisitions (Sentinel-1, SMOS) and data from shallow and deep wells in Southwestern France. *Sci. Rep.* **2019**, *9*, 8812. [CrossRef]
26. Burnol, A.; Foumelis, M.; Gourdier, S.; Deparis, J.; Raucoules, D. Monitoring of Expansive Clays over Drought-Rewetting Cycles Using Satellite Remote Sensing. *Atmosphere* **2021**, *12*, 1262. [CrossRef]
27. Deffontaines, B.; Kaveh, F.; Fruneau, B.; Arnaud, A.; Duro, J. Monitoring Swelling Soils in Eastern Paris (France) through Dinsar and psi Interferometry: A Synthesis. In *Engineering Geology for Society and Territory-Volume 5*; Springer: Berlin/Heidelberg, Germany, 2015; pp. 195–202.
28. Bai, L.; Jiang, L.; Wang, H.; Sun, Q. Spatiotemporal characterization of land subsidence and uplift (2009–2010) over wuhan in central china revealed by terrasar-X insar analysis. *Remote Sens.* **2016**, *8*, 350. [CrossRef]
29. Shi, X.; Zhang, S.; Jiang, M.; Pei, Y.; Qu, T.; Xu, J.; Yang, C. Spatial and temporal subsidence characteristics in Wuhan (China), during 2015–2019, inferred from Sentinel-1 synthetic aperture radar (SAR) interferometry. *Nat. Hazards Earth Syst. Sci.* **2021**, *21*, 2285–2297. [CrossRef]
30. Cigna, F.; Tapete, D. Satellite InSAR survey of structurally-controlled land subsidence due to groundwater exploitation in the Aguascalientes Valley, Mexico. *Remote Sens. Environ.* **2021**, *254*, 112254. [CrossRef]
31. Svigkas, N.; Loupasakis, C.; Papoutsis, I.; Kontoes, C.H.; Alatza, S.; Tzampoglou, P.; Tolomei, C.; Spachos, T. InSAR Campaign Reveals Ongoing Displacement Trends at High Impact Sites of Thessaloniki and Chalkidiki, Greece. *Remote Sens.* **2020**, *12*, 2396. [CrossRef]
32. Ilia, I.; Loupasakis, C.; Tsangaratos, P. Land subsidence phenomena investigated by spatiotemporal analysis of groundwater resources, remote sensing techniques, and random forest method: The case of Western Thessaly, Greece. *Environ. Monit. Assess.* **2018**, *190*, 623. [CrossRef]
33. Osmanoğlu, B.; Dixon, T.H.; Wdowinski, S.; Cabral-Cano, E.; Jiang, Y. Mexico City subsidence observed with persistent scatterer InSAR. *Int. J. Appl. Earth Obs. Geoinf.* **2011**, *13*, 1–12. [CrossRef]
34. Raspini, F.; Loupasakis, C.; Rozos, D.; Adam, N.; Moretti, S. Ground subsidence phenomena in the Delta municipality region (Northern Greece): Geotechnical modeling and validation with Persistent Scatterer Interferometry. *Int. J. Appl. Earth Obs. Geoinf.* **2014**, *28*, 78–89. [CrossRef]
35. Raspini, F.; Loupasakis, C.; Rozos, D.; Moretti, S. Advanced interpretation of land subsidence by validating multi-interferometric SAR data: The case study of the Anthemountas basin (Northern Greece). *Nat. Hazards Earth Syst. Sci.* **2013**, *13*, 2425–2440. [CrossRef]
36. Zhou, C.; Gong, H.; Chen, B.; Li, X.; Li, J.; Wang, X.; Gao, M.; Si, Y.; Guo, L.; Shi, M. Quantifying the contribution of multiple factors to land subsidence in the Beijing Plain, China with machine learning technology. *Geomorphology* **2019**, *335*, 48–61. [CrossRef]
37. Ziwen, Z.; Liu, Y.; Li, F.; Li, Q.; Ye, W. Land subsidence monitoring based on InSAR and inversion of aquifer parameters. *EURASIP J. Wirel. Commun. Netw.* **2019**, *2019*, 291. [CrossRef]
38. Malik, K.; Kumar, D.; Perissin, D.; Pradhan, B. Estimation of ground subsidence of New Delhi, India using PS-InSAR technique and Multi-sensor Radar data. *Adv. Space Res.* **2022**, *69*, 1863–1882. [CrossRef]
39. Tsangaratos, P.; Ilia, I.; Loupasakis, C. Land Subsidence Modeling Using Data Mining Techniques. In *Natural Hazards GIS-Based Spatial Modeling Using Data Mining Techniques*; Pourghasemi, H.R., Rossi, M., Eds.; Springer: Berlin/Heidelberg, Germany, 2018.
40. Khorrami, M.; Abrishami, S.; Maghsoudi, Y.; Alizadeh, B.; Perissin, D. Extreme subsidence in a populated city (Mashhad) detected by PSInSAR considering groundwater withdrawal and geotechnical properties. *Sci. Rep.* **2020**, *10*, 11357. [CrossRef]
41. Han, Y.; Zou, J.; Lu, Z.; Qu, F.; Kang, Y.; Li, J. Ground deformation of wuhan, china, revealed by multi-temporal insar analysis. *Remote Sens.* **2020**, *12*, 3788. [CrossRef]
42. Chaussard, E.; Amelung, F.; Abidin, H.; Hong, S.-H. Sinking cities in Indonesia: ALOS PALSAR detects rapid subsidence due to groundwater and gas extraction. *Remote Sens. Environ.* **2013**, *128*, 150–161. [CrossRef]
43. Cantillo, V.; Mercado, V.; Pájaro, C. Empirical correlations for the swelling pressure of expansive clays in the city of Barranquilla, Colombia. *Earth Sci. Res. J.* **2017**, *21*, 45–49. [CrossRef]
44. Van der Merwe, D. Contribution to Specialty Session B, Current Theory and Practice for Building on Expansive Clays. In Proceedings of the 6th Regional Conference for Africa on Soil Mechanics and Foundation Engineering, Durban, South Africa, 1 January 1975; pp. 166–167.
45. Yilmaz, I. Indirect estimation of the swelling percent and a new classification of soils depending on liquid limit and cation exchange capacity. *Eng. Geol.* **2006**, *85*, 295–301. [CrossRef]
46. Mowafy, M.; Bauer, G.E. Prediction of swelling pressure and factors affecting the swell behaviour of an expansive soils. *Transp. Res. Rec.* **1985**, *1032*, 23–28.
47. Rabba, S. Factors Affecting Engineering Properties of Expansive Soils. Master's Thesis, Al-Azhar University, Cairo, Egypt, 1975.
48. Zomeni, Z.; Koulermou, N. *Geohazard Description for Lefkosia*; Cyprus Geological Survey Department: Nicosia, Cyprus, 2013; p. 81.
49. Tzampoglou, P.; Loukidis, D. Investigation of the importance of climatic factors in COVID-19 worldwide intensity. *Int. J. Environ. Res. Public Health* **2020**, *17*, 7730. [CrossRef]
50. Constantinou, G.; Petrides, G.; Kyrou, K.; Chrysostomou, C. *Swelling Clays: A Continuous Threat to the Byilt Environment of Cyprus. UNOPS Project Final Report*; Technical Chamber of Cyprus: Nicosia, Cyprus, 2002.

51. Stavridakis, E.I.; Al-Rawas, A.; Goosen, Z. Assessment of Anisotropic Behavior of Swelling Soils on Ground and Construction Work. In *Expansive Soils: Recent Advances in Characterization and Treatment*; Taylor and Francis: London, UK, 2006; pp. 371–384.
52. Take, W.; Bolton, M. Seasonal ratcheting and softening in clay slopes, leading to first-time failure. *Géotechnique* **2011**, *61*, 757. [CrossRef]
53. Hobbs, P.R.N.; Loukaides, G.; Petrides, G. Geotechnical Properties and Behaviour of Pliocene Marl in Nicosia, Cyprus. British Geological Survey, Cyprus Geological Survey Department: Nicosia, Cyprus, 1986; p. 288.
54. Peel, M.C.; Finlayson, B.L.; McMahon, T.A. Updated world map of the Köppen-Geiger climate classification. *Hydrol. Earth Syst. Sci.* **2007**, *11*, 1633–1644. [CrossRef]
55. Thornthwaite, C.W. An approach toward a rational classification of climate. *Geogr. Rev.* **1948**, *38*, 55–94. [CrossRef]
56. Bateson, L.; Cuevas, M.; Crosetto, M.; Cigna, F.; Schijf, M.; Evans, H. *PANGEO: Enabling Access to Geological Information in Support of GMES: Deliverable 3.5 Production Manual*; Version 1; European Commission: Brussels, Belgium, 2012; p. 106.
57. Castellazzi, P.; Schmid, W. Interpreting C-band InSAR ground deformation data for large-scale groundwater management in Australia. *J. Hydrol. Reg. Stud.* **2021**, *34*, 100774. [CrossRef]
58. Erzin, Y.; Erol, O. Correlations for quick prediction of swell pressures. *Electron. J. Geotech. Eng.* **2004**, *9*, 476.
59. Nayak, N.V.; Christensen, R. Swelling characteristics of compacted, expansive soils. *Clays Clay Miner.* **1971**, *19*, 251–261. [CrossRef]

remote sensing

MDPI

Article

Kinematics of Active Landslides in Achaia (Peloponnese, Greece) through InSAR Time Series Analysis and Relation to Rainfall Patterns

Varvara Tsironi [1,2,*], Athanassios Ganas [1], Ioannis Karamitros [1], Eirini Efstathiou [1], Ioannis Koukouvelas [2] and Efthimios Sokos [2]

1 National Observatory of Athens, Institute of Geodynamics, Lofos Nymfon, Thission, 11810 Athens, Greece; aganas@noa.gr (A.G.); jkaram@noa.gr (I.K.); euirini@gmail.com (E.E.)
2 Department of Geology, University of Patras, 26504 Rio, Greece; iannis@upatras.gr (I.K.); esokos@upatras.gr (E.S.)
* Correspondence: vtsironi@noa.gr

Abstract: We studied the kinematic behaviour of active landslides at several localities in the area of Panachaikon Mountain, Achaia (Peloponnese, Greece) using Sentinel (C-band) InSAR time series analysis. We processed LiCSAR interferograms using the SBAS tool, and we obtained average displacement maps for the period 2016–2021. We found that the maximum displacement rate of each landslide is located at about the center of it. The average E-W velocity of the Krini landslide is ~3 cm/year (toward the east) and 0.6 cm/year downward. The line-of-sight (LOS) velocity of the landslide (descending orbit) compares well to a co-located GNSS station within (±) 3 mm/yr. Our results also suggest a correlation between rainfall and landslide motion. For the Krini landslide, a cross-correlation analysis of our data suggests that the mean time lag was 13.5 days between the maximum seasonal rainfall and the change in the LOS displacement rate. We also found that the amount of total seasonal rainfall controls the increase in the displacement rate, as 40–550% changes in the displacement rate of the Krini landslide were detected, following to a seasonal maximum of rainfall values at the nearby meteorological station of Kato Vlassia. According to our results, it seems that large part of this mountainous region of Achaia suffers from slope instability that is manifested in various degrees of ground displacement greatly affecting its morphological features and inhabited areas.

Keywords: InSAR; GNSS; landslide; rainfall; Achaia; Greece

Citation: Tsironi, V.; Ganas, A.; Karamitros, I.; Efstathiou, E.; Koukouvelas, I.; Sokos, E. Kinematics of Active Landslides in Achaia (Peloponnese, Greece) through InSAR Time Series Analysis and Relation to Rainfall Patterns. *Remote Sens.* **2022**, *14*, 844. https://doi.org/10.3390/rs14040844

Academic Editor: Fulong Chen

Received: 26 December 2021
Accepted: 4 February 2022
Published: 11 February 2022

1. Introduction

The movement of an active landslide can be efficiently captured by the InSAR measurements [1–4]. Through InSAR time series analysis, the velocity of the ground movement can be measured with an accuracy of a few mm/yr (e.g., [5]). This remote sensing technique can provide an accurate identification of the area affected by active landslides and can also assist through hillslope monitoring by detecting potential slope failures. We use the term "active landslide" for landslide bodies that have a long history of affecting the area above and around them, especially the road network that requires constant maintenance. As active landslide (according to [6]) is considered a landslide that is currently moving and includes first time movements and reactivations.

The InSAR time series method has been applied successfully in a range of landslide studies, not only to locate landslide bodies, but also to identify spatial-temporal patterns of movement [7]. The analysis of The InSAR-generated displacement time series has the potential to identify periods of accelerated landslide deformation and to evaluate possible correlations with different triggers (rainfall, earthquakes). Large landslides and debris flows form a frequently occurring geohazard posing significant risk to lives and livelihoods.

Landslide phenomena have been previously recognized and mapped in Greek territory using InSAR techniques. Recently, Vassilakis et al. [8] used interferometric processing of TerraSAR-X data to estimate ground displacement rates of a large landslide inside a coal mine in western Macedonia; Kontoes et al., [4] produced susceptibility maps for the landslides in western Greece using ERS/ENVISAT data; Elias et al. [5] used ERS/ENVISAT/Sentinel-1 data to monitor slope stability upstream from an artificial dam; Tsangaratos et al., [9] examined the identification of mass movements due to landslides on Lefkada island with InSAR time series analysis using the TERRAFIRMA data (produced by the persistent scatterers interferometry—PSI technique). Kyriou & Nikolakopoulos, [10] mapped the landslides south of Patras using interferometric processing (offset-tracking method); several other studies on critical locations used InSAR technology [11,12].

The Achaia prefecture in NW Peloponnese (Greece), is an area strongly affected by localized subsidence and landslides [7,13–18], especially the mountainous area near the villages of Krini, Pititsa and Sella (Figure 1; [15,19]). These villages are located on the north-facing slopes of the mountain Panachaiko (maximum elevation 1926 m; Figure 1a), and all of them are affected by active landslide phenomena [20]. The NW Peloponnese shows spatially and temporally diverse climatic conditions across its area, mostly due to irregular topography and different atmospheric circulation patterns. Overall, the climate is typical Mediterranean and exhibits large seasonal variations with mild, wet winters and hot, dry summers [21]. The wet period usually extends from October through May [17].

In this study we used Sentinel-1 (C-band) InSAR displacement data to map the active landslides and monitor their kinematics. We compared our InSAR results to a co-located GNSS station at one of the sites. We then cross-correlated our InSAR time series to the rainfall data from a nearby meteorological station of NOA. We demonstrated a robust correlation between maximum rainfall and an increase in displacement rates.

2. Study Area

The study area is close to the Gulf of Corinth which is considered to be a paradigm of an active rift system in Greece (Figure 1; [22]) and is among the fastest extending continental regions in the world [15,23–25]. This rift was formed by normal slip on large, E-W striking fault segments that extend the crust of central Greece in a N-S direction. The length of the Corinth rift is 130 km, and the width is 20–40 km. The deepest sea depth is ~900 m, and the major peaks of the mountains around the Gulf of Corinth are ~2500 m. The south coast of the Corinth rift is uplifting, whereas the north part is subsiding [15,22,25–28]. From space geodesy data (GNSS) we know that the Peloponnese (southern part) moves faster toward the southwest than the Greek mainland (with respect to the "stable" Europe; [23,25,29]). The net result from this movement is that these two areas move away from each other with an average speed of 1 cm/year while the rate of movement increases from east to west [23,24]. The onshore active faults have normal kinematics with the main active structures being north-dipping faults visible along the southern coast of the Gulf of Corinth, while the rest are located offshore in the central part of the rift [29–32]. Moreover, the study area has rugged relief, several rivers flowing in the general N-S direction (Figure 1), many narrow valleys and numerous other erosional landforms such as triangular faceted spurs on the footwall of active faults [33–36]. Moderate to strong earthquakes occur frequently, inducing additional geological phenomena such as slow and continuous ground displacements in the area [13,15,37].

The area of Achaia specifically has been the focus of many studies concerning landslide events and vulnerability in the last decade [13,14,21,37–44]. The lithological condition of this area is one of the most controlling parameters for landslide occurrence [38,39]. The higher density of landslide occurrence is in Pliocene and Pleistocene fluvio-terrestrial and clastic flysch formations [39]. The most common triggering mechanisms are the seismicity, the steepness of slopes, heavily fractured rocks in the source areas and heavy rainfall [37,45]. The artificial and natural parts of the terrain slopes along a road, as well as the hydrographic axes of river networks, are also considered important factors for landslide manifestation [39,46]. From these common factors Koukis et al. [13] proposed two main triggering mechanisms for large mass movements in the area, i.e., the excessive rainfall resulting in high pore pressure in the rocks during these events and strong earthquakes resulting in dynamic loading conditions at the failure surface. Therefore, the increased permeability of the rock formations produced by earthquakes (due to strong ground motion and/or fissuring) together with events or periods of intense rainfall are the main factors for the intensity of landslide events.

The bedrock geology of Achaia comprises mainly carbonates (Figure 1b). According to IGME (institute of Geological and Mineral Exploration) maps [47,48], the study area is part of the Olonos–Pindos geotectonic zone of the Hellenides and mainly comprises limestone, schists and cherts of the Triassic–Jurassic age, covered by thinly bedded limestones with radiolarites of the Cretaceous age. A transition zone overlays the radiolarites, including limestones, shales, cherts and marls, leading to typical flysch sequence sediments of the Upper Eocene [13]. The syn-rift rocks are Pliocene–Quaternary age sedimentary rocks such as marls, sandstones, conglomerates and alluvial fan deposits.

In this paper we focus our research on the northern / eastern slopes of Panachaiko mountain, with emphasis on well-known landslide activity near the villages of Krini, Pititsa and the monastery of Agia Eleoussa (Figure 1a). In particular, the area around Krini (elevation 775 m; Figure 1b) mostly consists of the Upper Cretaceous–Palaeocene flysch. The flysch mainly includes beds of sandstone and marls (rarely), alternating with thin pelagic limestone, about 50 m thick [47,48]. During 1985 an IGME investigation about a landslide observed at the southern part of Krini was conducted in this area. According to the field report [49], the sliding area close to the village consists of Neogene clay and marl sediments overlapped by a weathered material about 1.5–2 m thick. The Krini landslide is a large earth flow with a mainly E-W slope-parallel displacement trend, which in its eastern (lower) part converts to a translational slide (based on the categorization proposed by [6,50,51]). This is possibly due to changes in lithology. It can be seen in the geological map (Figure 1b), that the landslide extent changes from flysch and its weathered mantle (upper part towards west) to limestone. Krini also displays many characteristics of an earth flow as described by [51] with long periods of relative slow to very slow movement alternating with more rapid surges. This type of mass movement has also allowed settlements (such as the village of Krini) and roads to be built on top of the landslide, a common occurrence in many earthflows around the world [51,52]. During 2021 we visited this area and observed numerous landslides near the village, some of them affecting parts of the road network. Apart from the large slope failures, many small-scale shallow landslides were also observed. Those were affecting the weathered cover of both flysch and Neogene (marl and clay) units.

The village of Pititsa (elevation 650 m; Figure 1b) is also located inside the Upper Cretaceous–Palaeocene flysch, but in this locality the flysch is comprised of sandstone alterations with limestone intercalations and marls. North and east of the village, the flysch is overlapped by Pliocene–Pleistocene blue marls and sandy clays. The Pititsa landslide can also be categorized as an earth flow with the same characteristics as the main body of the Krini landslide but mostly maintaining the lithological homogeneity throughout. The main displacement here is also E-W. To the west of Pititsa, the village of Sella is also affected by active landslides, however, it is built upon terrestrial deposits (Figure 1b; [47,48]).

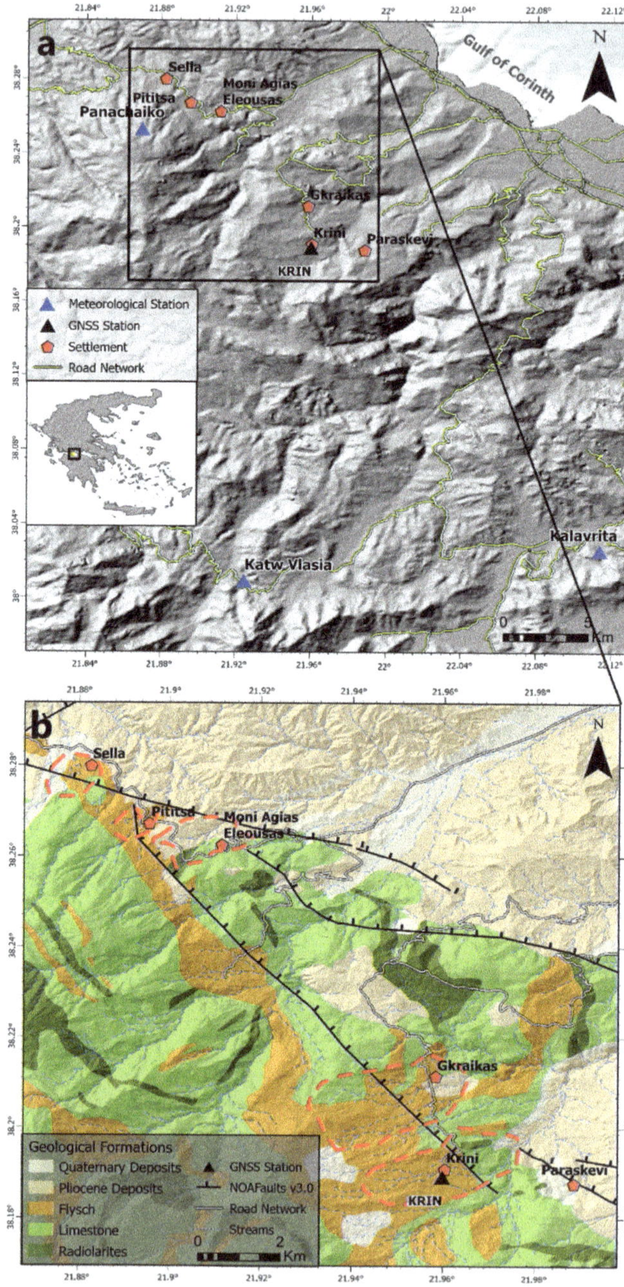

Figure 1. (**a**) shaded relief Map of northeastern Achaia showing the area of interest (box) including the villages where landslides have occurred; (**b**) simplified geological map [47,48]. The landslide of the village of Krini is located on flysch, and the Pititsa landslide is located on flysch and on Pliocene deposits. Active faults are shown with black lines, and red-dashed outlines represent the boundaries of the landslides (ticks on the downthrown side).

3. Data, Methods and Results

3.1. SAR Data Processing

The processing of InSAR time series analysis was completed using the LiCSBAS, an open-source package that integrates with LiCSAR products [53–55]. The processing chain and schematic description of data/methods used are shown in the flow chart of Figure S2. LiCSAR produced interferograms (wrapped and unwrapped) from SLC (single look complex) data of Sentinel-1 acquisitions. The interferograms were multilooked with a factor of 20×4 in range and azimuth (46×56 m spacing) and spatially filtered by a GAMMA adaptive power spectrum filter with an alpha value of 1.0 to decrease noise [56,57]. With the SNAPHU software, LiCSAR unwraps the phase in two dimensions using a statistical cost technique [58]. We employed unwrapped interferograms and coherence images that were geocoded with a pixel spacing of 0.001 degree (100 m) and converted to GeoTIFF format for our time series analysis. We used the frames with ID: 080D_05196_131104 for descending track and 175A_05184_121313 for ascending track, respectively. For ascending and descending track (see Figure S1), we used 346 and 317 available interferograms from the LiCSAR portal, respectively (out of 99 and 120 SAR images, respectively; see Table S1 for image dates).

The areas of interest are in the center-east of the frames ID 175A and 080D (Figure S1). The time span for the ascending track ranges was from 2015 to 2021 and for the descending from 2016 to 2021, respectively (see Table S1). Before the main processing, a tropospheric correction was applied on the unwrapped products using the generic atmospheric correction online service for InSAR (GACOS) data ([59]; pixel size 10×10 km). GACOS utilizes the iterative tropospheric decomposition (ITD) model to separate stratified and turbulent signals from tropospheric total delays. The tropospheric corrections were used in GeoTIFF format. To evaluate the success of the correction, the calculation of standard deviation (STD) of unwrapped phase before and after the GACOS correction and the reduction rate of the STD was performed through the LiCSBAS software [53]. The STD of unwrapped phases for each interferogram was typically lower (see correlation diagrams in Figures 2 and 3; see also reduction of STD through the interferograms in Figures S3 and S4), indicating that this tropospheric correction approach greatly reduced tropospheric noise. We see in the diagrams (Figures 2 and 3) that the reduction of STD for the dataset of ascending orbit is lower (reduction rate 29.57%) than the reduction of STD for the descending orbit (36.29%).

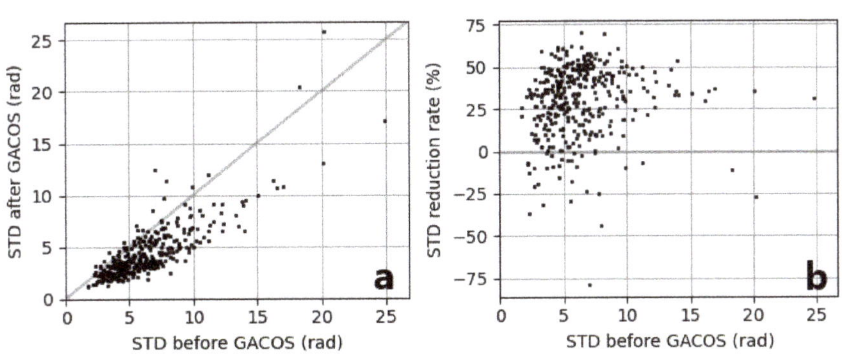

Figure 2. Correlation diagram showing standard deviation: (**a**), of unwrapped phases before and after the GACOS corrections and their reduction rates; (**b**), for ascending track 175. The gray line denotes a 1:1. The reduction of STD for the ascending orbit is moderate. The STD decreased from 6.35 rad to 4.41 rad on average with a mean reduction rate 29.57%.

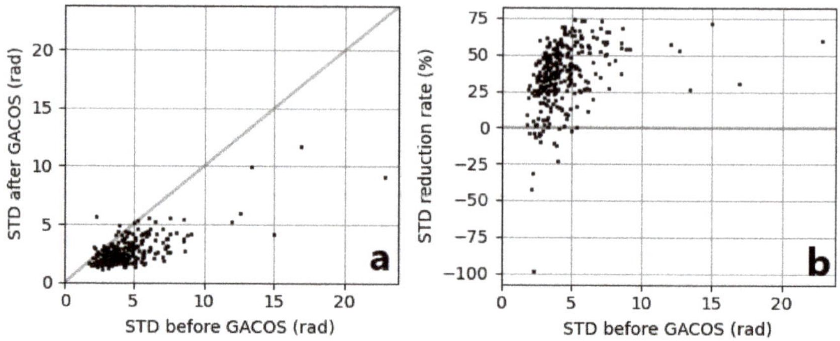

Figure 3. Correlation diagram showing standard deviation: (**a**), of unwrapped phases before and after the GACOS corrections and their reduction rates; (**b**), for descending track 80. The grey line denotes a 1:1. The STD decreased from 4.36 rad to 2.61 rad on average with a mean reduction rate 36.29%.

After the tropospheric correction, we performed the quality check function of LiCSBAS. Statistics such as average phase coherence (we applied a 0.3 threshold) and percentage of genuine unwrapped pixels are used to identify bad data. We also performed the network refinement by interferometric loop closure [60] and removed 100 interferograms (95 ascending and 5 descending orbit) from further processing. Unwrapped data may contain unwrapping mistakes, which can result in serious errors in the derived time series and should be eliminated or addressed prior to use. Moreover, after the quality checks, we applied the small baseline (SB) inversion on the network of interferograms (Figure 4). The least-squares approach was then used to get the mean displacement velocity (LOS) from the cumulative displacements. We used the NSBAS approach [61] to produce a more realistic time series of displacement even with a disconnected network (Figure 4). We selected the SBAS technique because it works well in the vegetated and rural areas such as the mountainous Achaia region, due to the use of distributed scatterers.

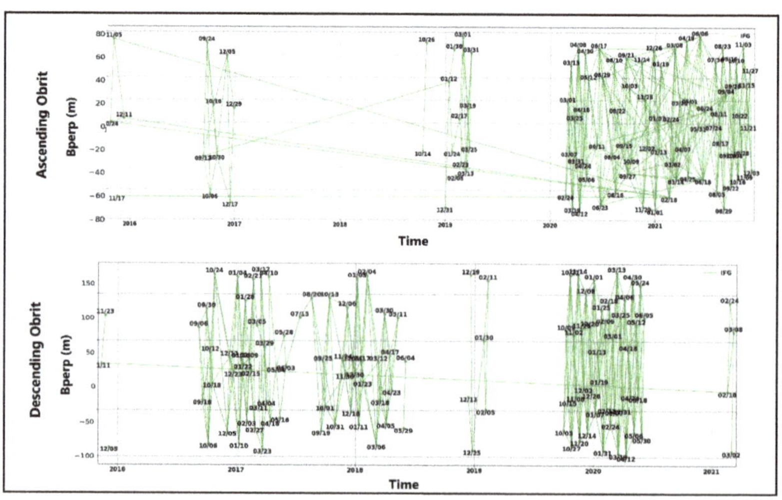

Figure 4. Perpendicular baseline configuration and network of the 346 small baseline interferograms formed from 99 Sentinel-1A images (ascending track, top panel) and network of the 317 SB interferograms formed from 120 Sentinel-1A images used in this study (descending track, bottom panel). Vertical lines indicate the network gaps.

We also used the percentile bootstrap method to calculate the LOS velocity's standard deviation from cumulative displacements. The next processing step was the masking of noisy pixels in the time series and the spatiotemporal filtering of time series. Finally, topography-correlated components (linear with elevation) were subtracted with deramping (see Figures S5 and S6), and we obtained the LOS velocity maps for each satellite track (Figure 5).

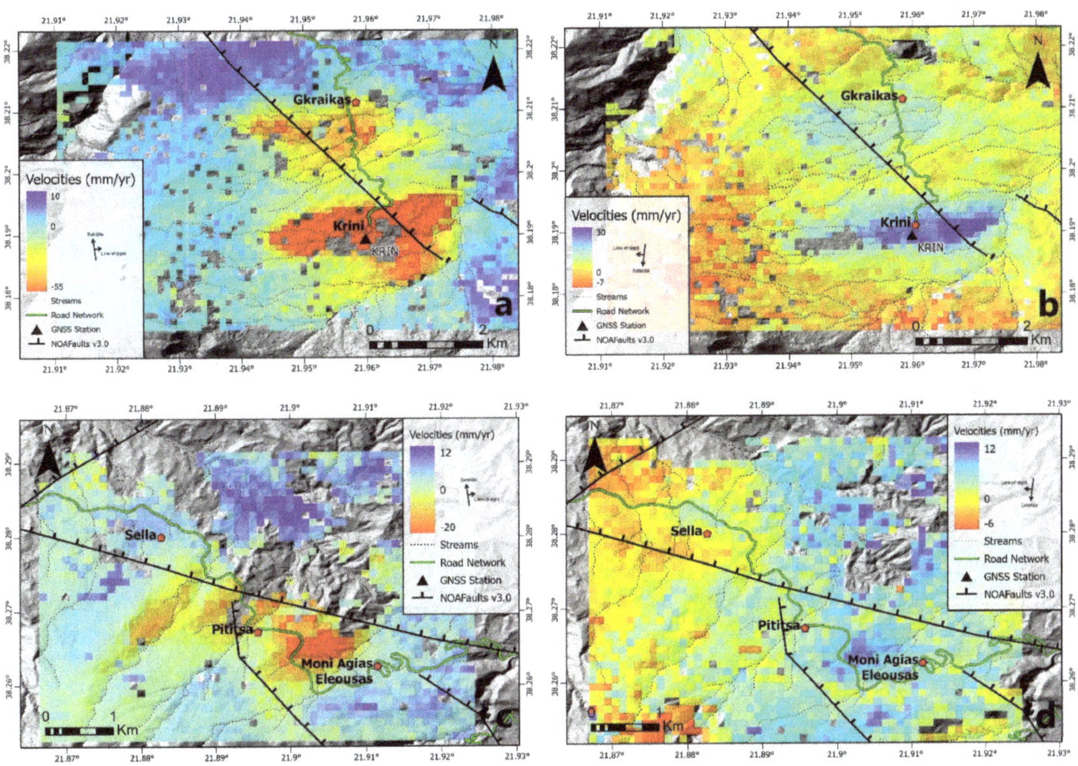

Figure 5. Displacement maps in LOS direction for the broader areas of Krini (**a**,**b**) and Pititsa villages (**c**,**d**) for ascending track and descending track (**a**) shows the displacement map for ascending track; and (**b**), for descending track around the village of Krini, respectively. The Figure 5c,d corresponds to displacement map for ascending and descending track around the villages of Pititsa and Sella, respectively.

After the main processing, we decomposed the LOS displacement data to east-west (E-W) and up-down (U-D) components assuming that the north-south (N-S) displacement is very small due to the track's imaging geometry of the Sentinel-1 satellite [62,63].

In the area around the village of Krini (Figure 1), an active landslide is present (Figure 5a,b). More active landslides are also located near the villages of Graikas, Pititsa and Sella and to the monastery of Agia Eleoussa (Figure 5c,d). In those locations we measured LOS velocities up to −55 mm/yr for the ascending track (Figure 5a). We obtained similar velocities, up to 30 mm/yr for the descending track (Figure 5b). Note that the values of LOS displacements are referenced to a local reference point which is located about 30 km south of the area of interest (see Figure S1 for location), and it is considered stable (free of gravitational or tectonic displacements).

Then, we produced maps of the east-west and up-down components, assuming that the north-south component was neglected due to the track's geometry of the Sentinel-1 satellite. We found that all active landslides were moving east and downward (subsidence, Figure 6a,b). All the affected areas are located on the east-facing slopes and in the foothills of Panachaiko mountain (Figure 1).

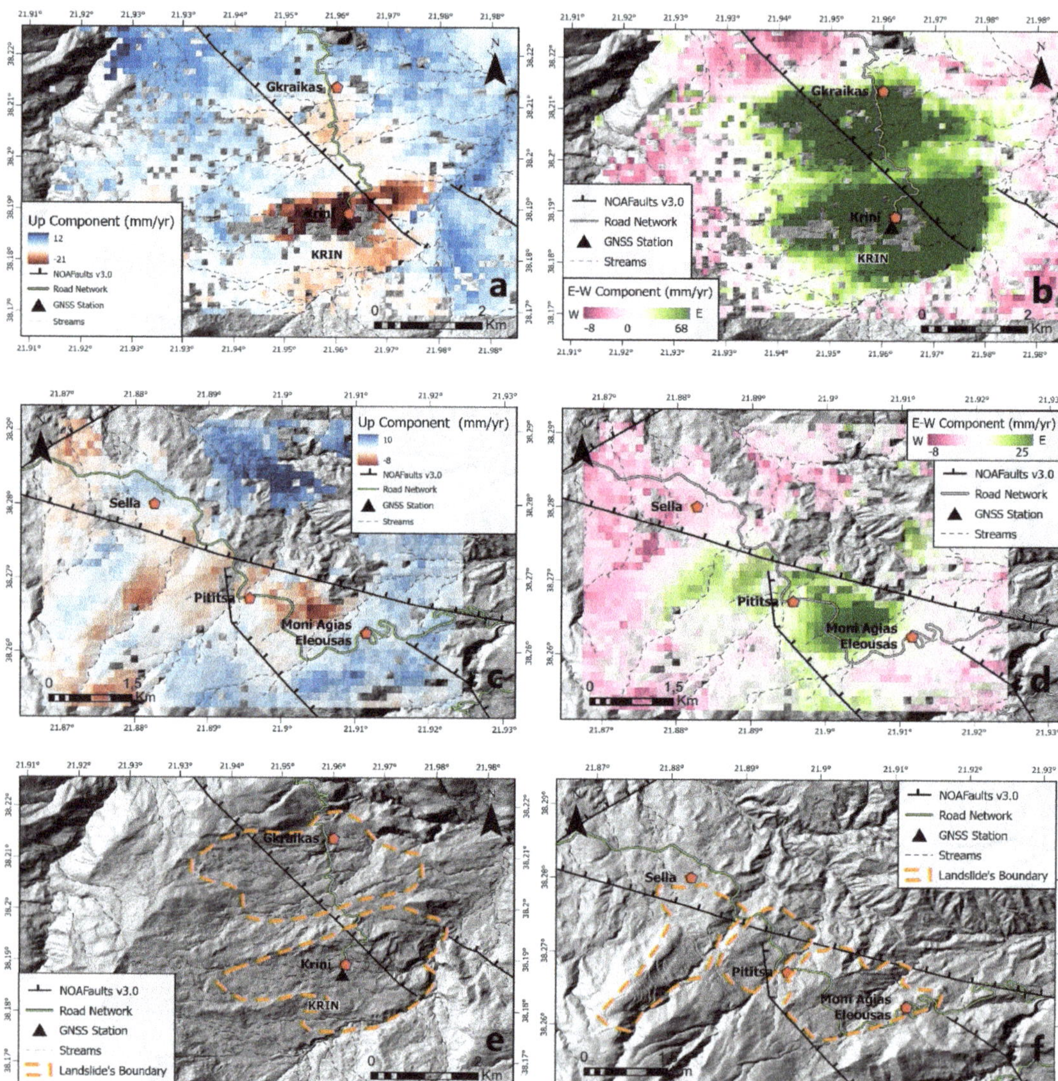

Figure 6. Displacement-rate maps in mm/yr: Up (**a**,**c**) and E-W (**b**,**d**), component after the decomposition of InSAR data for the broader areas of Krini and Pititsa villages, respectively; (**e**), showing the landslides' outlines for Krini and Graikas landslides; and (**f**), showing the landslides' outlines for the monastery Agia Eleoussa and the villages of Pititsa and Sella.

We traced the boundaries of the active landslides using the InSAR motion data on the E-W component (Figure 6e,f). We note that our motion data refer to the period 2016–2021 (Figure 4), so it is possible that landslides mapped in previous studies in the same area [15,20] have changed their motion patterns. In our study we measured the mean velocity of the larger landslides (Krini and Agia Eleoussa (see areas enclosed by orange lines in Figure 6e,f)) by a statistical approach and not by referring to a single pixel. The mean velocity of the Krini landslide was found to be 6 mm/yr downward (nearly 0.6 cm/year) and 28.7 mm/yr (nearly 3 cm/year) eastward. The mean velocity of the Agia Eleoussa landslide was found to be 1.8 mm/yr (~0.2 cm/year) downward and 7.7 mm/yr (~0.8 cm/year) eastward. We also computed the histograms of the velocities from both ascending and descending tracks and the Up and E-W components of the movement for Krini and Agia Eleoussa, respectively (Table 1, Figures 7 and 8). The area affected by the Krini landslide is 4,080,000 square meters (m^2). The area of the active landslide around the village of Pititsa is 800,000 square meters (m^2), and the area of the landslide near the monastery of Agia Eleoussa is 1,600,000 square m^2. It is probable that the two landslides are in fact part of one feature (mass movement). However, the uncertainty in our InSAR motion rates (~3 mm/yr LOS, see Figure 5c,d) is within the values we mapped in the area between the two landslides.

Table 1. Table of mean and median velocities of active landslides in the village of Krini and at the monastery of Agia Eleoussa. The landslide outlines are shown in Figure 6e,f. The period of observation was 2016–2020.

	Mean Velocity (mm/yr)	Median Velocity (mm/yr)
Krini Up Component	−6.0	−4.6
Krini E-W Component	28.7	24.5
Krini LOS (ascending)	−22.0	−18.7
Krini LOS (descending)	12.5	11.0
Agia Eleoussa Up Component	−1.8	−1.2
Agia Eleoussa E-W Component	7.7	7.0
Agia Eleoussa LOS (ascending)	−7.1	−6.3
Agia Eleoussa LOS (descending)	5.5	5.2

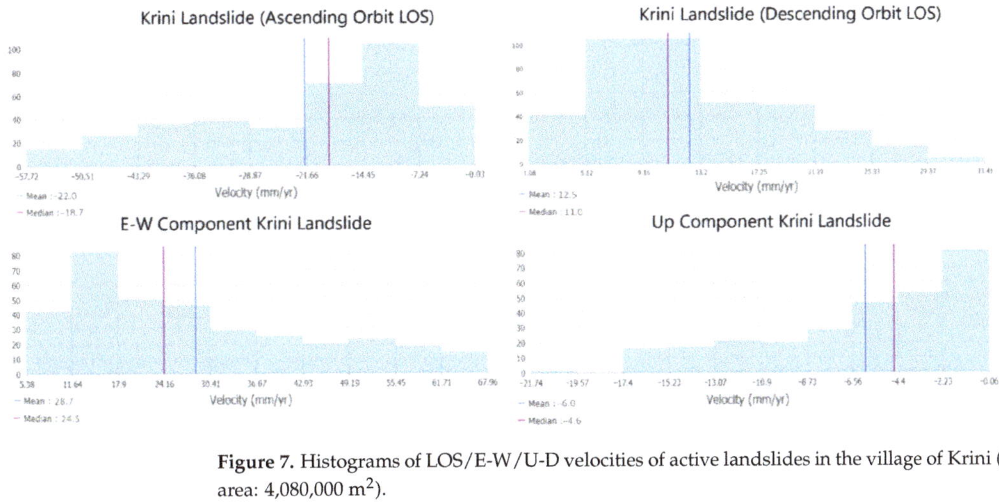

Figure 7. Histograms of LOS/E-W/U-D velocities of active landslides in the village of Krini (Affected area: 4,080,000 m^2).

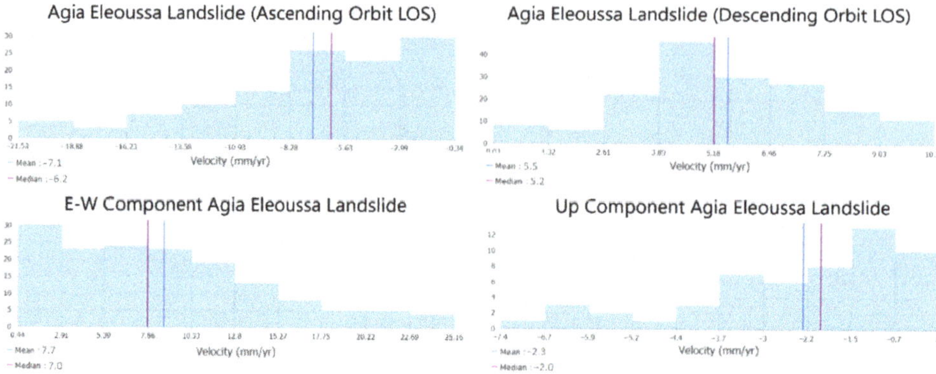

Figure 8. Histograms of LOS/E-W/U-D velocities of active landslides in the Agia Eleoussa area (Affected area: 1,600,000 m^2).

3.2. Rainfall Data

We also collected rainfall data of stations located in the Achaia prefecture (from the database of [64]) to identify possible temporal patterns of ground movement that could be correlated with the rainfall. The wet season lasts from October to May, during which the total rainfall accounts for 93% of the annual rainfall. The month with the highest precipitation is December, with a mean rainfall of 128.9 mm, followed by November (124.7 mm). August is the lowest with a mean rainfall of 7.0 mm, followed by July (8.8 mm) [17].

We used the daily rain data of three meteorological stations of NOA (Kalavrita, Panaxaiko and Kato Vlassia stations (Figure 1)) for the same time span (2015–2021) as with the InSAR and GNSS time series. These stations are located at similar elevations to and within 20 km from the landslides. The data are plotted in Figures 9–11.

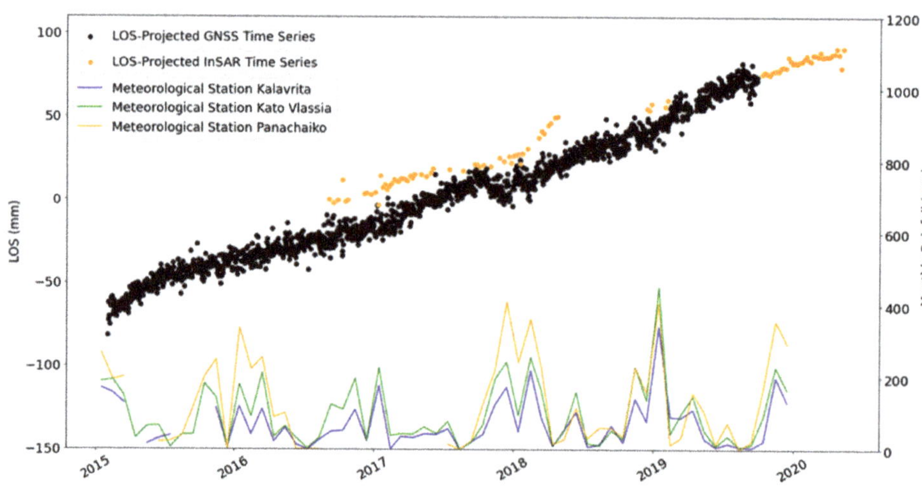

Figure 9. LOS-Projected GNSS time series (KRIN station: black points) and InSAR position time series (orange points: descending orbit) of the pixel which contains the GNSS station. Also, blue, green and orange lines represent the rainfall of the meteorological stations Kalavrita, Kato Vlassia and Panachaiko (see Figure 1 for station localities).

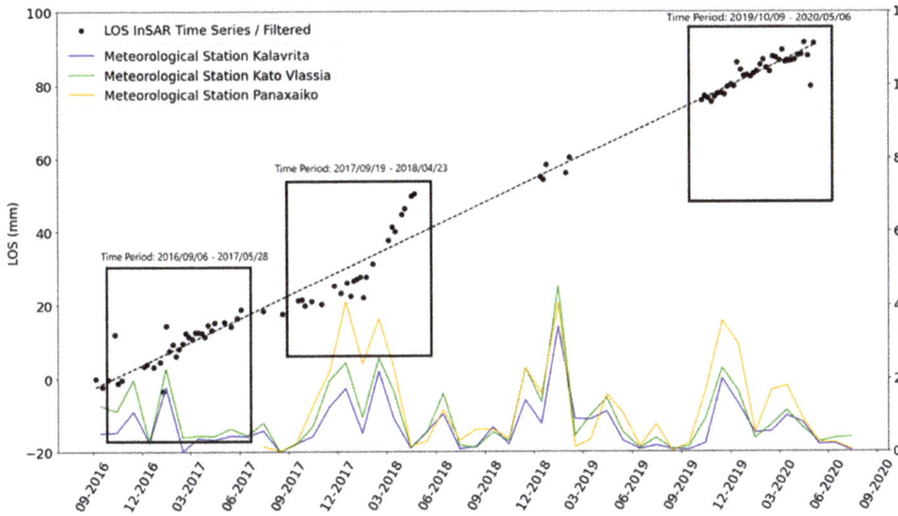

Figure 10. Graph showing LOS position time series for the Krini landslide descending track (080D). Blue, green and orange lines represent the average monthly rainfall of the meteorological stations Kalavrita, Kato Vlassia and Panachaiko, respectively (right axis). The three black boxes represent the time periods what are used for further processing (see Figure 11). The dashed line indicates the linear fit of the InSAR data for the whole period of observations (25 mm/yr).

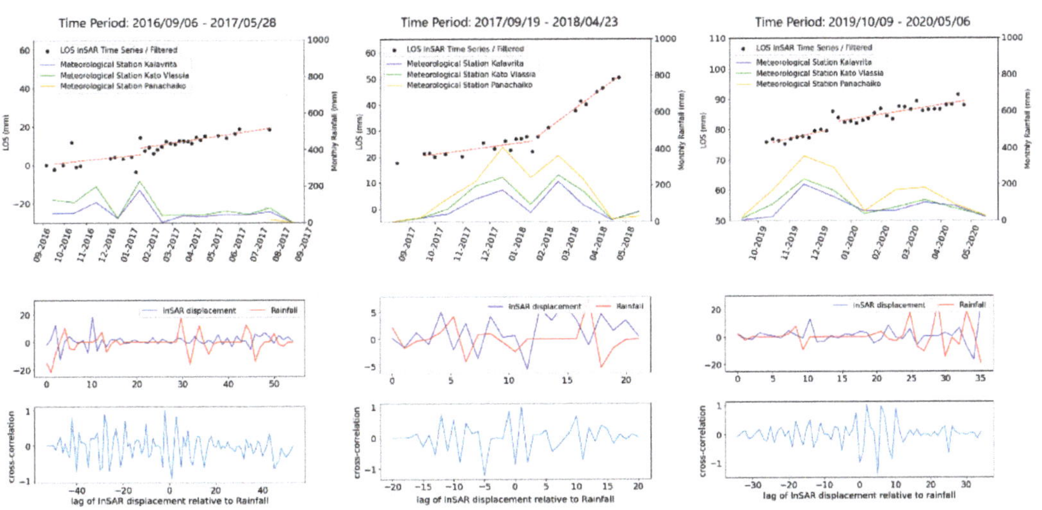

Figure 11. Graphs showing InSAR position time series and average monthly rainfall (Kato Vlassia station, top panel) for each time period corresponding graphs showing cross-correlation results (signal-converted waveforms, middle panel; correlation between the two waveforms, lower panel). The corresponding displacement rates are reported in Table 2.

Table 2. Displacement rate A corresponds to the linear displacement rate of the Krini landslide (descending orbit) before the maximum seasonal rainfall is reached, and linear displacement rate B corresponds to the displacement rate following the rainfall peak (see Figure 11 for trendlines). The total rainfall corresponds to the total rain during each time period. The displacement rate A & B values of the last line (period 2020–2021) refers to the ascending orbit data.

Time Period of Time Series	Displacement Rates A (mm/yr)	Displacement Rates B (mm/yr)	Total Rainfall (mm) (Kato Vlassia Station)	Time Lag between Rainfall Peak and InSAR Time Series (Days)
6 September 2016–28 May 2017	14.8	20.8	712	17.6
19 September 2017–23 April 2018	16.6	92.6	1041	12
9 October 2019–6 May 2020	19.1	26.6	704	11
24 September 2020–31 May 2021	30.0	99.8	935	12

3.3. Validation of InSAR Time Series with GNSS Data

To compare our InSAR displacements (velocities) with other geodetic data, we used the GPS observations of the permanent station KRIN of the Corinth Rift Laboratory (CRL) NFO of EPOS (http://crlab.eu/ (accessed on 20 December 2021).). Published GNSS data (30-s daily positions; [25]) were used. We visited the station in May 2021, and we confirmed the stability of the antenna (see Figure S7). We transformed the GNSS components into LOS through the equation [63]:

$$D_r = d_u \times \cos(inc) - (\sin(inc) \times (d_n \times \cos(ah\text{-}3 \times pi/2) + d_e \times \sin(Ah\text{-}3 \times pi/2))) \quad (1)$$

Ah-3 × Pi/2 corresponds to the angle to the azimuth look direction, which is perpendicular to the satellite heading; inc is the radar incidence angle; and d_u, d_n, d_e are the corresponding up, north and east displacements derived from GNSS solutions (Figure S8).

The GNSS time series and the InSAR time series analysis show slightly different displacement rates (Figure 9, descending orbit). The GNSS rate is slightly greater than the InSAR rate; the displacement rate from GNSS is 28 mm/yr and from InSAR time series it is 25 mm/yr (see Figure S9), providing a deviation of 3 mm/yr or 12% of the GNSS rate. This rate difference is expected as it is related to the difference in absolute terms between the point measurement of displacement, which corresponds to the GNSS time series (i.e., the motion of the antenna reference point or ARP), and the 'pixel' measurement on the ground, which corresponds to the InSAR time series. Note that the InSAR time series represent the mean value of displacement within a pixel size 100 × 100 m, i.e., enclosing the building where the GNSS antenna is located. This small deviation in rates may indicate: (a) the "smoothing" effect of InSAR; and (b) the possible occurrence of differential movements inside subareas which each SAR pixel covers. To complete the comparison—validation—we also projected the KRIN GNSS data into the LOS-ascending orbit geometry, confirming similar rates of motion (48 vs. 55 mm/yr, respectively; see time-series diagram in Figure S10).

4. Discussion

4.1. Landslide Motion and Rainfall Pattern

There are many publications concerning rainfall triggers for landslides for the areas of central and southern Europe (e.g., [65–67]). The most common parameters used for rainfall threshold definition is intensity (of both antecedent cumulated rainfall and event rainfall) and duration [67]. The effect of the antecedent rainfall depends heavily on the hydrological conductivity of the landslide mass, and only rainfall events with a large amount of precipitation are considered capable of triggering the increase of deformation rate of deep-seated landslides. The authors of [68] noted in their case that the time lag for peak precipitation and peak deformation rate may be up to 20 days, while [65] found a 12-day relation. Since proximal rainfall data is available for Achaia (Figure 1), we examined the correlation between the rainfall patterns and the seasonal movement of the Krini landslide because for this landslide the displacement rate is well constrained by the co-located GNSS station (Figure 9 and Figure S10). For this purpose, we used a signal processing method, the cross correlation [69]. We used this approach to convert our data to signal and compare them, so that we could determine if a time delay (time lag) exists between the maximum peaks of the time series of rainfall and of InSAR, respectively. The cross-correlation technique is used to measure the similarity of two time series as a function of the displacement of one relative to the other and to detect correlations among these two series [70]. We applied this technique to the daily rainfall dataset, which could play an important role in the movement of a landslide and to the InSAR time series. We used the InSAR time series of the descending track due to the completeness of its dataset during the period 2016–2020 rather than the time series from the ascending track (Figure 4). First, we resampled and interpolated the daily rainfall data to fit it to the time span of the InSAR time series. Because it was complete, we selected the data from the meteorological station Kato Vlassia (Figure 1a) and then performed the correlation for motions (displacements) of the Krini landslide.

We applied the cross correlation to the two time series, especially to three subperiods (see boxes in Figure 10). These time periods consisted of three sets of eight months of observations that had complete datasets. We selected to work with three separate time periods because our InSAR time series contains network gaps (Figure 4) which correspond to gaps in the time series. We used the following time periods: 6 September 2016–28 May 2017, 19 September 2017–23 April 2018 and 9 October 2019–6 May 2020. We calculated the displacement rate before and after the changes of the displacements through fitting the trendline to the data (Table 2). We observed in Figure 11 that when changes of the displacement rate were detected previously, a seasonal maximum of rainfall values occurred. Two of the three time periods showed an increase of displacement rate of about 40%, while the total seasonal rainfall (8 months) was quite similar (~700 mm, Table 2). The period September 2017–April 2018 showed an increase of the displacement rate of about 550%. This result was accompanied by a large amount of total rainfall (~1000 mm, Table 2). These findings allowed us to correlate the amount of total rainfall with the positive displacement rate (increase) of an active landslide. Through the cross-correlation method, we computed the time lag between the maximum peaks of InSAR and rainfall (Figure 11 lower panel; Table 2). The mean time lag (of the three periods) was 13.5 days between the maximum value of rainfall and the change in the InSAR displacement rate. The changes in InSAR displacement rates occurred on 28 January 2017, 23 January 2018 and 20 December 2019, respectively. Moreover, due to the large number of ascending orbit data available during the winter–spring period in 2020–2021 (Figure 4), we performed a separate cross correlation using the InSAR position time series (ascending orbit) against the daily rainfall data from the Kato Vlassia station (see Figure S11). Again, we found a 330% increase in the displacement rate of the landslide following a large interval of rainfall with a time lag of 12 days (see Table 2, last line). The change in the displacement rate of the Krini landslide occurred on 18 February 2021. Thus, whatever S1 orbit data we use in the case of Achaia, we always find this pattern, i.e., an increase in the landslide displacement rate following heavy rainfall.

4.2. Kinematic Characteristics of the Landslides

The Krini, Agia Eleoussa monastery and Pititsa are active landslides whose motion was measured by InSAR time series analysis for the period 2016–2021. First, we identified the moving pixels in the InSAR data (draped over the high-resolution DEM, Figure 6) and delineated the landslide boundary. We examined the decomposed velocities and found that the maximum displacement rate of each landslide is located at about the center of each landslide (Figure 12). Then, through InSAR mapping we identified more cases of mass movement than these three well-known landslides in the broader research area. In particular, we identified two additional active landslides, one around the village of Graikas (3 km north of Krini), and another around the village of Sella (Figure 5c,d and Figure 6c,d). Our results showed that a large part of this mountainous region suffers from slope instability that is manifested in various degrees of ground displacement that greatly affects its morphological features and inhabited areas. To validate these results, we provide field photographs of active landslides (taken on 7 May and 10 September 2021) which agree with our InSAR findings (Figure 12).

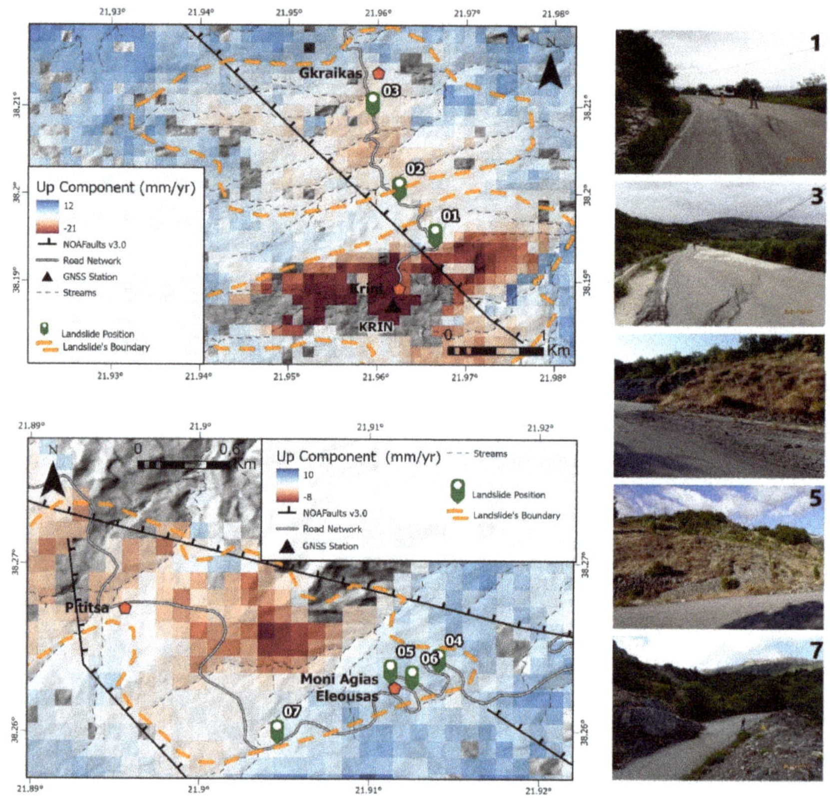

Figure 12. Maps showing U-D component of displacement rate (left panel, 2016–2021). Upper panel map includes the landslide of Krini and Graikas, and the bottom panel map includes the landslides of Pititsa and Agia Eleoussa, respectively. On the right side we provide field photos with numbers corresponding to the green markers on the maps.

In addition, the NOAFAULTs database of active faults of Greece [71,72] contains two active, north-dipping normal faults that affect the area of the active landslides. In particular, the Pititsa landslide seems to be located at the hanging wall of the Panachaiko fault segment, the Agia Eleoussa is situated between two faults (Panachaiko F. and Pititsa F.), while the Krini and Graikas landslides overlay the Panachaiko fault trace (Figure 1b). We could not find in the literature a fine-scale geological map showing the fault traces with high planimetric accuracy, so we had to rely on the coarse information provided by the NOAFAULTs database. In addition, there is no geodetic or geological evidence for any fault creep along these faults, so the tectonic influence of the landslide mobility is uncertain. However, we cannot exclude the possibility of a strong earthquake during the Holocene along these faults and so to further mobilize these landslides.

Furthermore, due to the lack of fine-scale geological mapping in this region, the lithological contacts are based on 1:50.000 scale maps where both cartographic accuracy and lithology detail are reduced. Nevertheless, the area around the Eleoussa monastery landslide consists mainly of flysch and unconformably overlying Neogene deposits, with small chert (Lower–Upper Cretaceous) and limestone (Jurassic) occurrences (Figure 12, point 7). In the area around Krini (Figures 1b, 12 and 13), we observed that the main body of the landslide is located near the contact between the "lower" flysch formation and the pelagic limestone of the Upper Cretaceous age. In addition, the main body of the landslide extends to areas consisting of Neogene sedimentary (clastic) formations. This comes in accordance with the observations made by [13,19,20] on the same region, where they attribute the highly sheared and weathered nature of flysch and Neogene sediments for contributing to the instability of the area. The weakened state of these clastic sediments influences the reactivation of ground motions when nearby seismic activity and heavy rainfall occurs.

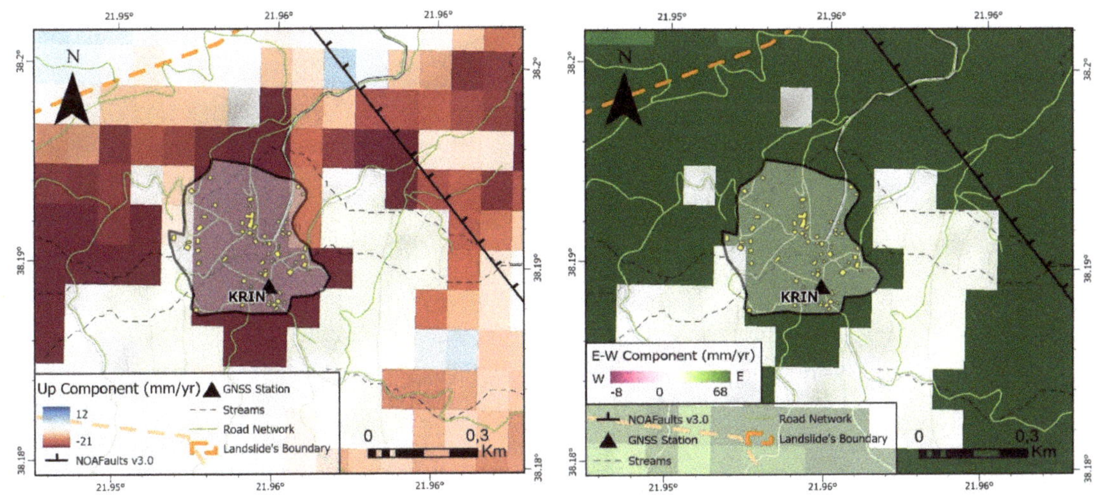

Figure 13. Maps showing InSAR velocities in the area of the village of Krini (shaded polygon: period 2016–2021). Up component (**left**), E-W component (**right**). Yellow boxes indicate houses, and black triangle shows the location of the GNSS station.

Inside the village of Krini itself, we mapped downward velocities up to 2.1 cm/year and eastward velocities up to 6.8 cm/year, respectively (Figure 13). The InSAR coverage of surface motion is discontinuous because of the scarcity of ascending orbit acquisitions (spring 2017–2018, Figure 4 and Table S1), yet there is robust geodetic (InSAR) evidence that the whole village as well as the road access to it, from both north and east directions, is affected by the deep-seated landslide.

Deep-seated landslides are characterized by long-term gradual deformation of millimeter to decimeter scale per year [73]. Their movement can be divided into slow and acceleration deformation phases, triggered by increased intensity in precipitation [74]. The deformation rate in deep-seated landslides is mainly controlled by hydrometeorological conditions. Their sliding behavior is a result of the relation of the shear strength of the soil to the shear (sliding) force applied by the gravitational forces acting on the landslide mass, a balance that the hydrological condition of the area can greatly affect [73]. This may be also the case for the Krini deep-seated landslide. We interpret its kinematic behavior as a result of seasonal changes in rainfall. For the first time, we determined the correlation between rainfall and movement of this landslide. Through the cross-correlation method, the maximum correlation between the two series were about 13.5 days. In all three time-periods studied there was an increase of displacement rate right after a period with rainfall. We suggest that the spatiotemporal pattern of movement is modulated by the seasonal rainfall which in turn, allows us to expect an increase of displacement rate of the landslide of Krini at the end of the rainy season and at the beginning of the dry period. In addition, a moderate to strong earthquake in this area could increase the displacement rate of the landslide, but such a case has not yet been demonstrated.

5. Conclusions

The main findings of this paper are:

i. The Krini, Agia Eleoussa monastery and Pititsa landslides are active landslides whose motion was measured by InSAR (C-band) time series analysis for the period 2016–2021.
ii. We processed LiCSAR interferograms using the SBAS tool and we obtained average displacement maps. The results indicate slow ground motions toward the east and downward (subsidence).
iii. The maximum displacement rate of each landslide is located at about the center of each landslide.
iv. Our results point that there is a correlation between rainfall and landslide motion. For the Krini landslide, we found the mean time lag to be 13.5 days between the maximum rainfall and the maximum of LOS displacement (descending orbit data).
v. The displacement rates of the Krini active landslide increase after a period of rainfall. Two of the three time periods examined showed an increase in the displacement rate by about 40% when the total rainfall was quite similar (~700 mm). The period September 2017–April 2018 showed an increase in the displacement rate by about 550%. This result was accompanied by a large amount of total rainfall (~1000 mm).
vi. Our findings suggest that the amount of total rainfall could control the amount of increase of the displacement rate of an active landslide.

Supplementary Materials: The following supporting information can be downloaded at: https://www.mdpi.com/article/10.3390/rs14040844/s1, Figure S1. ESRI map with the ID frames of LicSAR Portal (red outline is descending; blue outline is ascending frame). Green rhomb indicates the location of the reference point used in the InSAR analysis. Figure S2. Data processing flow chart used in this study. The arrows indicate input/output actions. The products are shown in orange. Figure S3. Unw_org corresponds to the interferogram before the GACOS correction, and Unw_cor corresponds to the interferogram after the GACOS correction. The standard deviation (STD) decreased from 3.2 rad to 2.4 rad. The reduction rate for this interferogram is 26.6%. The dates of the interferometric pair are 24 January 2019 and 17 February 2019. (Ascending Orbit). Figure S4. Unw_org corresponds to the interferogram before the GACOS correction, and Unw_cor corresponds to the interferogram after the GACOS correction. The standard deviation (STD) decreased from 5.2 rad to 2.2 rad. The reduction rate for this interferogram is 60.1%. The dates of the interferometric pair are 13 December 2016 and 4 January 2017. (Descending Orbit). Figure S5. Graph showing the time series of displacement of the village of Krini corresponding to the pixel enclosing the GNSS station (ascending orbit). Vel(1) indicates the velocity of this pixel (blue line) after the spatio-temporal filtering and deramping. Figure S6. Graph showing the time series of displacement of the village of Krini corresponding to the pixel enclosing the GNSS station (descending orbit). Vel(1) indicates the velocity of this pixel (blue line) after the spatio-temporal filtering and deramping. Figure S7. Field photograph showing the antenna of KRIN GNSS Station. View to the northeast. Photograph was taken on 7 May 2021. Figure S8. Graph showing the geometry of the LOS velocity vector. 34.58° is the incidence angle in the Krini study area. Figure S9. Graph showing the trend differences between InSAR and GNSS LOS displacements (descending orbit) of the same dates during the common period of observation (58 common dates). For comparison, the blue-line indicates a perfect (1:1) correlation. Figure S10. LOS-Projected GNSS Time series (KRIN station; black points) and InSAR position time series (orange points, ascending orbit) of the pixel which contains the GNSS station. Figure S11. Graphs showing position time series and average monthly rainfall (Kato Vlassia station, top panel) for the period September 2020–May 2021 and corresponding graphs (middle and lower panel) showing cross-correlation results. The corresponding displacement rates are reported in Table 2. Table S1. Dataset of Sentinel-1 SAR acquisitions of ascending and descending orbit used in this study.

Author Contributions: Conceptualization, A.G. and V.T.; methodology, V.T., A.G.; soft-ware, V.T.; validation, V.T., A.G. and I.K. (Ioannis Karamitros); formal analysis, V.T.; investigation, V.T., A.G.; resources, E.E., I.K. (Ioannis Karamitros); data curation, V.T., E.E. and I.K. (Ioannis Karamitros); writing—original draft preparation, V.T., A.G., I.K. (Ioannis Karamitros); writing—review and editing, V.T., A.G., I.K. (Ioannis Karamitros), E.E., I.K. (Ioannis Koukouvelas), E.S.; visualization, V.T.; supervision, A.G., I.K. (Ioannis Koukouvelas), E.S.; project administration, A.G.; funding acquisition, A.G. All authors have read and agreed to the published version of the manuscript.

Funding: This research was funded by the research project 'Platform of multidimensional monitoring with micro sensors of Enceladus of Hellenic Supersite-PROION' with Grant Number: MIS-5070928. https://proion-hellas.eu/ (accessed on 20 January 2022).

Institutional Review Board Statement: Not applicable.

Informed Consent Statement: Not applicable.

Data Availability Statement: The InSAR data are freely accessible through the LiCSAR—Copernicus portals; The rainfall data are available through www.meteo.gr (data accessed on 20 December 2021); GNSS observations (daily files) are available via the CNRS-CRL web portal.

Acknowledgments: We thank the CRL team which operated the GNSS station KRIN. Thanks to PROION partners for comment and discussions. We acknowledge concession for the use of ESRI products licensed to the Hellenic National Tsunami Warning Center, National Observatory of Athens, through the project "HELPOS—Hellenic Plate Observing System" (MIS 5002697). HELPOS is implemented under the Action "Reinforcement of the Research and Innovation Infrastructure", funded by the Operational Programme "Competitiveness, Entrepreneurship and Innovation" (NSRF 2014–2020) and co-financed by Greece and the European Union (European Regional Development Fund). We acknowledge KTIMATOLOGIO S.A. for kindly providing the digital elevation model. We also thank Vassileios Tsironis, Athina Psalta and Ioannis Fountoulakis for their help with cross-correlation techniques. We also thank Sotiris Valkaniotis and Helena Partheniou for comments and discussions.

Conflicts of Interest: The authors declare no conflict of interest. The funders had no role in the design of the study; in the collection, analyses, or interpretation of data; in the writing of the manuscript, or in the decision to publish the results.

References

1. Colesanti, C.; Wasowski, J. Investigating landslides with space-borne Synthetic Aperture Radar (SAR) interferometry. *Eng. Geol.* **2006**, *88*, 173–199. [CrossRef]
2. Aslan, G.; Foumelis, M.; Raucoules, D.; De Michele, M.; Bernardie, S.; Cakir, Z. Landslide Mapping and Monitoring Using Persistent Scatterer Interferometry (PSI) Technique in the French Alps. *Remote Sens.* **2020**, *12*, 1305. [CrossRef]
3. Solari, L.; Del Soldato, M.; Raspini, F.; Barra, A.; Bianchini, S.; Confuorto, P.; Casagli, N.; Crosetto, M. Review of Satellite Interferometry for Landslide Detection in Italy. *Remote Sens.* **2020**, *12*, 1351. [CrossRef]
4. Kontoes, C.; Loupasakis, C.; Papoutsis, I.; Alatza, S.; Poyiadji, E.; Ganas, A.; Psychogyiou, C.; Kaskara, M.; Antoniadi, S.; Spanou, N. Landslide Susceptibility Mapping of Central and Western Greece, Combining NGI and WoE Methods, with Remote Sensing and Ground Truth Data. *Land* **2021**, *10*, 402. [CrossRef]
5. Elias, P.; Valkaniotis, S.; Ganas, A.; Papathanassiou, G.; Bilia, A.; Kollia, E. Satellite SAR interferometry for monitoring dam deformations: The case of Evinos dam, central Greece. In Proceedings of the Eighth International Conference on Remote Sensing and Geoinformation of the Environment (RSCy2020), Paphos, Cyprus, 16–18 March 2020; Volume 11524, p. 115241I. [CrossRef]
6. Cruden, D.M.; Varnes, D.J. Landslide Types and Processes, Transportation Research Board. *US Natl. Acad. Sci. Spec. Rep.* **1996**, *247*, 36–75.
7. Singleton, A.; Li, Z.; Hoey, T.; Muller, J.P. Evaluating sub-pixel offset techniques as an alternative to D-InSAR for monitoring episodic landslide movements in vegetated terrain. *Remote Sens. Environ.* **2014**, *147*, 133–144. [CrossRef]
8. Vassilakis, E.; Foumelis, M.; Erkeki, A.; Kotsi, E.; Lekkas, E. Post-event surface deformation of Amyntaio slide (Greece) by complementary analysis of Remotely Piloted Airborne System imagery and SAR interferometry. *Appl. Geomat.* **2020**, *13*, 65–75. [CrossRef]
9. Tsangaratos, P.; Loupasakis, C.; Nikolakopoulos, K.; Angelitsa, V.; Ilia, I. Developing a landslide susceptibility map based on remote sensing, fuzzy logic and expert knowledge of the Island of Lefkada, Greece. *Environ. Earth Sci.* **2018**, *77*, 363. [CrossRef]
10. Kyriou, A.; Nikolakopoulos, K. Assessing the suitability of Sentinel-1 data for landslide mapping. *Eur. J. Remote Sens.* **2018**, *51*, 402–411. [CrossRef]

11. Papoutsis, I.; Kontoes, C.; Alatza, S.; Apostolakis, A.; Loupasakis, C. InSAR Greece with Parallelized Persistent Scatterer Interferometry: A National Ground Motion Service for Big Copernicus Sentinel-1 Data. *Remote Sens.* **2020**, *12*, 3207. [CrossRef]
12. Psychogyiou, C.; Papoutsis, I.; Kontoes, C.; Poyiadji, E.; Spanou, N.; Klimis, N. Multi-temporal Monitoring of slow moving Landslides in south Pindus mountain range, Greece. In Proceedings of the Fringe 2015 Workshop, Frascati, Italy, 23–27 March 2015; pp. 23–27.
13. Koukis, G.; Sabatakakis, N.; Ferentinou, M.; Lainas, S.; Alexiadou, X.; Panagopoulos, A. Landslide phenomena related to major fault tectonics: Rift zone of Corinth Gulf, Greece. *Bull. Eng. Geol. Environ.* **2009**, *68*, 215–229. [CrossRef]
14. Lebourg, T.; El Bedoui, S.; Hernandez, M. Control of slope deformations in high seismic area: Results from the Gulf of Corinth observatory site (Greece). *Eng. Geol.* **2009**, *108*, 295–303. [CrossRef]
15. Elias, P.; Briole, P. Ground Deformations in the Corinth Rift, Greece, Investigated Through the Means of SAR Multitemporal Interferometry. *Geochem. Geophys. Geosystems* **2018**, *19*, 4836–4857. [CrossRef]
16. Del Soldato, M.; Del Ventisette, C.; Raspini, F.; Righini, G.; Pancioli, V.; Moretti, S. Ground deformation and associated hazards in NW peloponnese (Greece). *Eur. J. Remote Sens.* **2018**, *51*, 710–722. [CrossRef]
17. Chen, W.; Chen, Y.; Tsangaratos, P.; Ilia, I.; Wang, X. Combining Evolutionary Algorithms and Machine Learning Models in Landslide Susceptibility Assessments. *Remote Sens.* **2020**, *12*, 3854. [CrossRef]
18. Sabatakakis, N.; Koukis, G.; Vassiliades, E.; Lainas, S. Landslide susceptibility zonation in Greece. *Nat. Hazards* **2012**, *65*, 523–543. [CrossRef]
19. Papadopoulos, G.A.; Ganas, A.; Koukouvelas, I. Landsliding phenomena in NW Peloponnese, Greece: A test-site of the EC LEWIS research project. *Geophys. Res. Abstr.* **2006**, *8*, 04402.
20. Papathanou, M. Small Scale Displacements within Landslides in the Region of Pititsa Using Earth Surveying and Photogrammetry. *Master's Thesis*, University of Patras: Rio-Patras, Greece; 97, *(unpublished)*.
21. Katrantsiotis, C. Holocene Environmental Changes and Climate Variability in the Eastern Mediterranean: Multiproxy Sediment Records from the Peloponnese Peninsula, SW Greece. Ph.D. Thesis, Department of Physical Geography, Stockholm University, Stockholm, Sweden, 2019.
22. Bernard, P.; Lyon-Caen, H.; Briole, P.; Deschamps, A.; Boudin, F.; Makropoulos, K.; Papadimitriou, P.; Lemeille, F.; Patau, G.; Billiris, H.; et al. Seismicity, deformation and seismic hazard in the western rift of Corinth: New insights from the Corinth Rift Laboratory (CRL). *Tectonophysics* **2006**, *426*, 7–30. [CrossRef]
23. Clarke, P.J.; Davies, R.R.; England, P.C.; Parsons, B.; Billiris, H.; Paradissis, D.; Veis, G.; Cross, P.A.; Denys, P.H.; Ashkenazi, V.; et al. Crustal strain in central Greece from repeated GPS measurements in the interval 1989-1997. *Geophys. J. Int.* **1998**, *135*, 195–214. [CrossRef]
24. Chousianitis, K.; Ganas, A.; Evangelidis, C.P. Strain and rotation rate patterns of mainland Greece from continuous GPS data and comparison between seismic and geodetic moment release. *J. Geophys. Res. Solid Earth* **2015**, *120*, 3909–3931. [CrossRef]
25. Briole, P.; Ganas, A.; Elias, P.; Dimitrov, D. The GPS velocity field of the Aegean. New observations, contribution of the earthquakes, crustal blocks model. *Geophys. J. Int.* **2021**, *226*, 468–492. [CrossRef]
26. Elias, P.; Kontoes, C.; Papoutsis, I.; Kotsis, I.; Marinou, A.; Paradissis, D.; Sakellariou, D. Permanent Scatterer InSAR Analysis and Validation in the Gulf of Corinth. *Sensors* **2009**, *9*, 46–55. [CrossRef]
27. Trikolas, C.; Koskeridou, E.; Tsourou, T.; Drinia, H.; Alexouli-Livaditi, A. Pleistocene marine deposits of the Aigialia region (N. Peloponnesus). *Bull. Geol. Soc. Greece* **2004**, *36*, 826–835. [CrossRef]
28. Palyvos, N.; Sorel, D.; Lemeille, F.; Mancini, M.; Pantosti, D.; Julià, R.; Triantaphyllou, M.; De Martini, P.M. Review and New Data on Uplift Rates at the W Termination of the Corinth Rift and the Ne Rion Graben Area (Achaia, Nw Peloponnesos). *Bull. Geol. Soc. Greece* **2018**, *40*, 412–424. [CrossRef]
29. Avallone, A.; Briole, P.; Agatza-Balodimou, A.M.; Billiris, H.; Charade, O.; Mitsakaki, C.; Nercessian, A.; Papazissi, K.; Paradissis, D.; Veis, G. Analysis of eleven years of deformation measured by GPS in the Corinth Rift Laboratory area. *Comptes Rendus Geosci.* **2004**, *336*, 301–311. [CrossRef]
30. Roberts, P.G.; Koukouvelas, I. Structural and seismological segmentation of the Gulf of Corinth fault system: Implications for models of fault growth. *Ann. Geofis.* **1996**, *39*, 619–646.
31. Roberts, G.; Ganas, A. Fault-slip directions in central and southern Greece measured from striated and corrugated fault planes: Comparison with focal mechanism and geodetic data. *J. Geophys. Res. Earth Surf.* **2000**, *105*, 23443–23462. [CrossRef]
32. Koukouvelas, I.; Stamatopoulos, L.; Katsonopoulou, D.; Pavlides, S. A paleoseismological and geoarchaeological investigation of Eliki fault, Gulf of Corinth, Greece. *J. Struct. Geol.* **2001**, *23*, 531–543. [CrossRef]
33. Verrios, S.; Zygouri, V.; Kokkalas, S. MORPHOTECTONIC ANALYSIS IN THE ELIKI FAULT ZONE (GULF OF CORINTH, GREECE). *Bull. Geol. Soc. Greece* **2004**, *36*, 1706–1715. [CrossRef]
34. Palyvos, N.; Pantosti, D.; De Martini, P.M.; Lemeille, F.; Sorel, D.; Pavlopoulos, K. The Aigion-Neos Erineos coastal normal fault system (western Corinth Gulf Rift, Greece): Geomorphological signature, recent earthquake history, and evolution. *J. Geophys. Res. Earth Surf.* **2005**, *110*, B09302. [CrossRef]
35. Tsimi, C.; Ganas, A.; Soulakellis, N.; Kairis, O.; Valmis, S. MORPHOTECTONICS OF THE PSATHOPYRGOS ACTIVE FAULT, WESTERN CORINTH RIFT, CENTRAL GREECE. *Bull. Geol. Soc. Greece* **2018**, *40*, 500–511. [CrossRef]
36. Verroios, S.; Zygouri, V. Geomorphological Analysis of Xilokastro Fault, Central Gulf of Corinth, Greece. *Geosciences* **2021**, *11*, 516. [CrossRef]

37. Zygouri, V.; Koukouvelas, I.K. Landslides and natural dams in the Krathis River, north Peloponnese, Greece. *Bull. Eng. Geol. Environ.* **2019**, *78*, 207–222. [CrossRef]
38. Rozos, D.; Bathrellos, D.; Skilodimou, D. Landslide Susceptibility Mapping of the northeastern part of Achaia Prefecture using analytical hierarchical process and GIS techniques. *Bull. Geol. Soc. Greece* **2010**, *43*, 1637–1646. [CrossRef]
39. Rozos, D.; Bathrellos, G.D.; Skillodimou, H.D. Comparison of the implementation of rock engineering system and analytic hierarchy process methods, upon landslide susceptibility mapping, using GIS: A case study from the Eastern Achaia County of Peloponnesus, Greece. *Environ. Earth Sci.* **2010**, *63*, 49–63. [CrossRef]
40. Kavoura, K. Landslide Inventory Map Using GIS Tools and Field Experience—The Case of Achaia Prefecture, Western Greece. Master's Thesis, University of Patras, Rio-Patra, Greece, 2013; p. 98.
41. Polykretis, C.; Ferentinou, M.; Chalkias, C. A comparative study of landslide susceptibility mapping using landslide susceptibility index and artificial neural networks in the Krios River and Krathis River catchments (northern Peloponnesus, Greece). *Bull. Eng. Geol. Environ.* **2015**, *74*, 27–45. [CrossRef]
42. Konstantopoulos, K.; Miliaresis, G.C.; Desk, S. Unsupervised landslide risk dependent terrain segmentation on the basis of historical landslide data and geomorphometrical indicators. *SDRP J. Earth Sci. Environ. Stud.* **2018**, *3*, 1–9. [CrossRef]
43. Tsangaratos, P.; Loupasakis, C.; Rozos, D.; Ilia, I. Landslide susceptibility assessments using the k-Nearest Neighbor algo-rithm and expert knowledge. Case study of the basin of Selinounda River 2015, Achaia County, Greece. In Proceedings of the SafeChania 2015. The Knowledge Triangle in the Civil Protection Service Center of Mediterranean Architecture, Chania, Greece, 10–14 June 2015; pp. 10–14.
44. Sabatakakis, N.; Tsiambaos, G.; Rondoyanni, T.; Papanakli, S.; Kavoura, K. Deep-Seated Structurally Controlled Landslides of Corinth Gulf Rift Zone, Greece: The Case of Panagopoula Landslide. In Proceedings of the 13th ISRM International Congress of Rock Mechanics, Montreal, QC, Canada, 10–13 May 2015; Paper Number: ISRM-13CONGRESS-2015-273. ISBN 978-1-926872-25-4.
45. Guzzetti, F.; Ardizzone, F.; Cardinali, M.; Rossi, M.; Valigi, D. Landslide volumes and landslide mobilization rates in Umbria, central Italy. *Earth Planet. Sci. Lett.* **2009**, *279*, 222–229. [CrossRef]
46. Tavoularis, N.; Papathanassiou, G.; Ganas, A.; Argyrakis, P. Development of the Landslide Susceptibility Map of Attica Region, Greece, Based on the Method of Rock Engineering System. *Land* **2021**, *10*, 148. [CrossRef]
47. Loftus, D.L.; Tsoflias, P. *Geological Map of Greece in scale 1:50000. NAFPAKTOS Map Sheet*; Institute of Geology and Mineral Exploration: Athens, Greece, 1971.
48. Tsoflias, P. *Geological Map of Greece in scale 1:50000. KHALANDRITSA Map Sheet*; Institute of Geology and Mineral Exploration: Athens, Greece, 1984.
49. Eleftheriou, A. *Geotechnical Report on the Areas of Elekistra–Argyra and Krini Villages of the Achaia District*; IGME: Athens, Greece, 1985; p. 13.
50. Varnes, D.J. Slope movement types and processes. *Spec. Rep.* **1978**, *176*, 11–33.
51. Hungr, O.; Leroueil, S.; Picarelli, L. The Varnes classification of landslide types, an update. *Landslides* **2014**, *11*, 167–194. [CrossRef]
52. Bertolini, G.; Guida, M.; Pizziolo, M. Landslides in Emilia-Romagna region (Italy): Strategies for hazard assessment and risk management. *Landslides* **2005**, *2*, 302–312. [CrossRef]
53. Morishita, Y.; Lazecky, M.; Wright, T.J.; Weiss, J.R.; Elliott, J.R.; Hooper, A. LiCSBAS: An Open-Source InSAR Time Series Analysis Package Integrated with the LiCSAR Automated Sentinel-1 InSAR Processor. *Remote Sens.* **2020**, *12*, 424. [CrossRef]
54. COMET-LiCS Sentinel-1 InSAR Portal. Available online: https://comet.nerc.ac.uk/COMET-LiCS-portal/ (accessed on 26 May 2021).
55. LiCSBAS: LiCSBAS Package to Conduct InSAR Time Series Analysis Using LiCSAR Products. Available online: https://github.com/yumorishita/LiCSBAS (accessed on 26 May 2021).
56. Werner, C.; Wegmüller, U.; Strozzi, T.; Wiesmann, A. Gamma SAR and interferometric processing software. In Proceedings of the ERS-ENVISAT Symposium 2000, Gothenburg, Sweden, 16–20 October 2000.
57. Wegmüller, U.; Werner, C.; Strozzi, T.; Wiesmann, A.; Frey, O.; Santoro, M. Sentinel-1 IWS mode support in the GAMMA software. In Proceedings of the 2015 IEEE 5th Asia-Pacific Conference on Synthetic Aperture Radar (APSAR), Singapore, 1–4 September 2015.
58. Chen, C.W.; Zebker, H.A. Phase unwrapping for large SAR interferograms: Statistical segmentation and generalized network models. *IEEE Trans. Geosci. Remote Sens.* **2002**, *40*, 1709–1719. [CrossRef]
59. Wang, Q.; Yu, W.; Xu, B.; Wei, G. Assessing the use of GACOS products for SBAS-INSAR deformation monitoring: A case in Southern California. *Sensors* **2019**, *19*, 3894. [CrossRef]
60. Biggs, J.; Wright, T.; Lu, Z.; Parsons, B. Multi-interferogram method for measuring interseismic deformation: Denali Fault, Alaska. *Geophys. J. Int.* **2007**, *170*, 1165–1179. [CrossRef]
61. López-Quiroz, P.; Doin, M.P.; Tupin, F.; Briole, P.; Nicolas, J.M. Time series analysis of Mexico City subsidence constrained by radar interferometry. *J. Appl. Geophys.* **2009**, *69*, 1–15. [CrossRef]
62. Wright, T.J.; Parsons, B.E.; Lu, Z. Toward mapping surface deformation in three dimensions using InSAR. *Geophys. Res. Lett.* **2004**, *31*, L01607. [CrossRef]
63. Motagh, M.; Shamshiri, R.; Haghshenas-Haghighi, M.; Wetzel, H.-U.; Akbari, B.; Nahavandchi, H.; Roessner, S.; Arabi, S. Quantifying groundwater exploitation induced subsidence in the Rafsanjan plain, southeastern Iran, using InSAR time-series and in situ measurements. *Eng. Geol.* **2017**, *218*, 134–151. [CrossRef]

64. Lagouvardos, K.; Kotroni, V.; Bezes, A.; Koletsis, I.; Kopania, T.; Lykoudis, S.; Mazarakis, N.; Papagiannaki, K.; Vougioukas, S. The automatic weather stations NOANN network of the National Observatory of Athens: Operation and database. *Geosci. Data J.* **2017**, *4*, 4–16. [CrossRef]
65. Guzzetti, F.; Peruccacci, S.; Rossi, M.; Stark, C.P. Rainfall thresholds for the initiation of landslides in central and southern Europe. *Meteorol. Atmos. Phys.* **2007**, *98*, 239–267. [CrossRef]
66. Guzzetti, F.; Peruccacci, S.; Rossi, M.; Stark, C.P. The rainfall intensity–duration control of shallow landslides and debris flows: An update. *Landslides* **2008**, *5*, 3–17. [CrossRef]
67. Segoni, S.; Piciullo, L.; Gariano, S.L. A review of the recent literature on rainfall thresholds for landslide occurrence. *Landslides* **2018**, *15*, 1483–1501. [CrossRef]
68. Wei, Z.-L.; Shang, Y.-Q.; Sun, H.-Y.; Xu, H.-D.; Wang, D.-F. The effectiveness of a drainage tunnel in increasing the rainfall threshold of a deep-seated landslide. *Landslides* **2019**, *16*, 1731–1744. [CrossRef]
69. Tomás, R.; Li, Z.; Lopez-Sanchez, J.M.; Liu, P.; Singleton, A. Using wavelet tools to analyse seasonal variations from InSAR time-series data: A case study of the Huangtupo landslide. *Landslides* **2015**, *13*, 437–450. [CrossRef]
70. Bracewell, R. *Pentagram Notation for Cross Correlation. The Fourier Transform and Its Applications*; McGraw-Hill: New York, NY, USA, 1965; pp. 46, 243.
71. Ganas, A. NOAFAULTS KMZ Layer Version 3.0 (2020 Update) (Version V3.0) [Data Set]. *Zenodo* **2020**. [CrossRef]
72. Ganas, A.; Tsironi, V.; Kollia, E.; Delagas, M.; Tsimi, C.; Oikonomou, A. Recent upgrades of the NOA database of active faults in Greece (NOAFAULTs). In Proceedings of the 19th General Assembly of WEGENER, Grenoble, France, 10–13 September 2018; pp. 10–13.
73. Natijne, A.L.; Lindenbergh, R.C.; Bogaard, T.A. Machine Learning: New Potential for Local and Regional Deep-Seated Landslide Nowcasting. *Sensors* **2020**, *20*, 1425. [CrossRef]
74. Villaseñor-Reyes, C.I.; Dávila-Harris, P.; Delgado-Rodríguez, O. Multidisciplinary approach for the characterization of a deep-seated landslide in a semi-arid region (Cañón de Yerbabuena, San Luis Potosí, Mexico). *Landslides* **2021**, *18*, 367–381. [CrossRef]

remote sensing

MDPI

Article

Landslides Triggered by the 2016 Heavy Rainfall Event in Sanming, Fujian Province: Distribution Pattern Analysis and Spatio-Temporal Susceptibility Assessment

Siyuan Ma [1,2], Xiaoyi Shao [3,4] and Chong Xu [3,4,*]

1 Institute of Geology, China Earthquake Administration, Beijing 100029, China; masiyuan@ies.ac.cn
2 Key Laboratory of Seismic and Volcanic Hazards, Institute of Geology, China Earthquake Administration, Beijing 100029, China
3 National Institute of Natural Hazards, Ministry of Emergency Management of China, Beijing 100085, China; xiaoyishao@ninhm.ac.cn
4 Key Laboratory of Compound and Chained Natural Hazards Dynamics, Ministry of Emergency Management of China, Beijing 100085, China
* Correspondence: chongxu@ninhm.ac.cn or xc11111111@126.com

Abstract: Rainfall-induced landslides pose a significant threat to the lives and property of residents in the southeast mountainous area. From 5 to 10 May 2016, Sanming City in Fujian Province, China, experienced a heavy rainfall event that caused massive landslides, leading to significant loss of life and property. Using high-resolution satellite imagery, we created a detailed inventory of landslides triggered by this event, which totaled 2665 across an area of 3700 km^2. The majority of landslides were small-scale, shallow and elongated, with a dominant distribution in Xiaqu town. We analyzed the correlations between the landslide abundance and topographic, geological and hydro-meteorological factors. Our results indicated that the landslide abundance index is related to the gradient of the hillslope, distance from a river and total rainfall. The landslide area density, i.e., LAD increases with the increase in these influencing factors and is described by an exponential or linear relationship. Among all lithological types, Sinian mica schist and quartz schist (Sn-s) were found to be the most prone to landslides, with over 35% of landslides occurring in just 10% of the area. Overall, the lithology and rainfall characteristics primarily control the abundance of landslides, followed by topography. To gain a better understanding of the triggering conditions for shallow landslides, we conducted a physically based spatio-temporal susceptibility assessment in the landslide abundance area. Our numerical simulations, using the MAT.TRIGRS tool, show that it can accurately reproduce the temporal evolution of the instability process of landslides triggered by this event. Although rainfall before 8 May may have contributed to decreased slope stability in the study area, the short duration of heavy rainfall on 8 May is believed to be the primary triggering factor for the occurrence of massive landslides.

Keywords: landslides; heavy rainfall; distribution pattern; spatiotemporal assessment; Sanming area; Fujian province

Citation: Ma, S.; Shao, X.; Xu, C. Landslides Triggered by the 2016 Heavy Rainfall Event in Sanming, Fujian Province: Distribution Pattern Analysis and Spatio-Temporal Susceptibility Assessment. *Remote Sens.* **2023**, *15*, 2738. https://doi.org/10.3390/rs15112738

Academic Editors: Ioannis Papoutsis, Konstantinos G. Nikolakopoulos and Constantinos Loupasakis

Received: 6 April 2023
Revised: 22 May 2023
Accepted: 22 May 2023
Published: 24 May 2023

1. Introduction

Rainfall-induced landslides are a type of slope instability that may occur in densely distributed soil and/or debris under heavy rainfall, producing a significant amount of sediments in river networks [1,2]. These landslides often result in catastrophic debris flows, which cause severe damage to agricultural crops, infrastructure and human lives [3,4]. Therefore, effective risk mitigation measures and early warning systems are urgently needed to minimize the detrimental impacts of these slope instabilities on both local and regional scales [5–7].

The southeast coastal area of China falls within the subtropical monsoon climate zone and is frequently affected by typhoons and rainstorms. The coastal areas are characterized by mountainous and hilly terrains, which cover approximately 75% of the total area. The climate in this region is warm and humid with abundant rainfall, leading to strong physical weathering of rocks [8]. The slope surfaces in this area are predominantly covered by residual soil and heavily weathered rocks. As a result, approximately 90% of landslides occur during the rainy seasons, which typically span from May to September. During these months, landslides triggered by heavy rainfall events are the primary cause of building damage and human casualties. In August 2005, a landslide triggered by Typhoon "Sudilo" caused direct economic losses amounting to USD 4.0 million in Zhejiang Province and posed a threat to 4616 local residents. On 10 June 2019, Longchuan County, Guangdong Province, China, was subjected to incessant heavy rainfall, which subsequently led to widespread landslides, collapses and debris flows. Among the 352 affected villages, Mibei village in Longchuan County was hit the hardest, with 1571 individuals affected, 120 buildings completely destroyed and over 100 houses sustaining damage of varying degrees. The direct economic loss of this event reached USD 15.4 million [9]. In June 2019, due to the influence of southwest airflow on the south side of low-level shear, heavy rainfall occurred in the western region of Fujian Province. Since 6 June, 75,900 people have been affected in 19 counties of Sanming, Nanping and Longyan cities, resulting in a direct economic loss of USD 37.9 million. Rainfall-induced landslides have seriously affected the life safety as well as economic development in the southeastern coastal areas. Therefore, the study of distribution characteristics and spatiotemporal prediction of such landslides has become a major demand for ensuring national security and social development.

For the analysis of regional landslides induced by a single extreme event (e.g., earthquake or heavy rainfall), a comprehensive landslide inventory is essential and often indispensable. Such data provide a crucial foundation for the distribution characteristics of landslides [10], susceptibility and risk assessment [11,12], landslide formation mechanism and geomorphological evolution [13,14]. In contrast to earthquake events, the database of landslides triggered by heavy rainfall events is still limited [15–17]. At present, only 16 public rainfall-event-based landslide databases are available worldwide, but most of these event-based databases are small in scale. There are only four landslide databases with more than 2000 landslides, including the Morakot Typhoon event in southern Taiwan on August 6–9, 2008 [18], the heavy rainfall event in the Teres ó polis region of Brazil on 11–13 January [17], as well as the long-term heavy rainfall event in Japan's Hiroshima region from 28 June to 9 July 2018 [19] and the Hurricane Maria event in Dominica region from 18–22 September 2017 [20,21]. Therefore, compared to earthquake-event-based landslide inventories, the landslide inventories associated with heavy rainfalls still need more in-depth investigations.

Assessing the spatial susceptibility of rainfall-induced landslides plays an important role in effective landslide prevention and control [22–25]. Currently, three primary methods are pervasively used for rainfall-induced landslide prediction: empirical models [26–28], data-driven models [24,29] and physically based models [30,31]. The empirical model is primarily based on the analysis of rainfall characteristics, such as rainfall intensity, duration and total amount. This type of model is utilized to determine the likelihood of landslides. Although this method is simple and easy to implement, it only considers rainfall as a single factor and ignores other important topographical, geological and hydrological factors. In addition, empirical models require abundant landslide and rainfall data to determine the empirical rainfall threshold. Due to the lack of landslide and rainfall data, it is difficult to develop an efficient rainfall threshold model in mountainous areas with severe landslide disasters [5,28]. The statistical model analyzes the relationship between various factors, such as elevation, hillslope gradient, slope aspect, average rainfall and vegetation coverage. Then, based on data-driven models and actual landslides, a rainfall-induced landslide assessment model is created [15,32]. However, this method cannot explain the physical mechanism of landslide occurrence and requires a large amount of landslide data; otherwise, because it requires sufficient landslide data to establish the susceptibility assessment model, this results in assessment results that frequently lag behind practical application and cannot serve the emergency assessment in a short time [33].

The physically based method does not use actual landslide data but rather simulates the physical process of rainfall-induced landslide occurrence by combining hydrological and infinite slope models. Physical models can reproduce the physical process of landslide occurrence, which is considered to be an effective method for landslide susceptibility analysis [34]. Furthermore, the GIS technology has facilitated the widespread use of physically based models in large regions. As a result, physically based models have been widely used in the prediction and early warning of rainfall-induced landslides [35,36]. Currently, commonly used physical models for landslide susceptibility assessment include the SHALSTAB [37] and SINMAP models [38] based on steady-state hydrological modeling, as well as the SLIP [2,39], CRESTSLIDE [40], HIRESSS [41,42] and TRIGRS models [43] based on transient physical modeling. At present, the TRIGRS model has been widely used worldwide including in Italy, the United States, China, South Korea and Southeast Asia [9,43–46], and is currently one of the most popular models for the spatiotemporal prediction of rainfall-induced landslides. However, the application of the TRIGRS model in China's southeast area is limited, so it is necessary to investigate the applicability of the model in the southeast mountainous area.

From 5 to 10 May 2016, Sanming City in Fujian Province, China, experienced an unprecedented heavy rainfall event with a maximum hourly rainfall of 56.6 mm and a maximum daily rainfall of 259.6 mm, breaking the local daily rainfall record of 178.2 mm set in 1961. This rainfall event triggered extensive landslides, resulting in significant loss of people's lives and property. During this rainfall event, a landslide occurred in the Chitan Village of Kaishan Township, leading to the burial of the office building and construction site dormitory of the expansion project of the Chitan hydropower plant, which belongs to the China Huadian Corporation. As of 1:00 pm on 10 May, the landslide had caused 35 deaths and one person was missing. To better understand the characteristics of the landslides induced by this fatal event, a comprehensive landslide inventory is needed. Therefore, the objectives of this study are twofold: (1) to establish a rainfall-induced landslide inventory through visual interpretation and to analyze the distribution pattern of landslides with relevant factors; and (2) to achieve physically based spatiotemporal susceptibility assessments using an open-source tool of MAT.TRIGRS (V1.0) and to back-analyze the rainfall process' response to changes in landslide stability. This study can provide a significant scientific basis for the formation mechanism and spatiotemporal prediction of rainfall-induced landslides in the southeast coastal area.

2. Study Area

The study area is located in the western part of Fujian Province, between 25°30′N–27°07′N latitude and 116°22′E–118°39′E longitude. It neighbors Fuzhou to the east, connects Jiangxi Province to the west and borders Nanping to the north. The landform of the study area is mainly hilly terrain, with medium- and low-elevation mountains. Geologically, the area is an erosion-dominated mountainous region with strong tectonic activity. The terrain is generally higher in the southeast and lower in the northwest, with elevations ranging from 130 m to 1847 m (Figure 1). In addition, the area features a subtropical monsoon climate with a typical mountain climate. The rainfall is abundant, with an annual average precipitation reaching up to 1700 mm. The rainfalls are mainly concentrated from March to August and the annual average temperature is about 19.9 °C.

Figure 2 illustrates the distribution of the main lithology exposed in the study area, which ranges from Quaternary loose deposits (Q) to Proterozoic granite (Pt-g). The main lithology is protozoic plagioclase hornblende and granulite (Pt-p), which are mainly distributed in the north of the study area. In addition, Jurassic variegated sandstone and glutenite and K-feldspar granite (Jg and Js-g) are distributed in the west and east of the study area. Cretaceous rhyolite porphyry and glutenite (Kr-g) are mainly exposed in the southwestern area. Sinian mica schist and quartz schist (Sn-s) and Cambrian quartz sandstone (∈) are observed in the middle part of the study area. The slope surface is mostly covered by residual soil and strongly weathered rock. Under the condition of rainfall, shallow landslides are mainly formed by loose deposits.

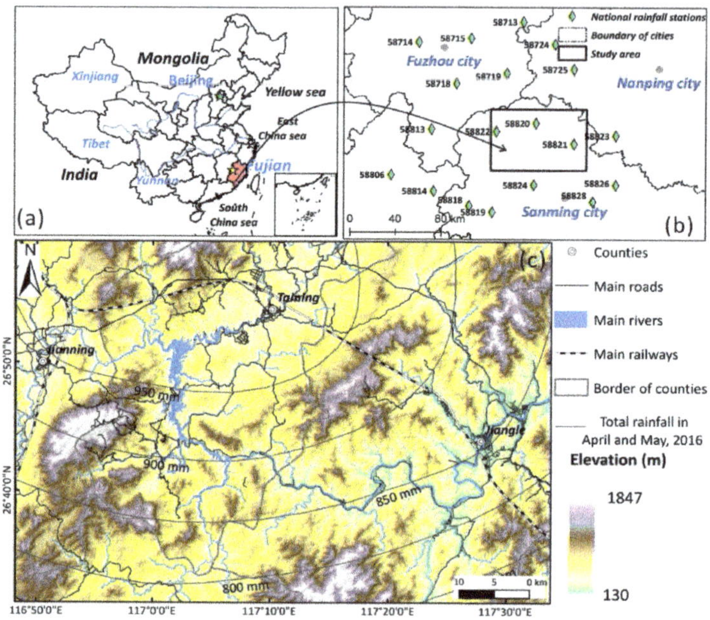

Figure 1. The geographical and topographical maps displaying the location and elevation distribution of the study area. (**a**) The location of Fujian Province; (**b**) the location of the study area and rainfall stations; and (**c**) the distribution of elevation, rainfall and water networks in the study area.

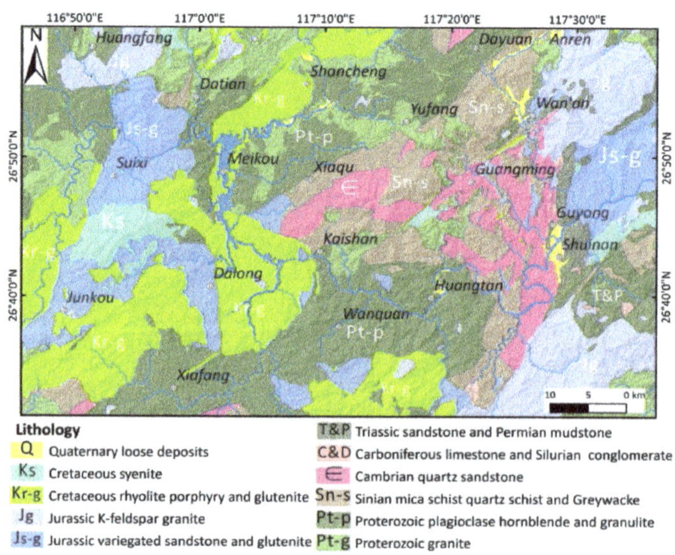

Figure 2. Lithology distribution of the study area. The geological map was created using China Geological Survey's 1:200,000 geological maps (http://dcc.cgs.gov.cn/, accessed on 5 April 2023).

3. Data and Methods

3.1. Landslide Mapping

A comprehensive rainfall-induced landslide inventory is of significance for studying the distribution pattern of landslides, landslide susceptibility and their impact on geomorphological evolution. In this study, we were able to conduct a detailed visual interpretation of landslides due to the availability of high-resolution satellite photographs on the Google Earth (GE) platform [47,48]. The satellite images used for landslide interpretation were all based on the GE platform, which provided a 100% coverage of high-resolution satellite images. Due to the high vegetation coverage in the study area, optical images can be used to better identify the landslide locations before and after the events. By comparing the pre- and post-rainfall images combined with field investigations, the landslide inventory associated with this rainfall event was ultimately established. Figure 3 shows the field photos of landslides triggered by this event.

Figure 3. Field photos of landslides triggered by this rainfall event.

3.2. Rainfall Data

We collected precipitation data in the Sanming area for the last two decades (from 2000 to 2020). According to these data, the average annual rainfall in the Sanming area has remained between 1200–2400 mm, with prominent fluctuation. In 2016, the annual rainfall exceeded 2200 mm, while in 2003 the annual rainfall was relatively low, less than 1200 mm (Figure 4a). After comparison, we found that the rainfall in May 2016 was more than the monthly average rainfall over the past 20 years. The precipitation in May was 300 mm, which is more than the average monthly rainfall in previous years (approximately 200 mm) (Figure 4b).

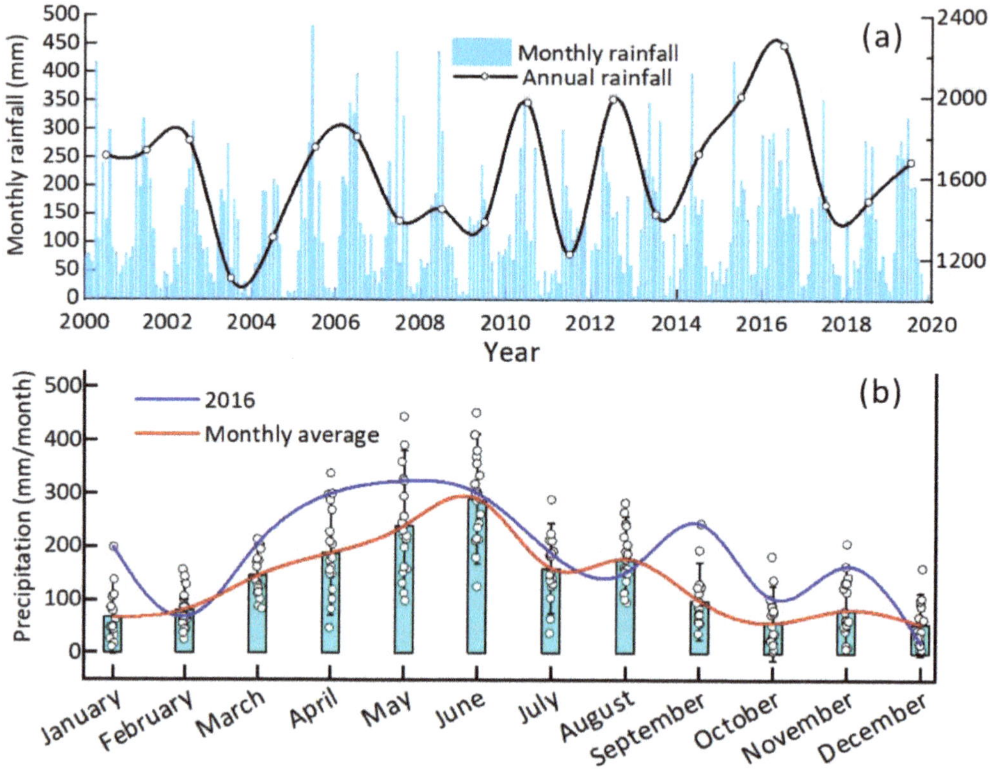

Figure 4. Monthly precipitation data of Sanming city over the past 20 years (2000–2020); (**a**) monthly and annual average precipitation data over the last 20 years; (**b**) comparison of the monthly rainfall in 2016 with the average precipitation over the last two decades.

We collected rainfall data from 19 rainfall stations of the China Meteorological Administration within a radius of 100 km in the study area. These stations recorded rainfall data every 12 h. Figure 5 shows the rainfall data from two stations located in the northern and western parts of the study area from 1 April to 30 May. Based on the rainfall data, we can observe that the rainfall event mainly occurred from 5 May to 10 May. The highest precipitation occurred on 8 and 9 May, reaching about 100–120 mm, which accounted for more than half of the total rainfall amount. The precipitation on the other four days was relatively low, averaging about 15–40 mm. Based on the above 19 rain gauges, the commonly used Kriging interpolation method was applied to obtain the distribution of the rainfall during different times of this event (Figure 6). The result indicates that the daily rainfall in the study area varies greatly, with a difference of around 160 mm on 8 May. In contrast, the spatial distribution of daily rainfall during other time periods shows relatively small variation, ranging from 10 mm to 40 mm.

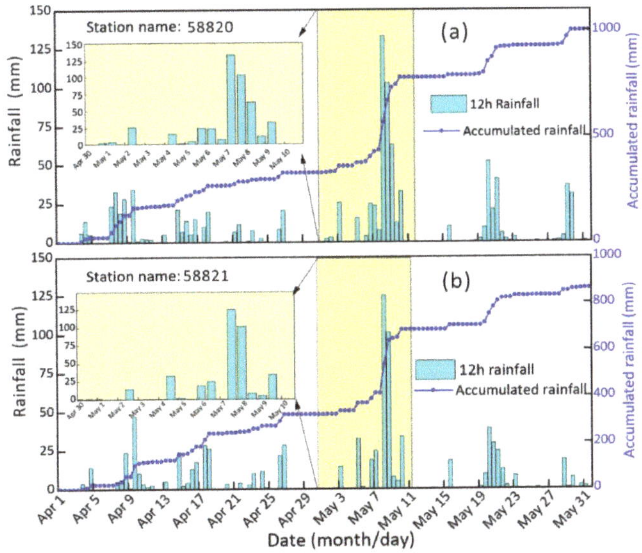

Figure 5. Precipitation data from every 12 h of two national stations in the study area from 1 April to 30 May; (**a**) the rainfall station (58,820) located in the north of the study area; (**b**) the rainfall station (58,821) located in the west of the study area.

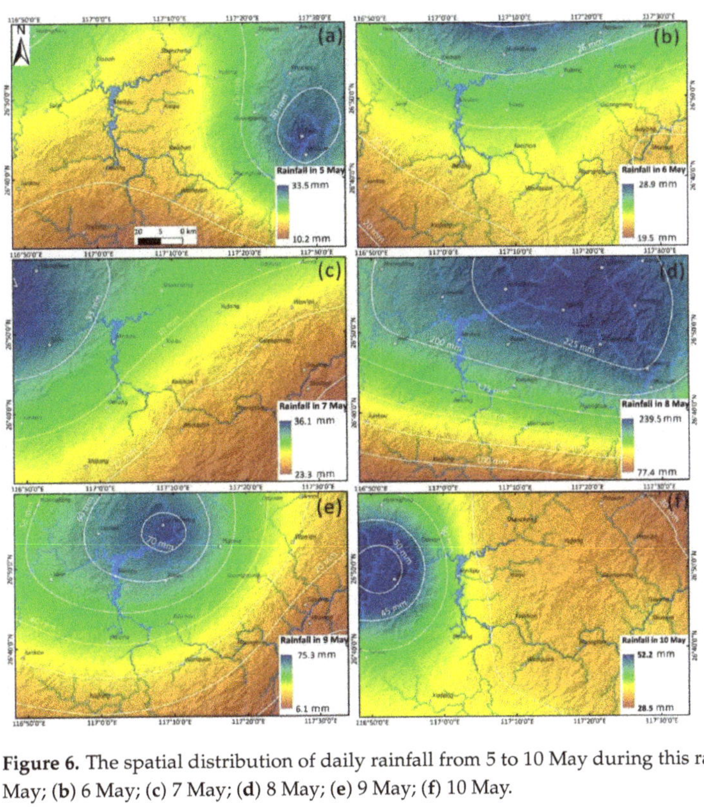

Figure 6. The spatial distribution of daily rainfall from 5 to 10 May during this rainfall event; (**a**) 5 May; (**b**) 6 May; (**c**) 7 May; (**d**) 8 May; (**e**) 9 May; (**f**) 10 May.

3.3. Data of Other Influencing Factors

Based on the distribution characteristics of landslides and influencing factors in the study area, as well as previous studies [9], we selected six factors, mainly including topography, geology, hydrology, land cover and rainfall. The ALOS PALSAR DEM data with a resolution of 12.5 m was used to calculate the hillslope gradient and slope aspect. We calculated the topographic relief based on the elevation range within a 1.0 km radius. TWI was calculated using GRASS GIS software and elevation data and drainages were derived from the DEM using ArcGIS software. The land use type data was derived from the 10 m resolution global land cover results [49]. Finally, all influencing factor layers were divided into 12.5 $\times$ 12.5 m grids and subjected to statistical analysis (Figure 7).

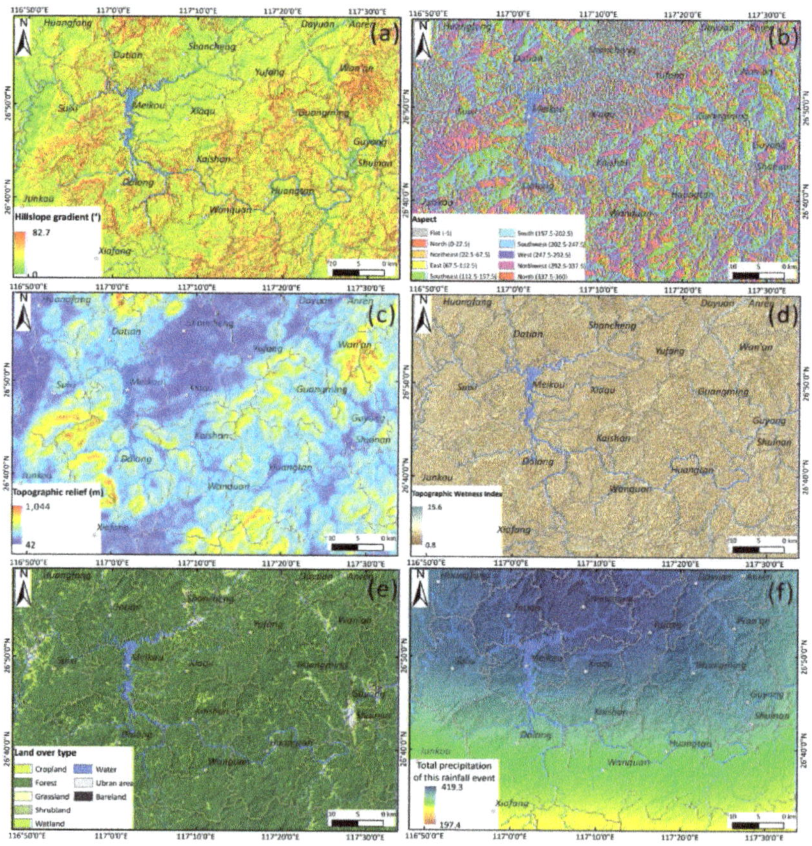

Figure 7. Map showing the spatial distribution of the influencing factors; (**a**) hillslope gradient; (**b**) aspect; (**c**) topographic relief; (**d**) topographic wetness index; (**e**) land over type; (**f**) total precipitation of this event.

3.4. TRIGRS Modelling

The TRIGRS model (transient rainfall infiltration and grid-based regional slope-stability model) is programmed by the USGS (United States Geological Survey) [50,51] and is widely used for evaluating shallow-rainfall-induced landslide susceptibility [52,53]. Specific input data are required such as rainfall, soil mechanics and hydrological characteristics of the study area [50]. After determining these parameters, using a GIS platform, the model calculates grid stability as a result of the change in transient pore water pressure of each grid during the rainfall period. Iverson [54] linearized the Richards equation solution, which serves as the foundation for infiltration models for moist beginning circumstances with steady and transient seepage components. The former is governed by the water table's initial depth and constant infiltration rate, which maintains slope stability. The latter refers to the increase in pore water pressure caused by rainfall, which can cause instability. The generalized solution in TRIGRS is

$$\psi(Z,t) = (Z-d)\beta + 2\sum_{n=1}^{N} \frac{I_{nz}}{K_s} H(t-t_n)[D_1(t-t_n)]^{\frac{1}{2}} \sum_{m=1}^{\infty}\left\{ierfc\left[\frac{(2m-1)d_{LZ}-(d_{LZ}-Z)}{2[D_1(t-t_n)]^{\frac{1}{2}}}\right] + ierfc\left[\frac{(2m-1)d_{LZ}+(d_{LZ}-Z)}{2[D_1(t-t_n)]^{\frac{1}{2}}}\right]\right\} - 2\sum_{n=1}^{N} \frac{I_{nz}}{K_s} H(t-t_{n+1})[D_1(t-t_{n+1})]^{\frac{1}{2}} \sum_{m=1}^{\infty}\left\{ierfc\left[\frac{(2m-1)d_{LZ}-(d_{LZ}-Z)}{2[D_1(t-t_{n+1})]^{\frac{1}{2}}}\right] + ierfc\left[\frac{(2m-1)d_{LZ}+(d_{LZ}-Z)}{2[D_1(t-t_{n+1})]^{\frac{1}{2}}}\right]\right\} \tag{1}$$

where ψ denotes pressure head; t is rainfall time; N is the number of rainfall time intervals; Z is the depth below the surface; d is the depth of water table;d_{LZ} indicates the impervious basement border depth; $\beta = cos^2\delta - (I_{ZLT}/K_s)$, δ is the hillslope gradient; I_{ZLT} is the constant surface flux; K_s is the saturated hydraulic conductivity; I_{nZ} is the nth time period, surface flux; $D_1 = D_0/cos^2\delta$, D_0 is the saturated hydraulic diffusivity and $H(t-t_n)$ is the Heaviside step function in which t_n is the time at the nth time interval in the rainfall sequence.

$$ierfc(\eta) = \frac{1}{\sqrt{\pi}}\exp\left(-\eta^2\right) - \eta erfc(\eta) \tag{2}$$

where $erfc(\eta)$ denotes the complementary error function.

The TRIGRS model computes infiltration (I) for each cell by adding precipitation (P) and any runoff from upslope cells (R_u). However, it is important to note that the saturated hydraulic conductivity (K_s) cannot be exceeded by infiltration. This ensures that the model accounts for the limitations of soil permeability.

$$I = P + R_u, if\ P + R_u \le K_s \tag{3}$$

$$I = K_s, if\ P + R_u > K_s \tag{4}$$

when $P + R_u$ surpasses K_s in a cell, the surplus is referred to as runoff (R_d) and it is directed to neighboring downslope cells.

$$R_d = P + R_u - K_s, if\ P + R_u - K_s \ge 0 \tag{5}$$

$$R_d = 0, if\ P + R_u - K_s < 0 \tag{6}$$

The TRIGRS model computes slope stability using an infinite-slope stability analysis (Equation (7)), as explained in Iverson [54]. In this analysis, it is the percentage of resistive basal Coulomb friction in the presence of gravitationally produced downslope. The instability of an infinite slope is characterized by basal driving stress [55]. The TRIGRS model calculates this ratio, known as the FoS, at depth Z by

$$FoS(Z,t) = \frac{tan\varphi'}{tan\delta} + \frac{c' - \psi(Z,t)\gamma_w tan\varphi'}{\gamma_s Z sin\delta cos\delta} \tag{7}$$

where c' is the soil cohesion; φ' is the friction angle of the soil; γ_s is the unit weight of soil and γ_w is the unit weight of groundwater.

To overcome the difficulties associated with manually updating several model parameters and sophisticated data processing in the standard TRIGRS model, Ma et al. [56]

posed a new TRIGRS model using Matlab® programming. This model can directly read grid data in TIF format as input and output prediction results, greatly simplifying data preparation and parameter configuration. It includes two script files, INPUT DATA.m and TRIGRS.m. The TIF input files are read by the INPUT DATA.m file, whereas TRIGRS.m is the executable program that calculates the pressure head and FoS. By computing the pressure head and FoS at various soil depths, the model provides the minimum FoS and accompanying pressure head in TIF format. More information may be found in [56]. The flow chart of this study is shown in Figure 8.

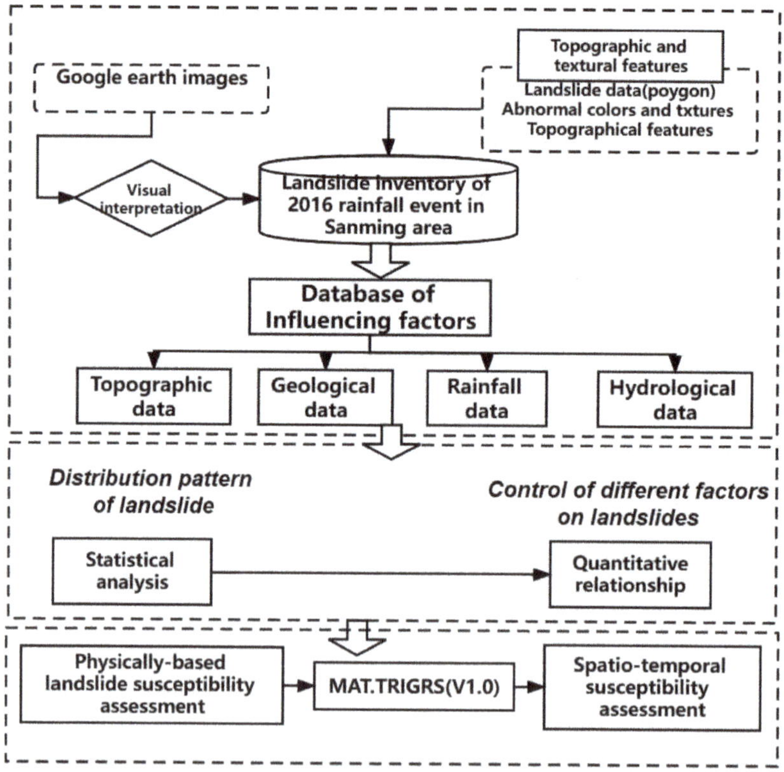

Figure 8. Flow chart of this study.

4. Results

4.1. Basic Characteristics of Rainfall-Induced Landslides

Based on the detailed landslide database of this rainfall event, it is clear that approximately 2665 landslides were triggered by this event (Figure 9a). The landslides were mostly small-scale shallow landslides with an elongated type. Among them, the largest landslide has an area about 50,000 m^2, while the smallest one is only 36 m^2, with an average area of 1070 m^2. The number of landslides with an area greater than 10,000 m^2 is 6. Approximately 21 landslides have an area between 5000–10,000 m^2 and 927 landslides have an area between 1000–5000 m^2. However, the majority of landslides (1711) possess an area of less than 1000 m^2. We used a moving window with a radius of 2.5 km and a Gaussian density kernel function to calculate the landslide number density (LND) in the study area (Figure 9b). The results show that the maximum LND reaches over 80/km^2. Spatially, the landslides are mainly concentrated in the northern part of Xiaqu town and the southern part of Yufang town.

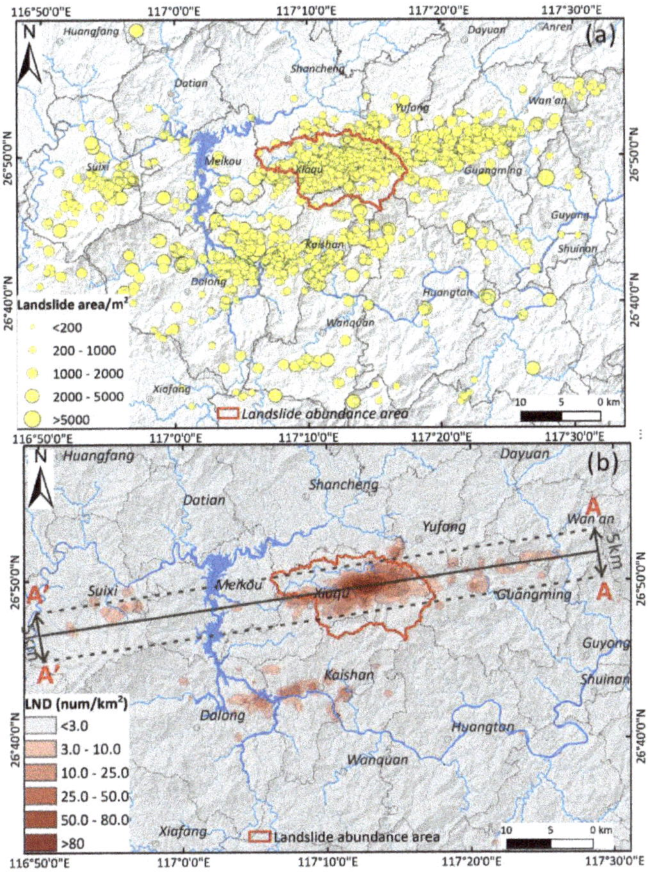

Figure 9. Spatial distribution of landslides and landslide number density (LND). (**a**) Inventory of the rainfall-induced landslides (**b**) landslide number density (LND); the red line delineates the landslide abundance area.

4.2. Correlation between Landslides and Influencing Factors

To analyze the relationship between the influencing factors and landslide occurrence, we conducted a statistical analysis of the frequency distribution of landslides and landscape (non-landslide) areas under different intervals, as well as landslide area density (LAD) under different intervals. Figure 10 shows the frequency density distribution of landslides and landscape areas under different influencing factors. Figure 11 shows the LAD distribution of six influencing factors under different intervals. Higher LAD values indicate the areas that are more prone to landslides. The results show that for elevation, most landslides are concentrated between 300–500 m. The landslide frequency density reaches a maximum of 0.22 in the elevation interval of 370–430 m. As for the hillslope gradient, most landslides are concentrated in the interval of 15–25°, with an average slope of 21.7°. Both landslide and non-landslide areas have the highest frequency density in the slope interval of 14–18° and the values are 0.15 and 0.16, respectively. The same distribution pattern can also be observed in TWI. Specifically, the landslide frequency density reaches its maximum near 4.5, with a value of 0.36. As for topographic relief, non-landslide areas are mainly concentrated between 300–500 m, while landslide areas are mainly concentrated between 150–200 m. For distance to a river, most landslides are concentrated within 800 m, which is the area most affected by river erosion. In the aspect of total rainfall, landslides are predominantly concentrated in the region with annual rainfall intervals of 320–330 mm and 380–400 mm, with landslide frequency densities of 0.1 and 0.36, respectively.

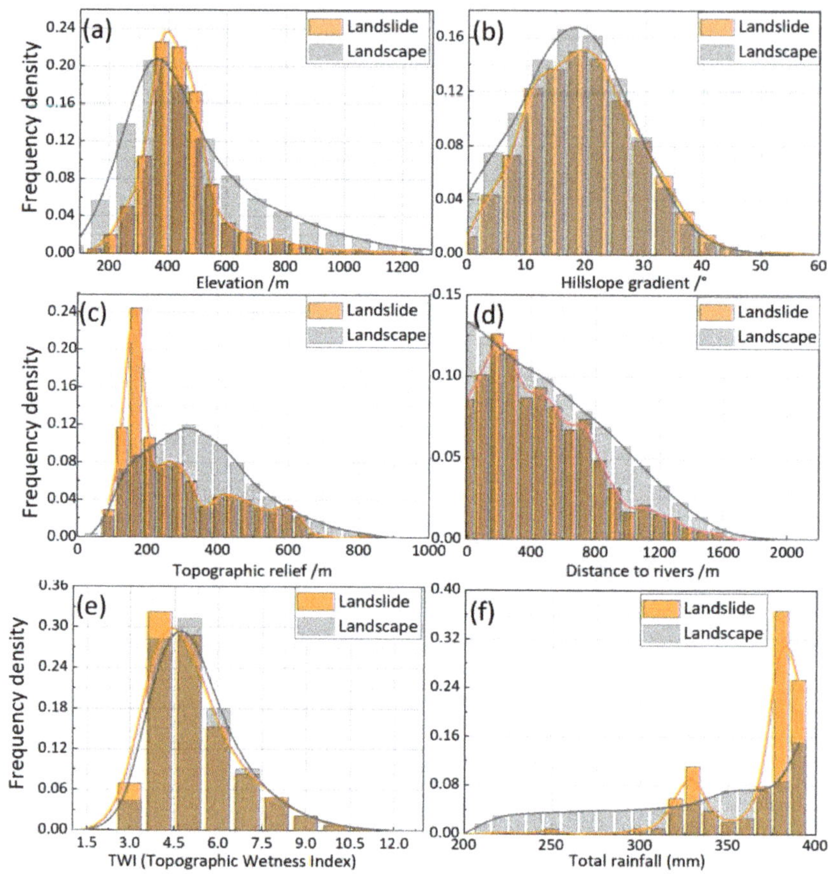

Figure 10. Frequency density estimates of landslide and landscape areas of six influencing factors; (**a**) elevation; (**b**) hillslope gradient; (**c**) topographic relief; (**d**) distance to river; (**e**) topographic wetness index (TWI); (**f**) total rainfall.

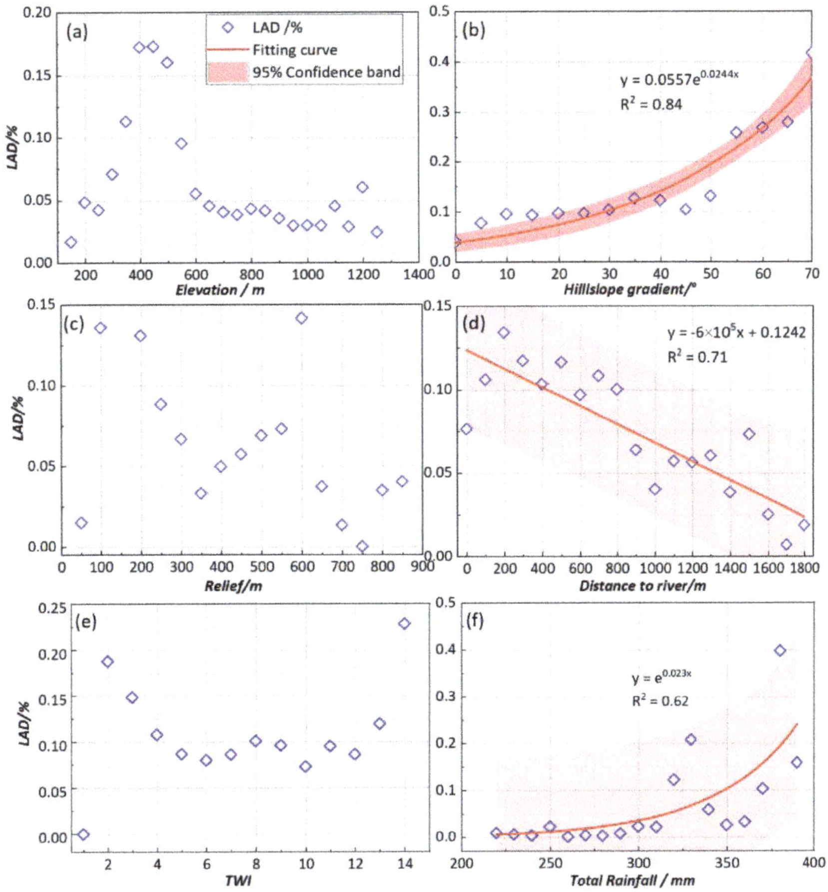

Figure 11. The relationship between the landslide areal density (LAD) and six influencing factors; (**a**) elevation; (**b**) hillslope gradient; (**c**) topographic relief; (**d**) distance to river; (**e**) topographic wetness index (TWI); (**f**) total rainfall.

Based on the statistical relationship between LAD and different influencing factors (Figure 11), it is observed that there is no significant correlation between elevation, relief, TWI and landslide abundance index. For elevation, the highest LAD (0.17%) is observed in the elevation range of 400–450 m, while the maximum LAD (0.13%) is observed in the relief range of 100–200 m and 550–650 m. By comparison, a tight correlation is seen between LAD and the other three influencing factors (i.e., hillslope gradients, distance to river and total rainfall). For hillslope gradient, the LAD increases with an increase in hillslope gradient and is described by an exponential relationship: $y = 0.0557e^{(0.0244x)}$, where x is the hillslope gradient and y is the LAD (Figure 11b). The equation indicates that with the increase in hillslope gradient, the possibility of landslide occurrence also raises. In terms of distance to a river, there is a negative linear relationship between the LAD and distance to a river: $y = -6 \times 10^5 x + 0.124$, where x represents the distance to a river and y is the LAD (Figure 11d), indicating that the LAD decreases with the increase in distance to rivers. For total rainfall, the LAD and total rainfall show an exponential relationship: $y = e^{0.023x}$, where x is the total rainfall and y is the LAD. Such a relationship demonstrates that landslides are more likely to occur in areas with high precipitation (Figure 11f).

Figure 12 shows the areal coverage (%) of various lithological types for landslide and landscape area, overlaid by the average landslide area and LAD estimated per unit. The result shows that the predominant lithology is protozoic plagioclase hornblende and granulite (Pt-p), which account for 25% of the study area, followed by Cretaceous rhyolite porphyry and glutenite (Kr-g), which account for more than 15% of the study area. Among all lithological types, Sinian mica schist and quartz schist (Sn-s) are the most prone to landslides, with over 35% of landslides occurring in 10% area. Furthermore, statistics on the average landslide area of different lithological units show that Kr-g and Pt-p have the largest average landslide area (>1400 m^2), followed by Cambrian quartz sandstone (∈), which has an average landslide area of 1200 m^2. Quaternary loose deposits (Q) have a small average landslide area, only 600 m^2.

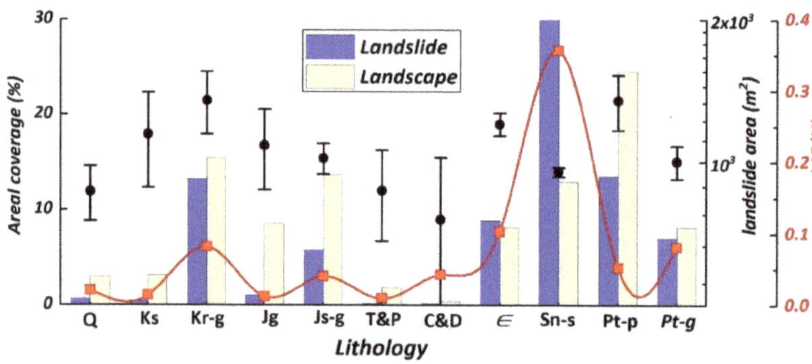

Figure 12. Areal coverage (%) of various lithological types for landslide and landscape, overlaid by average landslide area and LAD estimated per unit.

Based on the frequency density distribution and statistical analysis of landslide area density (LAD) across various aspects in both landslide and non-landslide areas, we found that the landslides on E-SE oriented slopes are highly developed, particularly within a slope aspect of 60–100°. The statistical results further revealed that landslide area density (LAD) was highest within the 50–110° range, reaching 0.17%. Overall, the eastern aspect of the area displayed a greater concentration of landslides, while the western slope exhibited lower levels of landslide development (Figure 13).

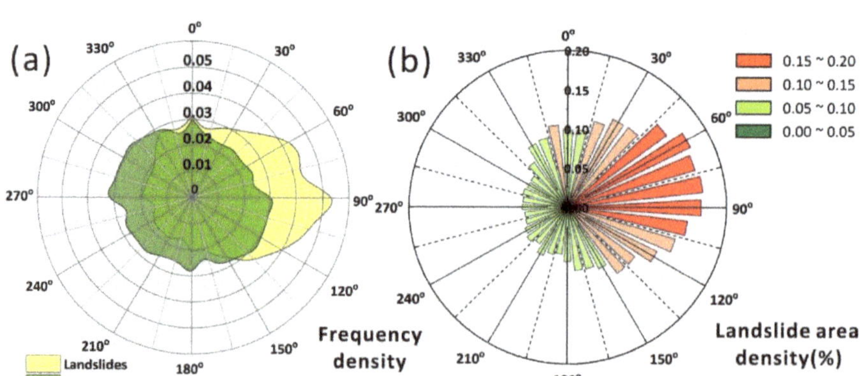

Figure 13. (**a**) The distribution of aspect within landslide and landscape areas; (**b**) correlations between aspect and LAD.

4.3. Spatio-Temporal Susceptibility Assessment

To achieve accurate landslide spatiotemporal susceptibility results using physically based models, obtaining sufficient and correct input parameter data is the foremost requirement [30,52,57,58]. The Z-model of Saulnier et al. [59] was used to evaluate the thickness of weathered soil mass. We assumed that the weathered soil mass covered by the upper layer of bedrock has a maximum thickness of 5 m and a minimum thickness of 0.5 m, as determined by previous studies [60,61]. Therefore, the estimation of soil thickness based on altitude can be calculated using Equation (8).

For landslide abundance areas, the lithology in the study area mainly includes Proterozoic granite (Pt-g), Sinian mica schist and quartz schist (Sn-s), Cambrian quartz sandstone (∈) and Quaternary loose deposits (Q). Based on previous studies [9,23,62] and rock engineering standards used in China [63], we assigned the corresponding values to hydrological and mechanical parameters, including soil cohesion (c'), internal friction angle (φ'), unit weight (γ_s) and saturated hydraulic conductivity (Ks) for different lithological types. Specific mechanical and hydrological parameter assignments for different lithologies can be found in Supplementary Materials Table S1. Otherwise, based on previous experience [44,64], saturated hydraulic diffusivity D_0 was set to $200K_s$ and the initial surface flux (I_{ZLT}) is generally less than Ks to one power or more and was often set to $I_{ZLT} = 0.01K_s$.

$$h_i = h_{max} - \left(\frac{Z_i - Z_{min}}{Z_{max} - Z_{min}}\right)(h_{max} - h_{min}) \quad (8)$$

where Z_{max} and Z_{min} refer to the maximum and minimum thicknesses of weathered soil mass, respectively. h_{max} and h_{min} are the maximum and minimum altitudes, respectively.

Figure 14 shows the predicted pictures of the FoS based on rainfall data over different time periods. From the results, it can be observed that prior to the rainfall, most of the study area had an FoS greater than 1.2. As the rainfall event began, unstable areas with an FoS less than 1.2 (shown in the red area) mainly appeared on both sides of the gullies and, by 8 am on 7 May, the unstable area within the study area rapidly increased. After 12 h of rainfall (reaching a rainfall amount of 102 mm) on 8 May at 8 am, the unstable area (shown in the red area) reached its critical value. Although several subsequent intermittent rainfall events occurred, the impact of rainfall on the change in the FoS was tiny and ignorable and the unstable area remained unchanged. We compared the prediction accuracy of the simulation results over different time periods based on the ROC curve. As shown in Supplementary Materials Figure S1, the prediction ability of the assessment results in different time periods varied between 0.68 and 0.72. Among them, the evaluation results on the 7 and 8 June had the highest prediction accuracy, around 0.72.

We conducted a statistical analysis on the temporal variations of the FoS results for different hillslope gradients during different time periods. Figure 15a shows the variations in the FoS of the grids with gradients less than 30°, where the majority of the FoS ranged from 1.8 to 2.4 with an average of approximately 2.1 before the onset of rainfall. As heavy rainfall occurred, the FoS gradually decreased and the average FoS reached around 1.95 at 8:00 on 7 May. Subsequently, the occurrence of intense rainfall on 8 May resulted in a significant decrease in the FoS for the grids with gradients less than 30°, with the average FoS maintained at around 1.85. Despite the persistent rainfall in the later period, the overall variation in the FoS for most grids was small, with an average of approximately 1.8. Similar phenomena were observed for the grids with gradients greater than 30° (Figure 15b). Prior to the rainfall, the FoS for these grids were basically distributed between 1.4 and 1.7, with an average of approximately 1.6. As rainfall increased, the FoS continued to decrease and after the intense rainfall event on 8 May, most of the shallow weathered soil layers became saturated. As a result, the FoS reached critical values and maintained between 1.1 and 1.4, with an average of approximately 1.3. Although the rainfall continued to occur, the impact of rainfall on the FoS was small due to the saturated state of the soil layers and the FoS for the grids remained stable.

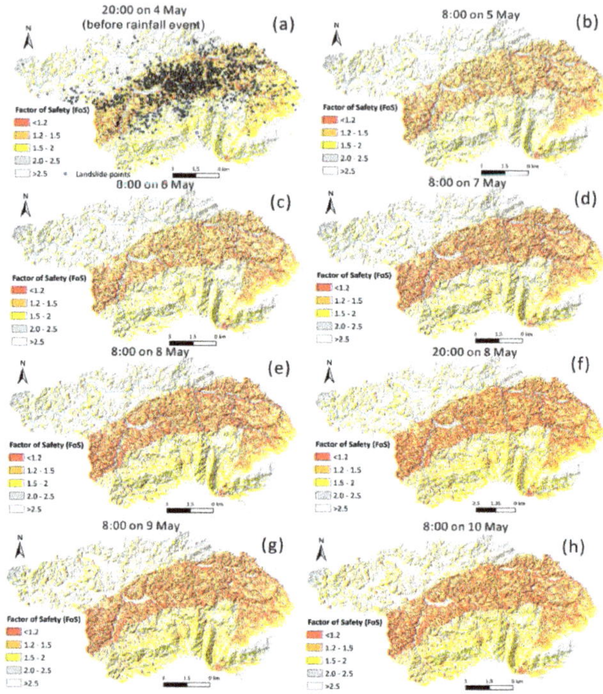

Figure 14. Conditions for slope stability, as measured by the factor of safety (FoS) at different times during the 2016 rainfall event; (**a**) 20:00 on 4 May (UTC + 8, before rainfall event); (**b**) 8:00 on 5 May (UTC + 8); (**c**) 8:00 on 6 May (UTC + 8); (**d**) 8:00 on 7 May (UTC + 8); (**e**) 8:00 on 8 May (UTC + 8); (**f**) 20:00 on 8 May (UTC + 8); (**g**) 8:00 on 9 May (UTC + 8); (**h**) 8:00 on 10 May (UTC + 8).

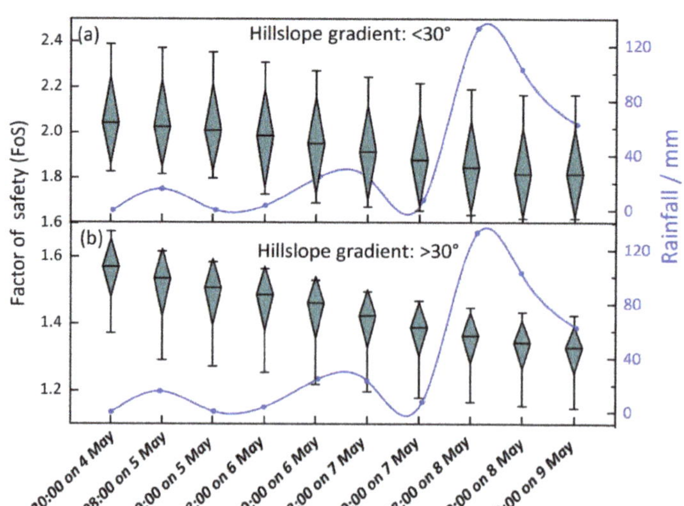

Figure 15. FoS results for various hillslope gradient intervals at different rainfall times; (**a**) hillslope gradient: <30°; (**b**) hillslope gradient: >30°.

5. Discussion

Topography, geological features and rainfall characteristics are considered to be essential factors influencing the occurrence of rainfall-induced landslides [65–67]. It is generally accepted that steeper terrain, weaker rock strength and higher rainfall amounts increase the occurrence likelihood of landslides [17,68,69]. To better understand the spatial distribution of the rainfall-induced landslides with different elevations, hillslope gradients, topographic relief, lithological types and rainfall characteristics, the swath profiles (EEN-WWS) with a width of 10 km are presented (Figure 16). The spatial distribution of landslides shows that the majority of the landslides are concentrated in Xiaqu town, which has the maximum landslide number density (100/km^2). In terms of topography, the Xiaqu area belongs to the transitional zone from high to low altitude, with an elevation generally ranging from 400 to 600 m. The hillslope gradients in this area are small, with a range from 10 to 30° and an average gradient of 20°. Regarding the rainfall distribution, the Xaiqu area is located in the region with the highest precipitation of nearly 400 mm. Thus, the spatial distribution of landslides is strongly controlled by the rainfall characteristics (Figures 16 and 10f). Otherwise, the primary rock type in the Xaiqu area is Sinian mica schist and quartz schist (Sn-s) and the statistical results show that over 35% of landslides are distributed in the area with this lithological type (Figure 12). We suggest that the schist belongs to metamorphic rock, which is influenced by geological structure, tectonics and mineral composition. When schist is in contact with water, characteristics such as creep, mechanical anisotropy, softening and deterioration may be observed. Due to these unique properties, schist in mountainous areas often experiences frequent landslides [70,71]. For the study area, mica schist and quartz schist have well-developed cleavage and high mica content. Meanwhile, due to the long-term physical weathering, the rock mass on the surface of the bedrock is fragmented, with highly developed fissures and locally filled mud. Therefore, this lithological type is prone to the development of weak structural layers after long-term immersion of rainfall, thereby providing a natural sliding surface for landslides [72]. Additionally, due to the development of cracks in the weathered rock and soil, long-term rainfall can penetrate through the cracks and increase pore water pressure in the rock mass and the decreased shear resistance caused by increased self-weight of the rock mass is also one of the main reasons for the occurrence of landslides.

In the southeast coastal area, orographic amplification of rainfall and the projection of rainfall-vector on hillslopes might result in greater rainfall in the windward hillslopes, resulting in a higher incidence of landslides on the hillslope scale [73]. During the summer months (June and July), the Sanming area is dominated by the southeast monsoon, which is influenced by the monsoon depression and tropical cyclone. The distribution of landslides during this rainfall event indicates that hillslopes facing southeast and east are more susceptible to collapse than those facing northwest-north (Figure 13). This phenomenon can be mainly attributed to the fact that the south-oriented slopes are predominantly windward, leading to greater rainfall and splash erosion.

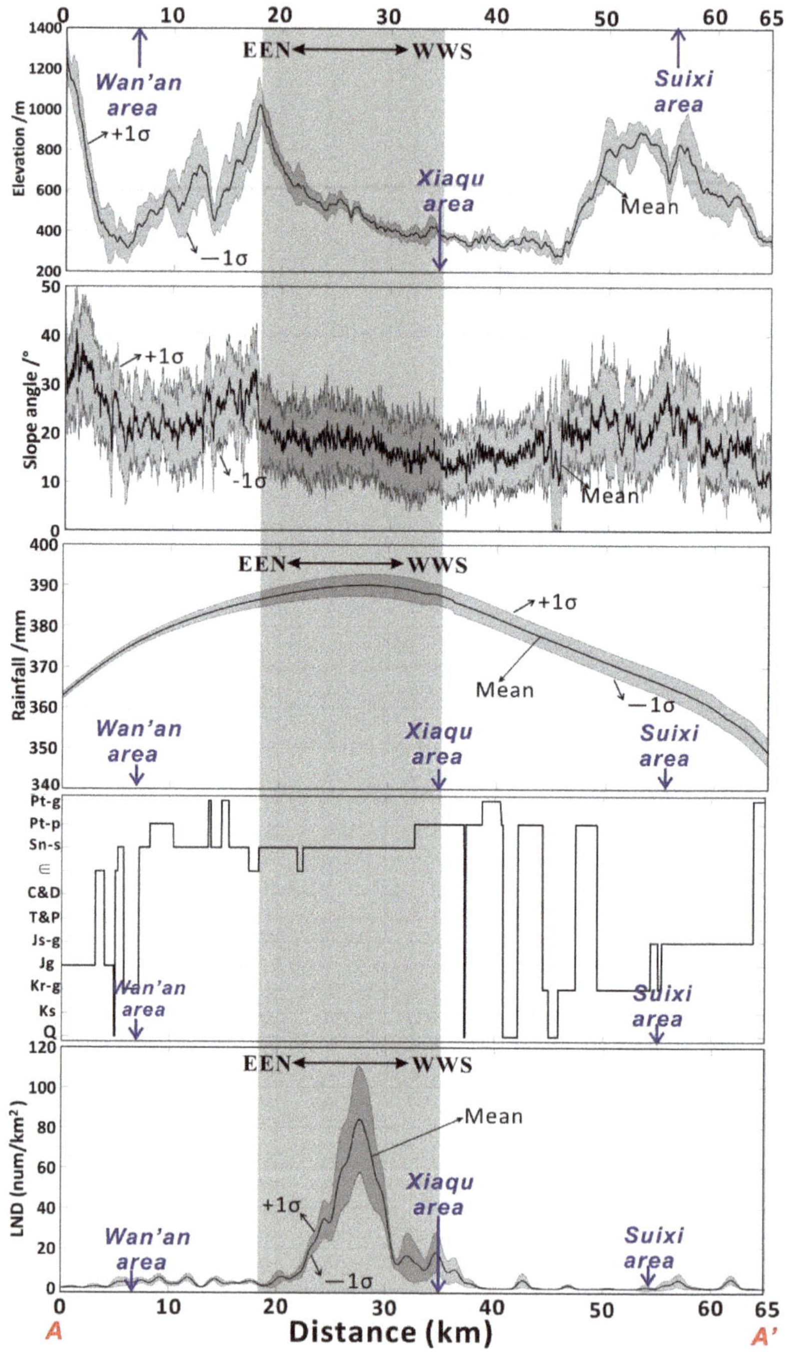

Figure 16. Comparison of longitudinal (EEN-WWS) swath profiles of elevation, slope angle, rainfall, lithology and LND; the location of the swath profile is shown in Figure 8.

In addition, we collected the landslide data and the corresponding population statistics from different towns (Figure 17). According to the statistical results, we found that the landslide size in each town was similar, with most landslides ranging from 600 to 1000 m^2. Hangtan, Guangming and Yunkou towns had relatively larger average landslide areas, reaching 1000 m^2 (Figure 17). In terms of the relationship between landslide numbers and population distribution in each town, we can observe that the towns with the most concentrated landslides were areas with the lowest populations (Figure 17). For example, more than 1000 landslides occurred in Xiaqu town but the local population is only 7000. Yufang town had nearly 600 landslides, but the population is only around 2700. On the other hand, areas with fewer landslides corresponded to regions with more concentrated populations, such as Guyong and Shancheng towns, which had populations of 50,000 and 45,000, respectively, but only around 50 landslides occurred in each town. By comparing the topography, geomorphology and rainfall characteristics of these towns, we found that Yufang and Xiaqu towns were located in an area with the highest rainfall intensity and the most susceptible strata for landsliding. In contrast, Guyong and Shancheng towns had gentler terrain, with most hillslope gradients being less than 15°, making them low-susceptibility areas for landsliding. Although these areas are considered low-susceptibility areas, the loss of people's lives and property is still possible due to the landslides triggered by typhoons and heavy rainfall events. Conversely, although Yufang and Xiaqu towns are considered high-susceptibility areas for landsliding, their low population density may result in a lower impact on residents and public facilities.

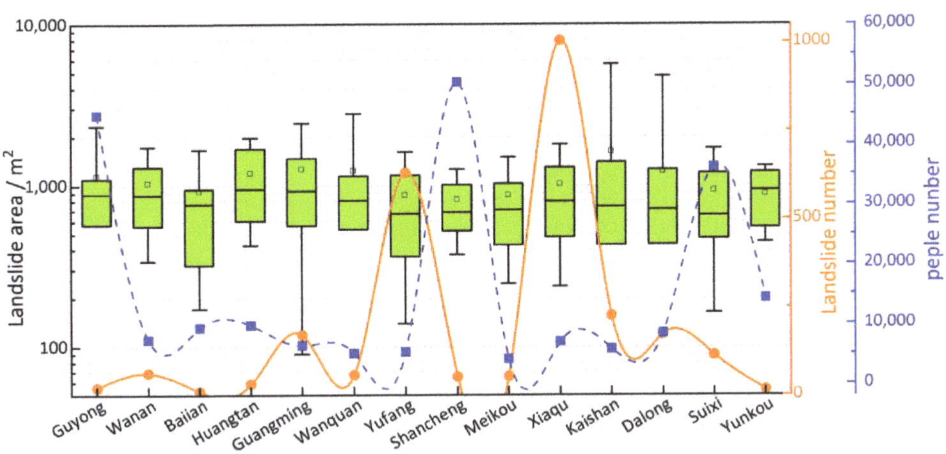

Figure 17. Landslide scale, number of landslides and the corresponding population distribution in different towns of the study area.

We calculated the antecedent precipitation index (API) of the study area with different statistical steps (5, 10, 15 days) based on the daily precipitation data (Figure 18). The relevant information about the API is detailed in the Supplementary Materials. From 1 April to 16 May, three API curves were all in the high-value range, with peaks occurring around 10 April, 20 April and 10 May, respectively. The first two peaks had relatively smaller API values (90 mm and 80 mm), while the largest peak appeared around 10 May. The Sanming area witnessed a strong rainstorm from 5 May to 10 May, with a total precipitation of 250 mm within 24 h and a maximum rainfall intensity of 56.6 mm/h. During this time, the API curves increased significantly and reached their pinnacle (250–280 mm). Subsequently, as the rainfall events ceased, the API rapidly decreased. In summary, the preceding rainfall events with low intensity (before 5 May) increased the soil water content and the short duration of heavy rainfall on 8 May caused the rapid saturation of soil water content, thus leading to the occurrence of massive landslides in the Sanming area.

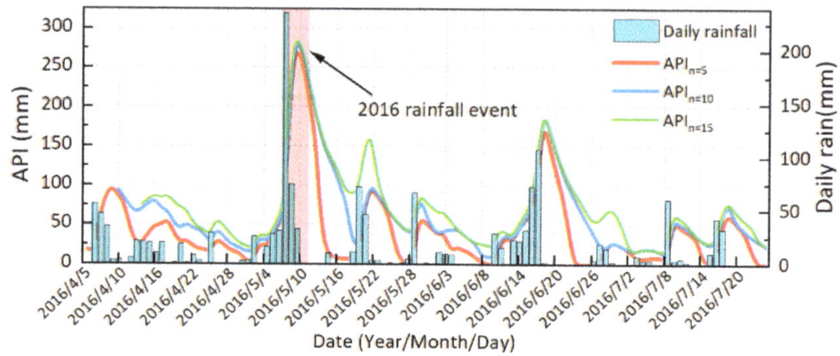

Figure 18. The rainfall data and calculated antecedent precipitation index (API) of this rainfall event.

In physically based models, accurate input information is essential to obtain accurate simulation results [58,74]. However, obtaining accurate input parameters for rock and soil is a challenging task in actual situations, so input data differ from the actual situation to some extent. For example, there is spatial heterogeneity in the mechanical parameters of different positions on slope body and the mechanical and hydrological parameters may be different at various depths [75]. In addition, the evaluation scale based on physically based models is a large area, which is limited by sampling manpower and material resources, resulting in a certain degree of uncertainty in the input parameters of the modelling [76,77]. The evaluation results based on MAT.TRIGRS (V1.0) show that the overall prediction accuracy is around 0.7 (Figure S1), which indicate that the prediction accuracy is reliable. However, due to the subjectivity of the input parameters, there are still some errors between the overall evaluation results and the actual landslides (Figure 14) and some areas with less landsliding are predicted as high-susceptibility areas. Therefore, how to obtain accurate input parameters on a regional scale remains an important constraint for rainfall-induced landslide susceptibility assessment based on physically based models.

6. Conclusions

The objective of this study was to examine the landslides that occurred during the heavy rainfall event from 5 to 10 May 2016 in the Sanming area and to identify the geological, geomorphological and hydrometeorological factors that contributed to landslide hazards. The rainfall event resulted in around 2700 landslides covering a total area of 2.8 km^2, predominantly in Xiaqu town. We analyzed the landslide distribution pattern and its relationship with various factors. Our findings suggest that elevation, relief and topographic wetness index (TWI) were not significantly correlated with the landslide abundance index, while hillslope gradient, distance to a river and total rainfall played a significant role in the occurrence of landslides. The Sinian mica schist and quartz schist (Sn-s) lithological types were found to be the most susceptible to landslides, with more than 35% of landslides occurring in only 10% of the area. This indicates that landslides are more likely to occur in Sn-s strata during rainfall events. Additionally, the study showed that southeast- and east-facing hillslopes are more susceptible to collapse than northwest-north-oriented hillslopes due to the influence of the summer southeast monsoon. To further examine the susceptibility of the area to landslides, we used the MAT.TRIGRS (V1.0) tool for spatio-temporal susceptibility assessment. The results showed that unstable areas mainly appeared on both sides of the gullies as the rainfall event began. By 8 am on 7 May, the unstable area within the study area rapidly increased. Our numerical simulations indicate that the preceding rainfall events with low intensity (before 5 May) increased the soil water content and the short duration of heavy rainfall on 8 May caused the rapid saturation of soil water content, thus leading to the occurrence of massive landslides in the Sanming area. Based on the findings, we recommend that more attention should be given to Sn-s lithological types in the region during the planning and implementation of landslide hazard mitigation measures. Additionally, our study highlights the importance of considering the influence of monsoons on landslide susceptibility in the area. Future studies could consider using high-resolution satellite images and other remote sensing techniques to monitor and map the landslide distribution in the study area, as well as to assess the effectiveness of the mitigation measures implemented.

Supplementary Materials: The following supporting information can be downloaded at: https://www.mdpi.com/article/10.3390/rs15112738/s1. Figure S1. Prediction curves of simulation results over different time periods; Table S1. Mechanical and hydrological parameters for different lithological types in landslide abundance areas. References [78–83] are cited in the Supplementary Materials.

Author Contributions: The research concept was proposed by C.X. who also contributed to the data curation and analysis. S.M. designed the research framework, processed the relevant data and drafted the manuscript. X.S. participated in the data analysis and contributed to the manuscript revisions. All authors have read and agreed to the published version of the manuscript.

Funding: This study was supported by the National Key Research and Development Program of China (2021YFB3901205).

Data Availability Statement: The source code of MAT.TRIGRS (V1.0) (https://doi.org/10.1016/j.nhres.2021.11.001, Ma et al., 2022 [56]) is available at https://github.com/Masiyuanlandslides/MAT.TRIGRS-V1.0- (accessed on 10 May 2023).

Acknowledgments: We thank Google Earth for the free-to-access satellite images used in this study. We would like to thank three anonymous reviewers for their comments, which are very useful for improving the quality of the MS.

Conflicts of Interest: The authors declare that they have no known competing financial interest or personal relationship that could have appeared to influence the work reported in this paper.

References

1. Salciarini, D.; Godt, J.W.; Savage, W.Z.; Conversini, P.; Baum, R.L.; Michael, J.A. Modeling regional initiation of rainfall-induced shallow landslides in the eastern Umbria Region of central Italy. *Landslides* **2006**, *3*, 181–194. [CrossRef]
2. Montrasio, L.; Valentino, R.; Losi, G.L. Rainfall-induced shallow landslides: A model for the triggering mechanism of some case studies in Northern Italy. *Landslides* **2009**, *6*, 241–251. [CrossRef]
3. Bordoni, M.; Vivaldi, V.; Lucchelli, L.; Ciabatta, L.; Brocca, L.; Galve, J.; Meisina, C. Development of a data-driven model for spatial and temporal shallow landslide probability of occurrence at catchment scale. *Landslides* **2020**, *18*, 1209–1229. [CrossRef]
4. Bălteanu, D.; Chendeş, V.; Sima, M.; Enciu, P. A country-wide spatial assessment of landslide susceptibility in Romania. *Geomorphology* **2010**, *124*, 102–112. [CrossRef]
5. Brunetti, M.; Peruccacci, S.; Rossi, M.; Luciani, S.; Valigi, D.; Guzzetti, F. Rainfall thresholds for the possible occurrence of landslides in Italy. *Nat. Hazards Earth Syst. Sci.* **2010**, *10*, 447–458. [CrossRef]
6. Petley, D. Global patterns of loss of life from landslides. *Geology* **2012**, *40*, 927–930. [CrossRef]
7. Wang, L.; Xiao, T.; Liu, S.; Zhang, W.; Yang, B.; Chen, L. Quantification of model uncertainty and variability for landslide displacement prediction based on Monte Carlo simulation. *Gondwana Res.* **2023**. [CrossRef]
8. Ma, T.; Li, C.; Lu, Z.; Bao, Q. Rainfall intensity–duration thresholds for the initiation of landslides in Zhejiang Province, China. *Geomorphology* **2015**, *245*, 193–206. [CrossRef]
9. Ma, S.; Shao, X.; Xu, C. Characterizing the Distribution Pattern and a Physically Based Susceptibility Assessment of Shallow Landslides Triggered by the 2019 Heavy Rainfall Event in Longchuan County, Guangdong Province, China. *Remote Sens.* **2022**, *14*, 4257. [CrossRef]
10. Marc, O.; Gosset, M.; Saito, H.; Uchida, T.; Malet, J.-P. Spatial patterns of storm-Induced landslides and their relation to rainfall anomaly maps. *Geophys. Res. Lett.* **2019**, *46*, 11167–11177. [CrossRef]
11. Broeckx, J.; Vanmaercke, M.; Duchateau, R.; Poesen, J. A data-based landslide susceptibility map of Africa. *Earth-Sci. Rev.* **2018**, *185*, 102–121. [CrossRef]
12. Bălteanu, D.; Micu, M.; Jurchescu, M.; Malet, J.-P.; Sima, M.; Kucsicsa, G.; Dumitrică, C.; Petrea, D.; Mărgărint, M.C.; Bilaşco, Ş.; et al. National-scale landslide susceptibility map of Romania in a European methodological framework. *Geomorphology* **2020**, *371*, 107432. [CrossRef]

13. Dai, L.; Scaringi, G.; Fan, X.; Yunus, A.P.; Liu-Zeng, J.; Xu, Q.; Huang, R. Coseismic Debris Remains in the Orogen Despite a Decade of Enhanced Landsliding. *Geophys. Res. Lett.* **2021**, *48*, e2021GL095850. [CrossRef]
14. Xiong, J.; Chen, M.; Tang, C. Long-term changes in the landslide sediment supply capacity for debris flow occurrence in Wenchuan County, China. *Catena* **2021**, *203*, 105340. [CrossRef]
15. Kirschbaum, D.; Stanley, T. Satellite-based assessment of rainfall-triggered landslide hazard for situational awareness. *Earth's Future* **2018**, *6*, 505–523. [CrossRef]
16. Emberson, R.; Kirschbaum, D.; Amatya, P.M.; Tanyas, H.; Marc, O. Insights from the topographic characteristics of a large global catalog of rainfall-induced landslide event inventories. *Nat. Hazards Earth Syst. Sci.* **2022**, *22*, 1129–1149. [CrossRef]
17. Marc, O.; Stumpf, A.; Malet, J.P.; Gosset, M.; Uchida, T.; Chiang, S.-H. Initial insights from a global database of rainfall-induced landslide inventories: The weak influence of slope and strong influence of total storm rainfall. *Earth Surf. Dyn.* **2018**, *6*, 903–922. [CrossRef]
18. Chang, K.-t.; Chiang, S.-h.; Chen, Y.-C.; Mondini, A. Modeling the spatial occurrence of shallow landslides triggered by typhoons. *Geomorphology* **2014**, *208*, 137–148. [CrossRef]
19. Hirota, K.; Konagai, K.; Sassa, K.; Dang, K.; Yoshinaga, Y.; Wakita, E.K. Landslides triggered by the West Japan Heavy Rain of July 2018, and geological and geomorphological features of soaked mountain slopes. *Landslides* **2019**, *16*, 189–194. [CrossRef]
20. Bessette-Kirton, E.; Cerovski-Darriau, C.; Schulz, W.; Coe, J.; Kean, J.; Godt, J.; Thomas, M.; Hughes, K. Landslides triggered by Hurricane Maria: Assessment of an extreme event in Puerto Rico. *GSA Today* **2019**, *29*, 4–10. [CrossRef]
21. Van Westen, C.J.; Zhang, J. Landslides and Floods Triggered by Hurricane Maria (18 September 2017) in Dominica, Digital or Visual Products, UNITAR-UNOSAT. Available online: https://www.unitar.org/unosat/node/44/2762 (accessed on 5 April 2023).
22. Saadatkhah, N.; Kassim, A.; Lee, L.M.; Rahnamarad, J. Spatiotemporal regional modeling of rainfall-induced slope failure in Hulu Kelang, Malaysia. *Environ. Earth Sci.* **2015**, *73*, 8425–8441. [CrossRef]
23. Chen, Y.-l.; Chen, D.-h.; Li, Z.-c.; Huang, J.-b. Preliminary studies on the dynamic prediction method of rainfall-triggered landslide. *J. Mt. Sci.* **2016**, *13*, 1735–1745. [CrossRef]
24. Merghadi, A.; Yunus, A.P.; Dou, J.; Whiteley, J.; ThaiPham, B.; Bui, D.T.; Avtar, R.; Abderrahmane, B. Machine learning methods for landslide susceptibility studies: A comparative overview of algorithm performance. *Earth-Sci. Rev.* **2020**, *207*, 103225. [CrossRef]
25. Guo, Z.; Torra, O.; Hürlimann, M.; Abancó, C.; Medina, V. FSLAM: A QGIS plugin for fast regional susceptibility assessment of rainfall-induced landslides. *Environ. Model. Softw.* **2022**, *150*, 105354. [CrossRef]
26. Melillo, M.; Brunetti, M.T.; Peruccacci, S.; Gariano, S.L.; Roccati, A.; Guzzetti, F. A tool for the automatic calculation of rainfall thresholds for landslide occurrence. *Environ. Model. Softw.* **2018**, *105*, 230–243. [CrossRef]
27. Guzzetti, F.; Gariano, S.L.; Peruccacci, S.; Brunetti, M.T.; Marchesini, I.; Rossi, M.; Melillo, M. Geographical landslide early warning systems. *Earth-Sci. Rev.* **2020**, *200*, 102973. [CrossRef]
28. Guzzetti, F.; Peruccacci, S.; Rossi, M.; Stark, C. Rainfall thresholds for the initiation of landslides in Central and Southern Europe. *Meteorol. Atmos. Phys.* **2007**, *98*, 239–267. [CrossRef]
29. Reichenbach, P.; Rossi, M.; Malamud, B.D.; Mihir, M.; Guzzetti, F. A review of statistically-based landslide susceptibility models. *Earth-Sci. Rev.* **2018**, *180*, 60–91. [CrossRef]
30. Baumann, V.; Bonadonna, C.; Cuomo, S.; Moscariello, M.; Biass, S.; Pistolesi, M.; Gattuso, A. Mapping the susceptibility of rain-triggered lahars at Vulcano island (Italy) combining field characterization, geotechnical analysis, and numerical modelling. *Nat. Hazards Earth Syst. Sci.* **2019**, *19*, 2421–2449. [CrossRef]
31. Tran, T.V.; Lee, G.; An, H.; Kim, M. Comparing the performance of TRIGRS and TiVaSS in spatial and temporal prediction of rainfall-induced shallow landslides. *Environ. Earth Sci.* **2017**, *76*, 315. [CrossRef]
32. Hong, Y.; Adler, R.; Huffman, G. Evaluation of the potential of NASA multi-satellite precipitation analysis in global landslide hazard assessment. *Geophys. Res. Lett.* **2006**, *332*. [CrossRef]
33. Nowicki Jessee, M.A.; Hamburger, M.W.; Allstadt, K.; Wald, D.J.; Robeson, S.M.; Tanyas, H.; Hearne, M.; Thompson, E.M. A global empirical model for near-real-time assessment of seismically induced landslides. *J. Geophys. Res. Earth Surf.* **2019**, *123*, 1835–1859. [CrossRef]
34. Fell, R.; Corominas, J.; Bonnard, C.; Cascini, L.; Leroi, E.; Savage, W. Guidelines for landslide susceptibility, hazard and risk zoning for land use planning. *Eng. Geol.* **2008**, *102*, 85–98. [CrossRef]
35. Sarkar, S.; Roy, A.K.; Raha, P. Deterministic approach for susceptibility assessment of shallow debris slide in the Darjeeling Himalayas, India. *Catena* **2016**, *142*, 36–46. [CrossRef]
36. Schiliro, L.; Montrasio, L.; Scarascia Mugnozza, G. Prediction of shallow landslide occurrence: Validation of a physically-based approach through a real case study. *Sci. Total Environ.* **2016**, *569–570*, 134–144. [CrossRef]
37. Montgomery, D.R.; Dietrich, W.E.; Torres, R.; Anderson, S.P.; Heffner, J.T.; Loague, K. Hydrologic response of a steep, unchanneled valley to natural and applied rainfall. *Water Resour. Res.* **1997**, *33*, 91–109. [CrossRef]
38. Pack, R.; Tarboton, D.; Goodwin, C.N. SINMAP 2.0—A Stability Index Approach to Terrain Stability Hazard Mapping, User's Manual. Produced in ArcView Avenue and C++ for Forest Renewal B.C. under Research Contract No: PA97537-0RE. Available online: https://digitalcommons.usu.edu/cee_facpub/16/ (accessed on 21 May 2023).
39. Montrasio, L.; Valentino, R.; Meisina, C. Soil saturation and stability analysis of a test site slope using the shallow landslide instability prediction (SLIP) model. *Geotech. Geol. Eng.* **2018**, *36*, 2331–2342. [CrossRef]

40. Zhang, K.; Xue, X.; Hong, Y.; Gourley, J.J.; Lu, N.; Wan, Z.; Hong, Z.; Wooten, R. iCRESTRIGRS: A coupled modeling system for cascading flood–landslide disaster forecasting. *Hydrol. Earth Syst. Sci.* **2016**, *20*, 5035–5048. [CrossRef]
41. Rossi, G.; Catani, F.; Leoni, L.; Segoni, S.; Tofani, V. HIRESSS: A physically based slope stability simulator for HPC applications. *Nat. Hazards Earth Syst. Sci.* **2013**, *13*, 151–166. [CrossRef]
42. Salvatici, T.; Tofani, V.; Rossi, G.; D'Ambrosio, M.; Tacconi Stefanelli, C.; Masi, E.B.; Rosi, A.; Pazzi, V.; Vannocci, P.; Petrolo, M.; et al. Application of a physically based model to forecast shallow landslides at a regional scale. *Nat. Hazards Earth Syst. Sci.* **2018**, *18*, 1919–1935. [CrossRef]
43. Baum, R.L.; Godt, J.W.; Savage, W.Z. Estimating the timing and location of shallow rainfall-induced landslides using a model for transient, unsaturated infiltration. *J. Geophys. Res. F Earth Surf.* **2010**, *115*. [CrossRef]
44. He, J.; Qiu, H.; Qu, F.; Hu, S.; Yang, D.; Shen, Y.; Zhang, Y.; Sun, H.; Cao, M. Prediction of spatiotemporal stability and rainfall threshold of shallow landslides using the TRIGRS and Scoops3D models. *Catena* **2021**, *197*, 104999. [CrossRef]
45. Lee, J.H.; Park, H.J. Assessment of shallow landslide susceptibility using the transient infiltration flow model and GIS-based probabilistic approach. *Landslides* **2015**, *13*, 885–903. [CrossRef]
46. Saadatkhah, N.; Kassim, A.; Lee, L.M. Hulu Kelang, Malaysia regional mapping of rainfall-induced landslides using TRIGRS model. *Arab. J. Geosci.* **2014**, *8*, 3183–3194. [CrossRef]
47. Sato, H.; Harp, E. Interpretation of earthquake-induced landslides triggered by the 12 May 2008, M7.9 Wenchuan earthquake in the Beichuan area, Sichuan Province, China using satellite imagery and Google Earth. *Landslides* **2009**, *6*, 153–159. [CrossRef]
48. Tian, Y.; Owen, L.A.; Xu, C.; Ma, S.; Li, K.; Xu, X.; Figueiredo, P.M.; Kang, W.; Guo, P.; Wang, S.; et al. Landslide development within 3 years after the 2015 Mw 7.8 Gorkha earthquake, Nepal. *Landslides* **2020**, *17*, 1251–1267. [CrossRef]
49. Gong, P.; Liu, H.; Zhang, M.; Li, C.; Wang, J.; Huang, H.; Clinton, N.; Ji, L.; Li, W.; Bai, Y.; et al. Stable classification with limited sample: Transferring a 30-m resolution sample set collected in 2015 to mapping 10-m resolution global land cover in 2017. *Sci. Bull.* **2019**, *64*, 370–373. [CrossRef]
50. Baum, R.L.; Savage, W.Z.; Godt, J.W. *TRIGRS—A Fortran Program for Transient Rainfall Infiltration and Grid-Based Regional Slope-Stability Analysis, Version 2.0*; US Geological Survey (USGS): Reston, VA, USA, 2008.
51. Alvioli, M.; Baum, R.L. Parallelization of the TRIGRS model for rainfall-induced landslides using the message passing interface. *Environ. Model. Softw.* **2016**, *81*, 122–135. [CrossRef]
52. Weidner, L.; Oommen, T.; Escobar-Wolf, R.; Sajinkumar, K.S.; Samuel, R.A. Regional-scale back-analysis using TRIGRS: An approach to advance landslide hazard modeling and prediction in sparse data regions. *Landslides* **2018**, *15*, 2343–2356. [CrossRef]
53. Tran, T.V.; Alvioli, M.; Lee, G.; An, H.U. Three-dimensional, time-dependent modeling of rainfall-induced landslides over a digital landscape: A case study. *Landslides* **2018**, *15*, 1071–1084. [CrossRef]
54. Iverson, R.M. Landslide triggering by rain infiltration. *Water Resour. Res.* **2000**, *36*, 1897–1910. [CrossRef]
55. Godt, J.; Baum, R.L.; Savage, W.Z.; Salciarini, D.; Schulz, W.; Harp, E.L. Transient deterministic shallow landslide modeling: Requirements for susceptibility and hazard assessments in a GIS framework. *Eng. Geol.* **2008**, *102*, 214–226. [CrossRef]
56. Ma, S.; Shao, X.; Xu, C.; He, X.; Zhang, P. MAT.TRIGRS (V1.0): A new open-source tool for predicting spatiotemporal distribution of rainfall-induced landslides. *Nat. Hazards Res.* **2021**, *1*, 161–170. [CrossRef]
57. Harp, E.; Michael, J.; Laprade, W. Shallow landslide hazard map of Seattle, Washington. *Rev. Eng. Geol.* **2008**, *20*, 67–82. [CrossRef]
58. Vieira, B.C.; Fernandes, N.F.; Augusto Filho, O.; Martins, T.D.; Montgomery, D.R. Assessing shallow landslide hazards using the TRIGRS and SHALSTAB models, Serra do Mar, Brazil. *Environ. Earth Sci.* **2018**, *77*, 260. [CrossRef]
59. Saulnier, G.-M.; Beven, K.; Obled, C. Including spatially variable soil depths in TOPMODEL. *J. Hydrol.* **1997**, *202*, 158–172. [CrossRef]
60. Bai, H.; Feng, W.; Yi, X.; Fang, H.; Wu, Y.; Deng, P.; Dai, H.; Hu, R. Group-occurring landslides and debris flows caused by the continuous heavy rainfall in June 2019 in Mibei Village, Longchuan County, Guangdong Province, China. *Nat. Hazards* **2021**, *108*, 3181–3201. [CrossRef]
61. Xu, Y. *Research on Evolvement Process and Failure Mechanism of Pseudo-Soil Slope under Heavy Rainfall Action*; China University of Geosciences: Wuhan, China, 2014.
62. Zhao, Y. *Study on the Disaster Mechanism and Susceptibility Prediction Evaluation of the "7•9" Mass Debris Flow in Xiyan Village, Pucheng County, Fujian Province*; Chengdu University of Technology: Chengdu, China, 2021.
63. Ministry of Water Resources of the People's Republic of China. *Standrad for Engineering Classification of Rock Masses GB/T 50218-2014*; Standards Press of China: Beijing, China, 2014. [CrossRef]
64. Park, D.W.; Nikhil, N.V.; Lee, S.R. Landslide and debris flow susceptibility zonation using TRIGRS for the 2011 Seoul landslide event. *Nat. Hazards Earth Syst. Sci.* **2013**, *1*, 2547–2587. [CrossRef]
65. Qi, T.; Zhao, Y.; Meng, X.; Chen, G.; Dijkstra, T. AI-Based Susceptibility Analysis of Shallow Landslides Induced by Heavy Rainfall in Tianshui, China. *Remote Sens.* **2021**, *13*, 1819. [CrossRef]
66. Saito, H.; Korup, O.; Uchida, T.; Hayashi, S.; Oguchi, T. Rainfall conditions, typhoon frequency, and contemporary landslide erosion in Japan. *Geology* **2014**, *42*, 999–1002. [CrossRef]
67. Abancó, C.; Bennett, G.L.; Matthews, A.J.; Matera, M.A.M.; Tan, F.J. The role of geomorphology, rainfall and soil moisture in the occurrence of landslides triggered by 2018 Typhoon Mangkhut in the Philippines. *Nat. Hazards Earth Syst. Sci.* **2021**, *21*, 1531–1550. [CrossRef]

68. Marin, R.; García, E.; Aristizábal, E. Effect of basin morphometric parameters on physically-based rainfall thresholds for shallow landslides. *Eng. Geol.* **2020**, *278*, 105855. [CrossRef]
69. Yang, H.; Yang, T.; Zhang, S.; Zhao, F.; Hu, K.; Jiang, Y. Rainfall-induced landslides and debris flows in Mengdong Town, Yunnan Province, China. *Landslides* **2020**, *17*, 931–941. [CrossRef]
70. Hong, Y.; Hiura, H.; Shino, K.; Sassa, K.; Suemine, A.; Fukuoka, H.; Wang, G. The influence of intense rainfall on the activity of large-scale crystalline schist landslides in Shikoku Island, Japan. *Landslides* **2005**, *2*, 97–105. [CrossRef]
71. Hong, Y.; Hiura, H.; Shino, K.; Sassa, K.; Fukuoka, H. Quantitative assessment on the influence of heavy rainfall on the crystalline schist landslide by monitoring system -case study on Zentoku landslide, Japan. *Landslides* **2005**, *2*, 31–41. [CrossRef]
72. Martha, T.R.; Roy, P.; Govindharaj, K.B.; Kumar, K.V.; Diwakar, P.G.; Dadhwal, V.K. Landslides triggered by the June 2013 extreme rainfall event in parts of Uttarakhand state, India. *Landslides* **2015**, *12*, 135–146. [CrossRef]
73. Chen, Y.-C.; Chang, K.-t.; Wang, S.F.; Huang, J.-C.; Yu, C.-K.; Tu, J.-Y.; Chu, H.-J.; Liu, C.-C. Controls of preferential orientation of earthquake- and rainfall-triggered landslides in Taiwan's orogenic mountain belt. *Earth Surf. Process. Landf.* **2019**, *44*, 1661–1674. [CrossRef]
74. Rosso, R.; Rulli, M.C.; Vannucchi, G. A physically based model for the hydrologic control on shallow landsliding. *Water Resour. Res.* **2006**, *42*. [CrossRef]
75. Carrara, A.; Crosta, G.; Frattini, P. Comparing models of debris-flow susceptibility in the alpine environment. *Geomorphology* **2008**, *94*, 353–378. [CrossRef]
76. Westen, C.J. Landslide hazard and risk zonation-why is it still so difficult? *Bull. Eng. Geol. Environ.* **2005**, *64*, 5–23. [CrossRef]
77. Wu, S.-J.; Hsiao, Y.-H.; Yeh, K.-C.; Yang, S.-H. A probabilistic model for evaluating the reliability of rainfall thresholds for shallow landslides based on uncertainties in rainfall characteristics and soil properties. *Nat. Hazards* **2017**, *87*, 469–513. [CrossRef]
78. Smolíková, J.; Blahut, J.; Vilímek, V. Analysis of rainfall preceding debris flows on the Smědavská hora Mt., Jizerské hory Mts., Czech Republic. *Landslides* **2016**, *13*, 683–696. [CrossRef]
79. Yan, Y.; Hu, S.; Zhou, K.; Jin, W.; Ma, N.; Zeng, C. Hazard characteristics and causes of the "7.22" 2021 debris flow in Shenshuicao gully, Qilian Mountains, NW China. *Landslides* **2023**, *20*, 111–125. [CrossRef]
80. Chung, J.; Lee, Y.; Kim, J.; Jung, C.; Kim, S. Soil Moisture Content Estimation Based on Sentinel-1 SAR Imagery Using an Artificial Neural Network and Hydrological Components. *Remote Sens.* **2022**, *14*, 465. [CrossRef]
81. Kohler, M.A.; Linsley, R. *Predicting the Runoff from Storm Rainfall*; Research Paper No.34; Weather Bureau, US Department of Commerce: Washington, DC, USA, 1951.
82. Cantarino, I.; Carrion, M.A.; Goerlich, F.; Martinez Ibañez, V. A ROC analysis-based classification method for landslide susceptibility maps. *Landslides* **2018**, *16*, 265–282. [CrossRef]
83. Swets, J.A. Measuring the accuracy of diagnostic systems. *Science* **1988**, *240*, 1285–1293. [CrossRef] [PubMed]

remote sensing

MDPI

Article

An Improved Multi-Source Data-Driven Landslide Prediction Method Based on Spatio-Temporal Knowledge Graph

Luanjie Chen [1,2], Xingtong Ge [1,2], Lina Yang [1,2,*], Weichao Li [1] and Ling Peng [1,2]

1 Aerospace Information Research Institute, Chinese Academy of Sciences, Beijing 100049, China; chenluanjie20@mails.ucas.ac.cn (L.C.); gexingtong21@mails.ucas.ac.cn (X.G.); liweichao@rsai.tech (W.L.); pengling@aircas.ac.cn (L.P.)

2 College of Resources and Environment (CRE), University of Chinese Academy of Sciences, Beijing 100049, China

* Correspondence: yangln@aircas.ac.cn

Abstract: Landslides pose a significant threat to human lives and property, making the development of accurate and reliable landslide prediction methods essential. With the rapid advancement of multi-source remote sensing techniques and machine learning, remote sensing data-driven landslide prediction methods have attracted increasing attention. However, the lack of an effective and efficient paradigm for organizing multi-source remote sensing data and a unified prediction workflow often results in the weak generalization ability of existing prediction models. In this paper, we propose an improved multi-source data-driven landslide prediction method based on a spatio-temporal knowledge graph and machine learning models. By combining a spatio-temporal knowledge graph and machine learning models, we establish a framework that can effectively organize multi-source remote sensing data and generate unified prediction workflows. Our approach considers the environmental similarity between different areas, enabling the selection of the most adaptive machine learning model for predicting landslides in areas with scarce samples. Experimental results show that our method outperforms machine learning methods, achieving an increase in F1 score by 29% and an improvement in processing efficiency by 93%. Furthermore, by comparing the susceptibility maps generated in real scenarios, we found that our workflow can alleviate the problem of poor prediction performance caused by limited data availability in county-level predictions. This method provides new insights into the development of data-driven landslide evaluation methods, particularly in addressing the challenges posed by limited data availability.

Keywords: landslide prediction; spatial–temporal knowledge graph; machine learning; multi-source remote sensing data

Citation: Chen, L.; Ge, X.; Yang, L.; Li, W.; Peng, L. An Improved Multi-Source Data-Driven Landslide Prediction Method Based on Spatio-Temporal Knowledge Graph. *Remote Sens.* **2023**, *15*, 2126. https://doi.org/10.3390/rs15082126

Academic Editors: Ioannis Papoutsis, Konstantinos G. Nikolakopoulos and Constantinos Loupasakis

Received: 31 December 2022
Revised: 12 April 2023
Accepted: 14 April 2023
Published: 17 April 2023

1. Introduction

A landslide is a process in which the soil or rock on a slope falls, dumps, slides, spreads or flows due to the influence of various causative factors [1]. In recent years, landslide hazards have caused serious losses of human life and property, severely constraining economic and social development on a worldwide scale. The scientific and accurate prediction of landslides is thus of primary importance.

The common methods of landslide prediction can be divided into knowledge-driven methods and data-driven methods. Knowledge-driven methods are based on an understanding of the mechanisms of landslide formation for susceptibility prediction. One of the most dominant approaches is to predict landslides by comprehending the physical mechanisms of landslide formation using physical equations and numerical simulation methods. Liu et al. [2] utilize physical modeling and various instruments to study the evolution and instability of a locked segment landslide under rainfall conditions and identify tilting deformation as a standard for landslide instability. Capparelli et al. [3] use a physical model, SUSHI, to simulate the role of subsurface hydrology in rain-induced landslides in Campania, Italy. The model enables a better understanding of rainfall infiltration and suction changes in the triggering mechanism of the phenomena. Additionally, some studies have predicted landslide susceptibility based on empirical or statistical methods that assign weights to each causative factor. Mandal et al. [4] applied the analytical hierarchy process (AHP) using geospatial tools to develop a landslide susceptibility map for the Lish River basin in the eastern Darjiling Himalaya. Akgun et al. [5] also produced landslide susceptibility maps for a landslide-prone area in Findikli District using likelihood frequency ratio (LRM) and weighted linear combination (WLC) models. The results showed that the WLC model performed better. However, knowledge-driven methods heavily rely on professional knowledge, and the results are greatly influenced by human expertise.

To overcome this shortcoming, remote sensing data-driven methods have been proposed for landslide prediction. Supervised machine learning methods are by far the most widely used data-driven approach applied to landslide prediction. Typically, machine learning models use remotely sensed images as the data source to generate landslide inventories [6], and then construct relationships between input and output variables based on these inventories [7]. The most commonly used machine learning methods include Support Vector Machine (SVM), Decision Tree (DT), Random Forest (RF), and Artificial Neural Net (ANN). For example, Chen et al. [8] compared kernel logistic regression and naive-Bayes tree and alternating decision tree models in landslide prediction in Taibai County (China). Tian et al. [9] adopted an artificial neural network (ANN) model to predict landslides in Minxian, China. Marjanović et al. [10] tested different kernel functions of the support vector machine (SVM), selected the most accurate kernel function as model parameters, and carried out landslide susceptibility mapping for Chittagong, Bangladesh. In addition, ensemble methods have been gradually applied to produce landslide susceptibility maps [11]. Pham et al. [12] combined a rotation forest and different machine learning classifiers to produce landslide susceptibility maps of India. Dou et al. [13] used SVM as the base learner to generate four classes of ensemble learning models to predict catastrophic rainfall-induced landslides. Hong et al. [14] used J48 decision tree to construct adaptive boosting (Adaboost), bootstrap aggregating (Bagging) and rotation forest models to conduct a comparative study of landslide susceptibility in Guangchang County, Fuzhou City, and the results showed that the rotation forest model has better spatial prediction. Although machine learning methods can predict landslides and achieve high accuracy, the prediction effectiveness of the model is closely related to the quantity of the dataset in the study area. For example, the study area may have problems such as the low spatial resolution of remote sensing data and noisy historical landslide data, which can result in a scarcity of available data and make it difficult to fit the model.

To date, some methods have addressed the problem of sample scarcity by introducing Adversarial Neural Networks (GANs). For example, Al-Najjar et al. [15] proposed a novel approach using GANs to correct imbalanced landslide datasets. Their research showed that integrating GANs with machine learning models can improve the effectiveness of landslide prediction. However, GANs' complex training procedures and lack of interpretability may limit their practicality and reliability for landslide prediction in real-world scenarios. Furthermore, some methods have addressed the problem of scarce environmental data by considering the environmental information of multiple regions. For instance, Zhu et al. [16] added an unsupervised representation learning module to form the underlying representations embedded in thematic maps, which improved the model's accuracy. Ai et al. [17] transferred features from a large dataset region, utilized a pre-trained model, and established a transfer-learning-based susceptibility assessment model to enhance landslide prediction in regions with limited samples. These methods involve multi-source remote sensing data, and as the number of research areas increases, the data scale sharply increases. Therefore, these remote sensing data need to be scientifically integrated and organized in practical applications to meet the requirements of effectiveness and efficiency. Based on well-organized data, it is necessary to establish a unified prediction process to ensure an accurate and fast landslide prediction analysis of the regions of interest according to a standardized procedure. However, existing methods seldom consider the difficulties of organizing environmental big data from multiple sources, which reduces the efficiency of data reuse. Additionally, the lack of a systematic workflow for transfer-learning-based methods leads to the need to establish different models in different fields, reducing the prediction efficiency of machine learning methods.

Knowledge graph is a modeling approach that uses symbols to describe entities, concepts, and relationships in the real world [18–20]. It has received increasing attention in recent years. In the domain of geoscience, knowledge graphs are used to obtain spatio-temporal knowledge and geographic knowledge from multi-source remote sensing data and textual data, also known as a spatio-temporal knowledge graph [21,22] or geographic knowledge graph [23,24]. The spatio-temporal knowledge graph is based on the graph structure for unified spatio-temporal data management, intelligent retrieval and inference analysis, which is an effective means to fuse, organize and compute the multi-source data involved in landslide prediction. In this paper, we propose a workflow for landslide prediction based on spatio-temporal knowledge graph, which not only alleviates the problem of landslide sample scarcity but also improves the efficiency of data usage and landslide prediction. On the one hand, the spatio-temporal knowledge graph is used to fuse remote sensing environmental data, models, and datasets that are closely related to landslide prediction, which makes multi-area environmental data under different conditions rapidly available. On the other hand, the applicability of the machine learning model is enhanced by designing semantic reasoning rules in the knowledge graph. The method extends the traditional machine-learning-based landslide prediction method by adding the process of extracting, storing, and analyzing environmental knowledge, which improves the landslide prediction under the condition of sample scarcity.

This paper has the following main contributions: (1) We propose a workflow for landslide susceptibility evaluation combining spatio-temporal knowledge graphs and machine learning model. (2) We propose a method for organizing remote sensing environmental data based on semantic structure, and improve the efficiency of remote sensing data usage by constructing schema. (3) We define inference rules for candidate model selection and environmental similarity analyses to reduce the impact of sample scarcity on landslide prediction results. (4) We incorporate the knowledge of environmental features in the remote sensing data-driven machine learning method to enhance the applicability of the model, and demonstrate the benefits of this method through experiments. In the following, we first explain our proposed workflow and introduce the construction method of spatio-temporal knowledge graph, and the details of predicting landslides using our method in Section 2. Then, the advantages of our method are demonstrated by experiments in Section 3 and the experimental results are analyzed in Section 4. Finally, our study is concluded in Section 5.

2. Materials and Methods

2.1. Workflow for Landslide Prediction

Generally, when using machine learning models for landslide prediction, it is necessary to first define the boundary of the area for landslide prediction. Secondly, data related to landslides, including historical landslide data and environmental data in the area, are collected by means of remote sensing techniques or fieldwork. Based on these data, datasets are created. The dataset contains the environmental features, i.e., causative factors, that need to be input to the model, and the landslide prediction results, i.e., labels, that are output from the model. Then, the parameters of the model are trained based on the dataset. After training, the optimal model is obtained, and the prediction performance is evaluated based on the model. Eventually, the landslide is predicted based on the model.

To improve the effectiveness of landslide prediction in areas with scarce samples, we introduce knowledge graph into the workflow of machine-learning-based landslide prediction. Firstly, as in the general workflow, we define the boundary of the area to be evaluated and collect historical landslide data and environmental data from the area. Secondly, environmental data are structured knowledge and imported into a knowledge graph, i.e., extracting knowledge from environmental data. Then, we evaluate whether the quantity of historical landslide data can support the training of the model. If it can, the subsequent steps are performed following general machine learning methods, including producing the dataset of the area, training the model, and predicting landslides. If the quantity of the historical landslide data cannot support the training of the model, the environmental similarity within the area is analyzed based on the knowledge graph. Finally, the model with the highest similarity to the study area is selected among the candidate models for landslide prediction. Figure 1 shows the difference between the general workflow and the workflow using the knowledge graph.

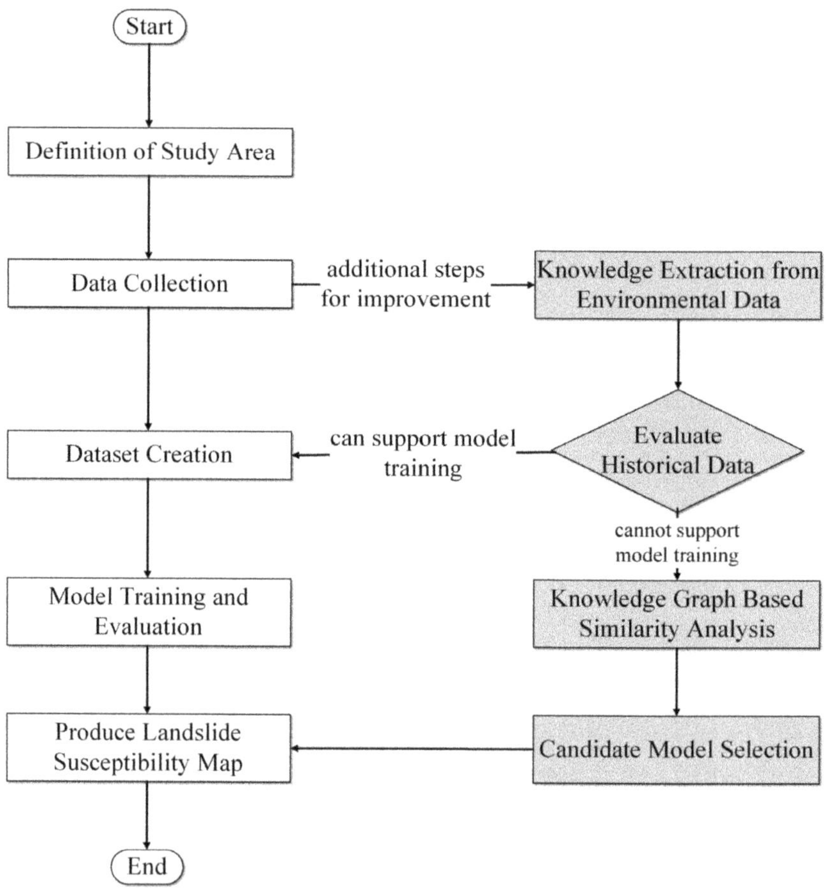

Figure 1. Landslide prediction workflow using machine learning (**left**), and additional steps for improvement using the spatio-temporal knowledge graph (**right**).

2.2. Design of Spatio-Temporal Knowledge Graph for Landslide Prediction

In this paper, we use the spatio-temporal knowledge graph to organize remote sensing environmental data, machine learning models and datasets of the study area. The spatio-temporal knowledge graph includes the schema layer and the data layer, as shown in Figure 2.

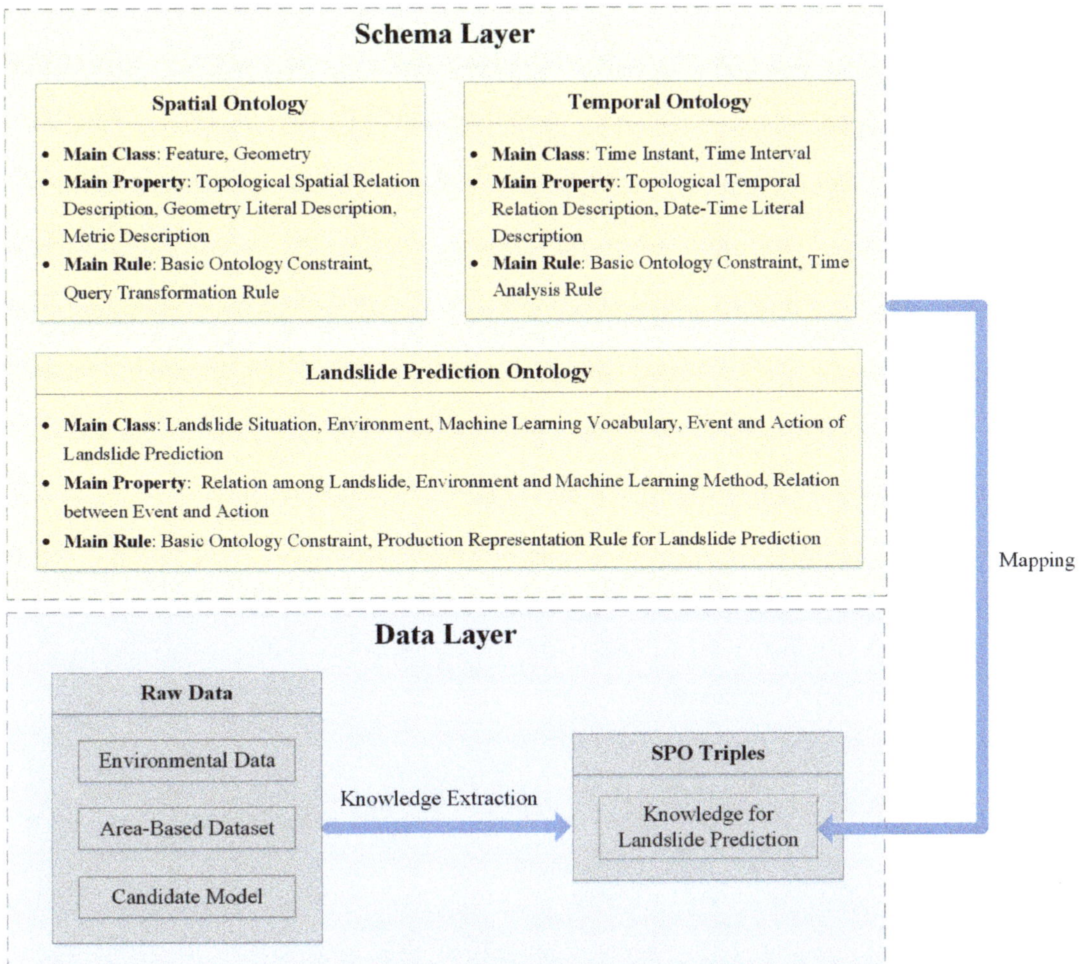

Figure 2. The structure of the spatio-temporal knowledge graph for Landslide Prediction.

2.2.1. Schema Layer

The schema of the knowledge graph is used to describe and organize the spatio-temporal data related to landslides and to define the rules for landslide prediction. We implement the schema using ontologies, which include a spatial ontology, a temporal ontology, and a landslide prediction ontology. Each ontology defines classes, properties, and rules. The classes and properties describe the concepts and their relations involved in landslide prediction, while the rules use classes and properties as symbols to describe the process of spatio-temporal analysis and landslide prediction. The structure of the schema, as well as the main ontologies, concepts, and attributes used, are shown in Figure 3.

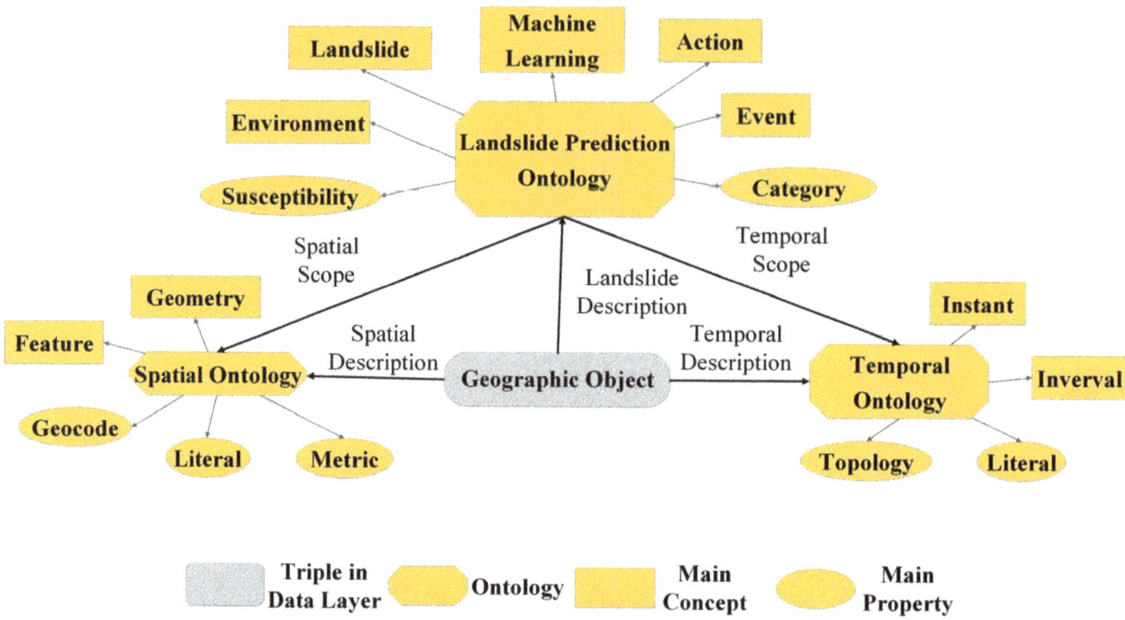

Figure 3. Schema structure, including the division of ontology, main concepts and attributes.

- Spatial ontology

The spatial ontology is used to describe the spatial information of geographic objects and is constructed based on the GeoSPARQL ontology [25]. Geocoding rules are designed in the spatial ontology to serve as the location index of the geographic object.

The classes of the spatial ontology mainly introduce two subclasses of geographic objects in the GeoSPQRQL ontology: the feature class and the geometry class. The spatial terms defined based on the feature and geometry classes can be helpful in modeling geospatial data.

The properties of the spatial ontology mainly define the topological spatial relations between geographic objects, as well as the geometry literal [26], which is the serialization standard used when generating geometry descriptions and the supported geometry types. In addition, the properties of the spatial ontology also include Metric [26], which are scalar spatial properties that describe the geographic object. The main rule of the spatial ontology includes basic ontology constraints for class and property, such as constraints on hierarchical relationships between classes and constraints on property values. The core rule includes rules defined in the GeoSPARQL ontology, such as the query transformation rule for computing spatial relations between geographic objects based on their geometries [25].

In addition, indexable location information helps to improve the efficiency of spatial analysis. However, remote sensing data describe spatial information with latitude and longitude coordinates, which cannot be objectified and indexed. To solve this problem, we designed a geographic tile-based spatial indexing rule, i.e., geocode. Figure 4 illustrates an example of geocoding.

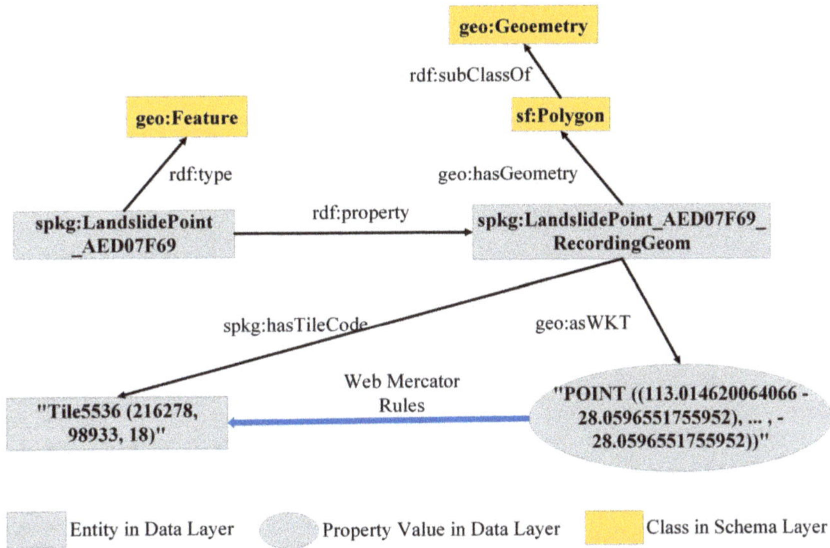

Figure 4. An example of a spatial description of a geographic object (landslide point), converting the coordinate property of the geographic object to a geocode property (Tile5536) so that the spatial information of the geographic object can be indexed.

Geocode converts the coordinate properties of geographic objects into tile-coded entities according to the Web Mercator rules [27], i.e., the tile number of the Web Mercator coordinate system is used, instead of the latitude and longitude coordinate system, as the unit to describe the location of the geographic entity. Each tile number consists of the horizontal coordinates, vertical coordinates and zoom level of the tile. The conversion rules are as follows:

$$x = \frac{lon + 180}{360} \cdot 2^{z} \tag{1}$$

$$y = (1 - \frac{\ln(\tan(lat \cdot \frac{\pi}{180}) + \frac{1}{\cos(lat \cdot \frac{\pi}{180})})}{\pi}) \cdot 2^{z-1} \tag{2}$$

where *lon* and *lat* denote the entered longitude coordinate and the entered latitude coordinate, x denotes the tile horizontal coordinate after conversion, y denotes the vertical coordinate after conversion, z denotes the zoom level of the tile. Each tile in the geocode represents a set of latitude and longitude coordinates, and tiles with different zoom levels contain different amounts of latitude and longitude coordinates. The higher the zoom level, the fewer the number of latitude and longitude coordinates in a tile, and the more accurate the spatial description of the geographic object.

- Temporal ontology

The temporal ontology is used to describe the temporal information of geographic objects, and we construct it based on the OWL-Time ontology [28].

The classes of the temporal ontology mainly define the instant and interval to describe the temporal position and duration of the geographic object. The properties of the temporal ontology mainly define the topological temporal relations between geographic objects, such as "meets", "overlaps" and "during", developed by Allen [29]. The temporal ontology also defines the Date–Time Literal, which is a serialization standard describing time. Similar to the spatial ontology, the main rule of the temporal ontology includes basic ontology constraints for classes and properties. Additionally, the main rule includes rules defined in the OWL-Time ontology. For example, OWL-Time defines time analysis rules for computing temporal relations between geographic objects based on their time instant and time interval [28].

- Landslide prediction ontology

The Landslide Prediction Ontology is used to describe the concepts needed for landslide prediction, the relations between concepts, and the reasoning process of landslide prediction.

The classes of the ontology define concepts that describe the landslide situation, such as the severity of the landslide and the phase it is in. Since the environment is the root cause of landslides, the classes also define concepts describing the environment, including the natural environment and the social environment. Additionally, concepts related to machine learning are defined in the classes, such as vocabulary to describe the features of models and datasets. Furthermore, the process of landslide prediction is divided into several events and actions; hence, we also need to define the events and actions involved in landslide prediction in the Landslide Prediction Ontology. For example, when selecting the best model for an area, candidate models are described in the ontology by defining classes.

The properties of the ontology mainly define the relations among landslides, the environment, and machine learning methods, such as describing which environmental factors are causative factors for landslides and which causative factors are used as features of the dataset. The Landslide Prediction Ontology also defines the relations between events and actions in the landslide prediction process. For example, when the environmental similarity between areas is calculated, the result triggers the action of model selection. The relation of this "trigger" is described as a property.

The main rule of the ontology includes basic ontology constraints for class and property. Meanwhile, based on the classes and properties of the Landslide Prediction Ontology, we use the production representation to define a series of rules to describe the process of remote sensing data-driven landslide prediction. This includes the calculation method of environmental similarity, the process of model selection, and the process of landslide prediction.

2.2.2. Data Layer

The data layer consists of subject–predicate–object (SPO) triples, where subjects and objects represent entities in the knowledge graph, and predicates denote the edges connecting them. The raw data include three types of independent data: environmental data, area-based dataset, and candidate model.

- Environmental Data: Environmental data record the causative factors of landslides in the area, with each type of environmental data corresponding to a specific causative factor. We extract both the environmental data and the environmental features in the area to generate SPO triples, which are used as the basis for analyzing environmental similarities between areas.
- Area-based Dataset: Area-based dataset refers to the dataset used for model training in specific areas. During the knowledge extraction process, we extract instances of dataset features to generate SPO triples. The features of the dataset include the number of samples, the sample area, and the statistical parameters of the causative factors contained in the samples.
- Candidate Model: Candidate model is the model trained based on the area-based dataset. We extract instances of model features to generate SPO triples, which include the name of the model, the address of the parameters, and the name of the samples used for model training.

After generating SPO triples in the data layer, the schema layer relates and organizes these triples to form knowledge that is useful in landslide prediction.

2.3. Landslide Prediction Using Spatio-Temporal Knowledge Graph

2.3.1. Knowledge Extraction and Storage

When processing environmental data, it is essential to extract and store knowledge while constructing a knowledge graph. After preprocessing the remote sensing monitoring data, they pass through the steps designed to generate knowledge. Simultaneously, data associated with the dataset and the machine learning model are produced by utilizing knowledge, and these data undergo a series of steps to generate corresponding knowledge that helps optimize the results of landslide prediction. The process of transforming data into knowledge is depicted in Figure 5.

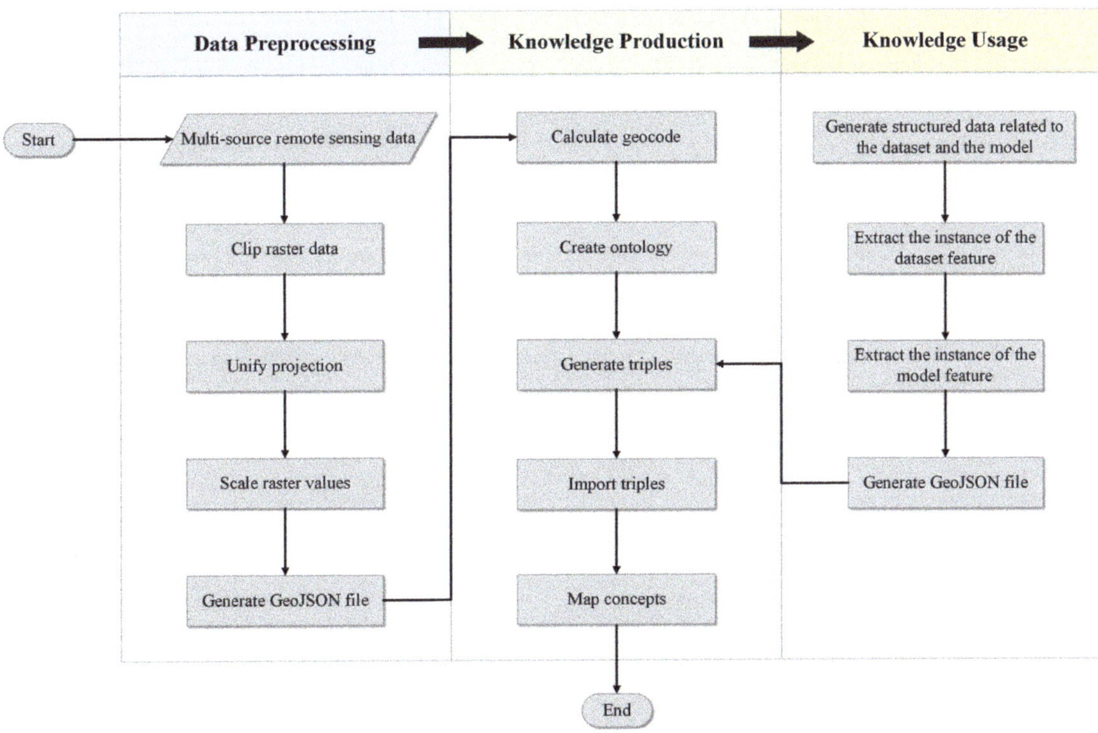

Figure 5. The process of producing knowledge from data in the knowledge graph approach.

- Data preprocessing

Remote sensing monitoring data quantify environmental elements by assigning a value to each pixel, such as the elevation in DEM data. Since knowledge in a knowledge graph is based on object representation, we convert discrete features in remote sensing data into attributes in objects. The GeoJSON file uses a feature object-based storage mode, which is more conducive to knowledge graph reading than remote sensing data. We preprocess the data and convert the original multi-source remote sensing data into a GeoJSON file that describes the distribution of environmental elements in the study area. After generating the GeoJSON file, we classify adjacent pixels in the remote sensing data with the same environmental element value into the same feature object.

Typically, remote sensing monitoring data of different environmental elements have different spatial ranges. Therefore, in the data preprocessing stage, we need to crop the original remote sensing data to obtain the remote sensing data within the spatial range of the research area. Additionally, remote sensing data from different sources may use different projected coordinate systems, so we must convert multi-source remote sensing data into the same projected coordinate system. We also scale the raster values, which may be decimal, to an integer by multiplying and rounding them. Finally, we generate a GeoJSON file by converting the raster data into vector data and then converting the vector data into the GeoJSON format. During the conversion process, we ensure that the original raster value is restored. The entire process can be automated using the GDAL library [30].

- Knowledge production

After data preprocessing, a GeoJSON file is generated, followed by the process of knowledge production. First, geocoding is calculated based on the spatial information of each object in the GeoJSON file. Next, SPO triples are generated to describe geographic entity attributes based on the GeoJSON file. The objects in the GeoJSON file correspond to the subjects in the SPO triples, the keys of the object properties correspond to the predicates in the SPO triples, and the values of the object properties correspond to the objects in the SPO triples. In this process, geocoding is also generated as an attribute of geographic entities in the form of SPO triples.

Next, we import the generated SPO triples into the knowledge graph. If the SPO triples are imported for the first time, the ontology needs to be created according to the ontology structure in the schema layer we designed. Spatial ontology and temporal ontology can be directly used as basic ontologies. For the landslide prediction ontology, we use Protégé [31] to define the Class, property, and rule in the ontology. Protégé is a tool that helps users quickly create and edit ontologies. The landslide prediction ontology edited with Protégé can be directly imported into the knowledge graph. In this paper, Virtuoso [32] is used to store ontologies and SPO triples. After importing the ontology and SPO triples, we map the SPO triples of the data layer and the ontology of the schema layer to generate semantic associations between data features to produce knowledge.

- Knowledge usage

During the process of knowledge usage, additional structured data related to the dataset and the machine learning model are generated, which also need to be extracted and stored in the knowledge graph. We extract instances of features from the dataset and the model, and write them into a GeoJSON file. The description objects of the dataset and the model are areas, and the characteristics of the dataset and the model in different areas are different. In the GeoJSON file, an area is defined as a feature object. The geometry of the feature describes the location of this area, and the properties of the feature represent the instance of the dataset feature and the model feature. After generating the GeoJSON file, we follow the steps of knowledge production to generate and import SPO triples. We map the SPO triples to schema layer ontologies to produce knowledge related to domain models and datasets.

2.3.2. Semantic Reasoning

Semantic reasoning is based on the production representation and recommends models for areas with sparse samples while following the main rule in the schema. It consists of two phases, similarity analysis and candidate model selection, each with several production rules. The general reasoning program automatically performs semantic reasoning as shown in Figure 6. A rule is triggered by an event object, and the corresponding action function is executed to generate a result based on the defined action object. The generated result then triggers the execution of other rules in the rule set until the phase is complete. Figure 7 shows the template defining this process.

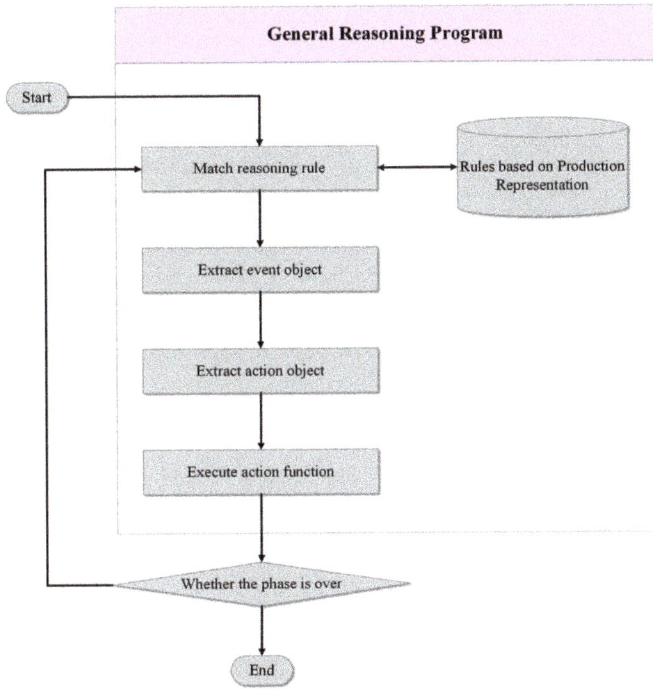

Figure 6. General reasoning program for semantic reasoning in each phase.

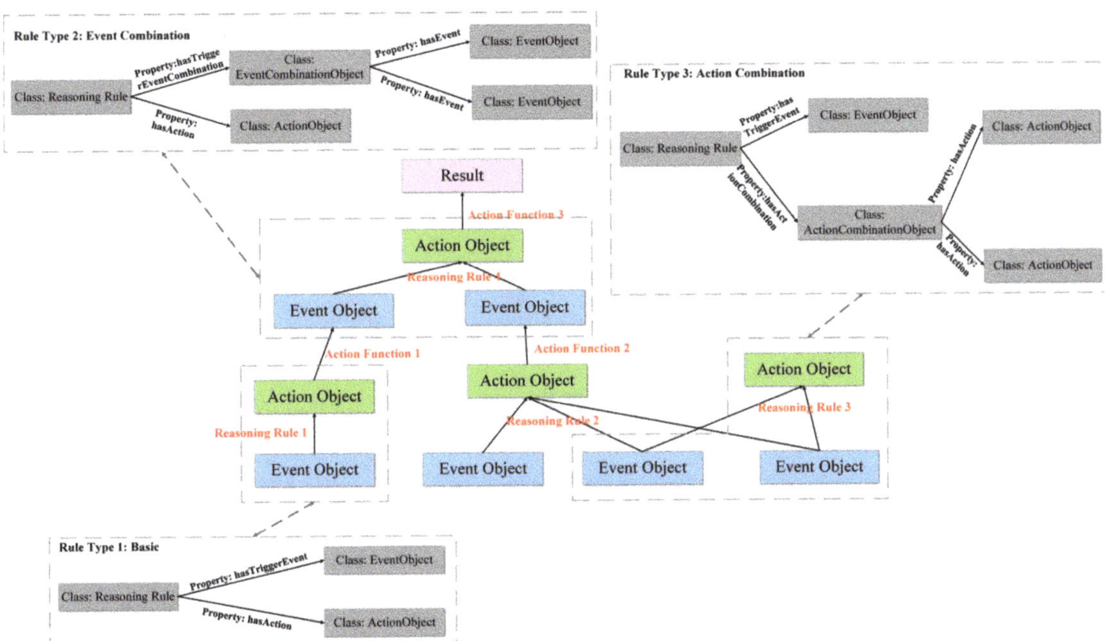

Figure 7. Template for completing phase reasoning from multiple events, and three types of rules, including one event triggering one action, event combination triggering one action, and one event triggering action combination.

The Jaccard index is used to evaluate the similarity of the environment between the area. The equation is as follows:

$$J(A,B) = \frac{|A \cap B|}{|A| + |B| - |A \cap B|} \tag{3}$$

where A and B denote the environmental feature collection of area A and area B, and the larger the Jaccard index, the more similar the environment of the two areas.

The Jaccard index essentially compares the number of environmental features that are similar between areas. For discrete environmental features, the mode of the feature values within the area is taken as the environmental feature value representing that area. If the environmental feature values representing two areas are equal, then this environmental feature is considered similar in the two areas. For continuous environmental features, the average of the feature values within the area is taken as the environmental feature value representing that area. For area A and area B, the similarity of the environmental features in the two areas is determined according to the following equation:

$$isSimilar = \begin{cases} Yes & |F_A - F_B| \leq \dfrac{F_{max} - F_{min}}{N_A + N_B} \\ No & |F_A - F_B| > \dfrac{F_{max} - F_{min}}{N_A + N_B} \end{cases} \tag{4}$$

where F_A and F_B denote the environmental feature values representing area A and area B, F_{max} and F_{min} denote the maximum and minimum values that can be obtained for the environmental feature, N_A denotes the number of the values of this feature in area A, and N_B denotes the number of the values of this feature in area B.

The statistical parameters calculated by the similarity analysis are stored as properties in the triples generated from the area-based dataset. When predicting landslides in study areas with sparse samples, the statistical parameters of the study area are first calculated. Then, a similarity analysis is performed based on the statistical parameters. Eventually, the area with the most similar environmental features to the study area is selected from the knowledge graph, and the model trained from the dataset of that area is obtained through a semantic query, i.e., the process of candidate model selection.

3. Experiment and Result

3.1. Study Area

We obtained historical landslide data for China from the Global Landslide Catalog [33]. China is one of the countries in the world with the highest frequency of landslide hazards, posing threats to both the ecological environment and the safety of people and their property. Furthermore, China is situated at the intersection of continental plates, and its mountainous areas account for nearly 70% of the land area, with a highly undulating terrain that provides natural conditions for landslides to occur.

To demonstrate the effectiveness of our method, we applied the DBSCAN algorithm [34] to cluster landslide points based on their spatial locations. Landslide points belonging to the same category are indicated in the same area, resulting in four simulated study areas, denoted as area 1, area 2, area 3, and area 4. Among them, area 3 and area 4 have the smallest sample sizes and can be simulated as sample scarcity cases. Then, environmental data were collected as causative factors for the training of machine learning models. Table 1 shows the sources and details of the experimental data, and Figure 8 depicts the process of obtaining samples from the experimental data for the four areas.

Additionally, although the selected simulated area can demonstrate the advantages of our method in terms of effectiveness, it is difficult to show the actual prediction results because the simulated area does not have clear boundaries. Therefore, we validated the practical effectiveness of our method using landslide data from Xiji County, located in the southern part of Ningxia Province, China. Xiji County has an area of approximately

1581.5 square kilometers, ranging from 35°35′ to 36°14′ north latitude and 105°20′ to 106°04′ east longitude. We obtained 82 landslide events, which mainly occurred in areas with broken topography and narrow ridges. We used the environmental data in Table 1 to create samples, but due to the scarcity of samples, it is difficult for conventional machine learning methods to make accurate predictions. The distribution of landslide points and the boundary of Xiji County are shown in the Figure 9.

Table 1. Sources and details of experimental data.

Type	Source	Spatial Resolution	Temporal Resolution	Acquisition Method or Sensor Used
Landslide	NASA Global Landslide Catalog [33]	Nationwide vector data	Acquired 1915-2021	Crowdsourcing
Terrain	Shuttle Radar Topography Mission DEM [35]	30 m × 30 m	Acquired 11-22 February 2000	STS Endeavour OV-105, SIR-C/X-SAR
Precipitation	Annual spatial interpolation dataset of Chinese meteorological elements [36]	1 km × 1 km	Update annual	Multi-element weather station
Lithology	Global Lithological Map [37]	0.5° × 0.5°; Rasterized at 250 m resolution	Released 2014	Assembled from existing regional geological maps
Landform	Global Landform classification from ESDAC [38]	500 m × 500 m	Released 2008	Applied two algorithms [39,40] on global DEM datasets
Land Cover	Landsat-derived annual land cover product of China [41]	30 m × 30 m	Update annual	Landsat
Road	OpenStreetMap [42]	Nationwide vector data	Update daily	Crowdsourcing
Normalized Difference Vegetation Index (NDVI)	China Annual NDVI Spatial Distribution Dataset [36]	1 km × 1 km	Update annual	SPOT/VEGETATION

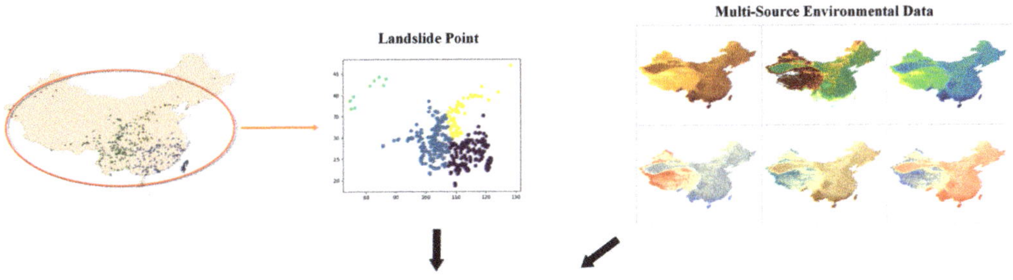

β	Area 1		Area 2		Area 3		Area 4	
	Positive Sample	Negative Sample	Positive Sample	Negative Sample	Positive Sample	Negative Sample	Positive Sample	Negative Sample
1.0	224	224	253	253	33	33	74	74
1.1	224	246	253	278	33	36	74	81
...	...	...	...	...	...	...	...	...
2.9	224	649	253	733	33	95	74	214
3.0	224	672	253	759	33	99	74	222

Figure 8. Template for completing phase reasoning from multiple events, and three types of rules, including one event triggering one action, event combination triggering one action, and one event triggering action combination.

Figure 9. Information of the Study Area: Xiji County.

3.2. Machine Learning Model

The performance of four methods, SVM [43], RF [44], KNN [45], and GCF [46], was compared based on landslide prediction research. To assess the landslide prediction, the landslide and non-landslide samples are both randomly divided into two parts: samples for model training and samples for performance testing.

- Support Vector Machine: The SVM algorithm is a supervised learning binomial classifier based on the risk minimization principle of structured architecture. It can accurately deal with complex nonlinear boundary models.
- Random Forest: The RF algorithm is a combination algorithm based on the classification and regression tree (CART) proposed by Breiman. By randomly selecting k samples from the training set and putting them back into the ground, a decision tree corresponding to the training samples is generated; thus, a random forest composed of k decision trees is generated. According to the prediction result of each tree, the final prediction result is obtained according to the category with the most votes.
- K-Nearest Neighbors: The KNN algorithm is a supervised machine learning classification algorithm. In the K-nearest neighbor method, the K value and distance measure are determined in advance, and the training set and test set are prepared in advance. Through the training set, the feature space is divided into subspaces, and every sample in the training set occupies a part of a space.
- Multigraded Cascade Forest: The GCF algorithm is a supervised ensemble learning method that combines the theory of random forests and a deep neural network. The GCF is composed of a multilevel random forest model, and each level of the random forest model contains many different types of random forests. This multilevel and multidimensional random forest processes the probabilistic eigenvector of the input data, and can effectively enhance the performance of the prediction algorithm for the input data and help to improve the prediction accuracy. Each stage uses the output of the upper stage and the original probability feature vector as its input; that is, it uses the feature information after the upper stage is processed, combined with the original probability feature vector. The new information is processed at this level and passed on to the next level.

3.3. Metrics

The process of landslide prediction based on machine learning is a binary classification process for landslide and non-landslide points. Several measures, including precision, recall and the F1 index, are employed to evaluate the overall landslide prediction accuracy for model comparisons. The equations of precision, recall, and F1 index are shown below:

$$Precision = \frac{TP}{TP + FN} \tag{5}$$

$$Recall = \frac{TP}{TP + FP} \tag{6}$$

$$F_1 = \frac{2PrecisionRecall}{Precision + Recall} \tag{7}$$

where TP denotes the number of true positives predicted as being in the positive category; TN denotes the number of true negatives predicted as negatives; FP denotes the number of true negatives predicted as positives; FN denotes the number of true positives predicted as negatives.

Additionally, the Receiver Operating Characteristic (ROC) curve and the Area Under the Curve (AUC) are used to evaluate the results. The horizontal and vertical axes of the ROC curve represent the false positive (FP) rate and true positive (TP) rate, respectively. The AUC is the area under the ROC curve. When the AUC exceeds 0.5, the model is considered to have positive discriminative ability. A higher AUC value, closer to 1, indicates a better predictive performance.

3.4. Experimental Results

3.4.1. Effectiveness of the Method

Table 2 presents a summary of the results obtained from predicting landslides in four areas using different candidate models. Initially, existing samples of each area were used to predict landslides, and candidate models numbered 1, 2, 3, and 4 were obtained from the training samples of areas 1, 2, 3, and 4, respectively. It was observed that models in areas with sparse samples were generally difficult to fit or had a poor performance after dividing the training and test sets. To address this issue, the environmental similarity (Jaccard Index) between sample-sufficient areas and sample-scarce areas was calculated based on the spatio-temporal knowledge graph's reasoning rules, and candidate models were selected for landslide prediction. When predicting landslides for area pairs with similar environmental features, using model 2 to predict the landslide of area 3, for instance, reduced the issue of sample scarcity in the prediction process for area 3. It was also noted that a larger Jaccard Index indicated that the environmental features of the two areas were more similar, and the model performed better. In terms of selecting model types, SVM and GCF showed better prediction performance for area 3, while KNN and GCF performed better for area 4. Table 3 presents the optimal performance achieved by predicting sample-sparse areas using the general workflow and the workflow incorporating knowledge graph, while Figure 10 displays the corresponding ROC curves. It is evident that incorporating the knowledge graph into the workflow enhances the predictive capability for sample-scarcity areas. However, it should be noted that the number of candidate models and the limitations of model training knowledge still leave room for improvements in the AUC.

Table 2. Results of using candidate models to predict landslide in different areas, including predictions for regular areas and predictions for scarce areas with similar environments.

Area Number	Model Number	Jaccard Index	Precision				Recall				F_1			
			SVM	RF	KNN	GCF	SVM	RF	KNN	GCF	SVM	RF	KNN	GCF
1	1	-	0.61	0.66	0.63	0.75	0.60	0.60	0.63	0.60	0.60	0.63	0.63	0.67
2	2	-	0.66	0.73	0.63	0.78	0.60	0.62	0.61	0.60	0.63	0.67	0.62	0.68
3	3	-	0.34	0.33	0.46	0.38	0.34	0.42	0.40	0.45	0.34	0.38	0.43	0.41
4	4	-	0.52	0.52	0.48	0.52	0.50	0.52	0.42	0.53	0.51	0.52	0.45	0.52
3	1	0.2	0.52	0.54	0.78	0.61	0.52	0.55	0.53	0.57	0.52	0.54	0.63	0.59
3	2	0.6	0.86	0.67	0.58	0.85	0.62	0.62	0.60	0.60	0.72	0.64	0.59	0.70
4	1	0.6	0.61	0.60	0.81	0.72	0.56	0.61	0.57	0.60	0.58	0.60	0.67	0.65
4	2	0.5	0.58	0.63	0.58	0.67	0.58	0.60	0.58	0.61	0.58	0.61	0.58	0.64

Table 3. Predicted performance of sample scarcity areas.

Sample Scarcity Area Number	F_1 of General Workflow	F_1 of Workflow Using Knowledge Graph
3	0.43 (Sample size too small to fit)	0.72
4	0.52	0.67

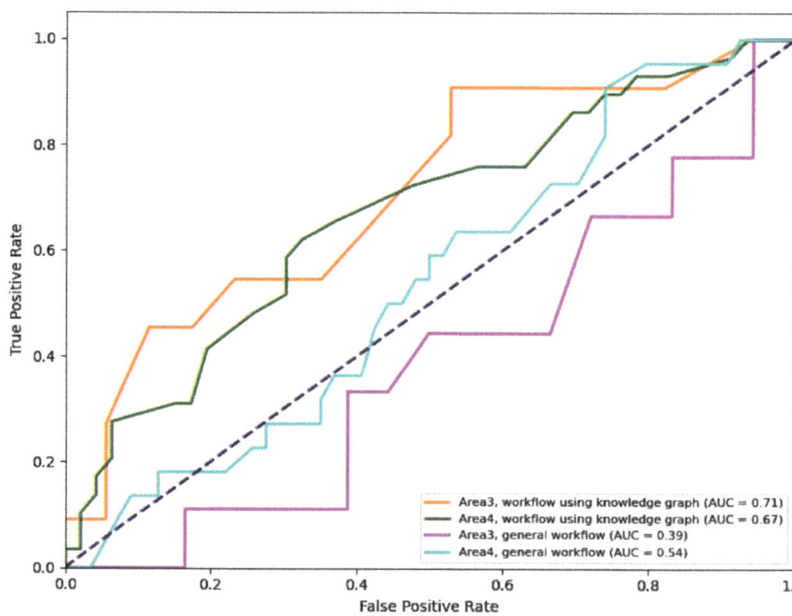

Figure 10. Comparison of Receiver Operating Characteristic Curves between General Workflow and Workflow Using Knowledge Graph in Area 3 and Area 4.

Moreover, we implemented two workflows for landslide prediction using machine learning. In areas with limited samples, the general workflow involves more manual steps. On the other hand, the workflow based on the knowledge graph offers the advantages of automation and faster computation. Table 4 provides a comparison of the two workflows.

Table 4. Effectiveness comparison of general workflow and workflow with additional knowledge graph steps.

Workflow	Tools	Manual Steps	Calculation Time
General workflow	ArcGIS, Anaconda platform, Scikit-Learn package	1. Collect, load and crop the data required for the study area. 2. Unify the data scale and grid the raster data. 3. Extract data features using raster calculation tools to generate datasets. 4. Train and test the model. 5. Calculation and find similar feature areas, and transfer the model to sample scarcity areas. 6. Eliminate outliers and obtain disaster prediction results. 7. Bridging process between automation steps.	17.5 accumulated hours
Workflow with additional knowledge graph steps	Virtuoso database, Anaconda platform, Scikit-Learn package	1. Collect the missing data of the study area in the knowledge graph. 2. Train the candidate models. 3. Design reasoning rules. 4. Bridging process between automation steps.	1.2 hours

3.4.2. Validation in Xiji

To further demonstrate the effectiveness of this method in real-world scenarios, we applied the knowledge graph-based workflow to produce a landslide susceptibility map in Xiji County. Our approach has shown promising results in preliminary studies and we sought to validate it in a practical setting. We first collected environmental data from Xiji County from the sources listed in Table 1. Next, we extracted knowledge from the data and stored it in the knowledge graph, following the data processing process outlined in Figure 5. The knowledge graph performed semantic reasoning to predict landslides, using similarity analysis and candidate model selection as detailed in Section 2.3.2. Based on Equations (3) and (4), the Jaccard index of Area 1 and Xiji is 0.6, and the Jaccard index of Area 2 and Xiji is 0.3. Therefore, the knowledge graph selected the model produced by Area 1 from the candidate models to generate the landslide susceptibility map. Among the candidate models, RF produced the best results for predicting landslide susceptibility in Xiji, with 100 trees in the forest, a minimum of 2 samples required to split an internal node, and a ratio of positive to negative samples of 1.7.

In addition, we followed the general machine learning method shown in Figure 1 to generate the susceptibility map and compared it with our method. The results are presented in Figure 11. Compared to the real landslide sites, the general machine learning method was unable to accurately evaluate the spatial distribution of susceptibility in Xiji County due to the lack of dataset. On the other hand, the method using the knowledge graph workflow mitigated the effect of sample scarcity on the results.

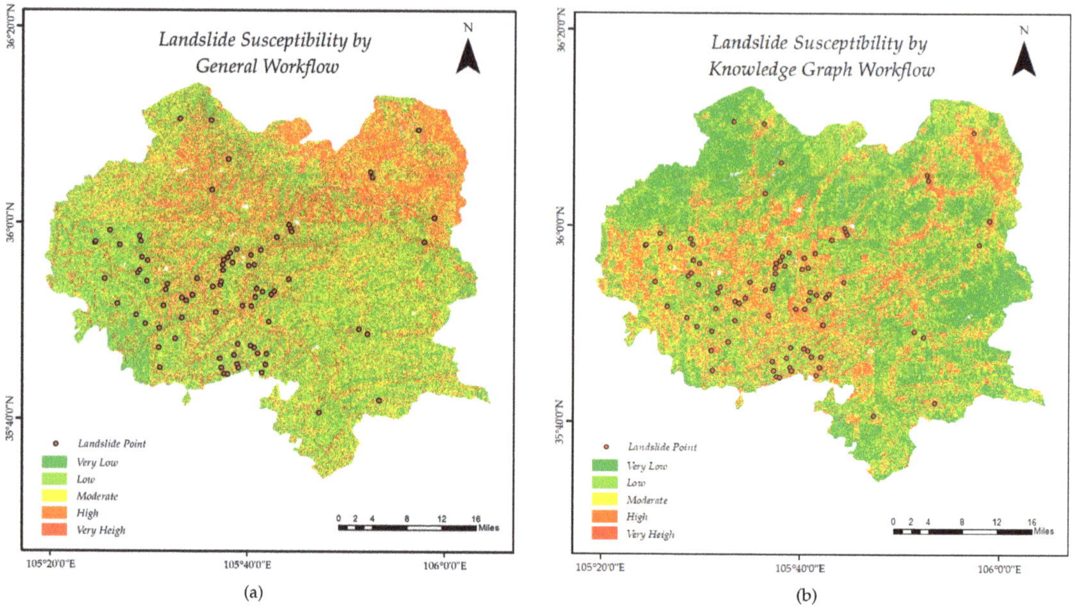

Figure 11. Comparison of Landslide Susceptibility Maps Produced by General Workflow and Workflow Using Knowledge Graph for Xiji Landslide

4. Discussion

In our experiments, we conducted both an effectiveness validation and a validation based on real scenarios. For the effectiveness validation, we divided the landslide dataset into four areas, including two sample-sufficient areas and two sample-scarce areas. We then used our proposed knowledge-graph-based method and general machine learning methods to predict the sample-scarce areas. Our method demonstrated several advantages over general machine learning methods, including better precision due to the use of similarity reasoning rules and environmental features stored in the spatio-temporal knowledge graph. The similarity analysis method we designed quantifies the similarity of geographical features, which improves prediction accuracy, as shown in our experiments. Additionally, the knowledge graph accelerated the prediction process by using automatic semantic reasoning rules and the storage advantages of the graph structure, providing a speed advantage over other methods.

Furthermore, for validation based on real scenarios in Xiji County, we further compared the effectiveness of our workflow and a general machine learning workflow to draw susceptibility maps. Our study demonstrated that the proposed workflow can mitigate the problem of poor prediction in sample-scarce areas. Among the candidate models, Random Forest performed the best, likely due to its ability to handle high-dimensional variables without variable deletion and reduce overfitting through the use of multiple trees, substitution methods, and random subset selection to split nodes.

However, it is important to acknowledge that our proposed method has limitations. Firstly, it is sensitive to prediction size, and larger study areas may require longer processing times and more storage space. Secondly, while our method shows promising results, the precision still needs improvement in real scenarios, which may be achieved by using higher-resolution environmental remote sensing data and more comprehensive landslide point records. Lastly, in future experiments, specific model training techniques could be incorporated as knowledge in the knowledge graph to standardize the comparison criteria, and the design of inference rules for model training should be carefully considered.

5. Conclusions

Data-driven methods usually require a sufficient number of samples to train the models. In areas where samples are limited, some studies employed prediction methods based on transfer learning or GANs. However, these methods face challenges in organizing multi-source remote sensing data or face difficulties training, making the munsuitable for disaster scenarios that require real-time prediction. Moreover, the lack of a systematic prediction process and the low level of automation in prediction resulted in low prediction efficiency for landslides. In this paper, we propose a novel approach to improve the performance of remote sensing data-driven landslide prediction, which makes the following main contributions:

- This paper proposes an efficient method for disaster analysis in the field of geohazard management by combining knowledge-driven and data-driven approaches.
- The problem of data-driven methods being over-sensitive to data is alleviated by semantic modeling and knowledge fusion.
- A novel paradigm is defined for the standardized integration of multi-source remote sensing resources, which helps to share and reuse formalized remote sensing resources and demonstrates the potential of spatio-temporal knowledge graphs in the field of remote sensing.

In future research, we will strive to improve the generalization ability of spatio-temporal knowledge graph. On the one hand, we should define the inference rules for machine learning training strategies in the spatio-temporal knowledge graph to improve the prediction accuracy of candidate models. We will also attempt to integrate other data-driven methods, such as representation learning. On the other hand, we will incorporate more disaster knowledge, such as exposure factors and other geological disaster concepts, into the model to assess the comprehensive risk of geological disasters. Additionally, we will pay more attention to the interpretability of landslide prediction methods. By leveraging the structural advantages of knowledge graph, modeling landslide disaster environments based on multi-source remote sensing data helps to explain the inherent features between causative factors and positively contributes to the prediction results of landslides.

Author Contributions: Conceptualization, L.C., L.P. and L.Y.; methodology, L.C., L.P. and W.L.; validation, L.C. and W.L.; resources, L.C.; data curation, L.C. and X.G.; writing—original draft preparation, L.C. and X.G.; writing—review and editing, L.C., X.G. and L.Y.; funding acquisition, L.P. All authors have read and agreed to the published version of the manuscript.

Funding: This work was supported by Ningxia Key R&D Program (2020BFG02013). This work was sponsored by Tianjin intelligent manufacturing special fund project (NO. 20201198).

Data Availability Statement: Data sharing not applicable.

Conflicts of Interest: The authors declare no conflict of interest.

References

1. Yilmaz, I. Landslide susceptibility mapping using frequency ratio, logistic regression, artificial neural networks and their comparison: A case study from Kat landslides (Tokat—Turkey). *Comput. Geosci.* **2009**, 35, 1125–1138. [CrossRef]
2. Liu, H.; Liu, J.; Chen, J.; Qiu, L. Experimental study on tilting deformation and a new method for landslide prediction with retaining-wall locked segment. *Sci. Rep.* **2023**, 13, 5149. [CrossRef] [PubMed]
3. Capparelli, G.; Versace, P. Analysis of landslide triggering conditions in the Sarno area using a physically based model. *Hydrol. Earth Syst. Sci.* **2014**, 18, 3225–3237. [CrossRef]
4. Mandal, B.; Mandal, S. Analytical hierarchy process (AHP) based landslide susceptibility mapping of Lish river basin of eastern Darjeeling Himalaya, India. *Adv. Space Res.* **2018**, 62, 3114–3132. [CrossRef]
5. Akgun, A.; Dag, S.; Bulut, F. Landslide susceptibility mapping for a landslide-prone area (Findikli, NE of Turkey) by likelihood-frequency ratio and weighted linear combination models. *Environ. Geol.* **2008**, 54, 1127–1143. [CrossRef]
6. Đurić, U.; Marjanović, M.; Radić, Z.; Abolmasov, B. Machine learning based landslide assessment of the Belgrade metropolitan area: Pixel resolution effects and a cross-scaling concept. *Eng. Geol.* **2019**, 256, 23–38. [CrossRef]

7. Ma, Z.; Mei, G.; Piccialli, F. Machine learning for landslides prevention: A survey. *Neural Comput. Appl.* **2021**, *33*, 10881–10907. [CrossRef]
8. Chen, W.; Xie, X.; Peng, J.; Wang, J.; Duan, Z.; Hong, H. GIS-based landslide susceptibility modelling: a comparative assessment of kernel logistic regression, Naïve-Bayes tree, and alternating decision tree models. *Geomat. Nat. Hazards Risk* **2017**, *8*, 950–973. [CrossRef]
9. Tian, Y.; Xu, C.; Hong, H.; Zhou, Q.; Wang, D. Mapping earthquake-triggered landslide susceptibility by use of artificial neural network (ANN) models: An example of the 2013 Minxian (China) Mw 5.9 event. *Geomat. Nat. Hazards Risk* **2019**, *10*, 1–25. [CrossRef]
10. Marjanović, M.; Kovačević, M.; Bajat, B.; Voženílek, V. Landslide susceptibility assessment using SVM machine learning algorithm. *Eng. Geol.* **2011**, *123*, 225–234. [CrossRef]
11. Tien Bui, D.; Shirzadi, A.; Shahabi, H.; Geertsema, M.; Omidvar, E.; Clague, J.J.; Thai Pham, B.; Dou, J.; Talebpour Asl, D.; Bin Ahmad, B.; et al. New ensemble models for shallow landslide susceptibility modeling in a semi-arid watershed. *Forests* **2019**, *10*, 743. [CrossRef]
12. Pham, B.T.; Prakash, I.; Dou, J.; Singh, S.K.; Trinh, P.T.; Tran, H.T.; Le, T.M.; Van Phong, T.; Khoi, D.K.; Shirzadi, A.; et al. A novel hybrid approach of landslide susceptibility modelling using rotation forest ensemble and different base classifiers. *Geocarto Int.* **2020**, *35*, 1267–1292. [CrossRef]
13. Dou, J.; Yunus, A.P.; Bui, D.T.; Merghadi, A.; Sahana, M.; Zhu, Z.; Chen, C.W.; Han, Z.; Pham, B.T. Improved landslide assessment using support vector machine with bagging, boosting, and stacking ensemble machine learning framework in a mountainous watershed, Japan. *Landslides* **2020**, *17*, 641–658. [CrossRef]
14. Hong, H.; Liu, J.; Bui, D.T.; Pradhan, B.; Acharya, T.D.; Pham, B.T.; Zhu, A.X.; Chen, W.; Ahmad, B.B. Landslide susceptibility mapping using J48 Decision Tree with AdaBoost, Bagging and Rotation Forest ensembles in the Guangchang area (China). *Catena* **2018**, *163*, 399–413. [CrossRef]
15. Al-Najjar, H.A.; Pradhan, B.; Sarkar, R.; Beydoun, G.; Alamri, A. A new integrated approach for landslide data balancing and spatial prediction based on generative adversarial networks (GAN). *Remote Sens.* **2021**, *13*, 4011. [CrossRef]
16. Zhu, Q.; Chen, L.; Hu, H.; Pirasteh, S.; Li, H.; Xie, X. Unsupervised feature learning to improve transferability of landslide susceptibility representations. *IEEE J. Sel. Top. Appl. Earth Obs. Remote Sens.* **2020**, *13*, 3917–3930. [CrossRef]
17. Ai, X.; Sun, B.; Chen, X. Construction of small sample seismic landslide susceptibility evaluation model based on Transfer Learning: A case study of Jiuzhaigou earthquake. *Bull. Eng. Geol. Environ.* **2022**, *81*, 116. [CrossRef]
18. Gutierrez, C.; Sequeda, J.F. Knowledge graphs. *Commun. ACM* **2021**, *64*, 96–104. [CrossRef]
19. Chen, X.; Jia, S.; Xiang, Y. A review: Knowledge reasoning over knowledge graph. *Expert Syst. Appl.* **2020**, *141*, 112948. [CrossRef]
20. Singhal, A. Introducing the knowledge graph: Things, not strings. *Off. Google Blog* **2012**, *5*, 16.
21. Nayyeri, M.; Vahdati, S.; Khan, M.T.; Alam, M.M.; Wenige, L.; Behrend, A.; Lehmann, J. Dihedron Algebraic Embeddings for Spatio-Temporal Knowledge Graph Completion. In Proceedings of the European Semantic Web Conference, Crete, Greece, 29 May–2 June 2022; Springer: Cham, Switzerland, 2022; pp. 253–269.
22. Ge, X.; Yang, Y.; Peng, L.; Chen, L.; Li, W.; Zhang, W.; Chen, J. Spatio-temporal knowledge graph based forest fire prediction with multi source heterogeneous data. *Remote Sens.* **2022**, *14*, 3496. [CrossRef]
23. Wang, S.; Zhang, X.; Ye, P.; Du, M.; Lu, Y.; Xue, H. Geographic knowledge graph (GeoKG): A formalized geographic knowledge representation. *ISPRS Int. J. Geo-Inf.* **2019**, *8*, 184. [CrossRef]
24. Yan, B.; Janowicz, K.; Mai, G.; Gao, S. From itdl to place2vec: Reasoning about place type similarity and relatedness by learning embeddings from augmented spatial contexts. In Proceedings of the 25th ACM SIGSPATIAL International Conference on Advances in Geographic Information Systems, Redondo Beach, CA, USA, 7–10 November 2017; pp. 1–10.
25. Battle, R.; Kolas, D. Enabling the geospatial semantic web with parliament and geosparql. *Semant. Web* **2012**, *3*, 355–370. [CrossRef]
26. Car, N.J.; Homburg, T. GeoSPARQL 1.1: Motivations, Details and Applications of the Decadal Update to the Most Important Geospatial LOD Standard. *ISPRS Int. J. Geo-Inf.* **2022**, *11*, 117. [CrossRef]
27. Battersby, S.E.; Finn, M.P.; Usery, E.L.; Yamamoto, K.H. Implications of web Mercator and its use in online mapping. *Cartogr. Int. J. Geogr. Inf. Geovis.* **2014**, *49*, 85–101. [CrossRef]
28. Time Ontology in OWL. Available online: https://www.w3.org/TR/owl-time/ (accessed on 8 August 2022).
29. Allen, J.F.; Ferguson, G. Actions and events in interval temporal logic. *J. Log. Comput.* **1994**, *4*, 531–579. [CrossRef]
30. Westra, E. *Python Geospatial Development*; Packt Publishing: Birmingham, UK, 2010.
31. Tudorache, T.; Noy, N.F.; Tu, S.; Musen, M.A. Supporting collaborative ontology development in Protégé. In Proceedings of the International Semantic Web Conference, Karlsruhe, Germany, 26–30 October 2008; pp. 17–32.
32. Virtuoso Universal Server. Available online: https://virtuoso.openlinksw.com (accessed on 8 August 2022).
33. Kirschbaum, D.; Stanley, T.; Zhou, Y. Spatial and temporal analysis of a global landslide catalog. *Geomorphology* **2015**, *249*, 4–15. [CrossRef]
34. Schubert, E.; Sander, J.; Ester, M.; Kriegel, H.P.; Xu, X. DBSCAN revisited, revisited: Why and how you should (still) use DBSCAN. *ACM Trans. Database Syst. (TODS)* **2017**, *42*, 1–21. [CrossRef]
35. LP DAAC—Homepag. Available online: https://lpdaac.usgs.gov (accessed on 8 August 2022).

36. Chinese Academy of Sciences Resource and Environmental Science Data Center. Available online: http://www.resdc.cn (accessed on 8 August 2022).
37. Global Lithological Map Database v1.0 (Gridded to 0.5° Spatial Resolution). Available online: https://doi.pangaea.de/10.1594/PANGAEA.788537 (accessed on 8 August 2022).
38. Panagos, P.; Van Liedekerke, M.; Borrelli, P.; Köninger, J.; Ballabio, C.; Orgiazzi, A.; Lugato, E.; Liakos, L.; Hervas, J.; Jones, A.; et al. European Soil Data Centre 2.0: Soil data and knowledge in support of the EU policies. *Eur. J. Soil Sci.* **2022**, *73*, e13315. [CrossRef]
39. Meybeck, M.; Green, P.; Vörösmarty, C. A new typology for mountains and other relief classes. *Mt. Res. Dev.* **2001**, *21*, 34–45. [CrossRef]
40. Iwahashi, J.; Pike, R.J. Automated classifications of topography from DEMs by an unsupervised nested-means algorithm and a three-part geometric signature. *Geomorphology* **2007**, *86*, 409–440. [CrossRef]
41. Yang, J.; Huang, X. The 30 m annual land cover dataset and its dynamics in China from 1990 to 2019. *Earth Syst. Sci. Data* **2021**, *13*, 3907–3925. [CrossRef]
42. Haklay, M.; Weber, P. Openstreetmap: User-generated street maps. *IEEE Pervasive Comput.* **2008**, *7*, 12–18. [CrossRef]
43. Cortes, C.; Vapnik, V. Support-vector networks. *Mach. Learn.* **1995**, *20*, 273–297. [CrossRef]
44. Breiman, L. Random Forests. *Mach. Learn.* **2001**, *45*, 5–32. [CrossRef]
45. Peterson, L.E. K-nearest neighbor. *Scholarpedia* **2009**, *4*, 1883. [CrossRef]
46. Zhou, Z.H.; Feng, J. Deep Forest: Towards An Alternative to Deep Neural Networks. In Proceedings of the IJCAI, Melbourne, Australia, 19–25 August 2017; pp. 3553–3559.

remote sensing

MDPI

Article

A Method for Predicting Landslides Based on Micro-Deformation Monitoring Radar Data

Weixian Tan [1,2], Yadong Wang [1,2], Pingping Huang [1,2,*], Yaolong Qi [1,2], Wei Xu [1,2], Chunming Li [1,2] and Yuejuan Chen [1,2]

1 College of Information Engineering, Inner Mongolia University of Technology, Hohhot 010051, China
2 Inner Mongolia Key Laboratory of Radar Technology and Application, Hohhot 010051, China
* Correspondence: hwangpp@imut.edu.cn; Tel.: +86-0471-360-1821

Abstract: Mine slope landslides seriously threaten the safety of people's lives and property in mining areas. Landslide prediction is an effective way to reduce losses due to such disasters. In recent years, micro-deformation monitoring radar has been widely used in mine slope landslide monitoring. However, traditional landslide prediction methods are not able to make full use of the diversified monitoring data from these radars. This paper proposes a landslide time prediction method based on the time series monitoring data of micro-deformation monitoring radar. Specifically, deformation displacement, coherence and deformation volume, and the parametric degree of deformation (*DOD*) are calculated and combined with the use of the tangent angle method. Finally, the effectiveness of the method is verified by using measured data of a landslide in a mining area. The experimental results show that our proposed method can be used to identify the characteristics of an imminent sliding slope and landslide in advance, providing monitoring personnel with more reliable landslide prediction results.

Keywords: landslide prediction; deformation monitoring; micro-deformation monitoring radar; coherence

Citation: Tan, W.; Wang, Y.; Huang, P.; Qi, Y.; Xu, W.; Li, C.; Chen, Y. A Method for Predicting Landslides Based on Micro-Deformation Monitoring Radar Data. *Remote Sens.* **2023**, *15*, 826. https://doi.org/10.3390/rs15030826

Academic Editors: Constantinos Loupasakis, Ioannis Papoutsis and Konstantinos G. Nikolakopoulos

Received: 4 January 2023
Revised: 29 January 2023
Accepted: 30 January 2023
Published: 1 February 2023

1. Introduction

As one of the most harmful natural disasters in the world, landslides can cause drastic loss of life and property every year [1–4]. In recent years, open-pit mining engineering activities have been expanding on a large scale along with the rapid development of mining technology and the continuous growth in global raw material demand, resulting in a sharp increase in the number of high and steep slopes in mining areas [5–8]. According to incomplete statistics, there are currently 10,100 non-coal metal mines and 73,548 non-metallic mines in China, and from 2001 to 2007, there were 1951 landslide accidents in metal and non-metal open-pit mines, with 3065 casualties [9–12]. The potential safety hazards on the rock slopes of open-pit mines represent the core problem of mine safety production. Accordingly, in 2018, China put forward the idea of "establishing an efficient and scientific natural disaster defense system and improving the ability of the whole society to prevent and control natural disasters" to achieve early identification, monitoring, and early warning of major geological disasters. In the future, organic collaborative means of air–space–ground monitoring will be used to ensure the safe production of mines [13–15].

The slopes of open-pit mining areas are so steep and abrupt that they can be rather dangerous to even observe. Large-scale and long-term real-time monitoring cannot be achieved using traditional slope measurement technologies such as GPS and total station, and it is also difficult to reflect the overall trend of slope deformation [16–21]. With the advancement of science and technology, micro-deformation monitoring radar has been widely used in open-pit mine slope landslide monitoring. It is capable of long-distance and all-weather real-time monitoring, using the phase change of electromagnetic waves to obtain the overall deformation information of the target area. As a result, it has attracted extensive attention worldwide [22–29]. In addition, coherence of the target deformation area can also be obtained using micro-deformation monitoring radar. A significant difference in the coherence of the target at two different times means that the target has a large deformation. The coherence will be reduced by rainfall, snowfall, and other factors, so good coherence is an important prerequisite for obtaining the deformation value [29–32]. In general, coherence gives an important and valuable reference and implication for deformation.

Since the 1960s, many scholars have carried out a series of studies on landslide prediction, wherein the Saito model, Fukuzono model, and Voight model are typical representatives of creep theory. They are suitable for short-term landslide prediction and temporary slip prediction [33–37] but are limited to gravity-type landslides only. Due to the complex failure mechanism of slopes, nonlinear models such as neural network models and collaborative prediction models always result in differences from the actual slope system [38–40]. In addition, Xu Qiang used the improved tangential angle method as an important indicator in early warning for landslides and achieved good results [41–43]. The existing methods can barely make the best of diverse data obtained by micro-deformation monitoring radar, despite its wide application. So, to take full advantage of these data, there is an urgent need to find a landslide prediction method suitable for micro-deformation monitoring radar.

In addition, the setting of the slope threshold is one of the key steps in landslide prediction. Once the corresponding parameters of a slope exceed the threshold, it is considered to be at risk of collapse. The early warning value is adjusted and continuously optimized in the process of data accumulation and application through the initially set empirical value so as to slowly approach the best early warning threshold applicable to different types of slopes [44]. However, the setting of the early warning value is not static, and there is no absolute universal early warning value. The traditional strategy usually uses a single threshold setting for early warning, which can be susceptible to the environment and other factors, resulting in false early warning.

In summary, based on the time series monitoring data of mine slopes obtained by micro-deformation monitoring radar, the degree of deformation (*DOD*) was calculated in this study by applying parameters such as deformation displacement, coherence, and deformation volume. Then, a comprehensive landslide early warning method combining the multi-threshold criterion and tangent angle was proposed. The effectiveness of the method in realizing early warning of slope landslides is verified through measured data.

2. Study Area

The experimental site, located in the northwest of China, is an open-pit mine on the slope of the low and middle part of the Altai Mountains. The altitude of the open-air stope is 1000~1300 m. The site is 1.2 km long in the northwest–southeast direction and 0.7 km long in the northeast–southwest direction; the terrain is high in the north and low in the east and south, with a relative height difference of 50~300 m; the bedrock slope was formed after the exploitation of the open-pit mine with the destruction of original terrain and landform. Accordingly, multiple platforms with noticeable elevation differences occur, leading to undulating terrain. Figure 1 shows a panorama of the open-pit mine from the radar perspective.

Geologically, the lithology of the study area can be divided into five categories: gneiss, schist, granulite, marble, and amphibolite. The rock mass inside the slope is relatively complete, and the integrity of the rock mass from the surface to the deep part gradually increases, except for the inner part of the slope due to the structural fracture zone caused by fragmentation of the rock mass, whereas the remaining rock masses are relatively complete and hard rock masses.

Figure 1. Open-pit mine slope from the radar perspective.

In this experiment, linear scanning micro-deformation monitoring radar was used for monitoring; the radar operates in the Ku band, with an azimuth resolution of 0.3° and a distance resolution of 0.2 m (Figure 2). The monitoring period was from 0:00 on 18 April 2021 to 10:00 a.m. on 21 April 2021, and a total of 225 radar images were acquired. A radar image obtained by the micro-deformation monitoring radar is shown in Figure 3. Since the monitoring scenes are mostly rock structures with strong scattering, the imaging results are fairly clear.

(**a**) (**b**)

Figure 2. (**a**) micro-deformation monitoring radar protection device; (**b**) micro-deformation monitoring radar.

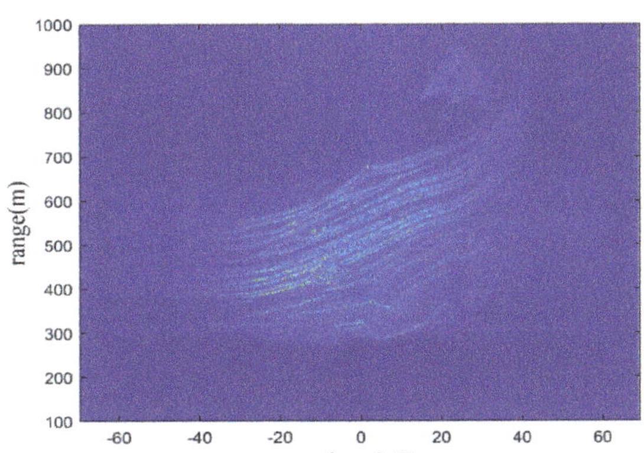

Figure 3. Radar image of monitored scene.

3. Method

In the 1960s, Saito proposed a landslide prediction model based on creep theory after a large number of landslide experiments, as shown in Figure 4, which divides the landslide process into three stages, and slope instability often occurs in the third stage, namely the accelerated deformation stage. In 2009, Xu Qiang proposed the early warning criterion of a tangent angle landslide based on the three-stage evolution characteristics of the Saito model curve, and the dimensions of the longitudinal and horizontal coordinates of the displacement time curve were found to be consistent through coordinate transformation [41,45]; the expression of the improved tangent angle is as follows:

$$\alpha = \arctan(V/V_a) \tag{1}$$

where V is the actual deformation rate of the landslide and V_a is the deformation rate of the uniform speed deformation stage of the landslide. When the tangent angle is greater than 45°, it marks the beginning of the landslide in the accelerated deformation stage, and the closer the tangent angle is to 90°, the closer the occurrence of a landslide (Figure 5).

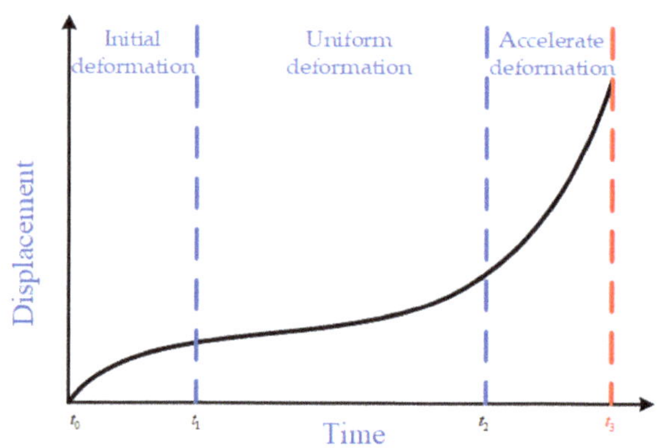

Figure 4. Saito curve model.

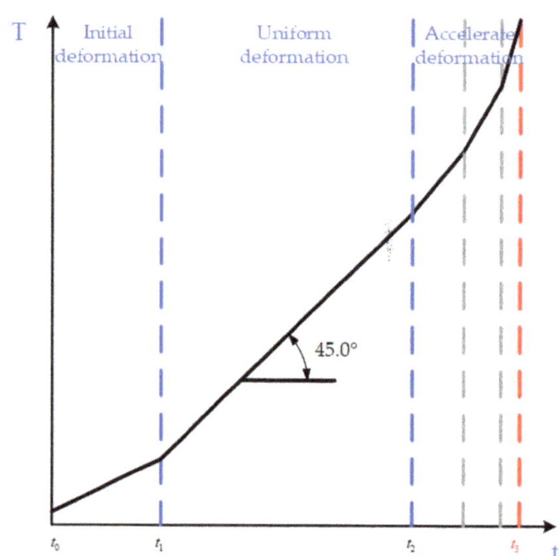

Figure 5. Tangent angle model.

3.1. Forecast Process

The flow chart of slope landslide prediction proposed in this paper is shown in Figure 6. First, all pixels of radar data are traversed, and deformed pixels are filtered out according to the cumulative deformation displacement threshold and deformation velocity threshold. Then, the filtered pixels are then connected by a connectivity algorithm to obtain multiple connected areas. When the deformation area is determined, the *DOD* (degree of deformation) of the target area is obtained by calculating the area, volume, and average coherence of the deformation area. Finally, the new tangent angle warning criterion is used in landslide prediction.

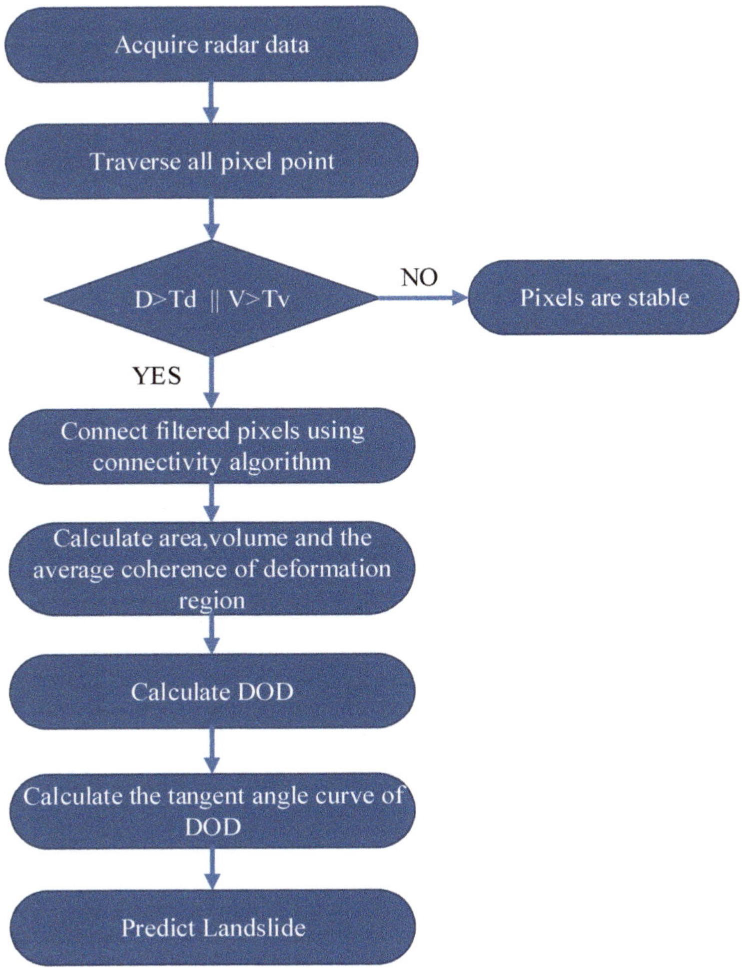

Figure 6. Flowchart of the landslide prediction method.

3.2. DOD

The azimuth resolution of the micro-deformation monitoring radar is expressed by the azimuth, which leads to an uneven size for the actual terrain corresponding to each pixel of the radar data. The farther the distance, the larger the pixel area. The pixel area can be calculated by the azimuth resolution, range resolution, and range sampling position of the micro-deformation monitoring radar [30]:

$$S_{single} = \frac{\alpha}{2}\left([n\beta + R]^2 - [(n-1)\beta + R]^2\right) = \alpha n\beta^2 + \alpha\beta R - \frac{\alpha}{2}\beta^2 \tag{2}$$

where S_{single} is the area of a single pixel, α is the azimuth resolution, β is the range resolution, n is the sampling position of the corresponding pixel in the range direction, and R is the radar monitoring close distance.

During the long-term monitoring process, the data may fluctuate due to certain uncontrollable factors in the monitoring environment, such as human activities, local vibrations caused by construction, and the noise of the system itself. If only a single pixel is used for region identification, pixels that fluctuate in value due to interference are mistaken for distorted pixels. Landslides, on the other hand, usually require a large deformation area to a certain extent and not just a few isolated pixels. In this paper, pixels are filtered according to the deformation velocity threshold T_v and cumulative displacement threshold T_d. The identified pixels are connected through the connectivity algorithm to identify the sensitive areas that need to be focused on. The landslide warning under multiple thresholds can have a good effect and avoid single-point false warnings. Not only should the actual situation of the monitoring site, including the influence of topography, landform, and human factors, be considered for the setting of the threshold, but also it needs to be continuously adjusted in the process of data collection. The area of the deformation area is the area of the contained pixels, and the deformation volume is calculated as follows:

$$B_v = \sum_{n=1}^{N}\left(D_{defo,n} \times S_n\right) \tag{3}$$

where B_v is the deformation volume, $D_{defo,n}$ is the deformation value corresponding to the nth pixel, S_n is the area of the nth pixel, and N is the total number of pixels in the deformation area.

The micro-deformation monitoring radar can obtain the coherence of the target, which is the similarity of the observation results of the target at two different times [46]. If the target does not change between two echoes, the target coherence is 1. If the coherence reaches 0, the target is completely incoherent. When the wavelength, angle of incidence, and time baseline of the radar are constant, there is a nonlinear inverse relationship between the deformation velocity and the coherence [32,47]. Qi Lin also demonstrated this conclusion through simulation [30]. The coherence is calculated as follows:

$$\gamma = \left| \frac{\sum_{i=1}^{n} A_i^2 A_{1,i} \exp\left(i\left[-\frac{4\pi}{\lambda}\Delta R + \Delta\beta\right]\right)}{\sqrt{\sum_{i=1}^{n} |A_i|^2 * \sum_{i=1}^{n} |A_i A_{1,i}|^2}} \right| \tag{4}$$

where γ is the coherence, A is the backscattering coefficient of the target, ΔR is the deformation value of the target, and $\Delta\beta$ is the scattering phase. Therefore, the average coherence of the deformation region is as follows:

$$\gamma_{average} = \left(\sum_{n=1}^{N} \gamma_{average}\right) \Big/ N \tag{5}$$

where $\gamma_{average}$ is the average coherence of the deformation region and N is the number of pixels in the deformation region.

Traditionally, the landslide is usually predicted according to a single pixel which can be easily disturbed by environmental factors during monitoring. Thus, the deformation value may be abrupt, leading to misjudgment of the situation of the slope in monitoring.

Therefore, from the point of view of slope, this paper proposes the parameter *DOD* for slope landslide prediction by using the nonlinear inverse relationship between deformation velocity and coherence. To describe the degree of deformation of the slope over a certain period of time, the *DOD* is calculated as follows:

$$DOD = \frac{B_v}{\gamma_{average}{}^2} \tag{6}$$

The variables used in the calculation of the *DOD* parameter presented in this paper are coherence and deformation volume, but the variables used in the displacement tangent angle method are displacement, so the variables used in the two methods are completely different. By applying the displacement tangent angle method, the diverse data obtained by micro-deformation monitoring radar could not be fully analyzed and studied. Therefore, the newly proposed method, *DOD* tangent angle, is in an attempt to improve this problem. In the actual monitoring process, multiple thresholds and *DOD* tangent angles are used to recognize which stage of accelerated deformation the landslide is currently in so that the landslide can be accurately predicted before it happens.

In summary, this paper proposes a landslide prediction method combining a multi-threshold criterion and the tangent angle, which provides a feasible method for the early warning of open-pit landslides. For a deformed mine slope, as the deformation accumulates, the calculated *DOD* begins to increase, and when the *DOD* increases to a certain extent, a landslide will occur. Usually, before the landslide, the *DOD* curve will increase significantly.

4. Result

The cumulative deformation displacement map within the time window of 0:00 on 18 April 2021 to 9:00 on 21 April 2021 is shown in Figure 7, and it can be clearly seen that the deformation value of pixels in region 1 changes greatly. The pixels are filtered according to the conditional threshold proposed in this article, and the obtained results are shown in Figure 8, where it is obvious that regions 1 and 2 are in the same position in the observation scene, and most of the pixels in region 1 are identified.

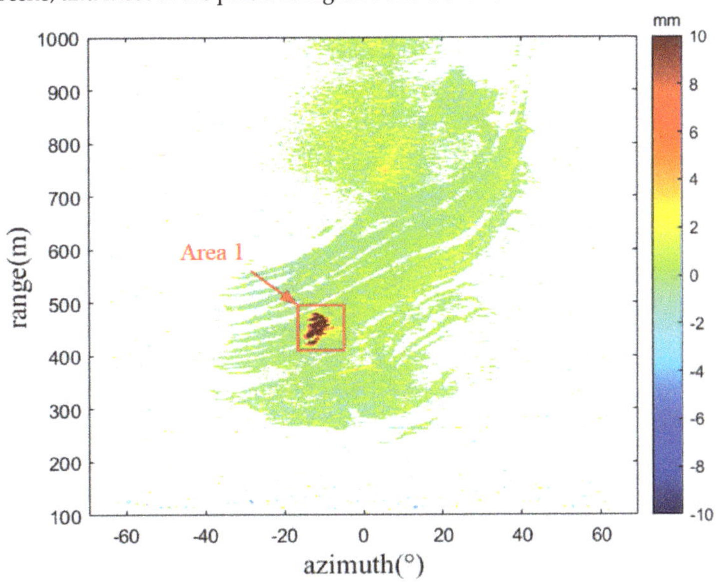

Figure 7. Cumulative deformation diagram.

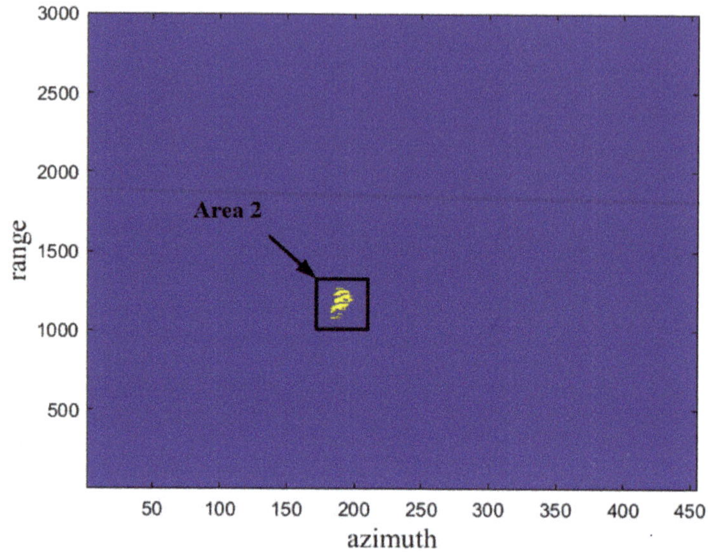

Figure 8. The deformed pixels are finally selected.

It can be seen from Figure 9 that the displacement in the monitoring area is continuously increasing, and the cumulative displacement curve basically conforms to the "three stages" of the Saito model landslide. From 22:54 on 18 April to about 2:00 on 19 April, the region was basically in a stable state without obvious deformation; from 3:00 on 19 April to around 19:00 on 20 April, the deformation in this area is in the stage of constant velocity deformation; from 19:00 on 20 April to around 8:50 on 21 April, the deformation in this area is in the accelerated deformation stage. The cumulative displacement of characteristic points increases sharply, and the deformation speed also continuously increases. The maximum cumulative displacement is 57 mm, and the maximum speed is 10.1 mm/h (Figure 10).

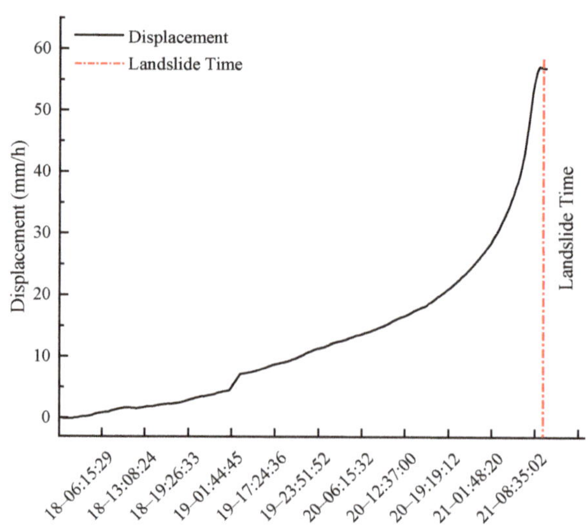

Figure 9. Displacement curve.

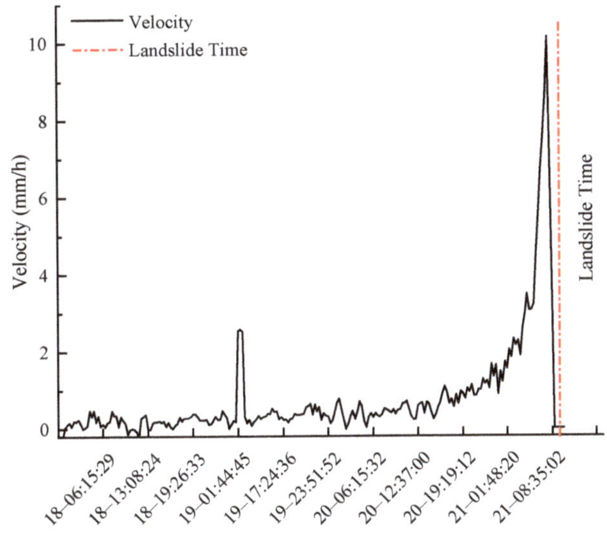

Figure 10. Velocity curve.

The displacement curve before sliding shows a horizontal trend, and the velocity curve shows a downward trend. Owing to rapid deformation, the significantly reduced coherence of the target region makes it impossible for the system to obtain valid deformation data (Figures 9 and 10). As shown in Figure 11, the coherence curve began to decrease overall on the morning of 20 April, and at about 8 a.m. on 21 April, coherence was less than 0.7 and began to drop sharply (based on monitoring experience, the threshold for coherence is usually 0.7). Changes in coherence before landslides provide us with a new perspective for monitoring slopes. Therefore, as shown in Figure 12, the change in coherence is added to the calculation not only to reproduce the deformation trend of the landslide area, but also for comparison with the velocity curve. The *DOD* curve is more concise and clear, and the starting point of rapid growth in *DOD* corresponds to the accelerated deformation stage in the Saito model. *DOD* increases sharply before slippage, and *DOD* obtains the maximum as the landslide occurs. After sliding, *DOD* continues to drop; then, there is a gradual stabilization of the target area and a gradual increase in coherence, and *DOD* continues to decrease. In summary, it is clear that the *DOD* curve reflects the overall trend of the slope in the accelerated deformation stage.

The most critical point of displacement tangent angle determination is the need to determine the uniform speed deformation stage of the displacement–time curve. Because of the interference from environmental factors and measurement errors, the deformation rate at each moment in the constant velocity deformation stage will fluctuate rapidly or slowly, thereby causing difficulties in determining the rate of the uniform speed deformation stage. Therefore, it should be continuously adjusted in line with the actual deformation (Figure 13). The calculation of the value for the parameter *DOD* proposed in this paper begins when the cumulative displacement and the deformation rate reach a certain threshold, at which the slope of the *DOD* curve shows great changes within a relatively short time window. The time window of the uniform deformation stage of *DOD* is narrower, making it easier to obtain the uniform deformation rate of *DOD* compared with the displacement rate. As can be seen from Figure 14, the *DOD* tangent angle curve is more intuitive and clear, the maximum tangent angle is reached at the slip moment, and the displacement tangent angle is determined by the fluctuation of the data. Although there seems to be an overall trend of a gradual increase in the tangent angle, the up and down fluctuations in the actual monitoring process make it easy to make misjudgments in monitoring.

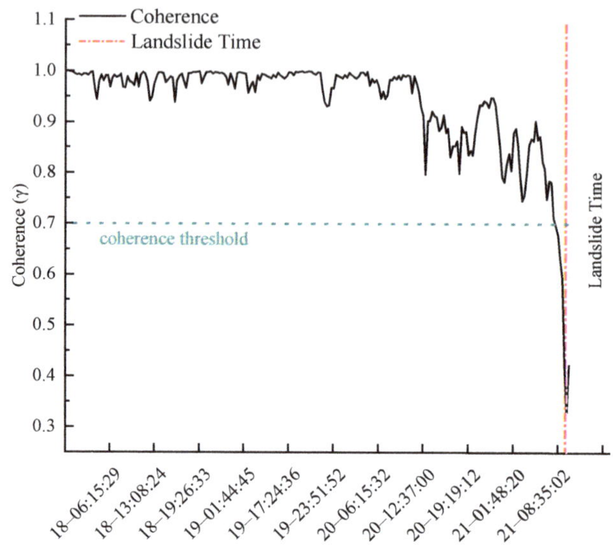

Figure 11. Coherence curve.

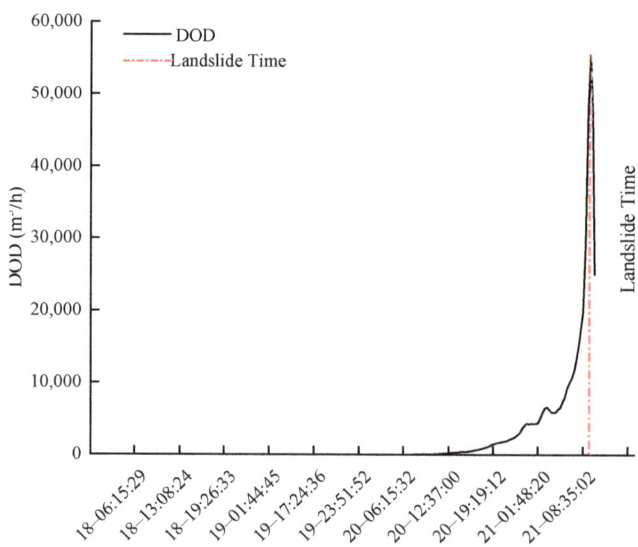

Figure 12. *DOD* curve.

Landslide prediction can be divided into medium and long-term prediction (more than 6–12 months), short-term prediction (3–6 months), and imminent sliding prediction. Generally, for a mine, prediction 2 h in advance is required. In addition, the instability of slope rock mass is unavoidable. On the premise of not affecting production, the mine will pursue the steepest slope angle with maximum benefit [11]. In summary, in order to more accurately judge the slippage trend of the slope, the velocity threshold, cumulative displacement threshold, and *DOD* tangent angle threshold of different deformation stages before the slope slippage of the open-pit mine were established based on the historical data of the previous slope landslide and the diversified data of the micro-deformation monitoring radar. In addition, a comprehensive early warning criterion, as shown in Table 1, was established. The acceleration is divided into three stages: the initial acceleration stage is a blue warning, the medium acceleration stage is an orange caution, and the critical sliding stage is a red alarm.

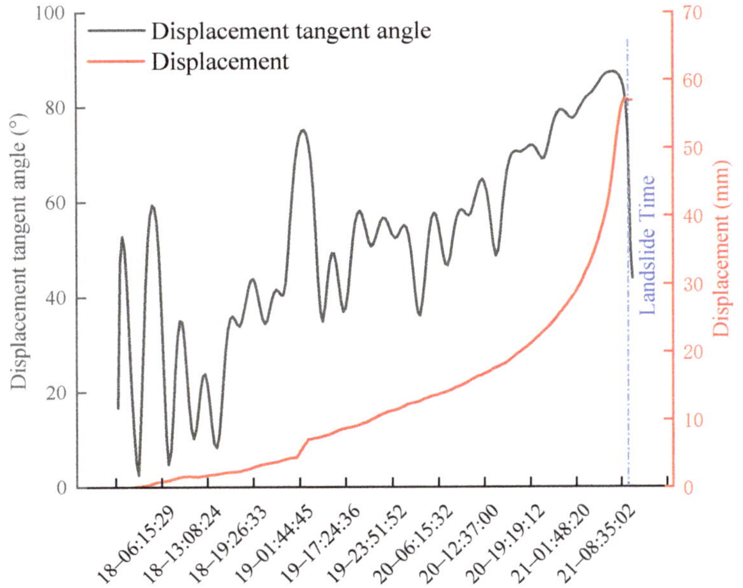

Figure 13. Displacement tangent angle.

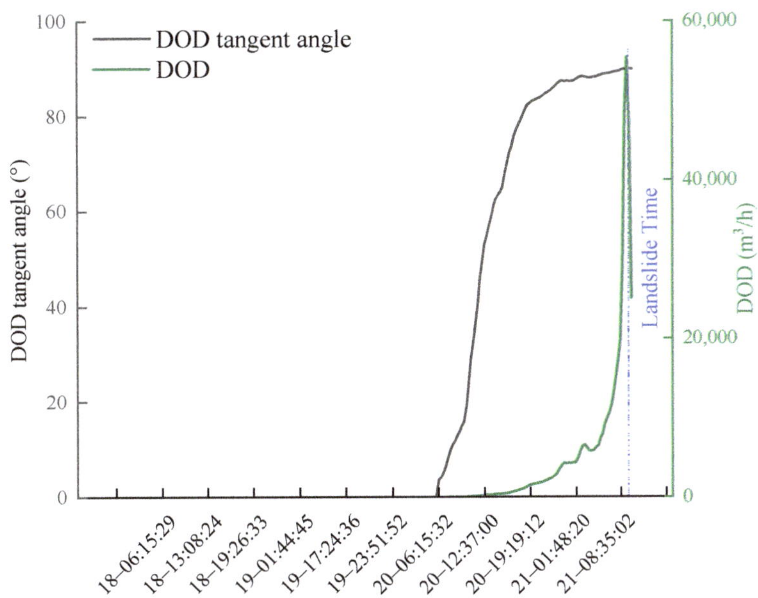

Figure 14. *DOD* tangent angle.

Table 1. Comprehensive early warning criterion at an accelerated stage.

Deformation Stage	Initial Acceleration	Medium Acceleration	Critical Sliding
Deformation velocity, V (mm/h)	$V > 0.5$	$2 < V < 0.5$	$V > 3$
Cumulative displacement, D (mm/day)	$D > 15$	$14 < D < 30$	$D > 30$
DOD tangent angle, α	$70° < \alpha < 85°$	$85° \leq \alpha < 89°$	$\alpha \geq 89°$
Warning level	Warning	Caution	Alarm

According to the landslide warning criterion proposed in this paper, the fractal deformation rate of 7:05 a.m. on 21 April 2021 was 7.5 mm/h, the cumulative deformation displacement was 47.9 mm, and the *DOD* tangent angle reached 89.13°, triggering the red warning, and at about 9:37 a.m. on 21 April 2021, the landslide was unstable, and the surveillance video of the mining area recorded the moment of the landslide (Figures 14 and 15). The displacement tangent angle criterion is that when the slope enters the slippery stage, the tangent angle is greater than 85°, and the corresponding moment is 04:45 a.m. on 21 April 2021 (Figures 12 and 13). Complying with the management requirements of the open-pit mine slope, the *DOD* tangent angle can give a red alarm 2 h in advance so that the mining area is insusceptible. Meanwhile, time is sufficient to transfer the mining facilities from the dangerous area in order to avoid the loss of property and casualties to the maximum extent. The scene after the landslide is shown in Figure 16.

Figure 15. Landslide collapse process.

Figure 16. The scene after the landslide.

The deformation process of the landslide area is shown in Figure 17. The deformation data from radar monitoring are combined with the digital terrain model (DSM) data such that the monitoring personnel can judge the large deformation in the monitoring area by observing the displacement and deformation cloud map and further improve the accuracy of landslide monitoring in combination with the slope landslide early warning system, greatly improving the effect of landslide prevention.

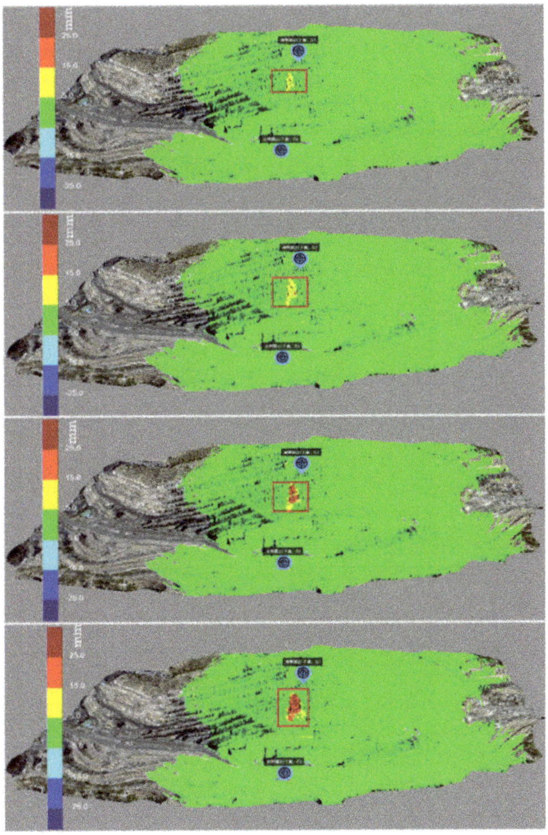

Figure 17. Slope radar monitoring cumulative deformation.

5. Discussion

The traditional threshold early warning strategy is usually used to set the threshold value of a single pixel point for early warning. This is not an ideal situation, because a single pixel point is vulnerable to interference from the external environment and other factors, resulting in the phenomenon of the monitoring data of a single pixel point having a sudden change or continuing to increase. Therefore, when the change in the monitoring data of only a single pixel point is used as the evaluation standard for landslide early warning, it is prone to false early warning. Using remote sensing monitoring means, monitoring is conducted in point cloud monitoring mode. At present, there is less research on the mode of landslide early warning based on the regional surface. Therefore, from the point of view of landslide surface and landslide mass, this paper proposes a landslide early warning index based on double thresholds of deformation velocity and cumulative displacement combined with the tangent angle model to propose a comprehensive early warning criterion for open-pit mine slope landslides.

The setting of a slope threshold is always a difficult problem, and few scholars have proposed effective thresholds. Therefore, the identification of rapidly growing data points is based on a set threshold, which is based on the analysis after a period of data accumulation and needs to be continuously optimized and adjusted in the process of gradual data collection and enrichment. This value varies for slopes of differing scales, formation conditions, and rock and soil types. The historical data analysis method is used in actual monitoring; it involves judging the maximum deformation rate in the past, which is used as the threshold for an unacceptable maximum deformation rate resulting in a landslide. When setting the actual threshold, according to an actual situation, there should be comprehensive consideration of the nature of rock and soil mass, the degree of artificial disturbance, the possible influence range, and other factors that increase the maximum deformation value by 10–20%. When the slope deformation speed exceeds this value at a certain time, it is necessary to focus on slope stability. If there is no landslide or signs of a landslide, the threshold value is updated according to the deformation speed at this time.

Based on the nonlinear inverse relationship between deformation volume and coherence, a new tangent angle early warning criterion is proposed according to monitoring experience. The target deformation will affect the scattering characteristics of the target. Similarly, weather factors such as rain and snow will also affect the radar echo. Ensuring good coherence is a prerequisite for accurate early warning. Added to that, it should be noted that *DOD* will be calculated only when the deformation speed and cumulative displacement reach a certain cumulative level. Accurate identification of the acceleration phase of the landslide requires substantial data accumulation and analysis.

6. Conclusions

The stability of high and large rock slopes in open-pit mines represents the core problem of safety production in open-pit mines, and the security of open-pit mines in China is still a serious issue. Micro-deformation monitoring radar is widely used in slope monitoring based on its characteristics of high spatial resolution, short time baseline, submillimeter monitoring accuracy, surface monitoring, and suitability for all-weather monitoring. In this paper, a new parameter, *DOD*, is proposed for measuring the degree of slope deformation based on the deformation volume and coherence value, and the tangent angle value of this parameter is combined with the dual threshold values of velocity and displacement in landslide prediction. With the introduction of deformation volume and coherence, this method can be used to provide timely and effective early warning before the sliding of a target and more accurate and reasonable prediction results for slope managers. Based on the measured data of a slope and landslide in a mining area, the method realizes timely warning of an impending landslide and allows the avoidance of property loss in the mining area.

Author Contributions: Conceptualization, W.T.; writing—original draft, Y.W.; methodology, W.T., Y.W. and Y.Q.; formal analysis, Y.W., P.H. and W.X.; supervision, P.H., C.L. and Y.C. All authors have read and agreed to the published version of the manuscript.

Funding: This work was supported in part by the National Natural Science Foundation of China (Grant Nos. 61971246 and 52064039), the Joint Funds of the National Natural Science Foundation of China (Grant No. U22A2010), the Science and Technology Planned Project of Inner Mongolia (Grant Nos. 2019GG139 and 2020GG0073), the Science and Technology Major Project of Inner Mongolia (Grant No. 2019ZD022), the Science and Technology Leading Talent Team of Inner Mongolia (Grant No. 2022LJRC0002), the Fundamental Research Funds for Universities of Inner Mongolia (Grant No. JY20220077), and the Natural Science Foundation of Inner Mongolia (Grant Nos. 2019MS04004, 2020ZD18 and 2021MS06004).

Data Availability Statement: Not applicable.

Acknowledgments: Many thanks to Inner Mongolia Pattern Technology Co., Ltd., for providing the field-measured data. Thanks to anonymous reviewers for their constructive comments to improve the quality of this paper.

Conflicts of Interest: The authors declare no conflict of interest.

References

1. Haque, U.; da Silva, P.F.; Devoli, G.; Pilz, J.; Zhao, B.; Khaloua, A.; Wilopo, W.; Andersen, P.; Lu, P.; Lee, J. The Human Cost of Global Warming: Deadly Landslides and Their Triggers (1995–2014). *Sci. Total Environ.* **2019**, *682*, 673–684. [CrossRef] [PubMed]
2. Lin, Q.; Wang, Y. Spatial and Temporal Analysis of a Fatal Landslide Inventory in China from 1950 to 2016. *Landslides* **2018**, *15*, 2357–2372. [CrossRef]
3. Ma, G.; Hu, X.; Yin, Y.; Luo, G.; Pan, Y. Failure Mechanisms and Development of Catastrophic Rockslides Triggered by Precipitation and Open-Pit Mining in Emei, Sichuan, China. *Landslides* **2018**, *15*, 1401–1414. [CrossRef]
4. Thiebes, B.; Bell, R.; Glade, T.; Jäger, S.; Mayer, J.; Anderson, M.; Holcombe, L. Integration of a Limit-Equilibrium Model into a Landslide Early Warning System. *Landslides* **2014**, *11*, 859–875. [CrossRef]
5. Whittle, D.; Brazil, M.; Grossman, P.A.; Rubinstein, J.H.; Thomas, D.A. Combined Optimisation of an Open-Pit Mine Outline and the Transition Depth to Underground Mining. *Eur. J. Oper. Res.* **2018**, *268*, 624–634. [CrossRef]
6. Amini, M. Stability Analysis of the North-Eastern Slope of Daralou Copper Open Pit Mine against a Secondary Toppling Failure. *Eng. Geol.* **2019**, *249*, 89–101. [CrossRef]
7. Obregon, C.; Mitri, H. Probabilistic Approach for Open Pit Bench Slope Stability Analysis—A Mine Case Study. *Int. J. Min. Sci. Technol.* **2019**, *29*, 629–640. [CrossRef]
8. Song, D.; Che, A.; Zhu, R.; Ge, X. Dynamic Response Characteristics of a Rock Slope with Discontinuous Joints under the Combined Action of Earthquakes and Rapid Water Drawdown. *Landslides* **2018**, *15*, 1109–1125. [CrossRef]
9. Wang, Y.; Zhang, N.N.; Sun, Z.; Fei, W. The major domestic natural disasters in 2017. *Disaster Reduct. China* **2018**, *5*, 38–41.
10. Tian, H.; Bai, R.; Zhao, H. Achievement and developing trend of open-pit mining in China. *Opencast Min. Technol.* **2019**, *34*, 1–9.
11. Yang, T.; Wang, H.; Dong, X. Current situation problems and countermeasures of intelligent evaluation of slope stability in open pit. *J. China Coal Soc.* **2020**, *45*, 2277–2295.
12. Jiao, R.; Wang, S.; Yang, H.; Guo, X.; Han, J.; Pei, X.; Yan, C. Comprehensive Remote Sensing Technology for Monitoring Landslide Hazards and Disaster Chain in the Xishan Mining Area of Beijing. *Remote Sens.* **2022**, *14*, 4695. [CrossRef]
13. Ge, D.; Dai, K.; Guo, Z.; Li, Z. Early Identification of Serious Geological Hazards with Integrated Remote Sensing Technologies: Thoughts and Recommendations. *Geomat. Inf. Sci. Wuhan Univ.* **2019**, *44*, 996–1007.
14. Xu, Q.; Dong, X.J.; Li, W.L. Integrated space-air-ground early detection, monitoring and warning system for potential catastrophic geohazards. *Geomat. Inf. Sci. Wuhan Univ.* **2019**, *44*, 957–966.
15. Li, Z.H.; Song, C.; Yu, C. Application of satellite radar remote sensing to landslide detection and monitoring: Challenges and solutions. *Geomat. Inf. Sci. Wuhan Univ.* **2019**, *44*, 967–979.
16. Qin, H.; Ma, H.; Yu, Z. Analysis Method of Landslide Early Warning and Prediction Supported by Ground-Based SAR Technology. *Geomat. Inf. Sci. Wuhan Univ.* **2020**, *45*, 1697–1706.
17. Bozzano, F.; Mazzanti, P.; Prestininzi, A.; Scarascia Mugnozza, G. Research and development of advanced technologies for landslide hazard analysis in Italy. *Landslides* **2010**, *7*, 381–385. [CrossRef]
18. Intrieri, E.; Gigli, G.; Casagli, N.; Nadim, F. Brief communication "landslide early warning system: Toolbox vand general concepts". *Nat. Hazards Earth Syst. Sci.* **2013**, *13*, 85–90. [CrossRef]
19. Tommaso, C.; Paolo, F.; Emanuele, I.; Kostas, B.; Nicola, C. On the monitoring and early-warning of brittle slope failures in hard rock masses: Examples from an open-pit mine. *Eng. Geol.* **2017**, *228*, 71–81.
20. Lopez-Vinielles, J.; Ezquerro, P.; Fernandez-Merodo, J. Remote analysis of an open-pit slope failure: Las Cruces case study, Spain. *Landslides* **2020**, *17*, 2173–2188. [CrossRef]
21. Carlà, T.; Intrieri, E.; Raspini, F.; Bardi, F.; Farina, P.; Ferretti, A.; Colombo, D.; Novali, F.; Casagli, N. Perspectives on the prediction of catastrophic slope failures from satellite InSAR. *Sci. Rep.* **2019**, *9*, 14137. [CrossRef]
22. Crosetto, M.; Monserrat, O.; Luzi, G.; Cuevas-González, M.; Devanthéry, N. A noninterferometric procedure for deformation mea-surement using GB-SAR imagery. *IEEE Geosci. Remote Sens. Lett.* **2013**, *11*, 34–38. [CrossRef]
23. Dick, G.J.; Eberhardt, E.; Cabrejo-Liévano, A.G.; Stead, D.; Rose, N.D. Development of an early-warning time-of-failure analysis methodology for open-pit mine slopes utilizing ground-based slope stability radar monitoring data. *Can. Geotech. J.* **2015**, *52*, 515–529. [CrossRef]
24. Zeng, T.; Deng, Y.K. Development state and application examples of ground-based differential interferometric radar. *J. Radars* **2019**, *8*, 154–170.
25. Liu, B.; He, K.; Han, M.; Hu, X.; Ma, G.; Wu, M. Application of UAV and GB-SAR in Mechanism Research and Monitoring of Zhonghaicun Landslide in Southwest China. *Remote Sens.* **2021**, *13*, 1653. [CrossRef]
26. Owerko, T.; Kuras, P.; Ortyl, Ł. Atmospheric Correction Thresholds for Ground-Based Radar Interferometry Deformation Monitoring Estimated Using Time Series Analyses. *Remote Sens.* **2020**, *12*, 2236. [CrossRef]
27. Hu, C.; Deng, Y.; Tian, W.; Zhao, Z. A Compensation Method for a Time–Space Variant Atmospheric Phase Applied to Time-Series GB-SAR Images. *Remote Sens.* **2019**, *11*, 2350. [CrossRef]
28. Huang, Z.; Sun, J.; Tan, W.; Huang, P.; Han, K. Investigation of Wavenumber Domain Imaging Algorithm for Ground-Based Arc Array SAR. *Sensors* **2017**, *17*, 2950. [CrossRef]
29. Gao, H.; Liao, M.; Liu, X.; Xu, W.; Fang, N. Source Geometry and Causes of the 2019 Ms6.0 Changning Earthquake in Sichuan, China Based on InSAR. *Remote Sens.* **2022**, *14*, 2082. [CrossRef]

30. Qi, L.; Tan, W.; Huang, P.; Xu, W.; Qi, Y.; Zhang, M. Landslide Prediction Method Based on a Ground-Based Micro-Deformation Monitoring Radar. *Remote Sens.* **2020**, *12*, 1230. [CrossRef]
31. Xiao, T.; Huang, W.; Deng, Y.; Tian, W.; Sha, Y. Long-Term and Emergency Monitoring of Zhongbao Landslide Using Space-Borne and Ground-Based InSAR. *Remote Sens.* **2021**, *13*, 1578. [CrossRef]
32. Zhang, H.; Wang, C. *Research on DInSAR Method Based on Coherent Target*; Science Press: Beijing, China, 2009; pp. 12–28.
33. Zheng, X.; Yang, X.; Ma, H.; Ren, G.; Zhang, K.; Yang, F.; Li, C. Integrated Ground-Based SAR Interferometry, Terrestrial Laser Scanner, and Corner Reflflector Deformation Experiments. *Sensors* **2018**, *18*, 4401. [CrossRef] [PubMed]
34. Fukuzono, T. A method to predict the time of slope failure caused by rainfall using the inverse number of velocity of surface displacement. *J. Japan Landslide Soci.* **1985**, *2*, 8–13. [CrossRef]
35. Voight, B. A method for prediction of volcanic eruptions. *Nature* **1988**, *332*, 125–130. [CrossRef]
36. Zhou, X.; Ye, T. Inverse-Square-Root-Acceleration Method for Predicting the Failure Time of Landslides. *Sci. China Technol. Sci.* **2021**, *64*, 1127–1136. [CrossRef]
37. Du, J.; Yin, K.; Lacasse, S. Displacement prediction in colluvial landslides, three Gorges reservoir, China. *Landslides* **2013**, *10*, 203–218. [CrossRef]
38. Du, W.; Wang, G. A one-step Newmark displacement model for probabilistic seismic slope displacement hazard analysis. *Eng. Geol.* **2016**, *205*, 12–23. [CrossRef]
39. Lian, C.; Zeng, Z.; Yao, W.; Tang, H. Multiple neural networks switched prediction for landslide displacement. *Eng. Geol.* **2015**, *186*, 91–99. [CrossRef]
40. Chousianitis, K.; Gaudio, V.; Kalogeras, I. Predictive model of Arias intensity and Newmark displacement for regional scale evaluation of earthquake-induced landslide hazard in Greece. *Soil Dyn. Earthq. Eng.* **2014**, *65*, 11–29. [CrossRef]
41. Xu, Q.; Zeng, Y.; Qian, J.; Wang, C.; He, C. Study on an improved tangential angle and the corresponding landslide pre-warning criteria. *Geol. Bull. China* **2009**, *28*, 501–505.
42. Xu, Q.; Peng, D.; He, C.; Qi, X.; Zhao, K.; Xiu, D. Theory and method of monitoring and early warning for sudden loess landslide—A case study atheifangtai terrace. *J. Eng. Geol.* **2020**, *28*, 111–121.
43. Xu, Q. Understanding the landslide monitoring and early warning:consideration to practical issues. *J. Eng. Geol.* **2020**, *28*, 360–374.
44. Intrieri, E.; Gigli, G.; Mugnai, F.; Fanti, R.; Casagli, N. Design and implementation of a landslide early warning system. *Eng. Geol.* **2012**, *147*, 124–136. [CrossRef]
45. Wang, J.; Zhang, Z. *System Engineering Geology Study on Typical Rapid Loess Landslides*; Sichuan Science & Technology Press: Chengdu, China, 1999.
46. Touzi, R.; Lopes, A. Coherence estimation for SAR imagery. *IEEE Trans. Geosci. Remote Sens.* **1999**, *37*, 135–149. [CrossRef]
47. Zebker, H.A.; Villasenor, J. Decorrelation in interferometric radar echoes. *IEEE Trans. Geosci. Remote Sens.* **1992**, *30*, 950–959. [CrossRef]

remote sensing

MDPI

Article

Timely and Low-Cost Remote Sensing Practices for the Assessment of Landslide Activity in the Service of Hazard Management

Aggeliki Kyriou, Konstantinos G. Nikolakopoulos * and Ioannis K. Koukouvelas

Department of Geology, University of Patras, 265 04 Patras, Greece
* Correspondence: knikolakop@upatras.gr

Abstract: Landslides are among the most dangerous and catastrophic events in the world. The increasing progress in remote sensing technology made landslide observations timely, systematic and less costly. In this context, we collected multi-dated data obtained by Unmanned Aerial Vehicle (UAV) campaigns and Terrestrial Laser Scanning (TLS) surveys for the accurate and immediate monitoring of a landslide located in a steep and v-shaped valley, in order to provide operational information concerning the stability of the area to the local authorities. The derived data were processed appropriately, and UAV-based as well as TLS point clouds were generated. The monitoring and assessment of the evolution of the landslide were based on the identification of instability phenomena between the multi-dated UAV and TLS point clouds using the direct cloud-to-cloud comparison and the estimation of the deviation between surface sections. The overall evaluation of the results revealed that the landslide remains active for three years but is progressing particularly slowly. Moreover, point clouds arising from a UAV or a TLS sensor can be effectively utilized for landslide monitoring with comparable accuracies. Nevertheless, TLS point clouds proved to be denser and more appropriate in terms of enhancing the accuracy of the monitoring process. The outcomes were validated using measurements, acquired by the Global Navigation Satellite System (GNSS).

Keywords: landslide mapping; landslide monitoring; UAV; terrestrial laser scanning; 3D point clouds

Citation: Kyriou, A.; Nikolakopoulos, K.G.; Koukouvelas, I.K. Timely and Low-Cost Remote Sensing Practices for the Assessment of Landslide Activity in the Service of Hazard Management. *Remote Sens.* **2022**, *14*, 4745. https://doi.org/10.3390/rs14194745

Academic Editor: Christian Bignami

Received: 18 August 2022
Accepted: 19 September 2022
Published: 22 September 2022

1. Introduction

Landslides are among the most dangerous and catastrophic natural disasters in the world. They usually occur suddenly and can be detrimental to the natural environment, the infrastructure or even human life itself. As climate change is unequivocal, model-based estimations propose that warming temperatures would lead to increased activity of landslides [1–3]. Over the years, several researchers dealt with the investigation of such phenomena and various methods focused on landslide vulnerability mapping, hazard zoning, risk assessment, or rock-fall simulations, have been developed [4,5]. However, landslide research has never been more urgent and important.

The increasing progress in remote sensing technology have made landslide observations more timely, systematic and less costly. Moreover, new possibilities for high-precision research of landslides located in inaccessible areas or extensive landslides have emerged. Novel methodologies, based on multiple remote sensing data, have already been indispensable tools for landslide assessment and risk prevention [6,7].

The growing use of Unmanned Aerial Vehicles (UAVs) has been a real milestone in Earth observation, and therefore, in landslide research [8,9]. The first approaches were based on the successful exploitation of UAVs with compact cameras for the rapid identification of landslides [10]. Subsequently, UAVs mounted with digital single-lens reflex (DSLR) cameras, were used for the documentation and monitoring of large earthflows [11]. An academic research team developed a UAV, equipped with a consumer-grade optical camera in order to generate 3D surface models for the more comprehensive characterization and monitoring of unstable areas [12].

Moreover, time series of UAV images were processed via the Structure from Motion (SfM) technique and the outputs were utilized for the quantification of the surface deformation, the measurement of the landslide volumetric change and the determination of the landslide's dynamics [13,14]. In a respective study, UAV data contributed effectively to the assessment of residual risk (post-landslide risk) on a medium- and long-term scale through the estimation of the evolution of the area [15]. UAV imagery along with digital photogrammetry has successfully assisted in the recording of slope conditions as well as in the enhancement of the understanding of landslide processes and the precise assessment of slope instabilities [16]. As time passes, new and more innovative approaches for landslide investigation are emerging based on the combined use of UAVs and machine learning algorithms for the extraction of landslide susceptibility maps and the monitoring of landslide risk areas [17].

Another common tool for many geotechnical studies and landslide investigations is Light Detection and Ranging (LiDAR) technology [18]. In particular, Airborne LiDARs have proven to be particularly effective in the detailed and accurate representation of the landslide's surface, the recognition of different types of mass movements and the monitoring of landslide dynamics as well as the classification of slow-moving landslides in densely vegetated areas [19–22]. In addition, a variety of studies have already been carried out regarding the utilization of LiDARs on rockfall mapping, rock mass characterization and rockfall susceptibility analysis [23,24]. However, landslide research through airborne LiDAR constitutes a quite expensive approach and thus continuous airborne monitoring is relatively limited.

On the contrary, Terrestrial LiDAR (TLS) surveys are more affordable, providing data with a higher temporal and spatial resolution. An overall overview of TLS acquisition and data processing concerning the characterization, volume estimation and monitoring of rock slopes has already been published [25]. Furthermore, a truly comprehensive study took place in Yosemite Valley where TLS and SfM were utilized for the detection of rockfalls over a 40-year period and the updating of the inventory database with more precise measurements (number, area, volume) [26]. In fact, the terrain models resulting from remote sensing techniques (TLS, SfM) were compared with the corresponding models derived from the processing of historical oblique photographs allowing the detection and quantification of surface changes and providing long-term monitoring.

The integrated use of a variety of remote sensing data is suggested as an alternative perspective for more comprehensive landslide research. Specifically, spaceborne satellite data (high resolution multispectral and radar images) were combined with UAV imagery and ground-based techniques, such as Ground-Based Interferometric SAR (GB-InSAR), TLS, etc., in order to identify, map and monitor landslides, which vary in their characteristics, failure mechanisms, evolution processes, spatial distribution and risk of instability [27]. Moreover, the analysis of the activity of the landslide and the estimation of its kinematic evolution can be achieved effectively either by the execution of repeated UAV campaigns along with Global Navigation Satellite Systems (GNSS) surveys [28] or by the utilization of UAV data in conjunction with (a) airborne LiDAR data [29] or (b) TLS surveys [30].

In the current study, we collected multi-dated data obtained by Unmanned Aerial Vehicle (UAV) and Terrestrial Laser Scanning (TLS) surveys for the accurate and immediate monitoring of a landslide aiming at the provision of operational information concerning the stability of the area to the local authorities. Specifically, a landslide occurred in an environmentally sensitive area, which is located in a steep and narrow valley. Thus, the monitoring of the area constitutes a particularly challenging task, as it must be achieved timely and accurately with as little environmental impact as possible and minimum costs. As both UAV and TLS field campaigns involve only transportation expenses to the study area, they have proved to be a reliable and cost affordable tool for continuous monitoring of small (ten of meters) to moderate (hundreds of meters) active landslides. The aforementioned challenge is the main objective of the current work. In this framework, the derived data were processed appropriately, and UAV-based as well as TLS point clouds were generated. The monitoring and assessment of the evolution of the landslide were based on the identification of instability phenomena between the multi-dated UAV and TLS point clouds using the direct cloud-to-cloud comparison and the estimation of the deviation between surface sections. Moreover, systematic measurements, obtained by the Global Navigation Satellite System (GNSS) were utilized for the verification of the results. Finally, research outcomes were communicated to local authorities in order to execute appropriate measures for the mitigation of the risk.

2. Landslide Area

Northern Peloponnese is characterized as one of the most tectonically and seismically active continental regions worldwide due to the existence of the fast-extending Corinthian rift [31–35]. The rift is dissecting the entire region, from the coastline in the north to the ridges inland since the Pliocene. Normal faulting, sea level changes, as well as the tectonic uplift of rift sediments, are recognized as dominant features in this process. The morphology of the wider area is obviously affected by the ongoing tectonics and it is evidenced by the development of deep and narrow valleys.

Our area of interest is located on the outskirts of the village of Kato Zachlorou within the Region of Western Greece (Figure 1). The first phenomena of instability at the specific site started as rock falls on 19 April 2019. A subsequent reactivation took place on 14 November 2019 through rockfalls and debris falls, while a more extensive event, including rock falls and debris falls occurred on 4 April 2020. Landslide material covered the area of the road in three different locations, contributing to the isolation of the local community from the surroundings (Figure 2).

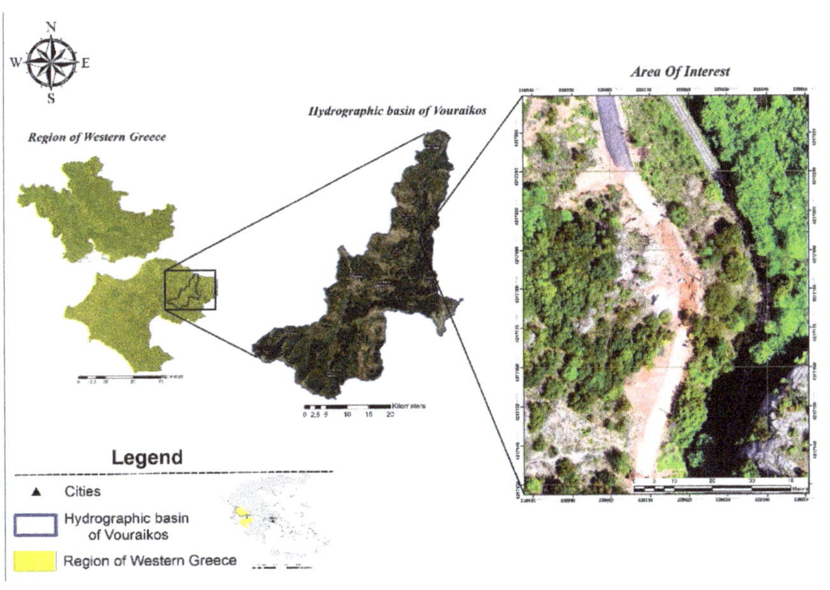

Figure 1. Area of Interest.

Figure 2. (**a**,**b**) UAV photos of the area of interest after the landslide on 4 April 2020. (**c**) Photo of the landslide material displaced on the area of the road.

The specific site is an outstanding natural heritage area, as well as a famous tourist destination. A historic rack railway, named "Odontotos" runs through the gorge of Vouraikos since 1896 and it is one of the most iconic touristic features of the wider area. In this context, the mitigation of landslide risk and the maintenance of human security are top priorities, especially in our area of interest, which is located in a particularly narrow area of the gorge and quite close to the railway lines (~30 m) (Figure 2a). Additionally, this environmentally sensitive area must be safeguarded and thus any monitoring should take place with as little environmental disturbance as possible.

3. Materials and Methods

3.1. Data Acquizition

The precise monitoring of our area of interest started after the occurrence of the extensive rock falls and debris fall on 19 April 2019 and is still in progress. Our datasets included repetitive representations of the study area acquired by either UAV or by TLS as well as high-precision GNSS measurements (Table 1). Repeated UAV/TLS surveys have been carried out at regular intervals for the immediate provision of stability information to the local authorities. UAV flights are operated within 1 h while each TLS survey lasts approximately 4 h.

A DJI Matrice 600 was utilized for the collection of the UAV imagery. The specific hexacopter is equipped with a Zenmuse X5 camera with a 15 mm F/1.7 lens and a 72-degree diagonal field of view. The camera operates with an electronic shutter, capturing images at 16 MP analysis, i.e., photo analysis of 4608 × 3456 pixels. The campaigns were executed once per month at an altitude of 70 m above the ground level, maintaining the same flight grid (Figure 3) and the corresponding photogrammetric characteristics throughout the monitoring period. The acquired UAV photos have 90% along-track and 75% across-track overlap and photogrammetric processing was carried out in Agisoft Metashape software.

TLS surveys were conducted using a Leica ScanStation P50, which allows the extremely fast scanning (1 million points per second) of large areas along with the extraction of high-quality 3D representations. The range accuracy of the specific laser scanner is estimated at about 1.2 mm for ranges varying from 120 m to 270 m. In addition, scanning can be performed under almost any weather conditions (exceptions include stormy winds and heavy rainfall) in a 360-degree horizontal and 270-degree vertical field of view. A Canon EOS 80D camera with 24 MP resolution is mounted onto Leica ScanStation P50 for the improvement of the sharpness of the obtained point clouds. The processing of the multi-dated point clouds took place in Leica Cyclone software.

Table 1. Dates of the repeated UAV, TLS and GNSS surveys.

Survey	Date	UAV	TLS	GNSS	Main Events
	19 April 2019				Rock falls
1	21 April 2019	X			
2	19 May 2019	X			
	14 November 2019				Rock falls and earth slides
	4 April 2020				Extensive rock falls and earth slides
3	11 April 2020	X			
4	7 May 2020	X			
5	10 June 2020	X	X		
	22 July 2020				Slope remediation
6	23 July 2020	X			
	24 July 2020				Construction of GNSS pillars
7	10 August 2020	X		X	
8	14 August 2020	X		X	
9	25 August 2020	X		X	
10	25 September 2020	X	X	X	
11	3 October 2020	X		X	
12	16 December 2020	X		X	
13	26 April 2021	X		X	
14	28 May 2021	X		X	
15	3 July 2021	X		X	
16	7 November 2021	X	X	X	

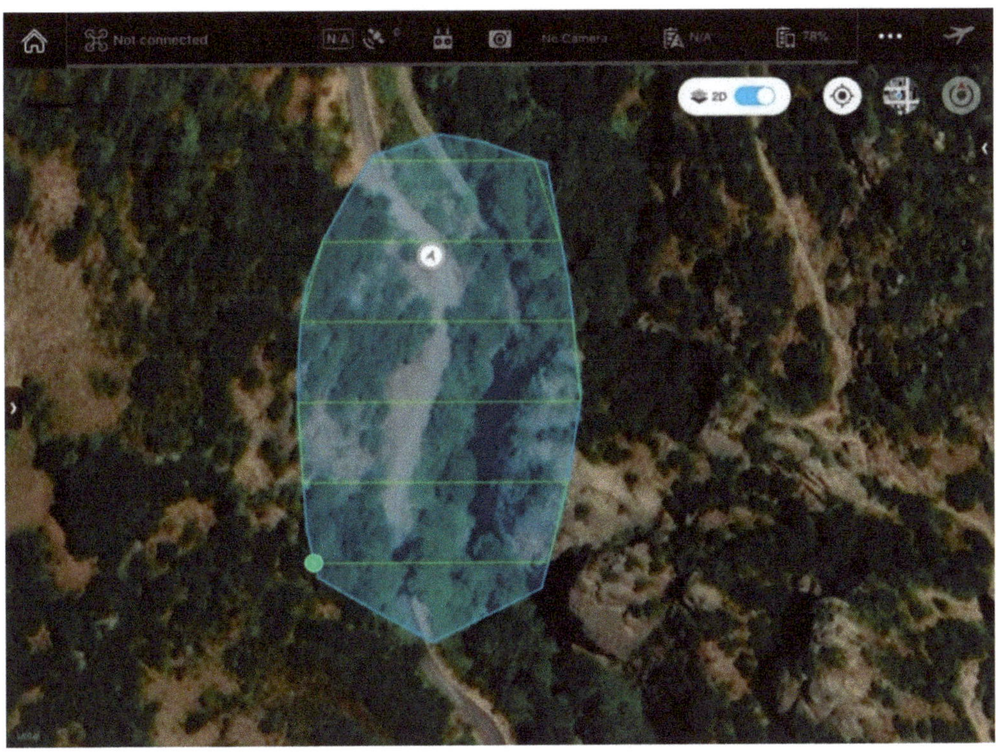

Figure 3. The flight grid of the photogrammetric UAV campaigns.

Moreover, several repetitive static GNSS measurements were executed utilizing a Leica GS08 GNSS Receiver. The adopted methodology for the construction of the permanent GNSS pillar and the and subsequent monitoring has already been described in detail [28]. The measurements were performed on permanent pillars (Figure 4), which are located at key points within the area of interest in order to guarantee the performance of each measurement exactly at the same position. In particular, three of these permanent pillars were constructed along the paved road, while two others were placed outside the landslide. GNSS measurements were utilized both to monitor instability phenomena and to verify the results of remote sensing approaches.

Figure 4. Permanent location on the road area for the execution of GNSS measurements.

Finally, square 4.5″ black and white targets were distributed throughout the area of instability (Figure 5) in each UAV campaign or TLS survey, aiming at minimizing georeferencing errors and enhancing the registration quality of the multi-dated outputs. The position of each target was measured using a Leica GS08 GNSS Receiver.

3.2. Methodology

The current research focuses on the accurate and timely monitoring of landslide activity using low-cost, repeatable remote sensing data obtained by UAV and TLS sensors, in an environmentally sensitive area. Our main purpose is to inform the local authorities about the stability of the area, within two days (maximum) from the field surveys, in order to respond immediately by planning the appropriate measures, in case of a possible emergency (future landslide). An overview of the adopted methodology is shown in the following flowchart (Figure 6). The validation of the results takes place through their comparative assessment with GNSS measurements, performed on permanent positions (Figure 7).

Figure 5. (**a**) Distribution of GCPs in the area of interest, (**b**) 4.5″ black and white target, (**c**) 4.5″ black and white target within the landslide area.

Figure 6. Flowchart depicting the applied methodology.

Figure 7. Distribution of the GNSS permanent pillars. The orthophoto presents the sliding area on 21 April 2019.

Specifically, the obtained UAV images were processed in Agisoft Metashape (v. 1.7.2., Agisoft LLC, St. Petersburg, Russia) according to the SfM photogrammetry. The technique transforms the overlapping, multi-view UAV images into a three-dimensional object model [36–38]. The origin of SfM is traced to the fields of computer vision and photogrammetry. In our case, the UAV images were aligned using the highest-quality option as described in the Agisoft Metashape manual [39]. The specific option is inextricably linked to the quality of the 3D reconstruction since the camera positions are calculated more accurately. In addition, the collected images were processed in their original size and at the same time, they were upscaled by a factor of 4. The dense point cloud generation was performed according to the ultra-high-quality setting which allows for the creation of more detailed and precise depth maps. Concerning the camera calibration and optimization, the default setting of Agisoft Metashape was selected. The SfM processing ended up in the extraction of point clouds, which were projected into the Hellenic Geodetic Reference System 1987.

The processing of the indicative TLS data was carried out in Leica Cyclone software. Six scan stations were required in each TLS survey to fully cover the area of interest. Therefore, the primary step of the processing was the correct registration of the scans obtained by the different scan positions. The procedure took place using Leica Cyclone REGISTER 360 through the identification of the 4.5″ black and white targets. The registration error in the specific step was measured at 6 mm. Afterward, the multi-dated point clouds were imported into Leica Cyclone REGISTER in order to be properly aligned. The alignment was executed through the detection of common points between the point clouds. The derived TLS three-dimensional representations along with the corresponding UAV-based point clouds were utilized for the monitoring of the landslide's evolution

4. Results

4.1. UAV Surveys

The systematic monitoring of mass movements across the investigated area using UAV imagery was based on change detection approaches. Specifically, high-resolution orthophotos were utilized for the visual identification of surface changes between the repetitive sliding episodes. Some typical relief changes, detected between the first mass movements in April 2019 (Table 1) and the more extensive ones in April 2020, are presented in Figure 8. Specifically, Figure 8a,b show area 1 between the two episodes, while Figure 8c,d correspond to area 2. In both cases, the crosshair is located in the same position. As can be observed, instabilities were evolving as evidenced by the displacement of large conglomerate boulders and soil towards the area of the road as well as vegetation changes.

Figure 8. Orthophotos: (**a**) Area 1 on 11 April 2020, (**b**) Area 1 on 21 April 2019, (**c**) Area 2 on 11 April 2020, (**d**) Area 2 on 21 April 2019.

Local authorities after the occurrence of the extensive mass movements (April 2020) decided to execute some stabilization measures in order to reduce the risk. Operations took place in late July 2020 and included scaling and trimming of the overhanging conglomerates. Therefore, the evolution of the topography of the investigated area from April 2020 to the end of 2021 is strongly related to man-made activities as well as natural processes.

The derived products of UAV photo processing were used for monitoring the topographic variation arising from both factors. In fact, multi-dated orthophotos, covering the investigated area, were exploited for the determination of the areas of instabilities through the monitoring of the vegetation evolution (Figure 9). The initial extent of the mass movements is displayed in blue, whilst the corresponding boundary of November 2021 is shown in red. The lines present significant changes, which are expressed by local vegetation variations and they are related to stabilization operations. Relief modifications related to the specific operations are equally obvious in surface profiles derived from UAV's DSMs (Figure 10). The surface profile of April 2019 is shown in magenta, while the respective profile after the remediation of the slope is depicted in red.

Figure 9. Evolution of the extent of the mass movements over time.

Concerning the monitoring of the instabilities using UAV point clouds, the processing relied on the direct cloud-to-cloud comparison, which was carried out in the Leica Cyclone 3DR extension. The results of this comparison, related to either manmade activities or natural processes are presented in Figure 11. In particular, UAV point clouds, obtained on 10 June 2020 and 25 September 2020 were exploited to detect relief variations that emerged from the stabilization measures (Figure 11a). The greatest changes were varying from 1.5 m to 1.7 m and were highlighted in reddish to magenta colors. Regarding the evolution of the mass movements influenced by natural processes, 3D point clouds acquired on 25 September 2020 and 7 November 2021 were compared (Figure 11b). Following this comparison, small, scattered surface variations were detected throughout the landslide. In the specific output, the topographic variations were overestimated concerning their distribution, which is associated with the sparser density of UAV point clouds and the temporal changes in vegetation. It is worth mentioning that in both periods, the surface changes depicted the dispersion area, despite the positive values that emerged from the processing procedure. Unfortunately, local authorities were removing the sliding materials daily in the area of the road, thus it is impossible to map the volume accumulated.

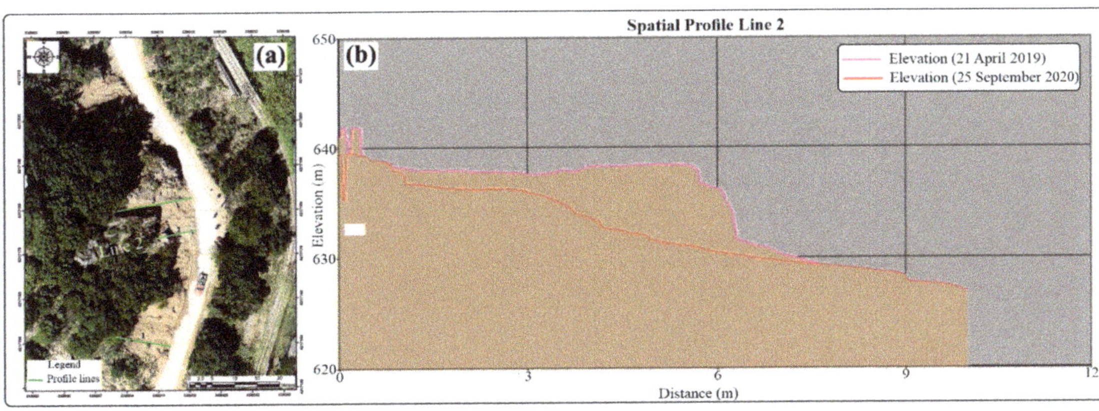

Figure 10. Surface profiles capturing relief modification after the stabilization operations. (**a**) Disribution of the profile sections within the area of interest. (**b**) Multitemporala surface profiles. The surface profile of April 2019 is shown in magenta, while the respective profile after the remediation of slope is depicted in red.

4.2. TLS Surveys

The detection of potential phenomena of instability within the area of interest, utilizing TLS surveys, was performed through cloud-to-cloud comparisons between the acquired multi-dated point clouds. In particular, TLS sensors are able to identify surface differences arising from manmade activities (Figure 12a) or natural processes (Figure 12b) in a more accurate way. The largest surface change associated with the stabilization measures was calculated at about 1.60 m and is close enough to the actual topographic modifications (Figure 12a). On the other hand, scattered relief changes related to erosion processes were noticed between the point clouds obtained on 25 September 2020 and 7 November 2021; they were estimated at about 0.5 m (Figure 12b). The procedure concerned the dispersion areas and not the areas of accumulation, as has already been noted.

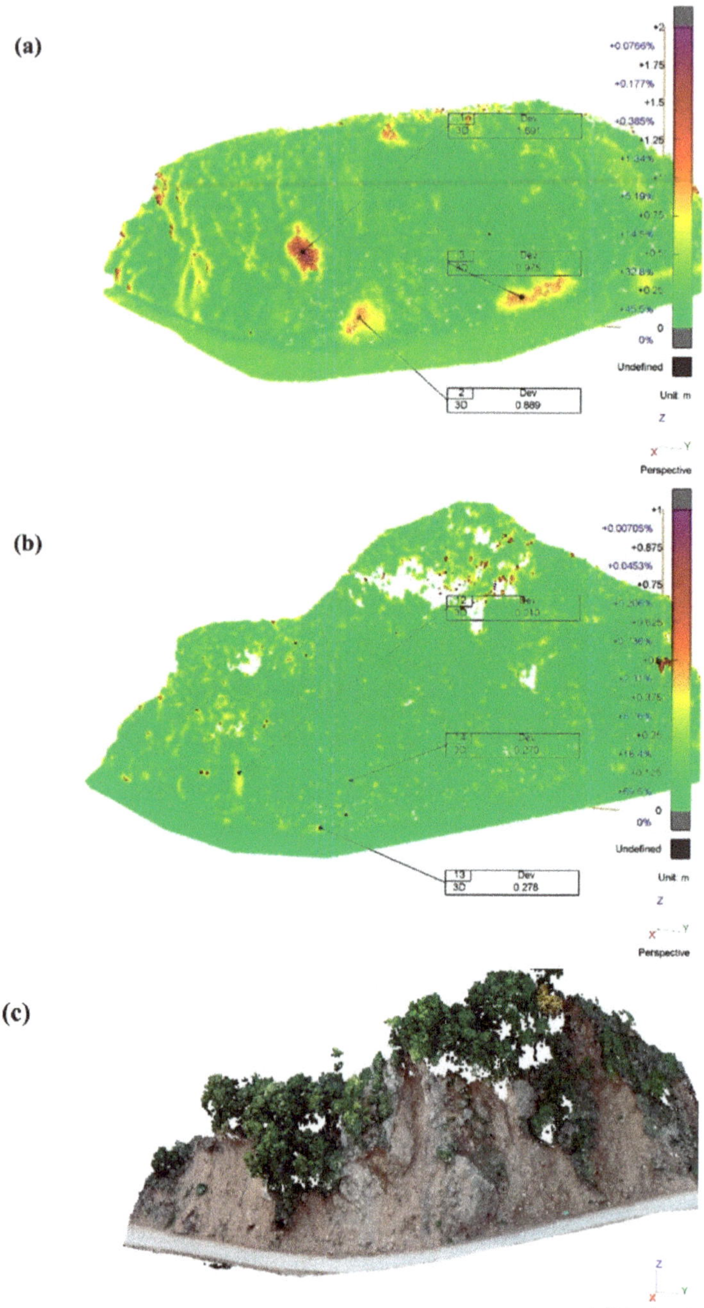

Figure 11. (**a**) Cloud-to-cloud comparison between the UAV point clouds, acquired on 10 June 2020 and 25 September 2020. (**b**) Cloud-to-cloud comparison between the UAV point clouds, acquired on 25 September 2020 and 7 November 2021. (**c**) UAV point cloud acquired on 25 September 2020 in RGB.

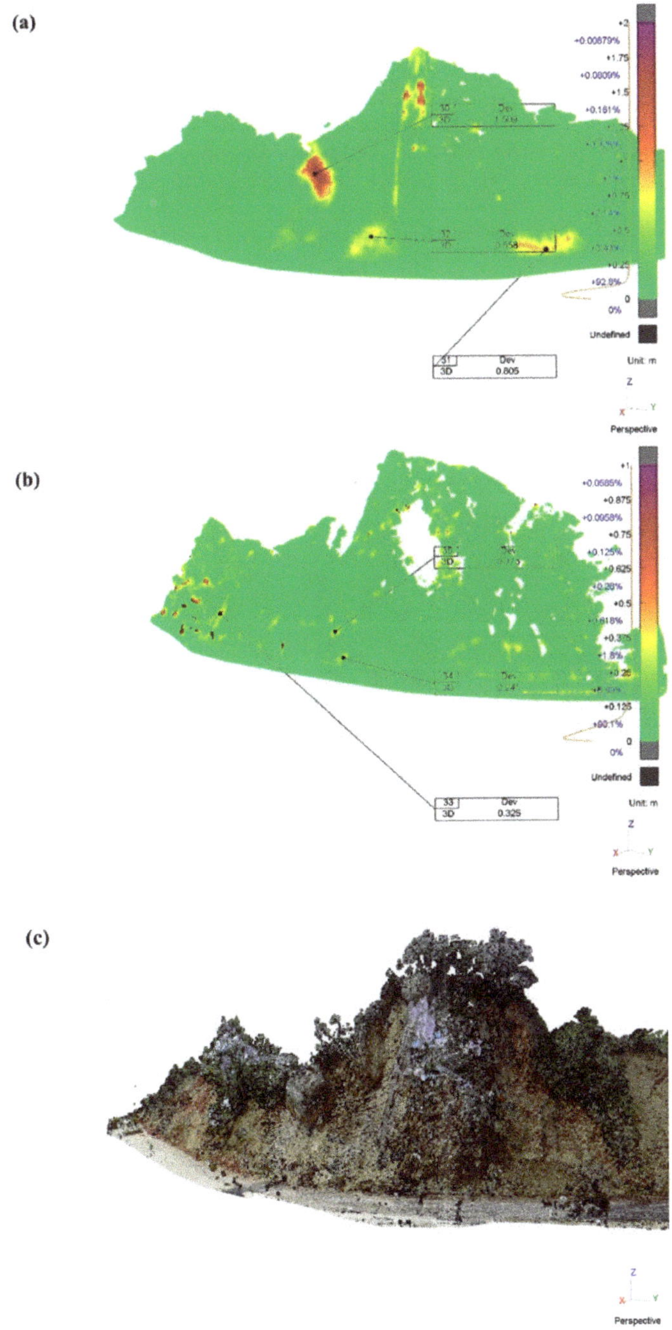

Figure 12. (**a**) Cloud-to-cloud comparison between the TLS point clouds acquired on 10 June 2020 and 25 September 2020. (**b**) Cloud-to-cloud comparison between the TLS point clouds, acquired on 25 September 2020 and 7 November 2021. (**c**) TLS point cloud acquired on 10 June 2020 in RGB.

Moreover, surface profiles of the multi-dated TLS 3D topographic representations were generated in the context of more comprehensive monitoring of the mass movements within the area of interest. Specifically, the intersection between the surface profiles of the TLS point clouds managed to capture accurately and efficiently the parts where the works for the remediation of the slope took place (Figure 13a). Red shades highlighted the areas where the greatest topographic changes have taken place due to the aforementioned operations. Although the output was expected, considering the accuracy of TLS data, it was truly surprising to discover that TLS sensors are able to detect micro-displacements associated with the particularly slow evolution of the topography of the area of interest (Figure 13b). The blue-green colors of the intersection depict the micro-displacements, while the road area seems to be lifting. In fact, these local surface changes, which emerged from the intersection of TLS surface profiles, were estimated at about 0.321 m and they have been confirmed by field observations and GNSS measurements (Table 2). These micro-displacements that indicate the ongoing activity of the slide, are easily observable in the contours of the multi-dated TLS representations, especially in the area of the road (Figure 14).

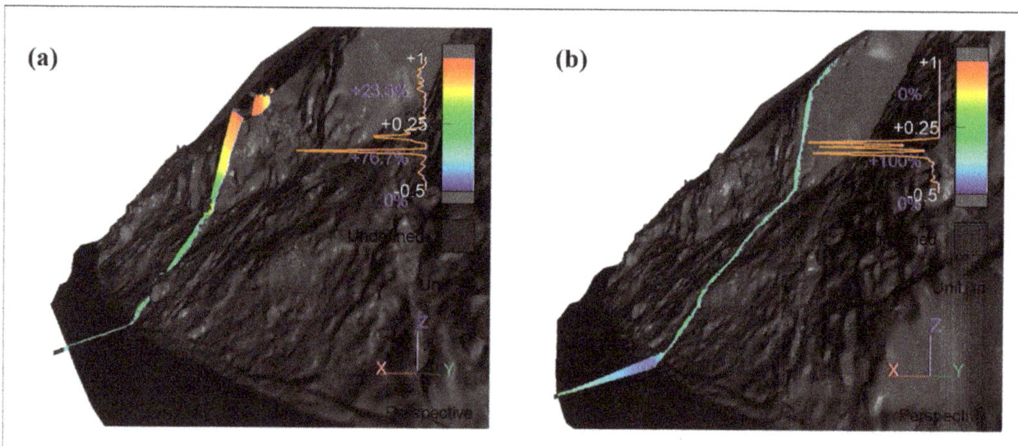

Figure 13. (**a**) Deviation between the surface sections of TLS point clouds acquired on 10 June 2020 and 25 September 2020. (**b**) Deviation between the surface sections of TLS point clouds, acquired on 25 September 2020 and 7 November 2021.

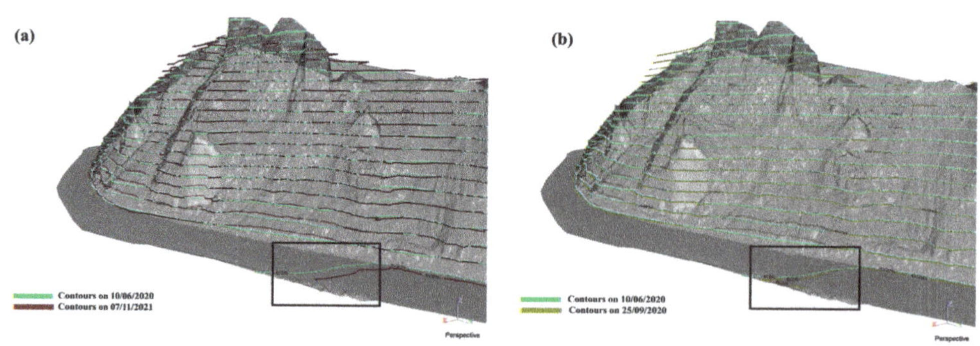

Figure 14. (**a**) Contours of the 3D TLS representations acquired on 10 June 2020 and 7 November 2021. (**b**) Contours of the 3D TLS representations acquired on 10 June 2020 and 25 September 2020. The black boxes surround the micro-movements along the road area.

Table 2. Surface deformation on permanent GNSS pillar 2.

Data Type	Surface Deformation (m)
GNSS measurements	0.334
UAV-based point clouds	0.203
TLS-based point clouds	0.312

4.3. Monitoring Overview and Computational Effort

As already mentioned, the main objective of this work is to provide information regarding the stability of the investigated area to the local authorities immediately, using low-cost and environmentally friendly approaches. The results demonstrated that UAVs, as well as TLS point clouds, are able to monitor the stability of the selected area precisely and effectively at regular intervals; local authorities have access to operational information about the selected site within 2 days.

In more detail, each UAV flight is performed in about 60 min and each scanning requires approximately 4 h for an 11,423 m^2 area with 120 m length (Table 3). The processing of the collected UAV and TLS data was performed in Agisoft Metashape and Cyclone software, respectively. The processing time for each approach is displayed in Table 3, while Table 4 contains the characteristics of the used computer. It is worth noting that TLS cloud density was particularly high between the repeated surveys, in contrast to the density of the UAV point clouds, which was obviously sparser. On the other hand, the increased and constant point density of the TLS point clouds is responsible for the good performance of the specific sensor even in the identification of micro-displacements. Furthermore, the operational cost of the applied approach is directly related only to the purchase of equipment and software or to repairing/upgrading issues.

Table 3. Characteristics and computational effort of UAV/TLS point clouds.

Sensor	Date	Point Cloud Density	Survey Time	Processing Time
UAV	10 June 2020	708.567 points	~60 min	~24 h
	25 September 2020		~60 min	~24 h
	7 November 2021		~60 min	~24 h
TLS	10 June 2020	6.000.000 points	~4 h	~12 h
	25 September 2020		~4 h	~12 h
	7 November 2021		~4 h	~12 h

Table 4. Characteristics of the used computer.

Processor	RAM	Disk	GPU
Intel Core i9 3.6 GHz	128 GB	SSD 1TB/HDD 2TB	NVIDIA GeForce RTX 3080

Meanwhile, we calculated the volume of these hanging rocks utilizing the point clouds obtained by either UAV or TLS, in the context of a comparison of the data used (Table 5). The calculation was carried out using the stockpile tools in Cyclone 3DR. Both approaches yielded similar results, which are slightly different from the assessment of the volume of the detached hanging rocks, as calculated by the local authorities. In particular, local authorities appointed staff from the Department of Geology of University of Patras, as consultants to monitor the instabilities within the area of interest [40]. In this framework, multi-dated UAV orthophotos and DSMs were imported into a GIS environment in order to calculate the volume of the hanging rocks. The estimation was based on simple mathematical operations between the digitized extents of the hanging rocks and the elevation differences. As can be observed, the specific estimation led to an overestimation of the volume, which is expected since the calculation took place in a rougher way. On the contrary, the proposed methodology, consisting of UAV and TLS surveys achieved a more realistic calculation of the volume with millimetric-scale accuracies.

Table 5. Estimation of the volume of the hanging rocks using UAV/TLS point clouds.

Data Type	Volume of the Hanging Rocks (m^3)
GIS methods	24.00
UAV-based point clouds	17.12
TLS-based point clouds	18.42

5. Discussion

Over the years, numerous researchers dealt with the investigation of landslides through the development of effective methodologies and tools for different aspects of landslide research, such as vulnerability mapping, risk assessment, etc. Nevertheless, proper landslide research has never been more urgent. New possibilities for landslide assessment and risk mitigation arose from the evolving development of technology and remote sensing. The aim of the current study is to provide the local authorities with useful and immediate operational information regarding the stability of an environmentally sensitive area while keeping at the same time operational costs as low as possible. On this basis, we performed multiple UAV flights and TLS surveys between April 2020 and November 2021. The collected UAV data were processed via SfM photogrammetry, resulting in the generation of multi-temporal point clouds. The accuracy assessment of the derived SfM products has already been examined in several studies using quantitative and qualitative methods along with a variety of reference data sets consisting of GNSS measurements, classical topographic measurements, etc. [41–43]. Most studies are emphasizing the effect of the number of GCPs and their distribution on the accuracy of the derived products [44–46]. Additionally, the registration and alignment of the TLS data led to the extraction of dense three-dimensional representations.

The monitoring and assessment of the landslide activity were carried out through the exploitation of the derived UAV-based point clouds and the respective TLS three-dimensional representations. TLS sensors have proven their effectiveness in the analysis of the spatio-temporal evolution of landslides in previous studies [47–49]. In this context, we tried to identify any surface change related to instability phenomena between the multi-dated UAV/TLS point clouds through the direct cloud-to-cloud comparison and the estimation of the deviation between surface sections. Surface variations emerging from either human activities or natural processes were successfully traced by both types of remote sensing data. The greatest surface modifications within the area of interest varied between 1.5 and 1.7 m; they were closely associated with the execution of the stabilization operations. Small, scattered surface displacements were detected throughout the area of interest during the last year as a result of erosion processes. Local authorities are constantly removing the newly induced sliding materials from the road area in order to keep the traffic flowing; however, this affects research in a way, as it only allows the mapping of the dispersion areas. In general, the outputs of the photogrammetric and the TLS processing were comparable; nevertheless, the evolution of the landslide area can be achieved more precisely (millimeter-scale) with the TLS-based point, as the corresponding UAV models were less accurate (centimeter-scale) due to the lower density of the point clouds (Table 3). Indeed, the micro-displacements which were observed in the TLS outputs (Figures 12–14), have been confirmed by field observations and the analysis of the GNSS measurements. In particular, Figure 15 depicted the x-, y-, z- coordinate variations of the permanent pillar 2 after the works for the remediation of the slope until now. The greatest differences vary between 0.1 m and 0.28 m. Therefore, it was demonstrated that the phenomena of instability within the area of interest were still ongoing until the end of 2021, but they are evolving slowly. We also have to mention that with the exception of the artificial boulder removal, all the other mass wasting over the slope is regularly distributed over the sliding area. These observations help the landslide classification as debris falls including minor conglomerate boulders. Similar methodologies based on cloud-to-cloud comparison have been applied towards the determination of the deformation of active landslides or in other emergencies [50–52].

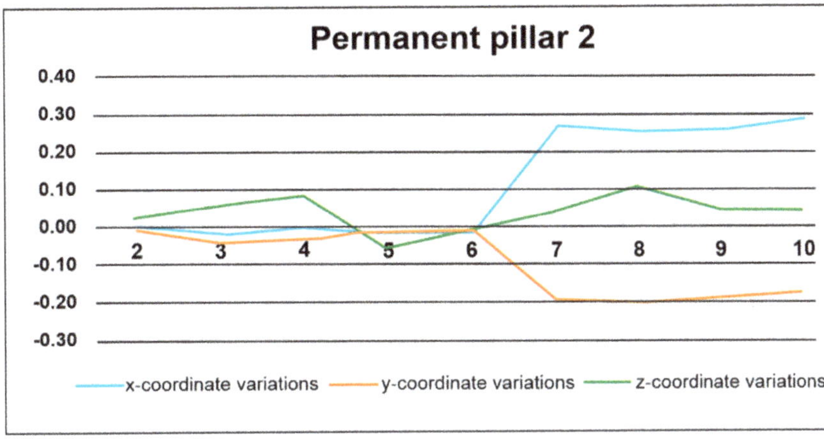

Figure 15. Variation of the coordinates of permanent pillar 2 over time. The *x*-axis corresponds to the period of GNSS measurements, while *y*-axis displays the observed displacements.

Many previous studies have examined the synergy of new technologies for landslide monitoring in conjunction with the minimization of operational costs. Synergistic use of TLS point clouds and GNSS measurements was proposed by [53] in order to map a landslide deformation in Puerto Rico. In order to reduce the high cost of purchasing commercial software, the specific study adopted the use of an open-source software package named Generic Mapping Tools for the processing of TLS data. Another study [54] proposes a combination of 12 low-cost single-frequency GNSS sensors, one seismometric station enhanced with another single-frequency GGNSS sensor and only one dual-frequency GNSS receiver established outside of the unstable area. The specific system provided daily information to the local stakeholders about the landslide movement. Another low-cost GNSS network was established by [55] composed of six sensors inside the landslide area and one outside to be used as a reference. Special attention was given to the logistic limitations (no electrical power, no Wi-Fi, etc.). The GNNS network results were analyzed with other data derived from boreholes, piezometers, inclinometers, crackmeters and meteorological stations. The supplementarity of TLS and UAV point clouds and the operational cost decrease with repeated UAV campaigns in areas with steep reliefs were also discussed [56]. Ref. [56] have emphasized the fact that in areas of very steep topography, the TLS data acquisition implies discontinuities in the point cloud and as a result, there is no homogeneous rendering over the broader area. This gap in point cloud data is filled with low-cost UAV campaigns. Similarly, it was mentioned that UAV campaigns at low altitudes and SfM photogrammetry can prevail over the visibility limitations that are present in land-based methodologies, such as TLS [57–59]. The TLS and UAV data combination for landslide monitoring in Brazil is presented in [60]. It is mentioned that the combination of those two methods presents great advantages over conventional, costly and time-consuming methods.

The current work proposes an effective methodology for the monitoring of challenging areas using low-cost data, acquired by UAVs and TLSs along with repetitive GNSS measurements. The total survey time (for both UAV and TLS) is only 5 h, while the processing of the collected data is carried out within 24 h. Thus, local authorities are informed about the stability of the landslide area, almost simultaneously with the monitoring procedure (within 2 days). The suggested methodology can be used as a guide for monitoring challenging sites or for the quick detection of surface/terrain changes in other emergencies. In more detail the main recommendations of the proposed methodology are:

- Monitoring of instabilities in environmentally sensitive areas can be implemented through repeated UAV and TLS surveys.
- Repeatability is determined by the activity of the instabilities.
- The installation of a permanent GNSS network is recommended. In particular, five permanent GNSS positions, installed in critical places, are sufficient for an area of approximately 1700 m^2. Generally, the number of permanent positions should be adjusted to the characteristics of the area under investigation.
- UAV surveys are able to detect topographic variation on the order of centimeters. On the other hand, TLS surveys can identify micro-displacements (millimetric-scale).
- The synergistic use of UAV and TLS data contributes to the enhancement of the spatial coverage and point density of UAV-based point clouds. This could be considered an ideal monitoring method for areas with complex topography.
- The presence of dense vegetation is an important challenge in the monitoring procedure. In the current research, we tried to reduce the influence of vegetation through the manual segmentation of UAV/TLS point clouds to contain as much topographical information as possible. Further research on the specific issue is needed.

6. Conclusions

The provision of operational information concerning the stability of an environmentally sensitive area to local authorities for the mitigation of the risk was the main objective of the current research. Based on this, the monitoring of the selected site was implemented using timely and low-cost remote sensing data. The appropriately processed multi-dated UAV-based point clouds and the respective TLS point clouds were submitted to comparison procedures aiming at the determination of the evolution of the landslide over time. Direct cloud-to-cloud comparison and the intersection between surface sections were exploited as change detection approaches. The overall evolution of the landslide is distinguished into two sub-periods: the one that was in line with human activities and the other emerging from natural processes, which are still in progress. Both types of data managed to trace the surface changes into the different sub-periods, proving that UAV and TLS data can be used effectively in emergencies and for the accurate landslide classification as debris falls. In fact, the outputs of the photogrammetric and the TLS processing were comparable. Specifically, the largest surface changes detected in UAV/TLS point clouds for the first monitoring period were varying between 1.5 m to 1.7 m. Additionally, the small local surface changes (0.20 m–0.30 m) which are identified on the sliding area during the second monitoring period—related to the natural processes—indicated that the landslide was still active until 11/2021, but it is evolving slowly. It is worth mentioning that the immediate and precise monitoring of the evolution of the landslide seems to be more efficient with the usage of the TLS-based point clouds, which are denser through the monitoring procedure. However, integrated use of UAV and TLS data could be suggested as a hybrid approach for the improvement of the spatial coverage and the point density of UAV-based point clouds. Eventually, this low-cost methodology can be utilized as a guide for the monitoring of challenging or sensitive sites or for the rapid detection of deformations arising from natural disasters or human activities.

Author Contributions: Conceptualization, K.G.N., I.K.K. and A.K.; methodology, K.G.N. and A.K.; software, K.G.N. and A.K.; validation, K.G.N., I.K.K. and A.K.; formal analysis, K.G.N., I.K.K. and A.K.; investigation, K.G.N., I.K.K. and A.K.; resources, K.G.N. and I.K.K.; data curation, K.G.N., I.K.K. and A.K.; writing—original draft preparation, K.G.N. and A.K.; writing—review and editing, K.G.N., I.K.K. and A.K.; visualization, K.G.N., I.K.K. and A.K.; supervision, K.G.N. and I.K.K.; project administration, K.G.N. and I.K.K.; funding acquisition, K.G.N. and I.K.K. All authors have read and agreed to the published version of the manuscript.

Funding: This research received no external funding.

Data Availability Statement: Data available on request due to restrictions. The data presented in this study are available on request from the corresponding author. The data are not publicly available due to public safety reasons.

Acknowledgments: The authors would like to thank Eidikos Logariasmos Kondilion kai Ereunas University of Patras for the support and the payment of the publication fees.

Conflicts of Interest: The authors declare no conflict of interest.

References

1. Gariano, S.L.; Guzzetti, F. Landslides in a changing climate. *Earth-Sci. Rev.* **2016**, *162*, 227–252. [CrossRef]
2. Huggel, C.; Khabarov, N.; Korup, O.; Obersteiner, M. Physical impacts of climate change on landslide occurrence and related adaptation. In *Landslides: Types, Mechanisms and Modeling*; Clague, J.J., Stead, D., Eds.; Cambridge University Press: Cambridge, UKI, 2012; pp. 121–133. [CrossRef]
3. Picarelli, L.; Lacasse, S.; Ho, K.K.S. The Impact of Climate Change on Landslide Hazard and Risk. In *Understanding and Reducing Landslide Disaster Risk*; Sassa, K., Mikoš, M., Sassa, S., Bobrowsky, P.T., Takara, K., Dang, K., Eds.; WLF 2020. ICL Contribution to Landslide Disaster Risk Reduction; Springer: Cham, Switzerland, 2021. [CrossRef]
4. Guzzetti, F.; Carrara, A.; Cardinali, M.; Reichenbach, P. Landslide hazard evaluation: A review of current techniques and their application in a multi-scale study, Central Italy. *Geomorphology* **1999**, *31*, 181–216. [CrossRef]
5. Mantovani, F.; Soeters, R.; Van Westen, C.J. Remote sensing techniques for landslide studies and hazard zonation in Europe. *Geomorphology* **1996**, *15*, 213–225. [CrossRef]
6. Lissak, C.; Bartsch, A.; De Michele, M.; Gomez, C.; Maquaire, O.; Raucoules, D.; Roulland, T. Remote Sensing for Assessing Landslides and Associated Hazards. *Surv. Geophys.* **2020**, *41*, 1391–1435. [CrossRef]
7. Zhong, C.; Liu, Y.; Gao, P.; Chen, W.; Li, H.; Hou, Y.; Nuremanguli, T.; Ma, H. Landslide mapping with remote sensing: Challenges and opportunities. *Int. J. Remote Sens.* **2020**, *41*, 1555–1581. [CrossRef]
8. Garnica-Peña, R.J.; Alcántara-Ayala, I. The use of UAVs for landslide disaster risk research and disaster risk management: A literature review. *J. Mt. Sci.* **2021**, *18*, 482–498. [CrossRef]
9. Gomez, C.; Purdie, H. UAV- based Photogrammetry and Geocomputing for Hazards and Disaster Risk Monitoring—A Review. *Geoenviron Disasters* **2016**, *3*, 23. [CrossRef]
10. Niethammer, U.; James, M.; Rothmund, S.; Travelletti, J.; Joswig, M. UAV-based remote sensing of the Super-Sauze landslide: Evaluation and results. *Eng. Geol.* **2012**, *128*, 2–11. [CrossRef]
11. Lindner, G.; Schraml, K.; Mansberger, R.; Hübl, J. UAV monitoring and documentation of a large landslide. *Appl. Geomat.* **2016**, *8*, 1–11. [CrossRef]
12. Rossi, G.; Tanteri, L.; Tofani, V.; Vannocci, P.; Moretti, S.; Casagli, N. Multitemporal UAV surveys for landslide mapping and characterization. *Landslides* **2018**, *15*, 1045–1052. [CrossRef]
13. Peppa, M.V.; Mills, J.P.; Moore, P.; Miller, P.E.; Chambers, J.E. Brief communication: Landslide motion from cross correlation of UAV-derived morphological attributes. *Nat. Hazards Earth Syst. Sci.* **2017**, *17*, 2143–2150. [CrossRef]
14. Turner, D.; Lucieer, A.; De Jong, S.M. Time Series Analysis of Landslide Dynamics Using an Unmanned Aerial Vehicle (UAV). *Remote Sens.* **2015**, *7*, 1736–1757. [CrossRef]
15. Godone, D.; Allasia, P.; Borrelli, L.; Gullà, G. UAV and Structure from Motion Approach to Monitor the Maierato Landslide Evolution. *Remote Sens.* **2020**, *12*, 1039. [CrossRef]
16. Rodriguez, J.; Macciotta, R.; Hendry, M.T.; Roustaei, M.; Gräpel, C.; Skirrow, R. UAVs for monitoring, investigation, and mitigation design of a rock slope with multiple failure mechanisms—A case study. *Landslides* **2020**, *17*, 2027–2040. [CrossRef]
17. Bernardo, E.; Palamara, R.; Boima, R. UAV and Soft Computing Methodology for Monitoring Landslide Areas (Susceptibility to Landslides and Early Warning). *Wseas Trans. Environ. Dev.* **2021**, *17*, 490–501. [CrossRef]
18. Jaboyedoff, M.; Oppikofer, T.; Abellán, A.; Derron, M.H.; Loye, A.; Metzger, R.; Pedrazzini, A. Use of LIDAR in landslide investigations: A review. *Nat. Hazards* **2012**, *61*, 5–28. [CrossRef]
19. Guo, C.; Xu, Q.; Dong, X.; Li, W.; Zhao, K.; Lu, H.; Ju, Y. Geohazard Recognition and Inventory Mapping Using Airborne LiDAR Data in Complex Mountainous Areas. *J. Earth Sci.* **2021**, *32*, 1079–1091. [CrossRef]
20. Mackey, B.H.; Roering, J.J.; McKean, J.A. Long-term kinematics and sediment flux of an active earthflow, Eel River, California. *Geology* **2009**, *37*, 803–806. [CrossRef]
21. Van Den Eeckhaut, M.; Kerle, N.; Poesen, J.; Hervás, J. Object-oriented identification of forested landslides with derivatives of single pulse LiDAR data. *Geomorphology* **2012**, *173–174*, 30–42. [CrossRef]
22. Ventura, G.; Vilardo, G.; Terranova, C.; Sessa, E.B. Tracking and evolution of complex active landslides by multi-temporal airborne LiDAR data: The Montaguto landslide (Southern Italy). *Remote Sens. Environ.* **2011**, *115*, 3237–3248. [CrossRef]
23. Fanos, A.M.; Pradhan, B. Laser Scanning Systems and Techniques in Rockfall Source Identification and Risk Assessment: A Critical Review. *Earth Syst Environ.* **2018**, *2*, 163–182. [CrossRef]
24. Guinau, M.; Tapia, M.; Pérez-Guillén, C.; Suriñach, E.; Roig, P.; Khazaradze, G.; Torné, M.; Royán, M.J.; Echeverria, A. Remote sensing and seismic data integration for the characterization of a rock slide and an artificially triggered rock fall. *Eng. Geol.* **2019**, *257*, 105113. [CrossRef]

25. Abellán, A.; Oppikofer, T.; Jaboyedoff, M.; Rosser, N.J.; Lim, M.; Lato, M.J. Terrestrial laser scanning of rock slope instabilities. *Earth Surf. Process. Landf.* **2014**, *39*, 80–97. [CrossRef]
26. Guerin, A.; Stock, G.; Radue, M.; Jaboyedoff, M.; Collins, B.; Matasci, B.; Avdievitch, N.; Derron, M.H. Quantifying 40 years of rockfall activity in Yosemite Valley with historical Structure-from-Motion photogrammetry and terrestrial laser scanning. *Geomorphology* **2020**, *356*, 107069. [CrossRef]
27. Casagli, N.; Frodella, W.; Morelli, S.; Tofani, V.; Ciampalini, A.; Intrieri, E.; Raspini, F.; Rossi, G.; Tanteri, L.; Lu, P. Spaceborne, UAV and ground-based remote sensing techniques for landslide mapping, monitoring and early warning. *Geoenviron. Disasters* **2017**, *4*, 9. [CrossRef]
28. Kyriou, A.; Nikolakopoulos, K.; Koukouvelas, I.; Lampropoulou, P. Repeated UAV Campaigns, GNSS Measurements, GIS, and Petrographic Analyses for Landslide Mapping and Monitoring. *Minerals* **2021**, *11*, 300. [CrossRef]
29. Pellicani, R.; Argentiero, I.; Manzari, P.; Spilotro, G.; Marzo, C.; Ermini, R.; Apollonio, C. UAV and Airborne LiDAR Data for Interpreting Kinematic Evolution of Landslide Movements: The Case Study of the Montescaglioso Landslide (Southern Italy). *Geosciences* **2019**, *9*, 248. [CrossRef]
30. Brook, M.S.; Merkle, J. Monitoring active landslides in the Auckland region utilising UAV/structure-from-motion photogrammetry. *Jpn. Geotech. Soc. Spec. Publ.* **2019**, *6*, 1–6. [CrossRef]
31. Avallone, A.; Briole, P.; Agatza-Balodimou, A.M.; Billiris, H.; Charade, O.; Mitsakaki, C.; Nercessian, A.; Papazissi, K.; Paradissis, D.; Veis, G. Analysis of eleven years of deformation measured by GPS in the Corinth Rift Laboratory area. *Comptes Rendus Geosci.* **2004**, *336*, 301–311. [CrossRef]
32. Fernández-Blanco, D.; de Gelder, G.; Lacassin, R.; Armijo, R. A new crustal fault formed the modern Corinth Rift. *Earth-Sci. Rev.* **2019**, *199*, 102919. [CrossRef]
33. Jolivet, L.; Labrousse, L.; Agard, P.; Lacombe, O.; Bailly, V.; Lecomte, E.; Mouthereau, F.; Mehl, C. Rifting and shallow-dipping detachments, clues from the Corinth Rift and the Aegean. *Tectonophysics* **2010**, *483*, 287–304. [CrossRef]
34. McKenzie, D. Active tectonics of the Alpine—Himalayan belt: The Aegean Sea and surrounding regions. *Geophys. J. Int.* **1978**, *55*, 217–254. [CrossRef]
35. Taylor, B.; Weiss, J.; Goodliffe, A.M.; Sachpazi, M.; Laigle, M.; Hirn, A. The structures, stratigraphy and evolution of the Gulf of Corinth rift, Greece. *Geophys. J. Int.* **2011**, *185*, 1189–1219. [CrossRef]
36. Micheletti, N.; Chandler, J.; Lane, S.N. Chapter 2—Structure from motion (SFM) photogrammetry. In *Geomorphological Techniques*; Section 2.2; British Society for Geomorphology: London, UK, 2015.
37. Smith, M.W.; Carrivick, J.L.; Quincey, D.J. Structure from motion photogrammetry in physical geography. *Prog. Phys. Geogr. Earth Environ.* **2016**, *40*, 247–275. [CrossRef]
38. Westoby, M.; Brasington, J.; Glasser, N.; Hambrey, M.; Reynolds, J. 'Structure-from-Motion' photogrammetry: A low-cost, effective tool for geoscience applications. *Geomorphology* **2012**, *179*, 300–314. [CrossRef]
39. Agisoft Metashape Professional Edition. Available online: https://www.agisoft.com/pdf/metashape-pro_1_7_en.pdf (accessed on 30 January 2022).
40. Nikolakopoulos, K.G.; Koukouvelas, I.K. Rockfalls systematic monitoring using UAVs: The case of Zachlorou village. In Proceedings of the Eighth International Conference on Remote Sensing and Geoinformation of the Environment (RSCy2020), Paphos, Cyprus, 16–18 March 2020. [CrossRef]
41. Brach, M.; Chan, J.C.-W.; Szymanski, P. Accuracy assessment of different photogrammetric software for processing data from low-cost UAV platforms in forest conditions. *iForest* **2019**, *12*, 435–441. [CrossRef]
42. Gindraux, S.; Boesch, R.; Farinotti, D. Accuracy Assessment of Digital Surface Models from Unmanned Aerial Vehicles' Imagery on Glaciers. *Remote Sens.* **2017**, *9*, 186. [CrossRef]
43. Nikolakopoulos, K.G.; Soura, K.; Koukouvelas, I.K.; Argyropoulos, N.G. UAV vs. classical aerial photogrammetry for archaeological studies. *J. Archaeol. Sci. Rep.* **2017**, *14*, 758–773. [CrossRef]
44. Agüera-Vega, F.; Carvajal-Ramírez, F.; Martínez-Carricondo, P. Assessment of photogrammetric mapping accuracy based on variation ground control points number using unmanned aerial vehicle. *Measurement* **2017**, *98*, 221–227. [CrossRef]
45. Akturk, E.; Altunel, A.O. Accuracy assessment of a low-cost UAV derived digital elevation model (DEM) in a highly broken and vegetated terrain. *Measurement* **2019**, *136*, 382–386. [CrossRef]
46. Ferrer-González, E.; Agüera-Vega, F.; Carvajal-Ramírez, F.; Martínez-Carricondo, P. UAV Photogrammetry Accuracy Assessment for Corridor Mapping Based on the Number and Distribution of Ground Control Points. *Remote Sens.* **2020**, *12*, 2447. [CrossRef]
47. Maurizio, B.; Fiani, M.; Lugli, A. Landslide monitoring using multitemporal terrestrial laser scanning for ground displacement analysis. *Geomat. Nat. Hazards Risk* **2015**, *6*, 398–418. [CrossRef]
48. Oppikofer, T.; Jaboyedoff, M.; Blikra, L.; Derron, M.-H.; Metzger, R. Characterization and monitoring of the Åknes rockslide using terrestrial laser scanning. *Nat. Hazards Earth Syst. Sci.* **2009**, *9*, 1003–1019. [CrossRef]
49. Rashidi, M.; Mohammadi, M.; Sadeghlou Kivi, S.; Abdolvand, M.M.; Truong-Hong, L.; Samali, B. A Decade of Modern Bridge Monitoring Using Terrestrial Laser Scanning: Review and Future Directions. *Remote Sens.* **2020**, *12*, 3796. [CrossRef]
50. Eker, R.; Aydın, A.; Hübl, J. Unmanned aerial vehicle (UAV)-based monitoring of a landslide: Gallenzerkogel landslide (Ybbs-Lower Austria) case study. *Env. Monit Assess* **2018**, *190*, 28. [CrossRef]

51. Fuad, N.A.; Yusoff, A.R.; Ismail, Z.; Majid, Z. Comparing the Performance of Point Cloud Registration Methods for Landslide Monitoring Using Mobile Laser Scanning Data. *ISPRS Int. Arch. Photogramm. Remote Sens. Spat. Inf. Sci.* **2018**, *4249*, 11–21. [CrossRef]
52. Martínez-Espejo Zaragoza, I.; Caroti, G.; Piemonte, A.; Riedel, B.; Tengen, D.; Niemeier, W. Structure from motion (SfM) processing of UAV images and combination with terrestrial laser scanning, applied for a 3D-documentation in a hazardous situation. Geomatics. *Nat. Hazards Risk* **2017**, *8*, 1492–1504. [CrossRef]
53. Wang GPhilips DJoyce, J.; Rivera, F. The Integration of TLS and Continuous GPS to Study Landslide Deformation: A Case Study in Puerto Rico. *J. Geod. Sci.* **2011**, *1*, 25–34. [CrossRef]
54. Zuliani, D.; Tunini, L.; Di Traglia, F.; Chersich, M.; Curone, D. Cost-Effective, Single-Frequency GPS Network as a Tool for Landslide Monitoring. *Sensors* **2022**, *22*, 3526. [CrossRef]
55. Šegina, E.; Peternel, T.; Urbančič, T.; Realini, E.; Zupan, M.; Jež, J.; Caldera, S.; Gatti, A.; Tagliaferro, G.; Consoli, A.; et al. Monitoring Surface Displacement of a Deep-Seated Landslide by a Low-Cost and near Real-Time GNSS System. *Remote Sens.* **2020**, *12*, 3375. [CrossRef]
56. Medjkane, M.; Maquaire, O.; Costa, S.; Roulland, T.; Letortu, P.; Fauchard, C.; Antoine, R.; Davidson, R. High-resolution monitoring of complex coastal morphology changes: Cross-efficiency of SfM and TLS-based survey (Vaches-Noires cliffs, Normandy, France). *Landslides* **2018**, *15*, 1097–1108. [CrossRef]
57. Cawood, A.J.; Bond, C.E.; Howell, J.A.; Butler, R.W.H.; Totake, Y. LiDAR, UAV or compassclinometer? Accuracy, coverage and the effects on structural models. *J. Struct. Geol.* **2017**, *98*, 67–82. [CrossRef]
58. Hugenholtz, C.H.; Whitehead, K.; Brown, O.W.; Barchyn, T.E.; Moorman, B.J.; LeClaire, A.; Riddell, K.; Hamilton, T. Geomorphological mapping with a small unmanned aircraft system (sUAS): Feature detection and accuracy assessment of a photogrammetrically-derived digital terrain model. *Geomorphology* **2013**, *194*, 16–24. [CrossRef]
59. Tonkin, T.N.; Midgley, N.G.; Graham, D.J.; Labadz, J.C. The potential of small unmanned aircraft systems and structure-from-motion for topographic surveys: A test of emerging integrated approaches at cwm Idwal, North Wales. *Geomorphology* **2014**, *226*, 35–43. [CrossRef]
60. Garcia, G.P.; Gomes, E.B.; Viana, C.D.; Grohmann, C.H. Comparing terrestrial laser scanner and UAV-based photogrammetry to generate a landslide dem. In *Anais do Xix Simpósio Brasileiro de Sensoriamento Remoto*; Santos. Anais eletrônicos. São José dos Campos, INPE, 2019. Proceedings of XIX Brazilian Symposium on Remote Sensing; ISBN 978-85-17-00097-3. Available online: https://proceedings.science/sbsr-2019/papers/comparing-terrestrial-laser-scanner-and-uav-based-photogrammetry-to-generate-a-landslide-dem?lang=en> (accessed on 31 July 2022).

remote sensing

MDPI

Article

Evaluation of SAR and Optical Data for Flood Delineation Using Supervised and Unsupervised Classification

Fatemeh Foroughnia [1], Silvia Maria Alfieri [1], Massimo Menenti [1,2,*] and Roderik Lindenbergh [1]

[1] Department of Geoscience and Remote Sensing, Faculty of Civil Engineering and Geosciences, Delft University of Technology, Stevinweg, 2628 CN Delft, The Netherlands; f.foroughnia@tudelft.nl (F.F.); s.m.alfieri-1@tudelft.nl (S.M.A.); r.c.lindenbergh@tudelft.nl (R.L.)

[2] State Key Laboratory of Remote Sensing Science, Aerospace Information Research Institute, Chinese Academy of Sciences, Beijing 100101, China

* Correspondence: m.menenti@tudelft.nl

Abstract: Precise and accurate delineation of flooding areas with synthetic aperture radar (SAR) and multi-spectral (MS) data is challenging because flooded areas are inherently heterogeneous as emergent vegetation (EV) and turbid water (TW) are common. We addressed these challenges by developing and applying a new stepwise sequence of unsupervised and supervised classification methods using both SAR and MS data. The MS and SAR signatures of land and water targets in the study area were evaluated prior to the classification to identify the land and water classes that could be delineated. The delineation based on a simple thresholding method provided a satisfactory estimate of the total flooded area but did not perform well on heterogeneous surface water. To deal with the heterogeneity and fragmentation of water patches, a new unsupervised classification approach based on a combination of thresholding and segmentation (CThS) was developed. Since sandy areas and emergent vegetation could not be classified by the SAR-based unsupervised methods, supervised random forest (RF) classification was applied to a time series of SAR and co-event MS data, both combined and separated. The new stepwise approach was tested for determining the flood extent of two events in Italy. The results showed that all the classification methods applied to MS data outperformed the ones applied to SAR data. Although the supervised RF classification may lead to better accuracies, the CThS (unsupervised) method achieved precision and accuracy comparable to the RF, making it more appropriate for rapid flood mapping due to its ease of implementation.

Keywords: SAR; optical; flood mapping; random forest; Otsu thresholding; unsupervised classification; supervised classification

Citation: Foroughnia, F.; Alfieri, S.M.; Menenti, M.; Lindenbergh, R. Evaluation of SAR and Optical Data for Flood Delineation Using Supervised and Unsupervised Classification. *Remote Sens.* **2022**, *14*, 3718. https://doi.org/10.3390/rs14153718

Academic Editors: Constantinos Loupasakis, Ioannis Papoutsis and Konstantinos G. Nikolakopoulos

Received: 3 June 2022
Accepted: 24 July 2022
Published: 3 August 2022

Publisher's Note: MDPI stays neutral with regard to jurisdictional claims in published maps and institutional affiliations.

1. Introduction

The literature on the detection and mapping of flooding events is rather abundant and documents a broad spectrum of methods, relying on multi-spectral (MS) and microwave remote sensing observations. Given the high dimensionality of flood mapping, we deemed it useful to evaluate in detail a limited number of combinations of methods and remote sensing signals. We chose for this experiment two extreme events in two different areas, which led to two completely different flooding patterns. We considered that by analyzing in detail a limited number of cases, it would be possible to understand the causes of the performance on flood mapping achieved in each case. This includes exploring the advantages and disadvantages of either signal, simple vs. advanced algorithms or single image vs. time series data sets. We conceived this study as an attempt to structure such options as an incremental approach where increasingly complex signals, data sets and algorithms are applied as needed, rather than going a priori for the most complex solution. The capability of Copernicus satellites to acquire remote sensing data at higher spatial and temporal resolution has improved surface water and flood mapping. Both optical and microwave sensors can be used for flood mapping, providing different capabilities and accuracy. MS imaging radiometers measure spectral radiance from the visible (VIS) through the shortwave infrared (SWIR) spectrum. The near-infrared (NIR) region is most suitable to distinguish water from dry surfaces due to the strong absorption of water [1]. Based on this, many simple spectral indices have been developed to delineate water areas using MS images.

The normalized difference vegetation index (NDVI), derived from Red and NIR ranges, has been widely used in water and flood mapping [2–5]. The NDVI, however, is a vegetation index, i.e., it is sub-optimal to capture information on a water surface [6]. The normalized difference water index (NDWI), calculated using Green and NIR spectral radiances, aims to maximize the spectral contrast between water and other terrestrial land covers in the Green and NIR regions [6]. NDWI has been extensively used to map inundated areas [7–12]. Although other indices were also successively developed [13–16], NDWI was demonstrated to provide higher performance in detecting water bodies [17]. The index ranges from −1 to +1, in which positive values are associated with surface water in the ideal situation of deep and clear water. The presence of dense vegetation, however, may easily lead to a higher NIR than Green reflectance, and NDWI values closer to values over land. In these cases, delineation of heterogeneous flooded areas using NDWI is not straightforward [8,18].

Microwave signals, on the other hand, benefit from good penetration through clouds, providing more efficient measurements in cloudy conditions than optical observations [19]. The difference in surface roughness is the main feature to detect surface water using synthetic aperture radar (SAR) data. Ideally, smooth open water exhibits specular reflection, i.e., away from the line of sight (LOS) of the SAR sensor, in strong contrast with the scattering of surrounding natural surfaces in dry conditions [20]. SAR backscattering is mainly influenced by soil roughness and the soil dielectric constant [21]. Specular reflectance can be affected by weather conditions, such as wind and precipitation, and also by ground-target types such as emergent vegetation, making the detection of open water difficult [22,23]. In addition, overestimation of the water extent using SAR backscatter is also frequent in sandy areas due to the similarity of radar backscatter over sand and water [24]. Notably, the quality of radar imaging of sandy regions is affected by the random reflection of the incident electromagnetic pulse which results in a loss of energy [25].

Various methodologies have been applied to delineate surface water from MS and SAR data. Water surfaces can be delineated by unsupervised [18,26–31] and supervised approaches [18,32–37] using single or multiple bands.

The literature reviewed above shows that in the case of an ideal situation, i.e., without any disturbance factors, unsupervised thresholding approaches provide a quick assessment of flooded areas. Thresholding methods, however, due to the presence of disturbance factors, which influence the real SAR backscatter and optical reflectance of the targets, may perform less effectively. When using MS data, NDWI thresholding may fail to detect standing water bodies beneath dense canopies and emergent vegetation due to the sensitivity of NIR reflectance to vegetation [38,39]. When using SAR data, flooded vegetation or forests appear bright due to the double and/or multi-bouncing effects, i.e., the interaction between the water surface and the vertical structure of stems and trunks [40–42]. Wind waves can also roughen the water surface, causing an increase in SAR backscatter to a similar/or even higher level than in surrounding non-flooded areas [43]. Moreover, the speckle noise inherent to all coherent imaging devices causes statistical fluctuations in the backscatter of pixels, which prevents stable estimates of threshold values [44].

Technically speaking, although histogram thresholding is one of the most rapid techniques in flood mapping, the selection of a suitable threshold value represents a critical step that strongly influences the outcome [45,46]. Essentially, a threshold-based method requires a bimodal histogram to binarize an image into the two semantic classes, target and background. However, since the water class only represents a small portion of the whole image in most flood cases, the histogram of the image values is often not obviously bimodal and it becomes difficult to separate the two classes [47]. To address this issue, some studies tried to divide the image into many sub-images and then apply the thresholding method to each sub-image to estimate a suitable threshold, where the histogram was bimodal [47,48]. An alternative is to merge all sub-images containing a sufficient number of flood pixels and to estimate one global threshold value which is then applied to all sub-images [49]. Other than pixel-based thresholding discrimination, image segmentation techniques, which gather connected homogeneous pixels into patches, can provide information at the object level. Furthermore, in the case of analyzing SAR data, image segmentation will reduce the speckle effect because both morphological and radiometric information is used.

The literature review led us to identify the following gaps in knowledge:

1. How to deal efficiently with challenges stemming from the heterogeneity and overlap of MS and SAR signatures of surface water types.
2. How to delineate fragmented flood water patches and estimate correctly the total flooded area.
3. How to identify an optimal combination of optical, SAR, and textural signatures as regards both accuracy and computational efficiency.
4. Comparative advantages of artificial intelligence (AI) algorithms above relatively simple thresholding and segmentation methods.

In this study, we addressed these gaps by developing and applying a stepwise approach to delineate surface water types and flooded areas, defined by the comparison of surface water areas before and during a flood event.

The research goal of this study was to evaluate alternate combinations of remote sensing data and delineation methods to determine flood extent. In this study, the flooded area is determined as the difference between the surface water area during a flooding event and the surface water area before it.

The approach applied in the study required multi-temporal image analysis. We have analyzed the MS and SAR signatures to identify a procedure to separate different water and non-water surfaces. In addition, both the MS and SAR signatures are likely to be rather heterogeneous due to the combined effects of terrain, vegetation, and sediments transported by flood water. For example, the optical signatures of emergent vegetation and turbid water were largely overlapping, but these surface types could be discriminated using SAR backscatter signatures. The optical signature of clear water, however, was very different from anything else and suggested the possibility of delineating this surface type using a simple spectral index. Hence, a classical thresholding procedure, i.e., with a predefined threshold, was not applicable to separate all water surface types. Therefore, we used a grid-based Otsu thresholding related to the distribution of threshold values in a set of heterogeneous sample areas. This approach, however, does not solve the problem of the fragmentation of flooded areas. To deal with fragmentation, we developed and applied a new unsupervised approach that benefits from the combination of thresholding and segmentation methods (CThS).

Given the heterogeneity of the water surfaces, we have experimented with AI algorithms to explore whether we could discover additional classification rules to classify different surface water types, which then could be aggregated to delineate the entire "surface water" area. The supervised classification method, random forest (RF), was applied to our datasets. This solution was suggested by its performance being less affected by outliers and noisy data, along with the easier parametrization and the absence of assumptions on data distribution [50]. Flood maps obtained with the RF classifier were explored to understand (a) the achievable improvements by using, first, either only SAR or MS data and, second, by combining both datasets for flood delineation and (b) which features are determinant in improving flood map accuracy. This has been done particularly focusing on the heterogeneous cases mentioned earlier. It should be noted that our first and second classification approaches are unsupervised while the third one is supervised. The solutions proposed in this study have been evaluated in two different case studies with highly heterogeneous water surfaces under different hydro-meteorological conditions.

The accuracy and precision of the methods were then evaluated using different reference datasets. A comparison between the three methods was performed and the difference in accuracy due to the use of different methodologies was evaluated.

2. Materials and Methods

2.1. Case Studies: Sesia and Enza Rivers

We carried out two case studies during extreme flood events in areas located in Northern Italy along the Sesia and Enza rivers (Figure 1). The extreme events were selected and characterized by [51] as part of a study on extreme hydro-meteoric events in the Emilia-Romagna region during the period 1989–2018.

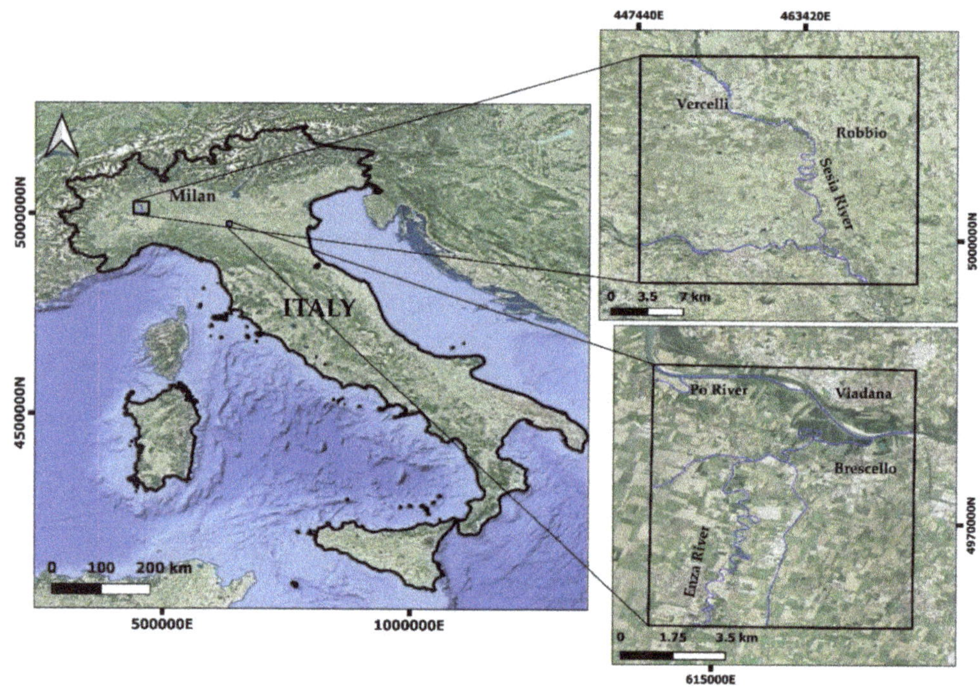

Figure 1. Location map of the two case studies in Italy. The lower and upper pictures show where the 2017 and 2020 events occurred, respectively. The blue lines represent rivers' routes. The footprints of the image tiles, i.e., the borders of the regions of interest, are indicated by black solid squares. The background is Google satellite imagery, available in the QGIS environment.

The Sesia is a left tributary of the Po River with the catchment entirely located in the Piemonte region. Its source is the Monte Rosa at 2500 m. It rapidly flows in the Valsesia valley where several smaller rivers flow into it, largely increasing its discharge. Between 2 and 3 October 2020, several flood events occurred in the Piemonte region. Among them, the one that occurred along the Sesia caused an embankment failure near Caresana, at the boundary with Pavia Province, leading to extensive flooding of agricultural fields and inundation of the municipalities of Borgosesia and Vercelli.

The Enza river, flowing in between the Parma and Reggio provinces (Emilia-Romagna region), is a right tributary of the river Po. Its source is in the Alpe di Succiso, in the northern Apennines at 1406 m. An extreme flood event occurred on 12 December 2017 when the Enza reached the maximum historical level of 12.44 a.m.s.l. [52] at Sorbolo in the Parma province. Consequently, the river broke up the embankment near Lentigione, inundating the entire urban area and forcing hundreds of residents to evacuate.

2.2. Datasets

The Sentinel-1 (S-1) and Sentinel-2 (S-2) satellites provide users with short revisit time data, good global coverage, and quick and free image delivery, and have good potential in land monitoring and emergency response [53,54].

According to the European Space Agency (ESA), the S-1 dual-polarized level-1 ground range detected high resolution (GRDH) products can be used in mapping affected flood areas. These datasets are acquired, multi-looked and projected to the ground range using an Earth ellipsoid model. The resulting product has approximately square pixels with a spatial resolution of 10 m with reduced speckle at the cost of worse spatial resolution. A number of (see Table 1) pre-processed (radiometrically calibrated and terrain corrected) GRD products with VV and VH polarizations were downloaded from the Google Earth Engine (GEE) server. S-1 SAR imagery in the GEE consists of Level-1 GRD scenes processed to backscatter coefficient (σ°) in decibels (dB) to ensure that images are statistically comparable [55]. The backscatter coefficient captures the target backscattering area (radar cross-section) per unit of ground area. Because this coefficient can vary by several orders of magnitude, it is usually converted to dB as $10{\cdot}\log_{10}\sigma^{\circ}$. The pre-processing of SAR GRD data in GEE includes applying orbit tracking, GRD border noise removal, thermal noise removal, radiometric correction and terrain correction using shuttle radar topography mission (SRTM) digital elevation data. In addition, some single look complex (SLC) data products were downloaded from the ESA Copernicus Open Access Hub [56] since we used the phase information as explained in Section 2.3.4. Level-1 SLC products consist of geo-referenced SAR data and are provided in zero-Doppler slant-range geometry. The data includes a single look in each azimuth and range direction using the full transmit signal bandwidth and contains complex samples preserving the phase information [56].

Table 1. Dates of GRDH Sentinel-1 (S-1), SLC S-1, and Sentinel-2 (S-2) data used. The explanation of the data products can be found in the text above. The flood and non-flood images are represented in blue and orange color respectively. The footprints of the images are also shown in Figure 1.

<table>
<tr><th colspan="3">Event 1
Enza River, 13 December 2017</th><th colspan="3">Event 2
Sesia River, 3 October 2020</th></tr>
<tr><th>Tiles</th><th>Processing Level</th><th>Image Date</th><th>Tiles</th><th>Processing Level</th><th>Image Date</th></tr>
<tr><td rowspan="13">Asc. track number 15</td><td rowspan="11">S-1 GRDH</td><td>2 October 2017</td><td rowspan="13">Asc. track number 88</td><td rowspan="8">S-1 GRDH</td><td>4 August 2020</td></tr>
<tr><td>8 October 2017</td><td>10 August 2020</td></tr>
<tr><td>14 October 2017</td><td>16 August 2020</td></tr>
<tr><td>20 October 2017</td><td>22 August 2020</td></tr>
<tr><td>26 October 2017</td><td>28 August 2020</td></tr>
<tr><td>1 November 2017</td><td>3 September 2020</td></tr>
<tr><td>7 November 2017</td><td>15 September 2020</td></tr>
<tr><td>13 November 2017</td><td>21 September 2020</td></tr>
<tr><td>19 November 2017</td><td>S-1 GRDH</td><td>27 September 2020</td></tr>
<tr><td>25 November 2017</td><td>S-1 SLC</td><td>3 October 2020</td></tr>
<tr><td>1 December 2017</td><td rowspan="3">S-1 GRDH</td><td>9 October 2020</td></tr>
<tr><td>S-1 GRDH</td><td>7 December 2017</td><td>15 October 2020</td></tr>
<tr><td>S-1 SLC</td><td>13 December 2017</td><td>27 October 2020</td></tr>
<tr><td rowspan="2">Granule T32TPQ</td><td rowspan="2">S-2 L2A</td><td>24 October 2017</td><td rowspan="2">Granule T32TMR</td><td rowspan="2">S-2 L2A</td><td>9 August 2020</td></tr>
<tr><td>13 December 2017</td><td>3 October 2020</td></tr>
</table>

The S-2 satellites carry the multi-spectral instrument (MSI) providing high spatial resolution multispectral imagery. MSI measures the Earth's reflected radiance in 13 spectral bands from VIS/NIR to SWIR with a spatial resolution ranging from 10 m to 60 m. Some S-2 Level-2A products (See Table 1) were downloaded from the ESA Copernicus Open Access Hub. S-2 Level-2A (MS) are provided after applying radiometric, geometric and atmospheric correction and were directly used in further processing. S-2 atmospheric correction (S2AC) is based on the algorithm Atmospheric/Topographic Correction for Satellite Imagery [57]. This algorithm allows retrieval of bottom-of-atmosphere (BOA) reflectance from top-of-atmosphere (TOA) reflectance images, available as Level-1C products. The method performs atmospheric correction based on the LIBRADTRAN radiative transfer model [58].

A complete list of the images applied in this study and the date of their acquisition is given in Table 1, while the footprints of the images are shown in Figure 1. A two-month time series per event, including 13 images (twelve non-flood and one flood image) in the same path (ascending with the track numbers of 15 and 88 for the events in 2017 and 2020, respectively), was used.

2.3. Methods

2.3.1. Overview of the Approach

The first step in the procedure adopted in this study is to evaluate SAR and MS signals. We identified, first, the main landscape units by interpreting true and false color composites, then analyzed the optical and microwave signatures of such units. The analyses of the signatures suggested that surface water types can be discriminated better by combining MS and SAR signatures. The characterization of landscape units is described in Section 3.1.

Second, a new stepwise workflow was developed as schematically illustrated in Figure 2 to delineate heterogonous surface water. In the first classification experiment, the Otsu method was applied to distinguish two classes with minimal intra-class variance and maximal inter-class difference. In the second experiment, we focused on improving the delineation of fragmented flood water patches by combining thresholding and segmentation. Based on the MS and SAR signatures, we labeled these classes as water and non-water, as required by these unsupervised classifications. The SAR backscatter and NDWI images were used for the unsupervised methods.

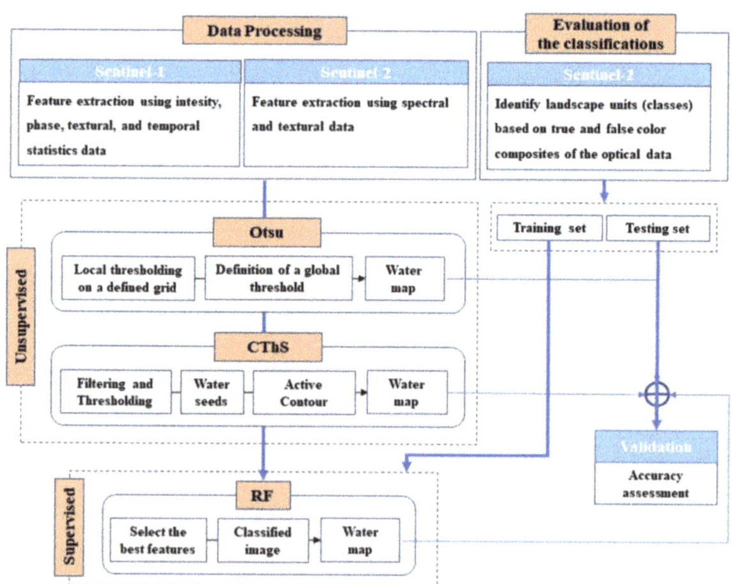

Figure 2. The work-flow of the approach, consisting of three different methods.

Flooded areas are better detectable with co-polarized SAR data rather than cross-polarized ones [59–61]. The S-1 data do not include HH backscatter data for our case studies, so we used the pre-processed S-1 backscatter data with only VV polarization, although the literature suggests that HH data may perform better [55,62,63]. Further detailed information about unsupervised methods can be found in Section 2.3.2. The RF approach was applied to multiple features obtained from SAR and MS data with a dual scope: first, to discover new classification rules to classify different surface water types and, second, to evaluate the classification performance when using either SAR or MS data only and when combining them. The RF supervised method is described in Section 2.3.3.

In this study, the flooded areas were determined as the difference between the water area during a flooding event and the permanent water areas before it. The accuracy and precision of water maps were evaluated by applying three different methods and two datasets.

There are four innovative elements in the proposed workflow: (1) the stepwise approach as an exploration of the capability of each dataset to distinguish landscape units starting from a simple method and simple data to increasingly complex algorithms and features to resolve ambiguities remaining at each step; (2) the combination of thresholding and segmentation; (3) the combination of optical and SAR derived features for RF classification and (4) the use of time-dependent features (anomalies) in the RF.

2.3.2. Unsupervised Methods

Basically, the global thresholding method assumes that image pixel intensity values follow a bimodal frequency distribution (histogram). The method tries to find a single intensity threshold that separates pixels into two classes, foreground and background. However, in most flood cases, the water feature covers only a small fraction of the scene, and the bimodality does not appear in the histogram. Furthermore, the abundance of two spectrally different features in the image, such as bare soil and vegetation, may give a threshold that is not appropriate to delineate water. To tackle the non-bimodality issue, we applied the simple Otsu thresholding method to sub-images and proposed a new unsupervised method (called CThS) based on the combination of histogram thresholding and active contour segmentation methods.

- Otsu thresholding method

The threshold in the Otsu method is determined by minimizing intra-class intensity variance, or equivalently, by maximizing inter-class variance [64]. To tackle the above-mentioned issue of non-bimodality, Otsu thresholding was applied to small sub-images of the original image. The entire image was subsampled into one hundred sub-images using a regular grid, and thresholds found were pooled to determine their frequency distribution (histogram). A unique threshold was determined from the histogram of sub-image thresholds. If the histogram of the thresholds is not bimodal, the threshold is identified first by a visual analysis to determine a threshold range that is not referring to water features and by excluding values in this range. Then, the maximum value (for SAR backscatter) or the minimum value (for NDWI) of the remaining thresholds is selected and applied to the entire image.

- The CThS method (combination of thresholding and segmentation)

The main idea of CThS is to find seeds that are definitely samples of water areas. To identify the water seeds a two-step procedure is applied using a textural feature, namely entropy, which maximizes the contrast between homogenous pixel samples. First, the entropy image is generated by applying a moving window. Then, a water and a non-water mask are constructed by applying a threshold to the moving window. The window size and the threshold are estimated by trial and error using the Otsu delineation of rivers and lakes in a pre-event image as a reference. By applying the mask to the input image at full resolution, the distribution of NDWI and VV backscatter values for water and non-water pixels is obtained. These distributions were applied to identify the water seeds. The histogram of all extracted seeds is reasonably bimodal so that a suitable threshold value can be determined by fitting a curve to the histogram to separate water and non-water pixels. The minimum turning point of the curve determines the threshold to extract water seed pixels. Having separated water seed points, in the second step an active contour segmentation method is used to delineate the full flood extent. The segmentation extends the initial seeds to fragmented patches. The active contour segmentation method has been widely employed for flood mapping [35,65,66]. Tong et al. (2018) showed that the Chan–Vase (C-V) active contour model [67] is computationally more efficient than the classical snake model, while it also performs better in weak boundary detection than the snake model. The snake model needs an initial set of boundary points, which are identified by applying the water and non-water gradient as a characteristic of boundary points. The gradient is estimated using an initial set of water patches. Because of the irregular and extensive distribution of the inundated areas, it is challenging to construct the initial set of water patches and estimate robust statistics on the gradient. Thus, the C-V model was applied in this study.

2.3.3. Supervised Random Forest Classification

RF is a machine learning-based method, which combines many weak classifiers, the individual decision trees, to obtain a strong classifier, the Random Forest, consisting of all decision trees together [68], and is, therefore, an example of a so-called ensemble method. The method takes a number of features as input, and, when applied to classification, is trained by a set of training feature vectors for which it is known to which class they correspond. All possible values of all features in the feature vectors together form the feature space. Each node of a decision tree in the random forest corresponds to a split of the feature space for one of its features, or one dimension. To partly decorrelate the decision trees, one individual tree is only built using a subset of the features, and each split in a decision tree is determined in the training phase such that it minimizes impurity, where the impurity is a measure of heterogeneity (entropy) of the two subsets of the feature space, generated by the split [69].

Each decision tree is created by selecting at random only two-thirds of the training feature vectors with replacement. The remaining one-third of the training samples are assigned as out-of-bag (OOB) data [68], which are used for inner cross-validation to evaluate the performance of RF. The importance of the input variables can be measured, which indicates their contributions to the classification accuracy [68]. Only two parameters need to be specified to parameterize the classifier, ntree, the number of decision trees making up the whole forest and, mty, the number of randomly selected features. In general, the OOB error decreases with the growth of ntree and the plot of OOB error vs. ntree is always necessary to see whether a given number of trees is sufficient to achieve the required performance in the grown forest [50].

A key functionality of the RF is the application of alternate criteria and metrics to rank candidate features on the basis of their importance. Gini importance or mean decrease impurity (MDI) is one of the methods to calculate the feature importance. For each feature, it is possible to assess how on average they decrease the impurity. The average over all trees in the forest is then the measure of the feature importance.

2.3.4. Feature Generation

S-2 bands 3 (*Green*) and 8 (*NIR*) were used to generate the *NDWI* image for both pre and in-flood dates. *NDWI* is mathematically expressed (Equation (1)) as a combination of *NIR* and *Green* bands [6]:

$$NDWI = \frac{Green - NIR}{Green + NIR}, \tag{1}$$

where, *Green* and *NIR* refer to the reflectance in the green and near-infrared bands of the MS data, respectively. The *NDWI* products from S-2 data are represented in Figure 3.

A total number of 32 features at pixel level were calculated as input for Random Forest Classification. These are based on statistics calculated with SAR amplitude, phase, temporal information and textural characteristics of S-1 data (Table 2).

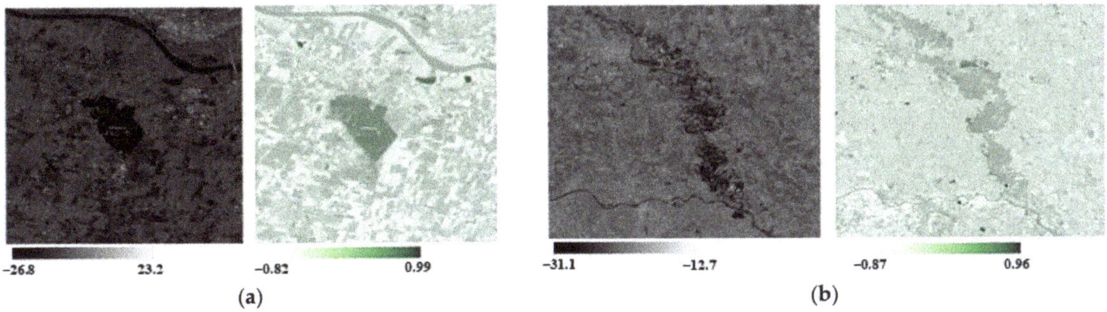

Figure 3. VV-derived SAR backscatter (**left**) and NDWI (**right**) images for the events in (**a**) 2017 and (**b**) 2020.

Table 2. Random forest SAR-derived features.

Data Type	Features	Description	No.
Intensity (from GRDH)	backscatter coefficients (VV, VH)	Log intensity in dB	2
Phase (from SLC)	Coherence (VV, VH)	Normalized cross-correlation coefficient between two interferometric images	2
	H/Alpha Dual Decomposition (VV + VH)	Scattering mechanism information	2
Texture (from GRDH)	GLCM: Contrast, Dissimilarity, Homogeneity, Angular Second Moment, Energy, Maximum, Entropy, GLCMMean, GLCMVariance, GLCMCorrelation, (VV, VH)	Gray Level Co-occurrence Matrix: second order textural features [70]	20
Temporal statistics (from GRDH)	Std (VV, VH)	Time-series standard deviation	2
	Z_Scores (VV, VH)	The number of standard deviations time-series pixels lie from the mean	2
	Anomalies (VV, VH)	Temporal Anomaly	2

The main parameter retrieved by SAR sensors is the backscatter coefficient, i.e., the amplitude squared of the SAR complex signal. SAR backscattering is mainly affected by soil roughness and soil dielectric constant [71]. Flooded areas appear darker, i.e., with lower backscatter coefficient values, than non-flooded areas due to specular reflectance of flood water surfaces when smooth, i.e., still and free of emergent vegetation. We did also benefit from VH available SAR polarization in line with the findings of [72]. The pre-processed VV (Figure 3) and VH SAR backscatter images were retrieved from the GEE server.

Additionally, some polarimetric and phase-based features can be extracted from the dual-polarized SLC data. SAR interferometry (InSAR), which provides information about the Earth's topography by processing two or more SAR data images, produces interferometric coherence. InSAR coherence is sensitive to the physical changes in the ground surface and is therefore useful for image segmentation and identification of geo-meteorological and hydrological features [73]. Coherence (γ) is defined as the normalized cross-correlation coefficient between two interferometric images I_1 and I_2:

$$\gamma = \frac{E(I_1 \cdot I_2^*)}{\sqrt{E\left(|I_1|^2\right)E\left(|I_2|^2\right)}}, \tag{2}$$

where E is the expectation operator, and the asterisk indicates complex conjugation.

An interferogram generated from two images before and after a flood represents flooded areas with uncorrelated phase information, thus more incoherent than non-flooded ones, since both dielectric constant and surface roughness of flooded patches can be different [73]. The SLC image pairs acquired on 7–13 December 2017 and 27 September–3 October 2020 were used to generate coherence maps for the first and second events, respectively. The short spatio-temporal baselines ensure that the least coherent areas most likely are flooded areas. The coherence processing was conducted using the Sentinel Toolbox software (SNAP) and consists of steps to apply orbit files, pre- and post-flood images co-registration, de-speckling (Refined Lee filter with the window size of 5 × 5), interferogram/coherence generation, Sentinel de-bursting, and multi-looking (to get square pixels). The coherence image finally was geocoded by correcting SAR geometric distortions using a digital elevation model (DEM), here the 1 Arc-Second SRTM DEM.

H/Alpha dual-polarimetric decomposition, which allows the separation of different scattering mechanisms, was also included in the RF classification. The H/Alpha decomposition of dual-polarization data uses an eigenvector analysis of the coherence matrix, which separates the parameters into scattering processes and their relative magnitudes [74]. Two parameters are extracted from the H/Alpha decomposition, entropy (H), and alpha (α). Entropy is calculated from the eigenvalue information and represents the heterogeneity of the scattering. Alpha (α) is calculated from the eigenvectors and represents a rotation that indicates the type of scattering mechanism.

Texture is one of the characteristics used in identifying objects or regions of interest in an image. The so-called gray-level co-occurrence matrix (GLCM) method is used to extract second and higher-order statistical texture features, considering the relationship between neighboring pixels. The GLCM function is an image texture indicator that works by computing the frequency of occurrence of pixel pairs with specific values and in a specific spatial relationship within an image. Fourteen textural features can be calculated from the probability matrix to derive the characteristics of texture statistics of images. Detailed definitions of the textural features can be found in [70]. We used, however, ten GLCM-derived features for each polarization, which are listed in Table 2.

At the same time, to improve the reliability of classification, some temporal SAR features, including standard deviation (Std), temporal Z-scores (Zs) [75] and normalized anomaly ($Anomaly$) of image pixels within our time-series images, were also calculated according to Equations (3), (4) and (5), respectively. Flooded pixels can be identified by using these indicators as features in the RF classification. The temporal Z-score, Zs, is a measure of the difference between the backscatter during the flood and the mean backscatter during the entire period of the observations (including pre-event data and co-event data). The anomaly is a measure of the difference between the backscatter during the flood and the mean backscatter during the non-flood period.

The time-series SAR backscatter data (including flood and non-flood images) was used to calculate the above-mentioned temporal features. Note that during these time periods, only the events analyzed in the current study occurred. These features were computed separately for both VV and VH polarizations.

$$Std = \sqrt{\frac{\sum_{n=1}^{N} (\sigma_n^\circ - \sigma_M)^2}{N-1}}, \tag{3}$$

where, σ_n° indicates the SAR backscatter coefficient of each pixel of Nth image within the time series. σ_M is the mean backscatter of each pixel along the whole stack of images. N is the number of images.

$$Zs = \frac{\sigma_F^\circ - \sigma_M}{Std}, \tag{4}$$

$$Anomaly = \frac{\sigma_F^\circ - \sigma_{M\_pre}}{\sigma_{Max}^\circ - \sigma_{Min}^\circ}, \tag{5}$$

where, σ_F° represents the SAR backscatter of the flood image. σ_{M_pre} is the mean backscatter of the pre-flood data stack only. σ_{Max}° and σ_{Min}° refer to the maximum and minimum of flood backscatter.

For MS data, the R, G, B, and NIR bands and the same GLCM-derived features (see Table 2) of RGB bands (at pixel level), as well as NDWI (15 features altogether for only the flooding date), are used for the classification.

2.3.5. Evaluation of the Classifications

To construct training and testing datasets, we first identified by expert interpretation the landscape units observable in both true color (R = S2 band 4, G = S2 band 3, and B = S2 band 2) and false color composites (R = S2 band 8; G and B as in the true color composite). Then, we analyzed the spectral and SAR signatures of such units to evaluate their separability. These signatures were estimated by sampling the MS and SAR images at locations identified as similar in both the true and false color composites. Specifically, the samples were randomly collected in small polygonal blocks so that all pixels within each polygon represented the same class. The training and testing samples were 70% and 30% of the total number of samples, respectively. The total number of samples was approximately 14,000 and 15,200 pixels for case studies in 2017 and 2020 respectively.

Three methods were applied to evaluate the accuracy of water maps, using different testing datasets. To compare the supervised and unsupervised results, the five classes mapped with the supervised RF classifications were aggregated into water and non-water classes. The testing dataset was used to calculate the producer accuracy related to the water class. The second reference dataset was obtained by delineating the water class in the entire scene and then estimating the fractional abundances (fraction of classified water pixels to the total number of pixels) of water on a regular 500 m resolution grid. The third evaluation was based on estimating the precision of each method by comparing each estimate within each grid cell with the median of all the estimates in the same cell.

3. Results

This section exhibits the results of different analyses in our research. First, the SAR and MS signatures of different landscape units are described and interpreted. The results of unsupervised and supervised methods are then presented. Having the accuracy and precision of water maps evaluated, the results on heterogeneous surface water which makes flood delineation more challenging are illustrated.

3.1. SAR and Multispectral Signatures of the Classes

The classes to be mapped by the supervised classification were identified by visual interpretation of the true and false color composites of the optical images acquired during the flood events. The two composites clearly identified similar land features, as shown in Figure 4, which were taken as a reference for the sampling and analysis of spectral signatures.

The first inspection (Figure 5) of the spectral profile of the land features led us to identify five different classes. According to the observed signatures, three water subclasses can be distinguished: emergent vegetation representing emergent vegetation (EV), turbid water (TW) (=defined here as flood) and clear water (CW). On the other hand, the signatures in Figure 5 identify two non-water classes: soil and vegetation. Due to the presence of sediments, most of the flooded areas had spectral characteristics different from clear water, which typically tends to have low reflectance values. Turbid water spectral reflectance is typically higher at increasing solid particle concentration [76,77]. The spectral reflectance of emergent vegetation was similar to turbid water (=flood), except the higher NIR and SWIR reflectance, while the spatial variability of the reflectance of turbid water was rather limited. The spectral profiles shown in Figure 5 suggest that the five classes might be discriminated by combining the SAR and MS signatures. The partial overlap of the MS signatures of emergent vegetation and flood water should be solved by using the VV and VH backscatter

(Figure 5b), which are clearly different. On the other hand, as expected, there is a clear overlap in the backscatter signatures of emergent and terrestrial vegetation, which can be tackled using the sharp contrast in the corresponding MS signatures.

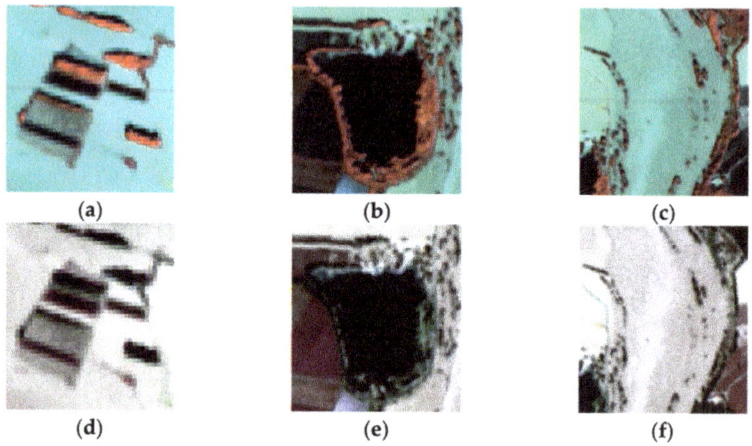

Figure 4. First row: Example of false color composites (R = S2 band 8(NIR), G = S2 band 3 (Green), B = S2 band 2 (Blue)) over (**a**) emergent vegetation (EV), (**b**) clear water (CW) and (**c**) turbid water (TW) classes. Second row: example of true color composite (RGB) over (**d**) emergent vegetation, (**e**) clear water, and (**f**) turbid water classes.

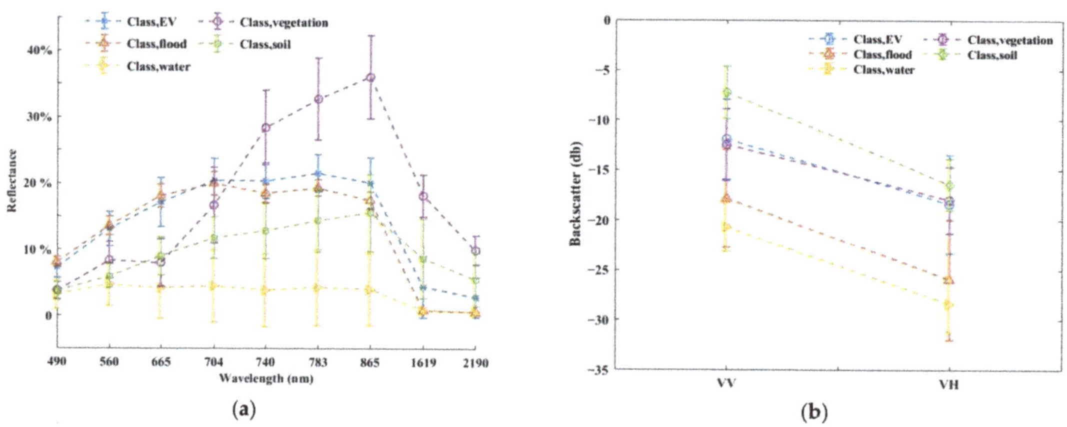

Figure 5. (**a**) Spectral profile (average and standard deviation) (**b**) VV and VH backscattering of the classes (EV: Emergent Vegetation, Flood: flood (turbid) water, Water: clear water, vegetation, and soil) included in the training dataset for Random Forest classification (the 2020 event).

3.2. Flood Maps Derived from Unsupervised Methods: Otsu and CThS Methods

The unsupervised classification methods were applied to flood SAR VV backscatter and optical images for each case study. Flood maps obtained by Otsu and CThS methods are shown in Figure 6. The RGB true color composite of the S-2 data for the flooding date is used in the background as a reference to provide visual support. To delineate only the area actually flooded during each event, pre-flood maps of permanent water bodies were generated and removed from the water maps obtained during the flood events. In both case studies, Otsu thresholding applied to MS data provided better delineation of flood areas than using SAR VV backscatter. In fact, it provided better-defined patterns compared to the SAR data, which gave flood maps more fragmented, i.e., some flood patterns with defined geometry, observable in the true color composite, were not well identified.

Results obtained by the CThS method (Figure 6e–h) showed that similar flood patterns were obtained by CThS compared to Otsu. The main difference between the two methods is that the CThS provided a less fragmented shape of flood areas than Otsu (see illustrations in Section 3.4. for a better visualization of the differences), as expected with the contour reconstruction. Overall, the total flooded areas were also comparable for the two 2017 and 2020 events and the given method (Figure 7). Differences between estimates based on SAR and MS were rather large, however. These differences are due to the partial overlap of the SAR signatures of TW and EV with vegetation and soil (Figure 5 and Table 3). The smaller difference between the SAR and MS total flooded area estimates for the 2020 event suggests, as expected, better performance of the CThS method when dealing with fragmented flooded areas.

3.3. Flood Maps with Supervised Methods: Random Forest Classification

The previous classification experiments highlighted a large variability in spectral signatures and the complexity of classifying water and non-water. To explore the potential advantages of using a larger number of features, we applied the RF classifier on either SAR and MS or both. Specifically, according to Section 2.3.4, we used 32 SAR (Table 2), 15 MS-derived features, and 47 SAR + MS features.

The optimal number of trees (ntree) can be determined by plotting the OOB error versus ntree, where the OOB error curve converged. Figure 8a,b exhibits the ranking of SAR feature importance we used for the classification of the flood event in 2017 (Enza), and a plot of the OOB error curve for the flood event 2020 (Sesia), respectively. The larger the number of trees is, the smaller the error is. The fluctuations around >400 trees were ignored. On the basis of the trend shown in the figure, we regarded the error as stabilized at around 400 trees, where the fluctuations became smaller than 0.001. Similarly, we regarded the contribution of additional features past the first six as negligible (Figure 8a), since the incremental contribution of each additional feature was smaller than 0.01. Therefore, in order to evaluate the reduction of the computational load, 400 trees have been selected and only the six most important features have been used to apply the RF classifier. On the other hand, a comparison between the use of all and the first six features' results provided differences in overall accuracy of less than three percent, confirming it was sufficient to use the best six features for our cases.

(a)

(b)

Figure 6. *Cont.*

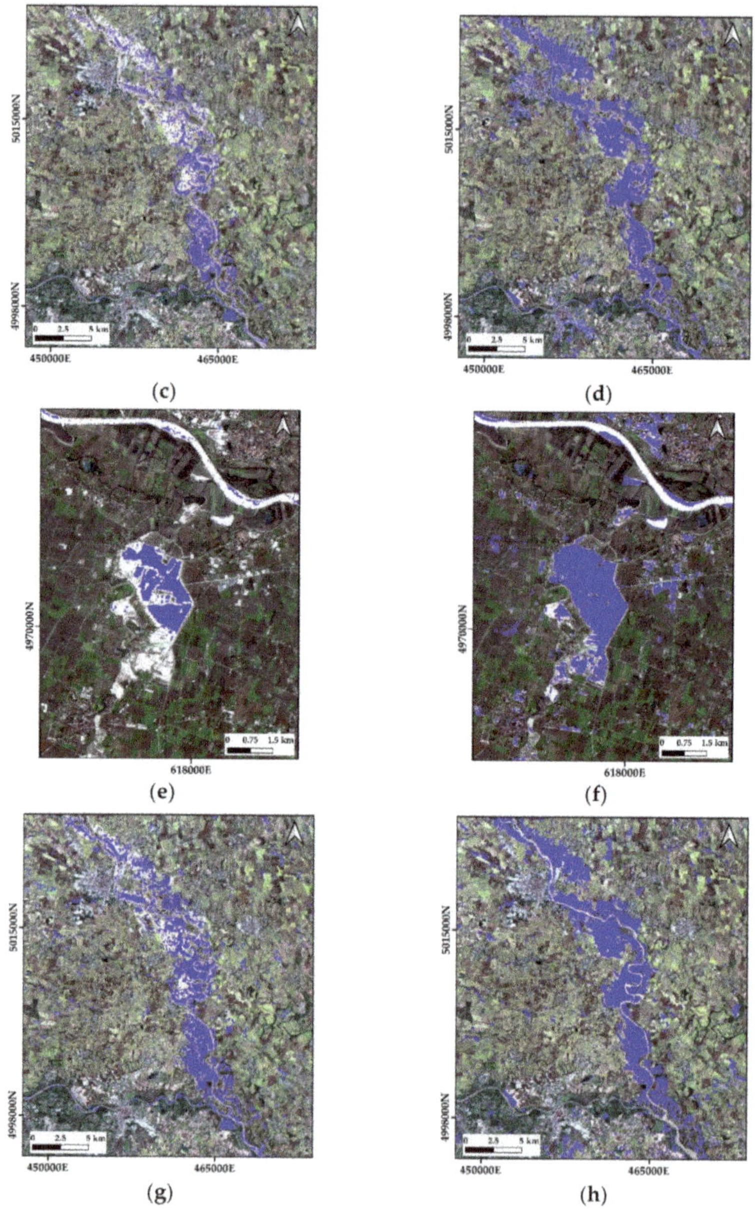

Figure 6. Flood maps obtained by unsupervised methods for the case studies Enza and Sesia rivers. The Otsu results of VV for the case study of (**a**) Enza river (in 2017) and (**c**) Sesia river (in 2020) and of NDWI for the case study of (**b**) Enza river (in 2017) and (**d**) Sesia river (in 2020). The CThS results of VV for the case study of (**e**) 2017 and (**g**) 2020 and of NDWI for the case study of (**f**) 2017 and (**h**) 2020. Pixels classified as water are displayed in blue color, overlaid onto the RGB true color composite of flooding date. The backscatter data on 7 December 2017 and 27 September 2020 as well as the MS data on 24 October 2017 and 9 August 2020 were used to map the permanent water body areas by each method separately.

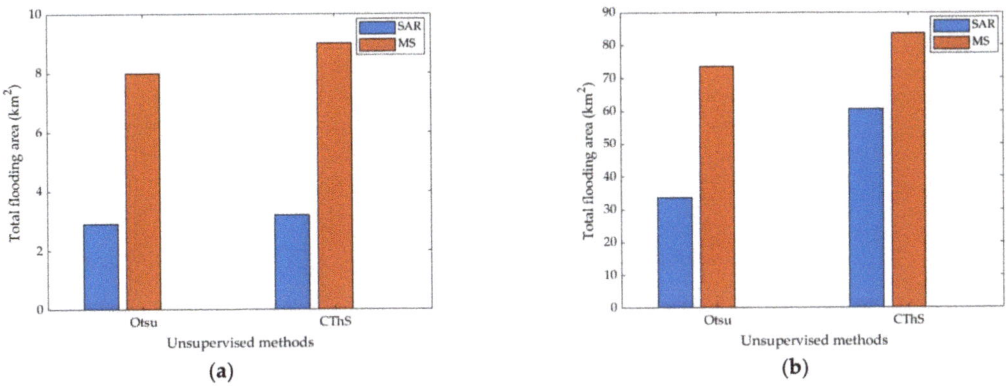

Figure 7. Total flooded areas estimated with Otsu and CThS methods using SAR and multi-spectral data for the events (**a**) 2017, Enza river and (**b**) 2020, Sessia river.

Table 3. Confusion matrices of the RF classification results of the flood event 2020 using SAR, MS, and the combination of them.

	Emergent Vegetation	Turbid Water	Clear Water	Vegetation	Soil
		SAR			
Emergent vegetation	324	319	11	511	304
Turbid water	117	2132	59	194	186
Clear water	0	158	492	9	1
Vegetation	366	640	24	4510	780
Soil	439	460	49	473	2588
		MS			
Emergent vegetation	791	63	5	82	192
Turbid water	365	3646	0	0	157
Clear water	0	0	628	0	0
Vegetation	0	0	0	5531	79
Soil	90	0	2	84	3431
		SAR + MS			
Emergent vegetation	768	84	6	17	321
Turbid water	396	3625	0	0	7
Clear water	0	0	628	0	0
Vegetation	0	0	0	5602	67
Soil	82	0	1	78	3464

Random Forest flood maps provided even better-defined flood patterns compared to unsupervised results (Figures 6 and 9). Well-defined geometric patterns were identified and mapped correctly when using SAR features. This is especially evident by looking at the event that occurred in 2020 (Figures 6 and 9).

The results obtained with the RF classifier using either SAR only or MS only or combined SAR and MS features for the event 2020 were compared by calculating confusion matrices (Table 3). The performance of classification was good when using MS features. In all cases, the percentage of misclassification of emergent vegetation as soil is rather high, with the worst case being SAR features. Likewise, the ambiguity between emergent vegetation and terrestrial vegetation was high in the case of SAR classification. This was slightly improved by combining MS's features with SAR's.

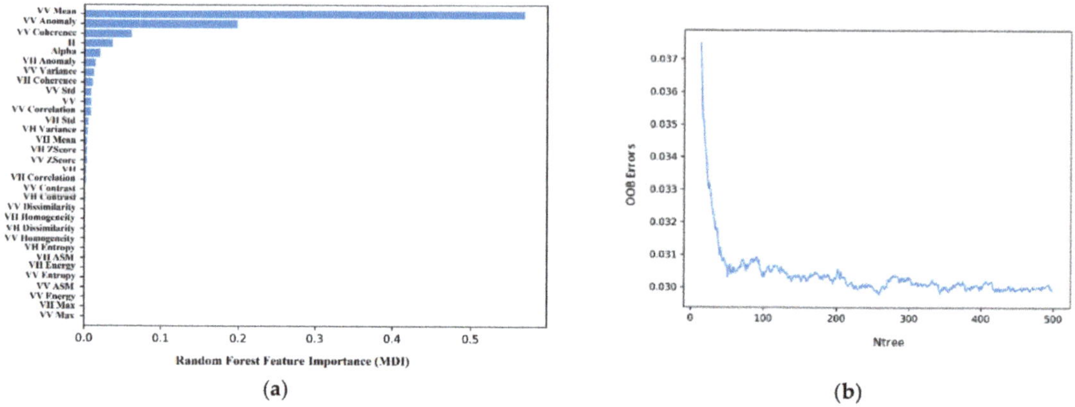

Figure 8. (**a**) Random Forest feature importance based on MDI for SAR-derived features, and (**b**) OOB error vs. number of trees for the best six SAR-derived features (event 2017).

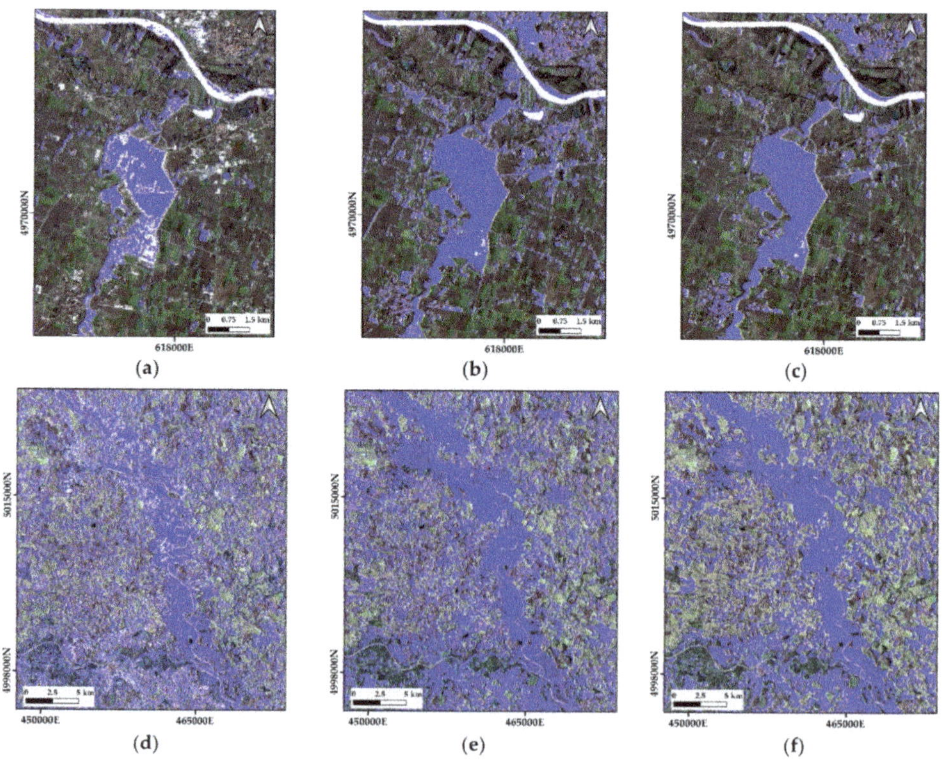

Figure 9. RF flood maps obtained for the two events in Enza (event 2017) and Sesia (event 2020) rivers by SAR (**a**) and (**d**), MS (**b**) and (**e**) and a combination of SAR and MS feature (**c**) and (**f**). Pixels classified as water are displayed in blue color, overlaid onto the RGB true color composite of flooding date. The backscatter data on 7 December 2017 and 27 September 2020 as well as the MS data on 24 October 2017 and 9 August 2020 were used to map the permanent water body areas by the Otsu thresholding method.

We have evaluated the composition of the lumped water class delineated by the unsupervised Otsu and CThS methods by using the information on water type in the testing dataset (Table 4). In addition, we have compared the EV, TW, and CW pixels classified by the supervised method on SAR and SAR + MS with the actual number of EV, TW, and CW pixels in the testing dataset. It appears that the SAR + MS RF classifier captured the largest fraction of the EV pixels, although still much lower than the total number of EV pixels in the testing dataset. As expected, the unsupervised methods captured almost all CW pixels in the testing dataset, but only part of the TW pixels and a small fraction of the EV pixels. The lumped water class delineated by the CThS method included a greater number of TW pixels than by the Otsu method and close to the number of TW pixels identified by the RF classifier applied to SAR data. The use of MS features in combination with the SAR improved the number of pixels that were classified correctly as emergent vegetation and turbid water.

Table 4. Otsu and CThS classifiers: number of pixels lumped in the water class, disaggregated into emergent vegetation (EV), turbid water (TW), and clear water (CW) according to the testing dataset. Number of pixels classified as EV, TW, and CW by the SAR RF and SAR + MS RF classifiers and actual number of EV, TW, and CW pixels according to the testing dataset.

	EV	TW	CW
SAR Otsu	58	1816	602
SAR CThS	149	2773	634
SAR RF	324	2132	492
SAR + MS RF	768	3625	628
Total testing samples	1246	3709	635

3.4. Evaluation of Flood Delineation

As defined in the methodology section, the water maps obtained by the different methodologies used in this work have been assessed in three ways. In the first evaluation, the producer accuracy of classified water (=percentage of correctly classified pixels divided by the total number of the testing samples, i.e., roughly 4000 and 4500 for the events 2017 and 2020, respectively) obtained by the different methodologies has been evaluated (Table 5).

Table 5. Water classification accuracies for the case studies of Enza (the event in 2017) and Sesia rivers (the event in 2020).

	Producer's Accuracy (%)	
	2017	2020
SAR Otsu	78	44
MS Otsu	88	89
SAR CThS	79	63
MS CThS	90	94
SAR RF	92	64
MS RF	99	98
SAR + MS RF	99	98

As mentioned before, the three water classes mapped with RF, i.e., emergent vegetation, turbid water, and clear water, were aggregated into the unique water class. The accuracy values of the water class confirmed a better performance with MS data compared to SAR for all the methods, as stated in the analysis of the flood maps provided in Sections 3.2 and 3.3. The CThS method outperformed the Otsu. The accuracy improvement ranges from about 1% (case 2017) to 19% (case 2020) when using SAR data and from 5% (case 2020) to 2% (case 2017) with MS-based classification. RF provided the highest accuracies with significant improvements in SAR based supervised classification compared to unsupervised (from 1% to 20%).

The second evaluation was performed by comparing the fractional abundance of water estimated with each classification against reference values obtained by delineating the water area by visual interpretation of the false color composites. For this purpose, we applied a regular 500 m resolution grid to sample the maps obtained as the results of the classifications. Then, the fractional abundance of water in each cell was plotted against the corresponding reference values (Figures 10 and 11). The plots show a greater dispersion of the results obtained with unsupervised methods compared to RF, where the classification exactly matches the reference values in most cases. The unsupervised methods generally underestimated flood extent, especially when using SAR, which gave the highest root mean square error (RMSE) values (from 17 to 34%) of the fractional abundance of water. Among the unsupervised classification, the CThS method provided a better delineation with lower RMSE values for both SAR and MS-based classification. The combination of SAR features with MS in RF classification did not give an improvement in terms of accuracy (Table 6) and flood delineation performance compared to considering only MS features. The Enza river case study (2017) showed a much better agreement between classification results and reference data (Figure 10).

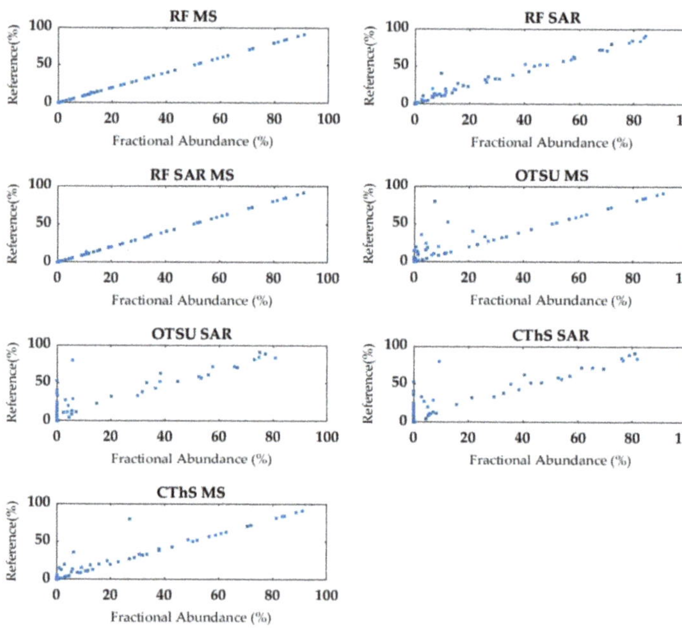

Figure 10. Plots of the fractional abundance of water (%) calculated over a regular 500 m resolution grid for the flood event 2017 (Enza river): results obtained by supervised and unsupervised methodologies (x-axis) against a reference flood delineation (y-axis).

In a further evaluation, the water maps were compared to assess the precision of the methods in each case study. This analysis was performed by evaluating the deviation of the fractional abundance calculated over the 500 m resolution grid from the median value of the results obtained from all the methods. Histograms of the deviations of fractional abundances are shown in Figures 12 and 13 for the events in 2017 and 2020, respectively.

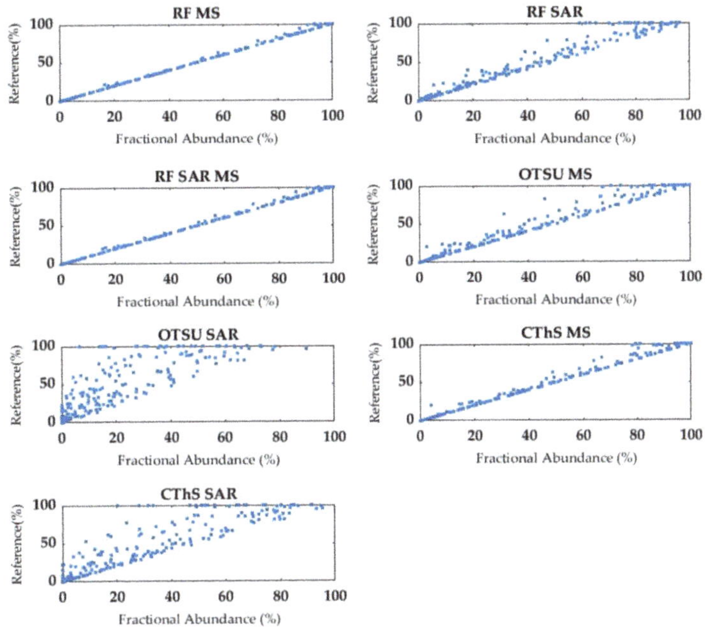

Figure 11. Plots of the fractional abundance of water (%) calculated over a regular 500 m resolution grid for the flood event 2020 (Sesia river): results obtained by supervised and unsupervised methodologies (*x*-axis) against a reference flood delineation (*y*-axis).

Table 6. Coefficient of determination (R2) and root mean square error (RMSE) calculated between the fractional abundance of water calculated from supervised and unsupervised classification methods vs. a reference flood delineation.

	Enza River, 2017		**Sesia River, 2020**	
	R2	**RMSE**	**R2**	**RMSE**
SAR Otsu	0.88	18.07	0.80	34.20
MS Otsu	0.91	13.23	1.00	8.10
SAR CThS	0.89	17.27	0.90	21.70
MS CThS	0.96	8.76	1.00	4.30
SAR RF	0.99	6.43	1.00	10.40
MS RF	1.00	0.23	1.00	1.40
SAR + MS RF	1.00	0.55	1.00	1.60

The results in Figure 12 indicate that the different methods provided similar estimates for the 2017 event (median deviations close to zero), with unsupervised methods giving in some areas smaller flood extent than RF. MS-based RF classification gave larger (positive) deviations from the other methods as well as larger dispersion of the deviations. Contrariwise, deviations were positive and larger with the SAR-based RF for the 2020 event, with a median value close to 30%. Positive deviations for RF classification in the 2020 event (Figure 14) were determined by the better performance, in terms of accuracy, compared to all the unsupervised methods (Table 5).

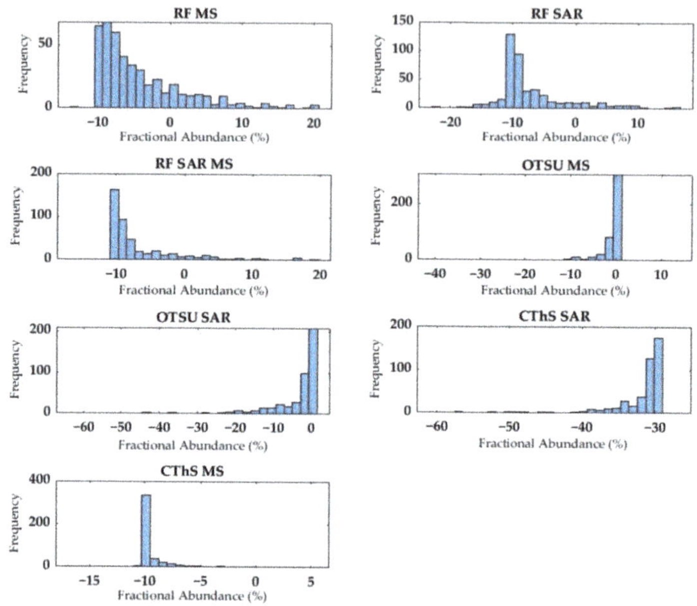

Figure 12. Deviation from the median value of the fractional abundance of water (event 2017) calculated over the nodes of a regular 500 m grid.

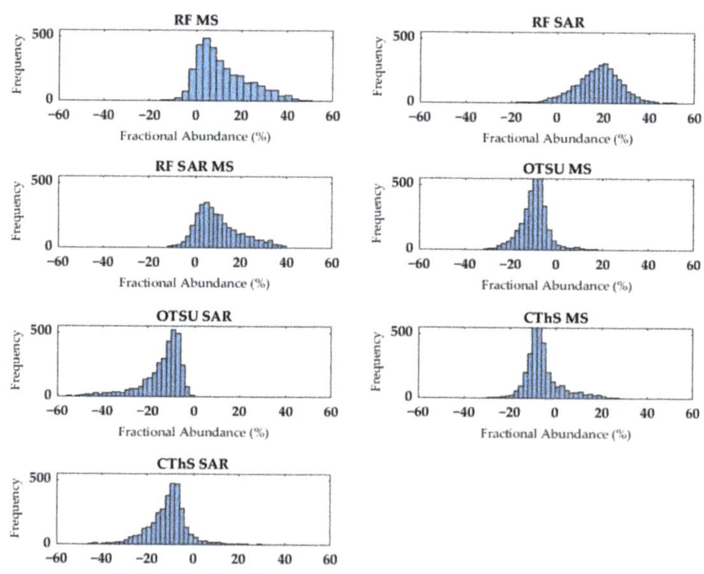

Figure 13. Deviation from the median value of the fractional abundance of water (event 2020) calculated over the nodes of a regular 500 m grid.

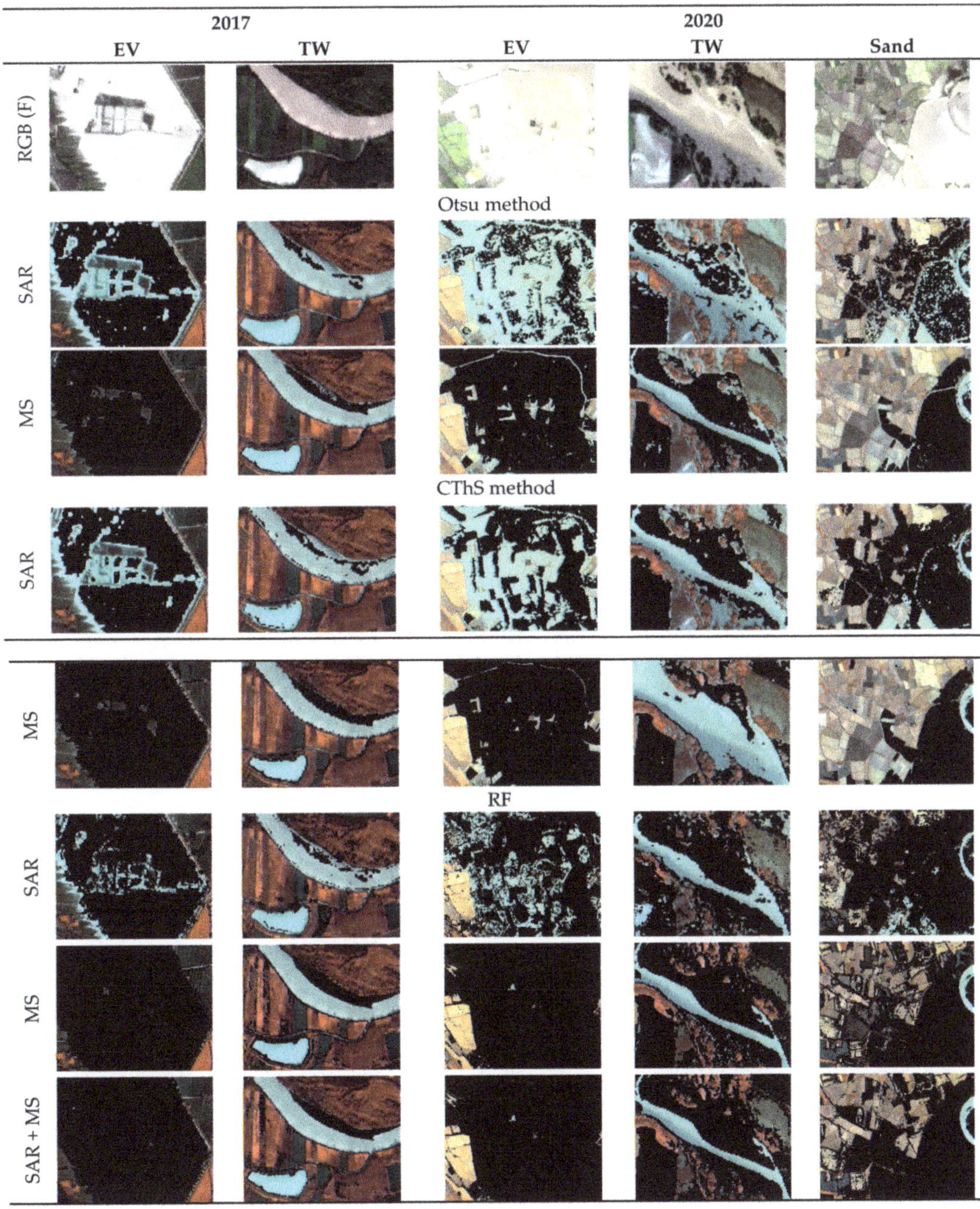

Figure 14. Flood maps of EV: emergent vegetation, TW: turbid water, and sand for all the experiments in 2017 and 2020. Pixels classified as water are displayed in black color, overlaid onto the NIRGB false color composite of the flooding dates. The light blue patches correspond to the river, which was masked out from the flooding water maps (Figure 2). The first row shows the pictures of RGB true color composite of flood conditions; 13 December 2017 and 03 October 2020. The backscatter data on 7 December 2017 and 27 September 2020 as well as the MS data on 24 October 2017 and 9 August 2020 were used to map the permanent water body areas by each method separately.

3.5. Sub-Cases: Emergent Vegetation, Sandy Areas, and Turbid Water

Based on the literature, there are some land cover types, including emergent vegetation, sandy areas, and turbid water, which make accurate flood mapping challenging. We randomly selected some areas with these land covers to investigate the differences between the flood maps derived using SAR and MS data (Figure 14).

The results indicated that the SAR-based unsupervised classifications did not capture completely the emergent/sub-merged vegetation observable in thc NIR, G and B false color composite. To understand this issue we compared the distribution of NDWI and backscatter (Figure 15) within the emergent vegetation (selected area in Figure 14) and a part of the river. The similarity in the distributions of the NDWI indicates that emergent vegetation is mapped as water while this cannot be achieved using backscatter, which has different distributions for emergent vegetation and river water. This implies that the unsupervised methods using NDWI provide a better delineation of the water class, as defined in Section 3.1.

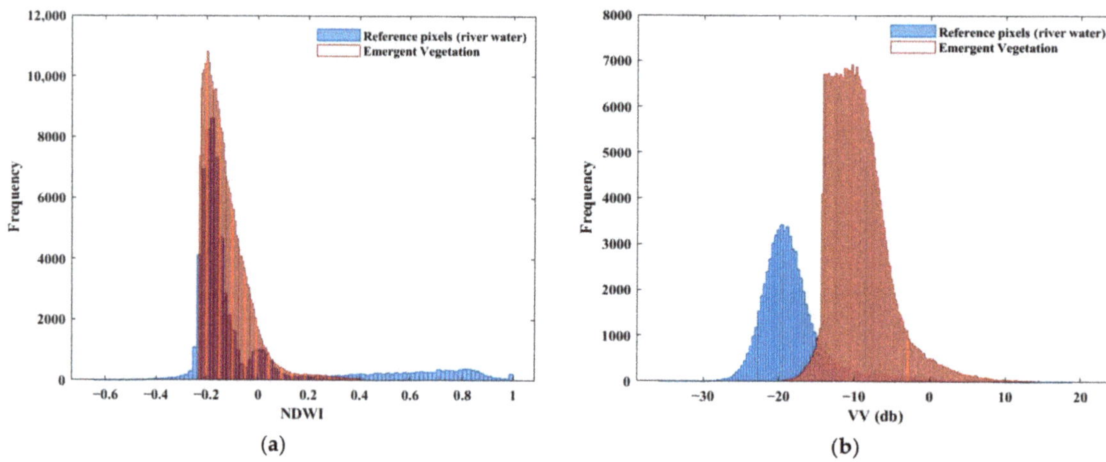

Figure 15. Signatures of emergent vegetation vs. reference water body observed during the 2020 flooding event: (**a**) NDWI from S2/MSI data and (**b**) VV backscatter (in dB) from S-1 SAR data.

We observed an improvement in the delineation of emergent vegetation when using SAR-based RF classification since even fragmented patches of emergent/sub-merged vegetation were correctly classified (see Figure 14, emergent vegetation sub-case, and Table 4). Random Forest achieved the highest performance in delineating emergent vegetation when using MS and a combination of MS and SAR signals.

Over sandy areas (based on the soil map of Regione Piemonte [78]), SAR data led to an over-estimation of water since the backscatter of sandy soil is similar to water, as shown by our analysis of the frequency distribution of backscatter (Figure 16) and confirmed in the literature [24]. The water extent was over-estimated in these sandy areas even when using RF with backscatter data. The distribution of NDWI suggests a clear threshold and good separability between water and sandy soil.

As illustrated in Figure 14, turbid water was mapped correctly by unsupervised classification since it was still possible to determine an appropriate threshold on NDWI. On the other hand, the pre-flood delineation of turbid water by the active contour segmentation when using MS data of the event 2020 was not completely accurate at the location indicated in Figure 14.

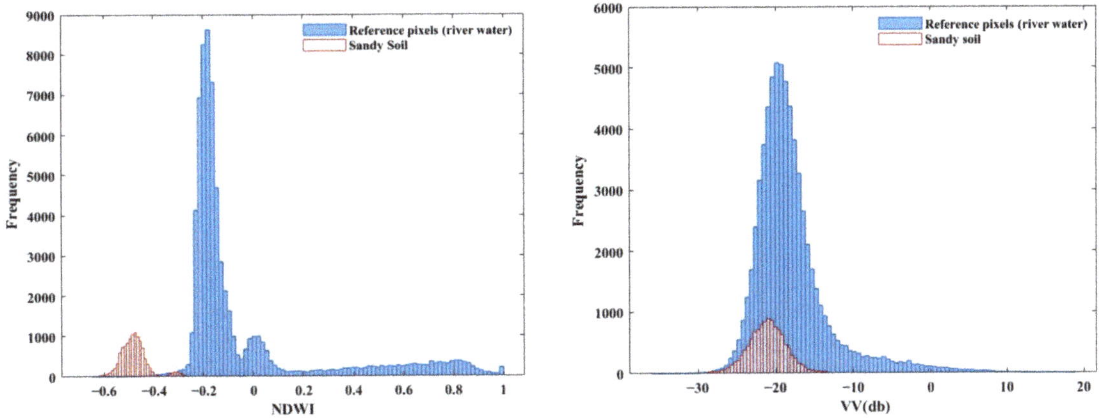

(**a**) (**b**)

Figure 16. Signatures of sandy soils vs. reference water body observed during the 2020 flooding event: (**a**) NDWI from S2/MSI data and (**b**) VV backscatter (in dB) from S1 SAR data.

4. Discussion

The results presented in the previous section provide answers to the research questions stated in the introduction as regards three main aspects:

1. Delineation of landscape units;
2. Spectral and backscatter features;
3. Classification methods.

1. Delineation of landscape units. Land cover, terrain and the depth of flood water concur in determining fragmented and heterogeneous patterns in floodwater. The high spatial resolution of S1/SAR and S2/MSI may capture small patches of emergent vegetation and of turbid water, which increases the heterogeneity of floodwater. The terrain in the two study areas is rather different, i.e., rather flat in the Sesia area and more heterogenous in the Enza area. Land cover is also very different with extensive rice, maze, and pastures, irrigated by flooding in the Sesia area and more fragmented agricultural land cover in the Enza river. In addition, hydro-meteorological conditions were quite different; the 2020 flood in the Sesia river was caused by an embarkment failure (see Section 2.1), rather than extreme rainfall and/or river water level. Contrariwise, the 2017 event in Enza river was caused by a record high river water level. Precipitation was slightly higher for the 2020 Sesia than for the 2017 Enza event. In other words, the combination of terrain, land cover and hydro-meteorological conditions led the 2020 event to be a rather complex flooding pattern, which explains the observed lower performance for this event.

A critical step in our approach was the delineation of landscape units to be mapped by interpreting the true and false color composites. The lack of calibration/validation data is a common problem when observing past extreme events associated with natural hazards. Under such circumstances, it is unlikely that concurrent in situ observations are available to analyze remote sensing data. Photo-interpretation of color composites is a widely used approach in these cases [35,79]. The identification of clearly different land units by photo-interpretation is still a challenge, however, and requires particular attention. Since our main interest was delineating water areas, we mainly focused on the correct identification of different surface water types, i.e., water–vegetation–sediment mixtures. Soil and vegetation classes, even showing intra-class heterogeneity in terms of spectral signature, could be easily identified as unique classes. The spectral signature of the classes is presented in Figure 5, where the mean values of the spectral reflectance confirm the overall separability of the defined classes. The slight overlap of the emergent vegetation and turbid water classes standard deviations suggests that few pixels may present a similar spectral signature. As regards the unsupervised methods, where the target classes are water and non-water, this similarity had no impact on the results obtained by thresholding of NDWI. Contrariwise, the thresholding of backscatter was not adequate to separate emergent (as water) from terrestrial vegetation (as non-water). The ambiguity of SAR backscatter data in the classification of the two classes, however, could be addressed by applying the RF classifier to the combined MS and SAR signatures. The five classes identified on the basis of the true and false color composites could be separated by applying the RF classifier to the combined MS and SAR signatures.

2. Spectral and backscatter features. The spectral and backscatter signatures of flooded areas are complex in two different ways. SAR backscatter is sensitive to the physical characteristics of the ground surface, i.e., roughness and the dielectric constant, making it more difficult to interpret. This concept is supported by the evidence in Table 3. Furthermore, the heterogeneity of the flooding pattern in both events implies that observed targets include rather different components, e.g., different vegetation types and water conditions, that can be better identified using MS spectral features. A flooded area is likely to include patches of turbid water and emergent vegetation which have different signatures from water. The spectral signatures in Figure 5 confirmed this hypothesis since the MS signatures of emergent vegetation and turbid water were roughly overlapping at shorter wavelengths, but slightly different beyond 740 nm. On one hand, the SAR signatures of emergent and terrestrial vegetation were completely overlapping. On the other hand, the combined MS and SAR signatures suggested that it was feasible to separate the five identified classes, as shown by the confusion matrices (Table 3) and by the frequency distributions (Figures 15 and 16). Emergent vegetation during the 2020 event had a spectral signature (NDWI) similar to river water (Figure 15a) and was classified correctly as a component of the water class. When using SAR to observe the same targets, however, the emergent vegetation appeared much brighter than water (Figure 15b) and was not classified correctly. Most likely this is due to the double bouncing effect that increases the backscatter, causing an under-estimation of water areas [41,42,80].

According to Figure 16, values of NDWI and VV SAR backscatter were also compared with reference (river) water in a sandy area, where we observed an overestimation of flooded areas with both supervised and unsupervised methods (Figure 14). The histograms prove that SAR backscatter (Figure 16b) mistakenly led to overestimating flood water by misclassification of sandy soil because of weak backscatter (Martone et al., 2014). As in the case of emergent vegetation, the sandy soil had a "drier" MS signature, i.e., negative NDWI, than water and was separated correctly (Figure 16a).

3. Classification methods. In general, the complexity of the landscape, as a consequence of the flooding pattern, makes it rather challenging both to estimate a reliable threshold in unsupervised methods and reliable signatures when applying the supervised method. As observed the flooding pattern in 2020 was more complex than in 2017, thus explaining the generally lower performance of all the methods evaluated in this study (Table 6).

Unsupervised methods demonstrated good overall performance. The grid-based estimation of the water/non-water thresholds gave satisfactory results when applying the Otsu approach to discriminate water from non-water. However, the accuracy analysis revealed better overall performance of the CThS method in delineating water extents compared to Otsu (Table 5). It generally improved the delineation of water extents by a better-defined geometric structure as it uses segmentation to grow the seed points to approach the optimal water boundaries (Figure 14 and Table 4). Nevertheless, there was an occurrence of misclassification of the water class in the pre-flood MS image (see Figure 14). This implies that a map of flood water extent beyond the boundary of the permanent water bodies was less accurate since it confused the bare soil around the river with the water class in the reference/pre-flood image (Figure 14, 2020 event using MS). As a result, when performing the change detection to remove the reference permanent water bodies, the flooded portion of the bare soil area was removed. On the other hand, the RF classification provided the highest accuracy in our flood mapping cases (Table 5). The advantage of RF appears when dealing with challenging cases, namely emergent vegetation which cannot be discriminated using SAR data alone, while acceptable results are obtained when using MS signatures. This is rather evident using SAR data alone. However, the CThS method provided, overall, precision and accuracy comparable to the supervised method and it is more appropriate for rapid flood mapping due to the easy implementation (Table 4).

Besides the complexity of constructing appropriate training and testing sets and defining efficient features for the supervised method, the computational complexity of RF is much higher than CThS. The computational complexity of RF is O(ntreexNxKxlogN), where N is the number of training samples, and K is the number of features [81], which gives in the more complex case of the 2020 event $O(400 \times 15{,}200 \times 6 \times \log 15{,}200) \approx O(152{,}553{,}654)$. The computational complexity of the CThS is mostly related to the active contour segmentation, which is O(MxN), where M and N refer to the image sizes [82]. Hence, the computational complexity of the CThS is equal to O(seeds number) $\approx$ O(52,000) in our complex case. The advantage of RF appears only when dealing with challenging cases, namely emergent vegetation which cannot be discriminated using SAR data alone, while acceptable results are obtained when using MS signatures.

The use of various input features instead of one, as well as the definition of the water classes on the basis of the signatures, increased the possibility of accurate class discrimination. The presence of emergent vegetation and sandy soil was the most problematic issue for flood mapping with the SAR data. Additionally, the overestimation of floods in non-water areas could also be due to the misclassification of vegetation with water. For the turbid water case, most MS-derived features (into the RF) were able to distinguish between turbid water and clear water class, leading to the most accurate delineation with RF. The detection of emergent vegetation by the SAR supervised method improved when compared to the unsupervised methods (Table 4). Both supervised and unsupervised methods overestimated flooded areas in sandy areas where the SAR backscatter signal is weak.

The confusion matrices in Table 3 indicated that SAR data could only discriminate clear water from all other classes. According to our experiments, described above, MS-derived features provided more reliable information on flooding than SAR. For example, the relatively small differences in reflectance beyond 470 nm between emergent vegetation and turbid water (Figure 5) were sufficient to mitigate the misclassifications between the two classes. The use of SAR in combination with MS resulted in more confusion in classifying emergent vegetation and soil compared to MS features alone. That was induced by the presence of sand and misclassification of sandy areas as water with SAR data. Furthermore, the use of MS features in combination with SAR data improved the separation of emergent vegetation from turbid water and from terrestrial vegetation compared with SAR only (Table 3).

The better performance of MS data for both supervised and unsupervised methods suggests that optical data should be preferred to the SAR. However, SAR data provides more efficient measurements in cloudy conditions than optical observations and increase the availability of data during flood events. Our study suggested that the combined use of SAR data and machine learning methods may lead to a better compromise in terms of data availability and method accuracy, providing performance improvements compared to unsupervised methods, notably in the case of the presence of emergent and/or sub-emergent vegetation. On the other hand, the CThS (unsupervised) method provided, overall, precision and accuracy comparable to the supervised method and is the most rapid technique to delineate flooded areas with acceptable performance.

5. Conclusions

Flood monitoring by remote sensing is a useful tool for rapid emergency response. The precise and accurate retrieval of flood maps is however a challenge mainly due to the heterogeneity of flooded and land areas. The use of multisource remote sensing imagery increases not only the chance of data availability at the time of extreme events but also precision and accuracy due to the different nature of signals. The goal of this paper was to evaluate the precision and accuracy of alternate combinations of classification methods and measurements of different and complementary natures (MS and SAR).

Flood mapping of two events in different regions of interest using S-1 (SAR) and S-2 (MS) datasets acquired during the 2017 and 2020 heavy precipitation was performed and evaluated. Two unsupervised methods, Otsu and CThS, as well as the RF supervised method, were applied. The results indicated that multi-spectral data provided more accurate flood maps using all methods compared to SAR data. Otsu-resulted maps exhibited more fragmented flooding areas, which was addressed by applying the CThS method. The CThS method takes the advantage of both thresholding and segmentation approaches. Consequently, better-defined patterns of inundated areas were obtained. Generally, the CThS resulted in more reliable water maps than Otsu.

There were some areas, like emergent vegetation and sandy soil, leading to misclassifications when using VV SAR backscatter data. The issue was tackled by applying supervised RF, in which different intensity-, phase-, texture- and temporal-based features were utilized to improve the SAR classification. An enhancement with emergent vegetation case was observed while some overestimations of water class over sandy soil still remained with the RF as well. In another experiment, the RF classifier was also applied to MS-derived features separately, as well as the combination of all SAR and MS features together. The highest accuracy in flood mapping was obtained by the supervised RF method in all the cases. Accuracies of 92%, 99%, and 99% were achieved for the 2017 event using SAR, MS, and SAR + MS, respectively. Similarly, high values were obtained for the 2020 event, i.e., 64%, 98%, and 98%. All the solutions evaluated in this study taken together, a better performance was achieved when using MS data, possibly due to the high heterogeneity of the two flooded areas because of the combined effect of terrain land cover and hydro-meteorological conditions in the 2017 and 2020 events.

Author Contributions: Conceptualization, S.M.A., M.M. and R.L.; methodology, F.F. and S.M.A.; software, F.F. and S.M.A.; validation, F.F., S.M.A. and M.M.; formal analysis, F.F. and S.M.A.; investigation, F.F. and S.M.A.; resources, F.F. and S.M.A.; data curation, F.F. and S.M.A.; writing—original draft preparation, F.F.; writing—review and editing, S.M.A., M.M. and R.L.; visualization, F.F. and S.M.A.; supervision, S.M.A., M.M. and R.L.; project administration, M.M. and R.L.; funding acquisition, M.M. All authors have read and agreed to the published version of the manuscript.

Funding: This work was carried out under the framework of OPERANDUM (OPEn-air laboRAtories for Nature baseD solUtions to Manage hydrometeorological risks) project, which is funded by the Horizon 2020 Program of the European Union under Grant Agreement No. 776848. M.M. acknowledges the support received from the MOST High-Level Foreign Expert program (Grant No. GL20200161002) and the Chinese Academy of Sciences President's International Fellowship Initiative (Grant No. 2020VTA0001).

Data Availability Statement: Not applicable.

Acknowledgments: The authors are grateful to B/Pulvirenti, F. Porcu and colleagues at the University of Bologna for discussions on flood events to be analyzed and vulnerable areas. We would also like to thank the European Space Agency (ESA) for providing, free of charge Sentinel 1 and 2 data.

Conflicts of Interest: The authors declare no conflict of interest.

References

1. Smith, L.C. Satellite remote sensing of river inundation area, stage, and discharge: A review. *Hydrol. Process.* **1997**, *11*, 1427–1439. [CrossRef]
2. Domenikiotis, C.; Loukas, A.; Dalezios, N. The use of NOAA/AVHRR satellite data for monitoring and assessment of forest fires and floods. *Nat. Earth Syst. Sci.* **2003**, *3*, 115–128. [CrossRef]
3. Fayne, J.V.; Bolten, J.D.; Doyle, C.S.; Fuhrmann, S.; Rice, M.T.; Houser, P.R.; Lakshmi, V. Flood mapping in the lower Mekong River Basin using daily MODIS observations. *Int. J. Remote Sens.* **2017**, *38*, 1737–1757. [CrossRef]

4. Powell, S.; Jakeman, A.; Croke, B. Can NDVI response indicate the effective flood extent in macrophyte dominated floodplain wetlands? *Ecol. Indic.* **2014**, *45*, 486–493. [CrossRef]
5. Zoffoli, M.L.; Kandus, P.; Madanes, N.; Calvo, D.H. Seasonal and interannual analysis of wetlands in South America using NOAA-AVHRR NDVI time series: The case of the Parana Delta Region. *Landsc. Ecol.* **2008**, *23*, 833–848. [CrossRef]
6. McFeeters, S.K. The use of the Normalized Difference Water Index (NDWI) in the delineation of open water features. *Int. J. Remote Sens.* **1996**, *17*, 1425–1432. [CrossRef]
7. Abazaj, F. SENTINEL-2 Imagery for Mapping and Monitoring Flooding in Buna River Area. *J. Int. Environ. Appl. Sci.* **2020**, *15*, 48–53.
8. Gao, B.-C. NDWI—A normalized difference water index for remote sensing of vegetation liquid water from space. *Remote Sens. Environ.* **1996**, *58*, 257–266. [CrossRef]
9. McFeeters, S.K. Using the normalized difference water index (NDWI) within a geographic information system to detect swimming pools for mosquito abatement: A practical approach. *Remote Sens.* **2013**, *5*, 3544–3561. [CrossRef]
10. Memon, A.A.; Muhammad, S.; Rahman, S.; Haq, M. Flood monitoring and damage assessment using water indices: A case study of Pakistan flood-2012. *Egypt. J. Remote Sens. Space Sci.* **2015**, *18*, 99–106. [CrossRef]
11. Thomas, R.F.; Kingsford, R.T.; Lu, Y.; Cox, S.J.; Sims, N.C.; Hunter, S.J. Mapping inundation in the heterogeneous floodplain wetlands of the Macquarie Marshes, using Landsat Thematic Mapper. *J. Hydrol.* **2015**, *524*, 194–213. [CrossRef]
12. Yang, X.; Zhao, S.; Qin, X.; Zhao, N.; Liang, L. Mapping of urban surface water bodies from Sentinel-2 MSI imagery at 10 m resolution via NDWI-based image sharpening. *Remote Sens.* **2017**, *9*, 596. [CrossRef]
13. Fisher, A.; Flood, N.; Danaher, T. Comparing Landsat water index methods for automated water classification in eastern Australia. *Remote Sens. Environ.* **2016**, *175*, 167–182. [CrossRef]
14. Shen, L.; Li, C. Water body extraction from Landsat ETM+ imagery using adaboost algorithm. In Proceedings of the 2010 18th International Conference on Geoinformatics, Beijing, China, 18–20 June 2010; IEEE: New York, NY, USA, 2010.
15. Wilson, E.H.; Sader, S.A. Detection of forest harvest type using multiple dates of Landsat TM imagery. *Remote Sens. Environ.* **2002**, *80*, 385–396. [CrossRef]
16. Xu, H. Modification of normalised difference water index (NDWI) to enhance open water features in remotely sensed imagery. *Int. J. Remote Sens.* **2006**, *27*, 3025–3033. [CrossRef]
17. Rokni, K.; Ahmad, A.; Selamat, A.; Hazini, S. Water feature extraction and change detection using multitemporal Landsat imagery. *Remote Sens.* **2014**, *6*, 4173–4189. [CrossRef]
18. Bangira, T.; Alfieri, S.M.; Menenti, M.; Van Niekerk, A. Comparing thresholding with machine learning classifiers for mapping complex water. *Remote Sens.* **2019**, *11*, 1351. [CrossRef]
19. Deijns, A.A.J.; Dewitte, O.; Thiery, W.; d'Oreye, N.; Malet, J.-P.; Kervyn, F. Timing landslide and flash flood events from SAR satellite: A new method illustrated in African cloud-covered tropical environments. *Nat. Hazards Earth Syst. Sci. Discuss.* **2022**, *172*, 1–38.
20. Bhatt, C.; Thakur, P.K.; Singh, D.; Chauhan, P.; Pandey, A.; Roy, A. Application of active space-borne microwave remote sensing in flood hazard management. In *Geospatial Technologies for Land and Water Resources Management*; Springer: Cham, Switzerland, 2022; pp. 457–482.
21. Santangelo, M.; Cardinali, M.; Bucci, F.; Fiorucci, F.; Mondini, A.C. Exploring event landslide mapping using Sentinel-1 SAR backscatter products. *Geomorphology* **2022**, *397*, 108021. [CrossRef]
22. Laugier, O.; Fellah, K.; Tholey, N.; Meyer, C.; De Fraipont, P. High temporal detection and monitoring of flood zone dynamic using ERS data around catastrophic natural events: The 1993 and 1994 Camargue flood events. In Proceedings of the third ERS Symposium, ESA SP-414, Florence, Italy, 17–21 March 1997.
23. White, L.; Brisco, B.; Dabboor, M.; Schmitt, A.; Pratt, A. A collection of SAR methodologies for monitoring wetlands. *Remote Sens.* **2015**, *7*, 7615–7645. [CrossRef]
24. Martinis, S.; Plank, S.; Ćwik, K. The use of Sentinel-1 time-series data to improve flood monitoring in arid areas. *Remote Sens.* **2018**, *10*, 583. [CrossRef]
25. Martone, M.; Bräutigam, B.; Rizzoli, P.; Krieger, G. TanDEM-X performance over sandy areas. In Proceedings of the EUSAR 2014, 10th European Conference on Synthetic Aperture Radar, Berlin, Germany, 3–5 June 2014; VDE: Berlin, Germany, 2014.
26. Ahmed, K.R.; Akter, S. Analysis of landcover change in southwest Bengal delta due to floods by NDVI, NDWI and K-means cluster with Landsat multi-spectral surface reflectance satellite data. *Remote Sens. Appl. Soc. Environ.* **2017**, *8*, 168–181. [CrossRef]
27. Amitrano, D.; Di Martino, G.; Iodice, A.; Riccio, D.; Ruello, G. Unsupervised rapid flood mapping using Sentinel-1 GRD SAR images. *IEEE Trans. Geosci. Remote Sens.* **2018**, *56*, 3290–3299. [CrossRef]
28. Kordelas, G.A.; Manakos, I.; Aragonés, D.; Díaz-Delgado, R.; Bustamante, J. Fast and automatic data-driven thresholding for inundation mapping with Sentinel-2 data. *Remote Sens.* **2018**, *10*, 910. [CrossRef]
29. Landuyt, L.; Verhoest, N.E.; Van Coillie, F.M. Flood mapping in vegetated areas using an unsupervised clustering approach on Sentinel-1 and-2 imagery. *Remote Sens.* **2020**, *12*, 3611. [CrossRef]
30. Liang, J.; Liu, D. A local thresholding approach to flood water delineation using Sentinel-1 SAR imagery. *ISPRS J. Photogramm. Remote Sens.* **2020**, *159*, 53–62. [CrossRef]
31. Zhang, Q.; Zhang, P.; Hu, X. Unsupervised GRNN flood mapping approach combined with uncertainty analysis using bi-temporal Sentinel-2 MSI imageries. *Int. J. Digit. Earth* **2021**, *14*, 1561–1581. [CrossRef]

32. Acharya, T.D.; Subedi, A.; Lee, D.H. Evaluation of Machine Learning Algorithms for Surface Water Extraction in a Landsat 8 Scene of Nepal. *Sensors* **2019**, *19*, 2769. [CrossRef]
33. Huang, M.; Jin, S. Rapid flood mapping and evaluation with a supervised classifier and change detection in Shouguang using Sentinel-1 SAR and Sentinel-2 optical data. *Remote Sens.* **2020**, *12*, 2073. [CrossRef]
34. Nandi, I.; Srivastava, P.K.; Shah, K. Floodplain mapping through support vector machine and optical/infrared images from Landsat 8 OLI/TIRS sensors: Case study from Varanasi. *Water Resour. Manag.* **2017**, *31*, 1157–1171. [CrossRef]
35. Tong, X.; Luo, X.; Liu, S.; Xie, H.; Chao, W.; Liu, S.; Liu, S.; Makhinov, A.; Makhinova, A.; Jiang, Y. An approach for flood monitoring by the combined use of Landsat 8 optical imagery and COSMO-SkyMed radar imagery. *ISPRS J. Photogramm. Remote Sens.* **2018**, *136*, 144–153. [CrossRef]
36. Benoudjit, A.; Guida, R. A novel fully automated mapping of the flood extent on SAR images using a supervised classifier. *Remote Sens.* **2019**, *11*, 779. [CrossRef]
37. Esfandiari, M.; Abdi, G.; Jabari, S.; McGrath, H.; Coleman, D. Flood hazard risk mapping using a pseudo supervised random forest. *Remote Sens.* **2020**, *12*, 3206. [CrossRef]
38. Ji, L.; Zhang, L.; Wylie, B. Analysis of dynamic thresholds for the normalized difference water index. *Photogramm. Eng. Remote Sens.* **2009**, *75*, 1307–1317. [CrossRef]
39. Townsend, P.A.; Walsh, S.J. Modeling floodplain inundation using an integrated GIS with radar and optical remote sensing. *Geomorphology* **1998**, *21*, 295–312. [CrossRef]
40. Chapman, B.; Russo, I.M.; Galdi, C.; Morris, M.; di Bisceglie, M.; Zuffada, C.; Lavalle, M. Comparison of SAR and CYGNSS surface water extent metrics over the Yucatan lake wetland site. In Proceedings of the 2021 IEEE International Geoscience and Remote Sensing Symposium IGARSS, Brussels, Belgium, 11–16 July 2021; IEEE: New York, NY, USA, 2021.
41. Moser, L.; Schmitt, A.; Wendleder, A.; Roth, A. Monitoring of the Lac Bam wetland extent using dual-polarized X-band SAR data. *Remote Sens.* **2016**, *8*, 302. [CrossRef]
42. Pulvirenti, L.; Pierdicca, N.; Chini, M.; Guerriero, L. Monitoring flood evolution in vegetated areas using COSMO-SkyMed data: The Tuscany 2009 case study. *IEEE J. Sel. Top. Appl. Earth Obs. Remote Sens.* **2013**, *6*, 1807–1816. [CrossRef]
43. Chaouch, N.; Temimi, M.; Hagen, S.; Weishampel, J.; Medeiros, S.; Khanbilvardi, R. A synergetic use of satellite imagery from SAR and optical sensors to improve coastal flood mapping in the Gulf of Mexico. *Hydrol. Process.* **2012**, *26*, 1617–1628. [CrossRef]
44. Refice, A.; Capolongo, D.; Pasquariello, G.; D'Addabbo, A.; Bovenga, F.; Nutricato, R.; Lovergine, F.P.; Pietranera, L. SAR and InSAR for flood monitoring: Examples with COSMO-SkyMed data. *IEEE J. Sel. Top. Appl. Earth Obs. Remote Sens.* **2014**, *7*, 2711–2722. [CrossRef]
45. Haralick, R.M.; Shapiro, L.G. Image segmentation techniques. *Comput. Vis. Graph. Image Process.* **1985**, *29*, 100–132. [CrossRef]
46. Druce, D.; Tong, X.; Lei, X.; Guo, T.; Kittel, C.M.; Grogan, K.; Tottrup, C. An optical and SAR based fusion approach for mapping surface water dynamics over mainland China. *Remote Sens.* **2021**, *13*, 1663. [CrossRef]
47. Chini, M.; Hostache, R.; Giustarini, L.; Matgen, P. A hierarchical split-based approach for parametric thresholding of SAR images: Flood inundation as a test case. *IEEE Trans. Geosci. Remote Sens.* **2017**, *55*, 6975–6988. [CrossRef]
48. Bangira, T.; Alfieri, S.M.; Menenti, M.; Van Niekerk, A.; Vekerdy, Z. A spectral unmixing method with ensemble estimation of endmembers: Application to flood mapping in the Caprivi floodplain. *Remote Sens.* **2017**, *9*, 1013. [CrossRef]
49. Bovolo, F.; Bruzzone, L. A split-based approach to unsupervised change detection in large-size multitemporal images: Application to tsunami-damage assessment. *IEEE Trans. Geosci. Remote Sens.* **2007**, *45*, 1658–1670. [CrossRef]
50. Feng, Q.; Liu, J.; Gong, J. Urban flood mapping based on unmanned aerial vehicle remote sensing and random forest classifier—A case of Yuyao, China. *Water* **2015**, *7*, 1437–1455. [CrossRef]
51. Porcù, F.; Leonardo, A. Data record on extreme events by OAL and by hazard. In *Open-Air Laboratories for Nature Based Solutions to Manage Hydro-Meteo Risks (OPERANDUM)*; University of Bologna: Bologna, Italy, 2019; pp. 1–117.
52. QN il Resto del Carlino. Meteo Reggio Emilia, la Piena del Fiume Enza sta Defluendo. 2017. Available online: https://www.ilrestodelcarlino.it/reggio-emilia/cronaca/meteo-fiume-enza-1.3603247 (accessed on 15 January 2021).
53. Drusch, M.; Del Bello, U.; Carlier, S.; Colin, O.; Fernandez, V.; Gascon, F.; Hoersch, B.; Isola, C.; Laberinti, P.; Martimort, P. Sentinel-2: ESA's optical high-resolution mission for GMES operational services. *Remote Sens. Environ.* **2012**, *120*, 25–36. [CrossRef]
54. Torres, R.; Snoeij, P.; Geudtner, D.; Bibby, D.; Davidson, M.; Attema, E.; Potin, P.; Rommen, B.; Floury, N.; Brown, M. GMES Sentinel-1 mission. *Remote Sens. Environ.* **2012**, *120*, 9–24. [CrossRef]
55. Martinis, S.; Rieke, C. Backscatter analysis using multi-temporal and multi-frequency SAR data in the context of flood mapping at River Saale, Germany. *Remote Sens.* **2015**, *7*, 7732–7752. [CrossRef]
56. ESA. 2014. Available online: https://scihub.copernicus.eu/ (accessed on 15 January 2021).
57. Richter, R.; Schläpfer, D. Atmospheric/topographic correction for satellite imagery. *DLR Rep. DLR-IB* **2005**, *438*, 565.
58. Mayer, B.; Kylling, A. The libRadtran software package for radiative transfer calculations-description and examples of use. *Atmos. Chem. Phys.* **2005**, *5*, 1855–1877. [CrossRef]
59. Evans, D.L.; Farr, T.G.; Ford, J.; Thompson, T.W.; Werner, C. Multipolarization radar images for geologic mapping and vegetation discrimination. *IEEE Trans. Geosci. Remote Sens.* **1986**, *GE-24*, 246–257. [CrossRef]
60. Wu, S.-T. Analysis of synthetic aperture radar data acquired over a variety of land cover. *IEEE Trans. Geosci. Remote Sens.* **1984**, *GE-22*, 550–557. [CrossRef]

61. Wu, S.-T.; Sader, S.A. Multipolarization SAR data for surface feature delineation and forest vegetation characterization. *IEEE Trans. Geosci. Remote Sens.* **1987**, *GE-25*, 67–76. [CrossRef]
62. Henry, J.B.; Chastanet, P.; Fellah, K.; Desnos, Y.L. Envisat multi-polarized ASAR data for flood mapping. *Int. J. Remote Sens.* **2006**, *27*, 1921–1929. [CrossRef]
63. Schumann, G.; Hostache, R.; Puech, C.; Hoffmann, L.; Matgen, P.; Pappenberger, F.; Pfister, L. High-resolution 3-D flood information from radar imagery for flood hazard management. *IEEE Trans. Geosci. Remote Sens.* **2007**, *45*, 1715–1725. [CrossRef]
64. Otsu, N. A threshold selection method from gray-level histograms. *IEEE Trans. Syst. Man Cybern.* **1979**, *9*, 62–66. [CrossRef]
65. Horritt, M.; Mason, D.; Luckman, A. Flood boundary delineation from synthetic aperture radar imagery using a statistical active contour model. *Int. J. Remote Sens.* **2001**, *22*, 2489–2507. [CrossRef]
66. Mason, D.C.; Speck, R.; Devereux, B.; Schumann, G.J.-P.; Neal, J.C.; Bates, P.D. Flood detection in urban areas using TerraSAR-X. *IEEE Trans. Geosci. Remote Sens.* **2009**, *48*, 882–894. [CrossRef]
67. Chan, T.F.; Vese, L.A. Active contours without edges. *IEEE Trans. Image Process.* **2001**, *10*, 266–277. [CrossRef] [PubMed]
68. Breiman, L. Random forests. *Mach. Learn.* **2001**, *45*, 5–32. [CrossRef]
69. Archer, K.J.; Kimes, R.V. Empirical characterization of random forest variable importance measures. *Comput. Stat. Data Anal.* **2008**, *52*, 2249–2260. [CrossRef]
70. Haralick, R.M.; Shanmugam, K.; Dinstein, I.H. Textural features for image classification. *IEEE Trans. Syst. Man Cybern.* **1973**, *SMC-3*, 610–621. [CrossRef]
71. Nico, G.; Pappalepore, M.; Pasquariello, G.; Refice, A.; Samarelli, S. Comparison of SAR amplitude vs. coherence flood detection methods-a GIS application. *Int. J. Remote Sens.* **2000**, *21*, 1619–1631. [CrossRef]
72. Carreño Conde, F.; De Mata Muñoz, M. Flood monitoring based on the study of Sentinel-1 SAR images: The Ebro River case study. *Water* **2019**, *11*, 2454. [CrossRef]
73. Dellepiane, S.; Bo, G.; Monni, S.; Buck, C. SAR images and interferometric coherence for flood monitoring. In Proceedings of the IGARSS 2000. IEEE 2000 International Geoscience and Remote Sensing Symposium. Taking the Pulse of the Planet: The Role of Remote Sensing in Managing the Environment. Proceedings (Cat. No.00CH37120), Honolulu, HI, USA, 24–28 July 2000.
74. Cloude, S.R.; Pottier, E. An entropy based classification scheme for land applications of polarimetric SAR. *IEEE Trans. Geosci. Remote Sens.* **1997**, *35*, 68–78. [CrossRef]
75. DeVries, B.; Huang, C.; Armston, J.; Huang, W.; Jones, J.W.; Lang, M.W. Rapid and robust monitoring of flood events using Sentinel-1 and Landsat data on the Google Earth Engine. *Remote Sens. Environ.* **2020**, *240*, 111664. [CrossRef]
76. Doxaran, D.; Froidefond, J.-M.; Lavender, S.; Castaing, P. Spectral signature of highly turbid waters: Application with SPOT data to quantify suspended particulate matter concentrations. *Remote Sens. Environ.* **2002**, *81*, 149–161. [CrossRef]
77. McLaughlin, J.; Webster, K. *Effects of a Changing Climate on Peatlands in Permafrost Zones: A Literature Review and Application to Ontario's Far North*; Ontario Forest Research Institute: Sault Ste. Marie, ON, Canada, 2013.
78. Piemonte, R. Carta dei Suoli. 2021. Available online: https://www.regione.piemonte.it (accessed on 15 January 2021).
79. Goffi, A.; Stroppiana, D.; Brivio, P.A.; Bordogna, G.; Boschetti, M. Towards an automated approach to map flooded areas from Sentinel-2 MSI data and soft integration of water spectral features. *Int. J. Appl. Earth Obs. Geoinf.* **2020**, *84*, 101951. [CrossRef]
80. Pulvirenti, L.; Pierdicca, N.; Chini, M.; Guerriero, L. An algorithm for operational flood mapping from synthetic aperture radar (SAR) data based on the fuzzy logic. *Nat. Hazard Earth Syst. Sci.* **2011**, *11*, 529–540. [CrossRef]
81. Kuppili, A. What Is the Time Complexity of a Random Forest, Both Building the Model and Classification? 2015. Available online: https://www.quora.com/What-is-the-time-complexity-of-a-Random-Forest-both-building-the-model-and-classification (accessed on 21 October 2021).
82. Cohen, R. *The Chan-Vese Algorithm*; Technion, Israel Institute of Technology: Haifa, Israel, 2010.

remote sensing

MDPI

Article

Using a Lidar-Based Height Variability Method for Recognizing and Analyzing Fault Displacement and Related Fossil Mass Movement in the Vipava Valley, SW Slovenia

Tomislav Popit *, Boštjan Rožič, Andrej Šmuc, Andrej Novak and Timotej Verbovšek

Department of Geology, Faculty of Natural Sciences and Engineering, University of Ljubljana, Aškerčeva 12, SI-1000 Ljubljana, Slovenia; bostjan.rozic@ntf.uni-lj.si (B.R.); andrej.smuc@ntf.uni-lj.si (A.Š.); andrej.novak@ntf.uni-lj.si (A.N.); timotej.verbovsek@ntf.uni-lj.si (T.V.)

* Correspondence: tomi.popit@ntf.uni-lj.si; Tel.: +386-14-704-601

Abstract: The northern slopes of the Vipava Valley are defined by a thrust front of Mesozoic carbonates over Tertiary flysch deposits. These slopes are characterized by a variety of different surface forms, among which recent and fossil polygenetic landslides are the most prominent mass movements. We used the height variability method as a morphometric indicator, which proved to be the most useful among the various methods for quantifying and visualizing fossil landslides. Height variability is based on the difference in elevations derived from a high-resolution lidar-derived DEM. Based on geologic field mapping and geomorphometric analysis, we distinguished two main types of movements: structurally induced movement along the fault zone and movements caused by complex Quaternary gravitational slope processes. The most pronounced element is the sliding of the huge rotational carbonate massif, which was displaced partly along older fault structures in the hinterland of fossil rock avalanches and carbonate blocks. In addition to the material properties of the lithology, the level of surface roughness also depends on the depositional processes of the individual sedimentary bodies. These were formed by complex sedimentary events and are intertwined in the geological past. The sedimentary bodies indicate two large fossil rock avalanches, while the smaller gravity blocks indicate translational–rotational slides of carbonate and carbonate breccia.

Keywords: slope process; surface roughness; rock avalanche; geomorphometric analysis; geological setting; deep-seated rotational and translational slides

Citation: Popit, T.; Rožič, B.; Šmuc, A.; Novak, A.; Verbovšek, T. Using a Lidar-Based Height Variability Method for Recognizing and Analyzing Fault Displacement and Related Fossil Mass Movement in the Vipava Valley, SW Slovenia. *Remote Sens.* **2022**, *14*, 2016. https://doi.org/10.3390/rs14092016

Academic Editors: Constantinos Loupasakis, Ioannis Papoutsis and Konstantinos G. Nikolakopoulos

Received: 1 March 2022
Accepted: 20 April 2022
Published: 22 April 2022

1. Introduction

Vipava valley (SW Slovenia) is located between the Karst plateau on the southwest side and the Nanos Mountain range in the northeast. The N and NE slopes of the valley are defined by a thrust front of Mesozoic carbonates over Tertiary flysch deposits [1–3]. This overthrusting has resulted in steep slopes and fracturing of the rock, producing highly weathered carbonates and large amounts of scree deposits in the upper part of the valley. In the lower part, these slopes are characterized by a variety of different surface forms, among which recent and fossil polygenetic landslides are the most prominent mass movements. Superficial deposits range from large-scale, deep-seated rotational and translational slides to shallow landslides, slumps, and sedimentary gravity flows in the form of debris or mudflows reworking the carbonate scree and flysch material [4–8]. The influence of tectonic fractures on mass movements is a common phenomenon in Slovenia, for example, the Ciprnik complex landslide in the Tamar Valley in northwestern Slovenia [9]. Due to tectonic stresses in the hinterland of the Ciprnik landslide, the initially highly bedded rocks were additionally fractured. This intense fracturing caused an increase in the effective porosity and a decrease in the strength of the material [9,10]. The relationship between tectonics and gravitational movement in the Vipava Valley and similar extreme cases in the Alps and Dinarides point to the need for a complex study of geologic processes [11].

In this paper, we present the morphometrical analysis of two large sedimentary bodies of fossil rock avalanches, Podrta gora and Gradiška gmajna, and a few smaller detached and translationally moved carbonate blocks named Stara baba, Veliki strel, and Klapačiše in Zagriža. As a morphometric indicator, we used a variation of roughness index; the variability of surface elevation was used, using five different methods [12]. Quantification of the variability of surface roughness was based on the height variability method (HV) [4]. This method is based on the differences in elevations obtained from a digital elevation model with a high spatial resolution (1 m × 1 m), derived from lidar scanning.

Based on the geomorphometric analyses of surface roughness in conjunction with geomorphological and geological mapping, two main types of displacements were analyzed in the considered work: structurally conditioned displacements (at the fault zone) and Quaternary displacements caused by gravitational processes. By studying fossil and (sub)recent landslides, we were able to identify the shape of the crown, main, and lateral edges, as well as the geomorphometric characteristics at the top of the single sedimentary body of rock avalanche.

2. Materials and Methods

2.1. Geological Setting

The general topography of the Vipava Valley is determined by thrust fronts of the Trnovo and Hrušica nappes composed of Mesozoic shallow-marine limestone and dolomites. Carbonates are thrust over gentle slopes of strongly folded Paleocene and Eocene basinal clastics deposits (flysch), which are composed of alternating sandstone, shale, and marl beds (Figure 1). The Mesozoic carbonate rocks are highly fractured along the thrust contacts and are cut by large NW–SE striking, Neogene, dextral, strike-slip faults characterized by fault zones up to 300 m wide [11,13,14].

The structural contact is expressed morphologically by the difference in lithology between the steep carbonate rocks in the massif and the gentler slopes of the underlying flysch in the lower part of the slope. This lithologic boundary, formed by the thrust contact, is covered by a variety of Quaternary slope deposits, broadly divided into two groups. The first group is lithified and unlithified scree deposits covering the upper part of the slope, while the second group is partially cemented complex Quaternary slope deposits covering the lower part of the slope. The latter represents a series of composite, fan-shaped sedimentary bodies with different compositions, internal structures, and textures, indicating a complex depositional history and polyphase genesis [4,6,15–17]. Quaternary slope deposits are moderately sorted and consist of gravel to medium boulder-sized clasts. Rarely, very large (approximately several meters in a linear direction) individual boulders are also present. In addition, carbonate megablocks reaching more than 100 m in length are found at the lower parts of the slopes (e.g., in the Lokavec area near Ajdovščina). They were detached from the stable carbonate karst plateau and were transported up to 2 km by translational and rotational slope movements [18] (Figure 2).

The structural and lithological settings also determine the hydrogeological conditions; therefore, most springs originate close to the contacts between the limestone and flysch (springs of the Vipava, Lijak, and Hubelj rivers) [14]. Many smaller springs also emerge within the gravel layers and lenses. These are very permeable and allow rapid infiltration of rainwater, which then flows within them and encounters significantly poorer permeable rocks in weathered flysch. In this area, the infiltrated rainfall water emerges in springs upon contact between the flysch below and the limestone gravel above and continues to flow superficially. The described geological structure and related hydrogeological conditions also influence the complex depositional processes of the slope deposit [3,4,15,19–23].

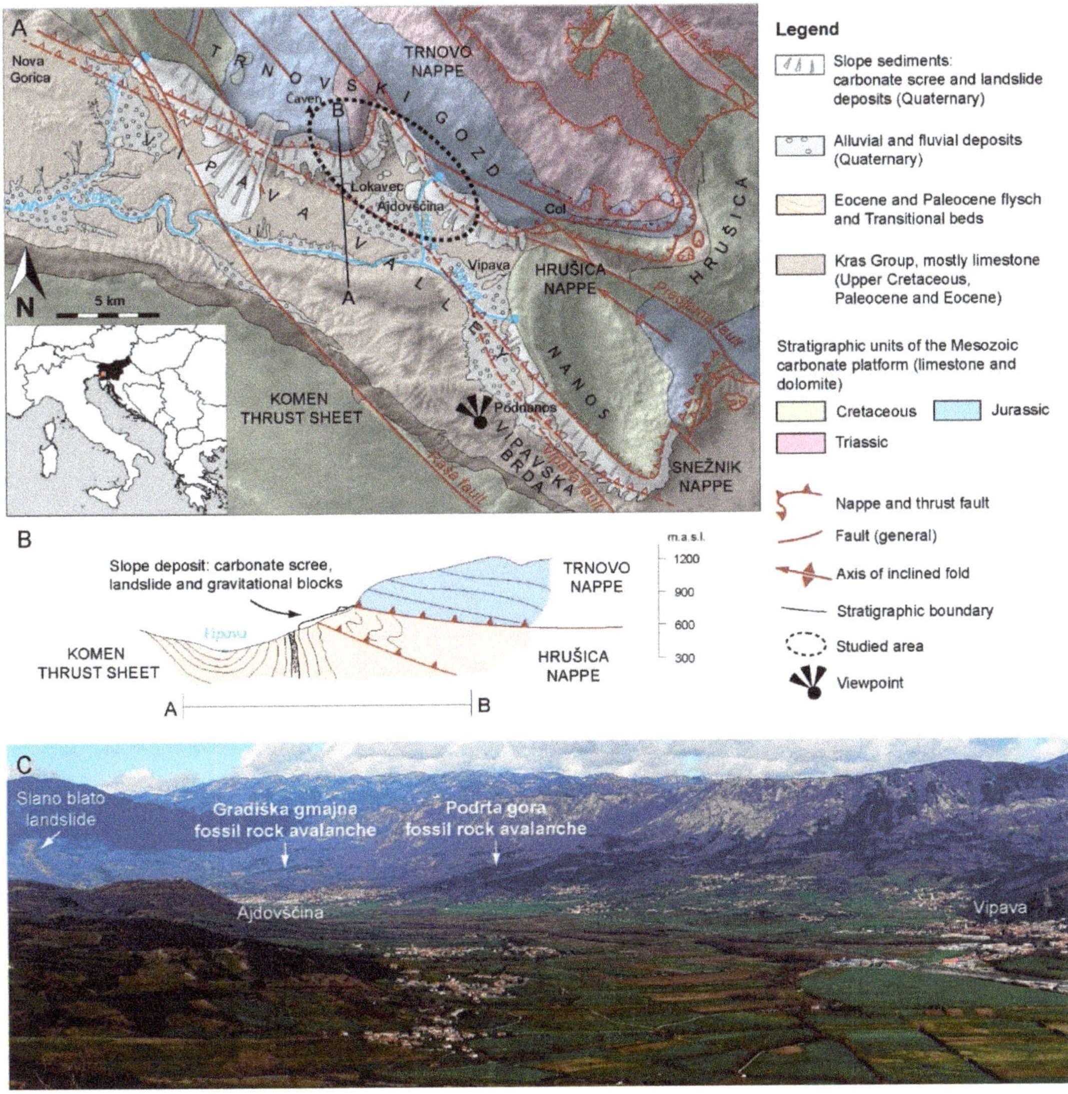

Figure 1. (**A**) Geological map of the extended area of the Vipava Valley; (**B**) cross-section through the Trnovski gozd and Vipava Valley; (**C**) panoramic view from Sveti Socerb to the Vipava Valley and Trnovski gozd in the hinterland (geological map modified after [1,3,11,14,24–27].

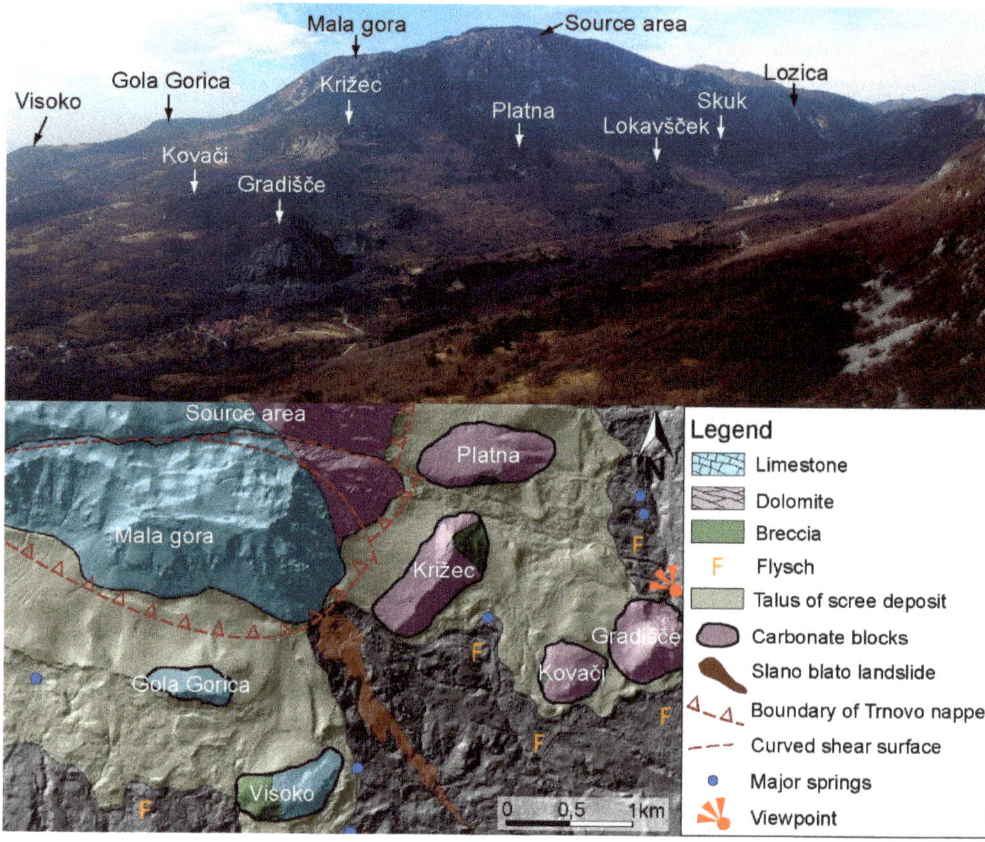

Figure 2. Photograph of carbonate gravitational blocks (upper figure) from Navrše hill (view towards W) and lithology of the wider source area and carbonate blocks (lower figure). Adapted and modified after [18]. Reprinted and adapted with permission from Ref. [18]. Copyright 2019, Založba ZRC SAZU, Geografski inštitut Antona Melika".

2.2. Geomorphological Analysis

A detailed digital hillshaded terrain model (DTM) was obtained from an openly available Airborne laser scanning (ALS) dataset of Slovenia [28]. The ALS point cloud was rasterized to a 1 m × 1 m resolution and later, by a combination of filtering and removal of non-ground base points by adaptive triangulated irregular network densification [12]. Relief Visualization Toolbox (Version 1.1 [29]) was used to aid the visual inspection of the hillshaded DTM to emphasize the positive and negative geomorphological anomalies, which simplifies the geomorphological interpretation of the studied area [12,30,31].

For quantitative geomorphological analysis of the rock avalanche surfaces, we have used several surface roughness methods which, apart from the surface curvature analyses, proved to be useful parameters for the investigation and detection of fossil landslides [4,12,32–35] and recent landslides [36–40].

A quantitative analysis of the surface roughness of the studied sedimentary bodies of fossil rock avalanches was performed in the program ArcGIS using the height variability (HV) method, which proved to be the most useful among the several methods for the quantification and visualization of landslide parts with different sedimentary composition and genesis [12]. The height variability method is performed on a raster elevation surface with, first, a resampling of the original 1 m × 1 m lidar resolution elevation data into a coarser 3 m × 3 m resolution (performed by the replacement of the center cell with the average value in the 3 m × 3 m moving window with the ArcGIS Focal Statistics tool) and, later, by the replacement of the center cell with the difference between the highest and lowest elevation (hence the name of the method) in the same window size. The original resolution turned out to be too detailed (too noisy) for further analyses. Larger moving windows were contrarily too coarse due to lost surface details.

The HV method is very similar to the method of slope variability cf. [41]; the only difference is that it uses the difference between the maximum and minimum elevation instead of the slope difference: DTV_{max}–DTV_{min} [12]. The results are given as the elevation difference (range) in meters, with a more variable ("rough") surface inside the search window giving the higher numerical values. The advantage of this method is a very distinguishable visualization of areas with different roughness, which correspond well to different sedimentary processes on the landslide body. This method was tested on the Quaternary slope sedimentary bodies of (i) Podrta gora and Gradiška gmajna fossil rock avalanches; (ii) the adjacent gravity carbonate megablocks of Stara baba, Veliki strel, Klapačiše, and Zagriža on the slopes of the Vipava Valley; and (iii) on the structural elements in the hinterland high karst plateau.

We also suggest using the additional approach for geomorphometric analysis of the surface using the VAT method [24,42]. This method was originally named visualization for archaeological topography, and although the name comes from its primary use in archeology, it can be used to explore small-scale topographic variations also in geology or in any field of geomorphology, generally. It is based on the analysis of surface elevation data, and it combines several DEM-derived input layers: hillshaded relief, positive openness [43], slope, and sky-view factor [31], and blends them to combine the information into a single image (VAT) suitable for visual or quantitative inspection of the surface morphology features, helping with the interpretation of the surface changes (landslide, erosion, construction works, etc.). The VAT method is a part of the Relief Visualization Toolbox (RVT) described in [31,42]. We have tested the VAT method and found that it is extremely useful when analyzing smaller landslide features on fossil rock avalanches [24]. However, in the presented case, in which we analyzed a much bigger (regional-scale) area several kilometers in size, the use of the VAT method did not contribute useful results to help with the interpretation of the terrain analysis.

3. Results

The HV values in the N and NE areas of the Vipava Valley reflect the differences between the surfaces of the flysch base and the surfaces of the fossil rock avalanche, gravity blocks built up of carbonate gravels, and structural elements in the hinterland of sedimentary bodies. Rock avalanches and carbonate blocks have high HV and a high degree of surface roughness at the edges, in contrast to flysch rocks, which have low HV and a low degree of surface roughness (Figure 3). The exceptions are flysch-cut ravines, where greater erosion of flysch rocks occurs and which represent areas of greater slope inclination than the surrounding area [44] and high height variability. The structural elements can be identified by linear changes in height variability.

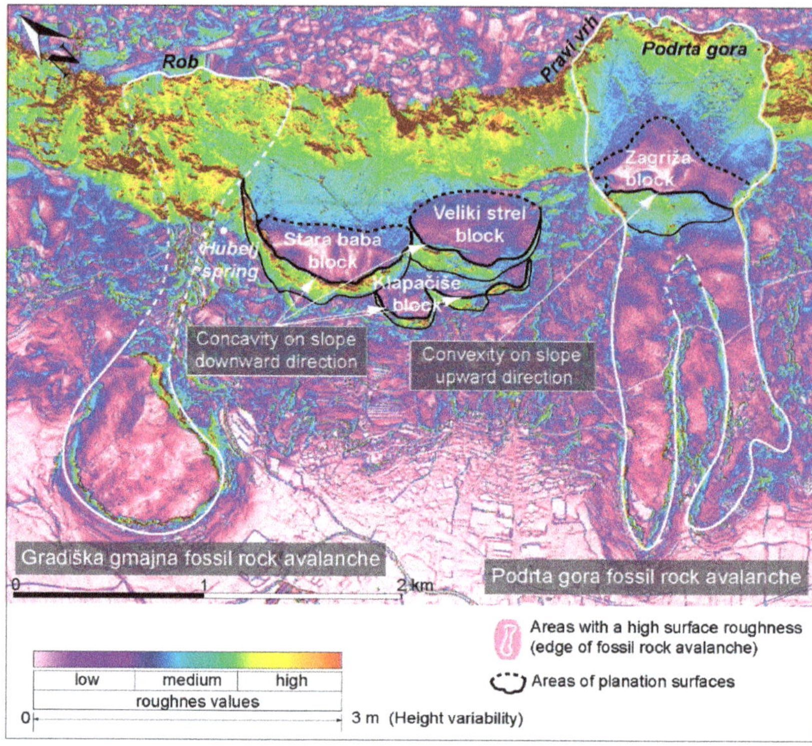

Figure 3. Map of HV with marked individual sedimentary bodies of Gradiška Gmajna and Podrta gora rock avalanches and larger leveling area of Stara baba, Veliki strel, Klapačiše, and Zagriža gravity carbonate blocks.

3.1. Boundary of Individual Sedimentary Bodies and Their Source Area

The values of HV in the Podrta gora and Gradiška gmajna fossil rock avalanches are shown in Figure 3. In narrow bands along the edges of the bodies, the variability is medium and rarely high, while the central part of the sedimentary bodies is mostly an area of low HV or low surface roughness. Medium to high HV is also observed in the narrow bands in the lower (fan-shaped) part of the Gradiška gmajna rock avalanche and in two tongue-shaped bodies of the Podrta gora rock avalanche (Figure 3). Almost the entire narrow area of the margins of both sedimentary bodies is dominated by medium to large surface roughness, which is in sharp contact with the smooth surface. The boundary is partially blurred only in the SW part of the Podrta gora rock avalanche, where the difference in surface roughness is not strongly pronounced, and in the central part of the Gradiška gmajna rock avalanche, where the boundary between the sedimentary body and its surroundings is not definable. On the NE side of the Gradiška gmajna rock avalanche, the lateral edge is clearly visible, as the gorge of the Hubelj River is cut next to it (Figure 3).

Aell-recognized geomorphological element is also the upper edge of the rock face in the hinterland of the Podrta gora and Gradiška gmajna rock avalanches (Figure 3). Below the upper edge, the values of HV (in the form of a jagged convex edge) are very high and indicate a sharp boundary between the carbonate rock face and the karst surface at the top of Trnovski gozd in the area of Rob, Pravi vrh, and especially of Podrta gora (Figure 3).

3.2. Boundary of Individual Sedimentary Bodies and Their Source Area

In both the Podrta gora and Gradiška gmajna rock avalanche areas and in the area between them, there are extensive areas with an extremely low degree of surface roughness

(Figure 3). These areas are the locations of the gravitational carbonate blocks named the Stara baba, Veliki strel, and Klapačiše sedimentary bodies and the Zagriža sedimentary body, located within the Podrta gora rock avalanche. The mentioned areas in the lower part, towards the SW, are adjacent to the areas of high surface roughness. The contacts between the smoothed and rough areas at the Stara Baba, Veliki Strel, and Klapačišče bodies are clearly concave in the downward direction, while the contact at the Zagriža body (within the Podrta gora rock avalanche) is approximately flat or slightly convex in the downward slope (Figure 3).

3.3. Identification of Structural Elements in the Hinterland of Sedimentary Bodies

The HV method is also useful for identifying faults and fracture zones. Fracture deformation and crack structures are most evident in the area of high karst plateaus, where faults can be identified in line-by-line changes in the values of height variability. In some places, the value of surface roughness increases; in other places, the response of the rock mass to the fracture zone is just the opposite, and the HV decreases. The most pronounced are the northern part of the Predjama fault, which runs across the plateau above Rob, and the fault on Mala gora above the Slano blato (Figure 3, cf. [45]. Linearly distributed alterations are evident in the carbonate rock face in height variability, mostly oblique to the edges of the rock face. Most likely, this is a morphological reflection of fracture zones running at different angles between the main fault systems [3,14].

Generally, the middle and upper parts of the rock face of carbonate rock (yellow and red color), extending over the whole Vipava Valley, have the greatest HV values (Figure 3). In the lower parts of the rock face in the foothills, the values of HV are medium (Figure 3). The transition to the upper karst plateau of Trnovski gozd is morphologically pronounced, as medium values and sometimes low values of HV begin to predominate in the sharp line. This transition is related to the erosional frontal retraction of the overthrust fronts and to the structural features of the area. The lower boundary, at the base of the rock face, is directly related to the thrust or fault contact with the carbonate rocks and the flysch bedrock. The sharpness of the boundary is obviously determined by the thickness of the carbonate gravel or scree deposit. In places where individual sedimentary bodies with a lot of carbonate gravel are observed, higher values of surface roughness are obtained, while in areas with thin gravel, the value of HV is low.

4. Discussion

In general, landslides are complex and consist of parts with different geomorphological characteristics [46,47]. Using the visual interpretation of the digital evaluation model (DEM), and the calculated surface roughness indicator, we were able to identify the surface properties of the individual sediment bodies very well in most cases. Based on a combination of geomorphometric indicators, we conclude that the sedimentary bodies have a very complex structure formed by different Quaternary sedimentation processes. The analysis of the typical morphological elements found in the sedimentary bodies of the Podrta gora and Gradiška gmajna rock avalanches and on the rotational blocks of Stara baba, Veliki strel, Klapačiše, and Zagriža is presented in Figure 4, cf. [47,48]. It has been shown that the degree of surface roughness is most strongly influenced by various sedimentation processes, in addition to the characteristics of the sedimentary material. Similarly, Grohmann et al. [49,50] determined that different surface roughness values are attributed to landslide formation processes (recent and fossil) and time elapsed since the surface formation, in addition to material characteristics.

The high contrast between the degree of surface roughness occurring at the margins of each body and the surface belonging to the surrounding and base of the sedimentary deposit reflects the difference between the two distinct lithologic units. The carbonate gravel or breccia belongs to the Gradiška gmajna and Podrta gora rock avalanches and to the Stara baba, Veliki strel, Klapačiše, and Zagriža gravity blocks. These blocks are characterized by a high degree of roughness, while flysch bedrock forms the base and immediate surroundings of sedimentary bodies, creating a relatively smooth surface.

4.1. Podrta Gora and Gradiška Gmajna Fossil Rock Avalanches

4.1.1. Crowns and Main Scarp

Based on the surface roughness, the upper edge above the Podrta gora and Gradiška gmajna landslides is clearly visible in the area of Rob, Pravi vrh, and Podrta gora (Figures 3 and 4). The values of HV in this area are very high in the form of a convex edge. The areas of fossil rock avalanches determined by remote sensing correspond with the results of geological mapping. High values of surface roughness in this area indicate that this geomorphometric element represents the crowns and main scarps of the Podrta gora and Gradiška gmajna rock avalanches (Figure 4A, points 1 and 2). The main direction of elongation of the crown and main scarp is perpendicular to the direction of mass transport, which is also one of the typical characteristics of crowns and can be detected at least in the case of the Podrta gora fossil rock avalanche, cf. [47]. In the immediate vicinity of the main scarp is also an area of carbonate gravel, representing a recent scree deposit (Figure 4A, point 7).

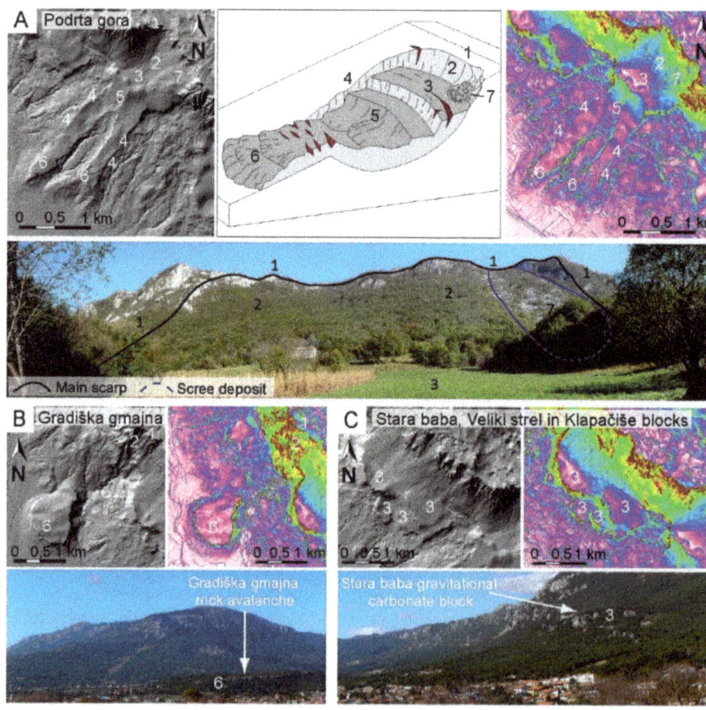

Morphometric elements	Properties
1. Crown	The convex edge is adjacent to the highest parts of the main scarp. The main direction is perpendicular to the transport direction (southwest).
2. Main (upper) scarp	A steep, semicircular, concave surface at the upper edge of the landslide with high surface roughness. Main direction is perpendicular to flow direction.
3. Head	Planation surfaces in the upper parts of the landslide along the contact between the displaced material and the main (upper) scarp.
4. Flank	Higher roughness at the edge of the landslide compared to the surrounding area (the lower part is clearly visible).
5. Depletion area	Lower surface roughness compared to the surroundings.
6. Accumulation area	The fan- and tongue-shaped body is convex in plan and profile with higher surface rougneses et the edge of the body.

Figure 4. Geomorphometric elements of sedimentary bodies: (**A**) Podrta gora fossil rock avalanche; (**B**) Gradiška gmajna fossil rock avalanche; and (**C**) Stara baba, Veliki strel, and Klapačiše carbonate blocks, based on the classification of the schematic representation of the landslide. Adapted from [46–48].

4.1.2. Minor (Lateral) Scarp of Fossil Rock Avalanches

At the border of sedimentary bodies, especially in the lower part, sharp transitions of HV values occur. These locations are the steep boundaries between the flysch bedrock and the sedimentary bodies of the Podrta gora and Gradiška gmajna fossil rock avalanches, which are made of carbonate gravel, mostly lithified in a slope breccia (Figure 5). Lateral scarps are areas of medium to high surface roughness, while the flysch bedrock is primarily a smooth area with an extremely low degree of HV (Figures 3 and 4). The lateral scarps of sedimentary bodies are approximately parallel to the main direction of transport, which is one of the characteristic elements of landslides [48]. The sharp transitions of HV at the periphery of the sedimentary bodies are also affected by the erosion of carbonate gravels and breccias. This is especially marked in the eastern part of the Gradiška gmajna fan-shaped sedimentary body, where the Hubelj River erodes part of the fan and changes its original shape (Figure 3. Shulz [51], for example, explained the lower reliability in detecting the lateral scarp and the toe of the Gradiška gmajna body precisely with the reworked surface of the fossil rock avalanches. The toe of the Gradiška gmajna rock avalanche with a high degree of surface roughness has also been eroded.

Figure 5. Up to a 10 m high wall of carbonate gravel, partly strongly lithified to breccia, in the lower part of the lateral scarp of the Podrta gora fossil rock avalanche in the abandoned Apnenec quarry, above the village of Kožmani.

4.1.3. Geomorphometry of the Central Part of Sedimentary Bodies

The surface roughness is low in the interior of the Podrta gora and Gradiška gmajna fossil rock avalanche. It is particularly low in the central part of the tongue- or fan-shaped areas of the body. Similarly, Glenn et al. [37] recognized the high surface roughness at the head scarp and the toe of rock avalanches. Surface roughness is different in different parts of a single landslide; namely, it is high in the areas of erosion and low in the body of the landslide [36,37].

The central parts of the sedimentary bodies were well identified at the Podrta gora rock avalanche, but this was the least accurate compared to other geomorphological elements. This is due to erosional processes that change the shape of the fan and increase the surface roughness. This can be seen in the eastern part of the Gradiška gmajna rock avalanche, where carbonate gravels occur at quite high elevations, at the source of the Hubelj, and the riverbed is cut to the flysch base (Figure 3. Habič [52] even stated that the water of the Hubelj River caused the sliding of the Gradiška gmajna breccia material, and the Gradišče carbonate breccia was displaced and transported when the breccia had already formed. The presence of the Hubelj karst spring [53] even before the lithification of the older gravel into the breccia indicates that this carbonate gravel had dammed the karst spring for a long time, and such a dam could only form when larger quantities of gravel poured into the original riverbed in a relatively short time during stronger earthquakes [52].

In the upper part of the Podrta gora rock avalanche is a large area of an accumulation of carbonate rocks with low surface roughness, most likely representing a huge rotational block that slid on the weathering flysch bedrock or muddy sediment at its base (Figure 3). The block, which is divided into three parts, consists of strongly cracked carbonate rocks, while carbonate breccias and slope gravels occur only in the hinterland. Similar gravity blocks were recorded near the Lokavec slide in a combination of translational and rotational block-type slope movements [18]. The gravitational block of Podrta gora most likely represented the first transport phase of the complex Podrta gora rock avalanche, from which a huge gravel landslide further developed and was transported in the form of a rock avalanche in the Vipava Valley. The two phases of the Podrta gora rock avalanche mass movement are also evidenced by the forms of the secondary scarp (convexity in the downward direction) in the Zagriža area (Figure 3). The latest proposed classification system by Hungr et al. [54], modified after Varnes [55] and Cruden and Varnes [48], classifies two-phase landslides in the class of complex landslides, and their transport complexity is referred to as a two-phase event [56]. In contrast to the two-phase Podrta gora rock avalanche, the two-phase transport process is not observed in the case of the Gradiška gmajna rock avalanche. The large area of the main scarp and the well-defined upper crown, as well as the large fan-shaped body in the lower part of the avalanche, may indicate that Gradiška gmajna represents a huge rockfall in the initial (first) phase, which further developed into a debris avalanche.

4.2. Gravitational Blocks

4.2.1. Planation Surface Area

Based on the analyses, we have identified many areas of extremely low surface roughness (Stara baba, Veliki strel, and Klapačiše), which spread upwards to individual scree deposits in the foothills. Individual planation surfaces were formed by large rotational slides, where individual blocks of the carbonate breccias rotated along the sliding surface at the contact of the gravel with the underlying flysch bedrock. The blocks of breccia also tilted towards the slope as they slid (Figure 6). At the outer edges, there was even a reverse tilt of the breccia blocks and the formation of steep rock faces with an extremely high degree of surface roughness and smooth areas in the hinterland depression. Similar geomorphometric features in the Rebrnice area were also recognized [6,19].

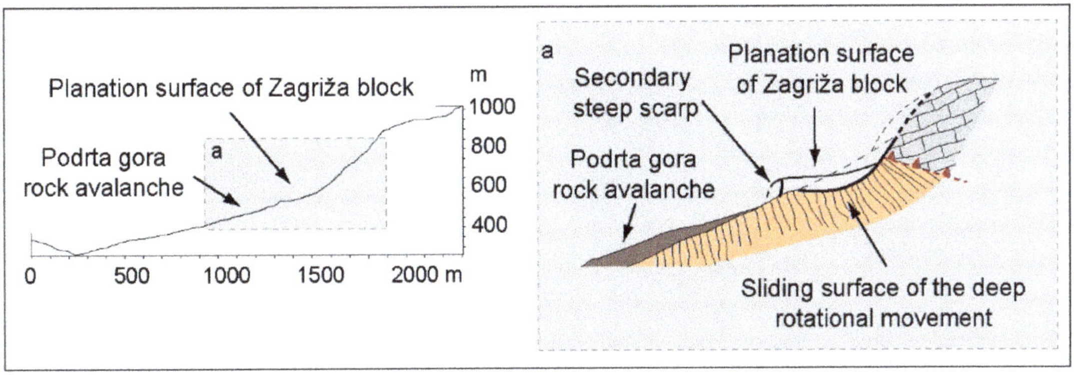

Figure 6. Rotational Zagriža carbonate block, which is part of the complex two-phase Podrta gora rock avalanche. The detail of rotational Zagriža carbonate block is on the right side of the image (area a).

4.2.2. Structurally Conditioned Movement

By analyzing the geomorphometric features, we can recognize the structurally induced movement. Today, the blocks are generally inactive and are important mainly because of their influence on the geological structure of the area [14,26,45,57]. The contact between the flysch and carbonate bedrock in the northwestern part of the area (north of the Gradiška gmajna rock avalanche) is at significantly higher elevations than in the central part. Thus, we find the highest flysch outcrops in the area of Gosta meja at 475 m and at the source of Hubelj at 240 m above sea level [14]. The structure in this area is a depressional synclinal bend of the overthrust surface of the Trnovo nappe with an axis in a northeast–southwest direction [57]. In addition, a regionally significant Avče fault was explored in the area, which would be at least a partially displaced thrust fault in this area [26,57]. In the Hubelj Spring area, detailed geological mapping identified a complex, highly branched NW–SE oriented fault system, one segment of which merges with the northern branch of the Predjama fault [14]. Thus, in the studied area, we are dealing with a complex structure, within which we cannot determine the structural significance of a single segment of the contact between the carbonates and flysch on the basis of the outcrops alone due to the overlap of the outcrops. Indeed, it can be assumed that the structures triggered a significant reduction of the thrust fault and, thus, influence the formation and hydrogeological characteristics of the area. At the same time, the diversity and intensity of the slope processes in this part of the examined area have significantly increased.

On the nearby Mala gora on the other side of the valley of the Lokavšček stream (Figure 7A), Placer et al. [45] suggested so-called structural landslides, reportedly also known in the Rebrnice area [58]. The exposed carbonate massif of Mala gora slides in the form of a large (deep-seated) rotational slide about 300 m down the slope toward the valley, including the flysch layers in the slide. The sliding surfaces (main scarps) in the carbonate massif represent the shape of normal faults in the hinterland of the Mala gora block [45].

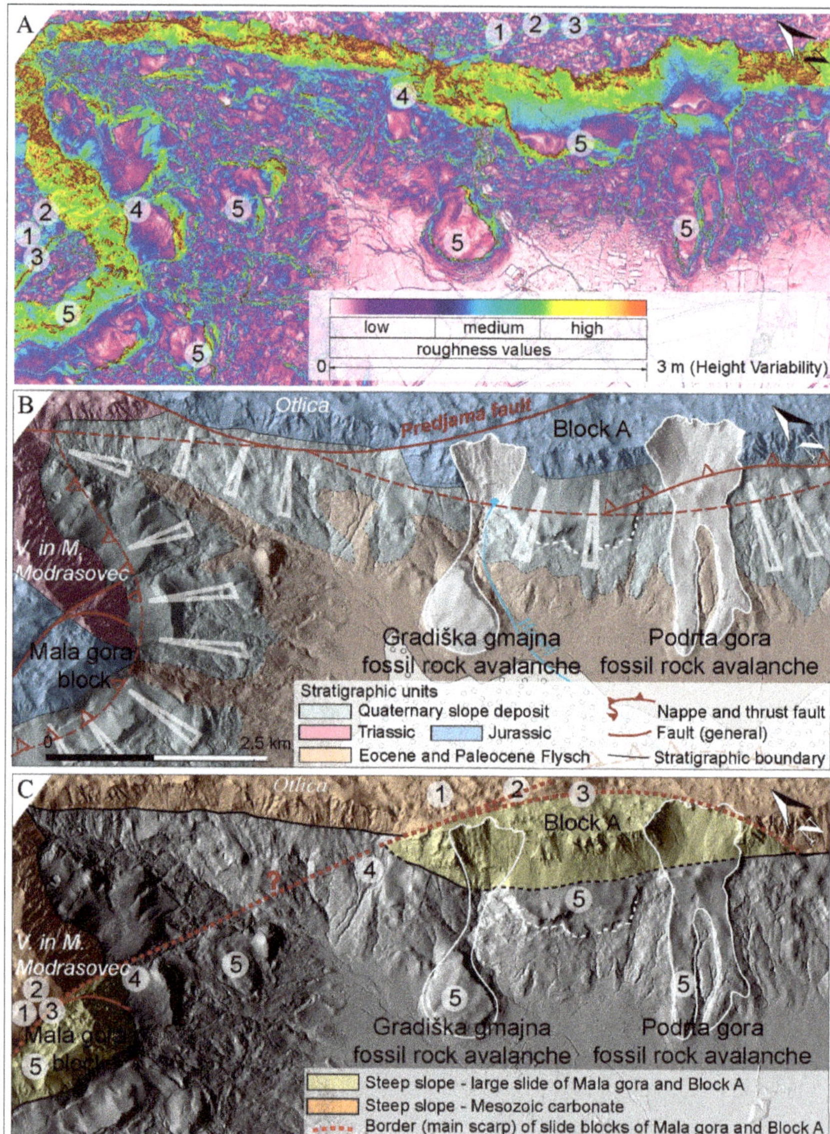

Figure 7. Interpretation of the processes in the area around Ajdovščina, where the whole area (Block A) above the hinterland of the fossil rock avalanche of Gradiška gmajna and Podrta gora could be classified as part of a large slide block by analogy with Mala gora. (**A**) HV map, (**B**) geological map, and (**C**) map with individual geomorphometric elements marked: 1 = linear distribution of change in the degree of HV along the northern branch of the Predjama fault is comparable to the main scarp of Mala gora; 2 = areas north to northeast of both lines have greater HV than southwestern areas; 3 = arcuate curvature of lines; 4 = southwestern carbonate massifs in the concave part of the lines, which are subsided; 5 = slopes below the concave margin where recent and sub-recent gravitational processes are intense.

A comparison of the results of the geomorphometric characteristics of the whole study area with the situation in Mala gora provides some details, namely:

(1) The linear distribution of the change in the degree of HV values along the northern branch of the Predjama fault is comparable to the main ridge of the Mala gora carbonate block (Figure 7A,C, point 1);
(2) Areas north of both lines show greater HV values than those southwest (Figure 7A,C, point 2);
(3) The lines are arcuately curved (Figure 7A,C, point 3);
(4) The southern carbonate massifs (Mala gora and Block A) in the concave part of the lines are subsided (Figure 7A,C, point 4);
(5) Intense recent and sub-recent gravitational movements are observed on the lower slopes in the concave part of these lines (Figure 7A,C, point 5).

Based on these results, we propose that the whole area between Rob and Podrta gora, analogous to the Mala gora massif, could be part of a large carbonate block, settled relative to the hinterland. The movement of the carbonate block (Block A; Figure 7A,C) occurred along the fracture surface of the Predjama fault in the western part and most likely along roughly parallel fracture zones in the eastern part. Consequently, the displacement of Block A also affected the increased intensity and diversity of gravitational movements in the lower part of the slopes, such as the large fossil rock avalanches Podrta gora and Gradiška gmajna. Based on the observed similarities, two questions regarding the geological history of the area are still unanswered: whether the lowered thrust fault in the area under consideration is only an influencing factor or if it is an active participant in the gravitational movement of rock masses. The other question is if the linear geomorphometric elements indicate a connection with the Mala gora main scarp and the fault zone of the northern branch of the Predjama fault. It is possible that both displacements could belong to a large-scale structural movement of Mala gora and Block A (Figure 7C).

5. Conclusions

Based on the geomorphometric analyses of surface roughness, we have roughly distinguished two main types of displacements: structurally induced displacements (along the fault zone) and displacements caused by Quaternary gravitational slope processes. Quaternary slope deposits were studied geomorphometrically on two larger sedimentary bodies, the Podrta gora and Gradiška gmajna fossil rock avalanches, and on some smaller gravitational carbonate bodies—Stara baba, Veliki strel, Klapačiše, and Zagriža. It turns out that the quantitative parameter of surface roughness proved to be very useful in the studies of fossil and recent or sub-recent rock avalanches or landslides, generally. Specifically, we were able to detect very well the shapes of the main and minor scarps, as well as the geomorphometric characteristics of the deposits within individual bodies. In addition, structural elements that influence mass movements have been successfully identified.

The degree of surface roughness depends mainly on various deposition processes, in addition to material properties. The visualization of roughness values in a GIS environment allowed us to understand the two phases of complex avalanches that evolved from a sliding rotational landslide, in the case of Podrta gora, or a large rockfall, in the case of Gradiška gmajna, to a rock avalanche. The cases of the Stara baba, Veliki strel, and Klapačiše gravity blocks indicate large translational–rotational slides. In this sense, we strongly recommend the use of surface roughness analysis in future research of mass movements induced by various displacement causes. In addition, the geomorphometric analyses also revealed some peculiarities in the structural observation, the most pronounced element being the sliding of the huge carbonate block on the Predjama fault in the hinterland of the fossil rock avalanches and carbonate blocks. A comparison of the geomorphometric elements of the carbonate block (A) with the Mala gora rotational block above Lokavec shows that the whole area may be a part of a major rotational slide of the carbonate massif that was displaced partly along older fault structures.

Author Contributions: Conceptualization, T.P., B.R., A.Š., A.N. and T.V.; methodology, T.V. and T.P.; writing—original draft preparation, T.P., B.R., A.Š., A.N. and T.V.; writing—review and editing, T.P., B.R., A.Š., A.N. and T.V.; visualization, T.P. and T.V.; funding acquisition, T.P. and T.V. All authors have read and agreed to the published version of the manuscript.

Funding: This research was partly funded by the Slovenian Research Agency's core funding (No. P1-0195—Geoenvironment and Geomaterials).

Acknowledgments: The work was carried out in the framework of the International Consortium on Landslides (ICL), project number IPL-216 (entitled Diversity and Hydrogeology of Mass Movements in the Vipava Valley, SW Slovenia), and the ICL Adriatic–Balkan Network (ABN). We appreciate the constructive comments and suggestions made by the anonymous reviewers.

Conflicts of Interest: The authors declare no conflict of interest. The sponsors had no role in the design execution, interpretation, or writing of the study.

References

1. Placer, L. Geologic structure of south-western Slovenia. *Geologija* **1981**, *24*, 27–60.
2. Placer, L. Contribution to the macrotectonic subdivision of the border region between Southern Alps and External Dinarides. *Geologija* **1998**, *41*, 223–255. [CrossRef]
3. Popit, T.; Rožič, B.; Šmuc, A.; Kokalj, Ž.; Timotej, V.; Košir, A. A lidar, GIS and basic spatial statistic application for the study of ravine and palaeo-ravine evolution in the upper Vipava Valley, SW Slovenia. *Geomorphology* **2014**, *204*, 638–645. [CrossRef]
4. Popit, T.; Verbovšek, T. Analysis of surface roughness in the Sveta Magdalena paleo-landslide in the Rebrnice area. *RMZ-Mater. Geoenviron.* **2013**, *60*, 197–204.
5. Verbovšek, T.; Košir, A.; Teran, M.; Zajc, M.; Popit, T. Volume determination of the Selo landslide complex (SW Slovenia): Integrating field mapping; ground penetrating radar and GIS approaches. *Landslides* **2017**, *14*, 1265–1274. [CrossRef]
6. Popit, T. Origin of planation surfaces in the hinterland of Šumljak sedimentary bodies in Rebrnice (upper Vipava valley, SW Slovenia) = Nastanek reliefnih izravnav v zaledju sedimentnih teles Šumljak na Rebrnicah (zgornja Vipavska dolina, SW Slovenija). *Geologija* **2017**, *60*, 297–307. [CrossRef]
7. Jemec Auflič, M.; Jež, J.; Popit, T.; Košir, A.; Maček, M.; Logar, J.; Petkovšek, A.; Mikoš, M.; Calligaris, C.; Boccali, C.; et al. The variety of landslide forms in Slovenia and its immediate NW surroundings. *Landslides* **2017**, *14*, 1537–1546. [CrossRef]
8. Verbovšek, T.; Popit, T. GIS-assisted classification of litho-geomorphological units using Maximum Likelihood Classification, Vipava Valley, SW Slovenia. *Landslides* **2018**, *15*, 1415–1424. [CrossRef]
9. Šmuc, A.; Janecka, K.; Lempa, M.; Kaczka, R.J. The spatio-temporal dynamics of the Ciprnik complex landslide, Tamar Valley, Julian Alps, Slovenia. *Stud. Geomorphol. Carpatho-Bal.* **2015**, *49*, 35–54. [CrossRef]
10. Novak, A.; Popit, T.; Šmuc, A. Sedimentological and geomorphological characteristics of Quaternary deposits in the Planica-Tamar Valley in the Julian Alps (NW Slovenia). *J. Maps* **2018**, *14*, 382–391. [CrossRef]
11. Placer, L. Vipavski prelom. *Geologija* **2008**, *51*, 101–105. [CrossRef]
12. Popit, T.; Supej, B.; Kokalj, Ž.; Verbovšek, T. Primerjava metod za geomorfometrične analize hrapavosti površja na primeru Vipavske doline = comparison of methods for geomorphometric analysis of surface roughness in the Vipava valley. *Geod. Vestn.* **2016**, *60*, 227–240. [CrossRef]
13. Čar, J.; Gospodarič, R. Geološka zgradba in nekatere hidrološke značilnosti bruhalnika Lijaka = Geological Setting and some Hydrological Properties of Lijak Effluent. *Acta Carsologica* **1988**, *17*, 13–32.
14. Janež, J.; Čar, J.; Habič, P.; Podobnik, R.; Hlad, B. *Vodno Bogastvo Visokega Krasa: Ranljivost Kraške Podzemne Vode Banjščic, Trnovskega Gozda, Nanosa in Hrušice*; Geologija d.o.o: Idrija, Slovenia, 1997; ISBN 961-90568-0-9.
15. Popit, T.; Košir, A. Pleistocene debris flow deposits in the Vipava Valley, SW Slovenia. In *Proceedings of the 22nd IAS Meeting of Sedimentology: E-Abstracts Book*; Vlahovič, I., Ed.; Institute of Geology: Zagreb, Croatia, 2002.
16. Popit, T.; Košir, A.; Šmuc, A. Sedimentological Characteristics of Quaternary Deposits of the Rebrnice Slope Area (SW Slovenia). In *Knjiga Sažetka, Proceedings of the 3rd Znastveni Skup Geologija Kvartara u Hrvatskoj s Međunarodnim Sudjelovanjem, Zagreb, Croatioa, 21–23 March 2013*; HAZU: Zagreb, Croatia, 2013.
17. Novak, A.; Verbovšek, T.; Popit, T. Heterogeneously composed Lozice fossil landslide in Rebrnice area; Vipava valley = Heterogeni fosilni plaz Lozice na območju Rebrnic v Vipavski dolini. *Geologija* **2017**, *60*, 145–155. [CrossRef]
18. Kocjančič, M.; Popit, T.; Verbovšek, T. Gravitational sliding of the carbonate megablocks in the Vipava Valley, SW Slovenia. *Acta Geogr. Slov.* **2019**, *59*, 7–22. [CrossRef]
19. Jež, J. Vzroki in mehanizem zemeljskega plazenja na Rebrnicah v Vipavski dolini = Reasons and mechanism for soil sliding processes in the Rebrnice area, Vipava valley, SW Slovenia. *Geologija* **2007**, *50*, 55–63. [CrossRef]
20. Popit, T.; Košir, A. Kvartarni paleoplazovi na Rebrnicah. In *Povzetki in Ekskurzije, Proceedings of the 3rd Slovenski Geloški Kongeres, Bovec, Slovenia, 16–18 Septmber 2010*; Znanstvenoraziskovalni center SAZU; Inštitut za raziskovanje krasa; Paleontološki inštitut Ivana Rakovca: Postojna, Slovenia, 2010.

21. Košir, A.; Martín-Pérez, A.; Popit, T. Zakaj je Slano blato slano? In *Proceedings of the 21st Posvetovanje Slovenskih Geologov, Ljubljana, Slovenia*; Geološki Zbornik: Ljubljana, Slovenia, 2013; p. 22.
22. Popit, T.; Jernej, J.; Košir, A.; Fifer Bizjak, K.; Ribičič, M. Kvartarni pobočni sedimenti severnega in vzhodnega roba Vipavske doline (Quaternary slope sediments of northern and eastern border of the Vipava Valley). In *Abstracts and Field Trips, Proceedings of the 4th Slovenian Geological Congress, Ankaran, Slovenia, 8–10 October 2014*; Rožič, B., Verbovšek, T., Vrabec, M., Eds.; Faculty of Natural Sciences and Engineering: Ljubljana, Slovenia, 2014; pp. 98–103.
23. Ribičič, M. Zemeljski plazovi. In Vipavska dolina. Neživi svet; rastlinstvo; živalstvo; zgodovina; umetnostna zgodovina; gmotna kultura; gospodarstvo. In *Naravovarstvo*; Pavšič, J., Ed.; Slovenska Matica: Ljubljana, Slovenia, 2013; ISBN 978-961-213-242-2.
24. Verbovšek, T.; Popit, T.; Kokalj, Ž. VAT Method for Visualization of Mass Movement Features: An Alternative to Hillshaded DEM. *Remote Sens.* **2019**, *11*, 2946. [CrossRef]
25. Buser, S. Basic geological map of SFRJ 1:100,000. In *Guide for Sheet Gorica*; Federal Geological Survey: Belgrade, Yugoslavia, 1973; p. 50.
26. Buser, S. Basic geological map of SFR Yugoslavia 1:100,000. In *Gorica Sheet*; Federal Geological Survey: Belgrade, Yugoslavia, 1968.
27. Placer, L.; Vrabec, M.; Celarc, B. The bases for understanding of the NW Dinarides and Istria Peninsula tectonics. *Geologija* **2010**, *53*, 55–86. [CrossRef]
28. ARSO Lidar Slovenia. Available online: http://gis.arso.gov.si/atlasokolja/profile.aspx?id=Atlas_Okolja_AXL@Arso (accessed on 28 February 2022).
29. Zakšek, K.; Oštir, K.; Pehani, P.; Čotar, K. *Relief Visualization Toolbox*; Version 1.1. Manual; Research Centre of the Slovenian Academy of Sciences and Arts: Ljubljana, Slovenia; Available online: iaps.zrc-sazu.si/en/rvt (accessed on 28 February 2022).
30. Kokalj, Ž.; Zakšek, K.; Oštir, K. Application of sky-view factor for the visualization of historic landscape features in lidar-derived relief models. *Antiquity* **2011**, *85*, 263–273. [CrossRef]
31. Zakšek, K.; Oštir, K.; Kokalj, Ž. Sky-view factor as a relief visualization technique. *Remote Sens.* **2011**, *3*, 398–415. [CrossRef]
32. Staley, D.M.; Wasklewicz, T.A.; Blaszczynski, J.S. Surficial patterns of debris flow deposition on alluvial fans in Death Valley; CA using airborne laser swath mapping data. *Geomorphology* **2006**, *74*, 152–163. [CrossRef]
33. Cavalli, M.; Marchi, L. Characterisation of the Surface Morphology of an Alpine Alluvial Fan Using Airborne LiDAR. *Hazards Earth Syst. Sci.* **2008**, *8*, 323–333. [CrossRef]
34. Cavalli, M.; Tarolli, P.; Marchi, L.; Dalla Fontana, G. The Effectiveness of Airborne LiDAR Data in the Recognition of Channel-bed Morphology. *Catena* **2008**, *73*, 249–260. [CrossRef]
35. Jaboyedoff, M.; Oppikofer, T.; Abellán, A.; Derron, M.-H.; Loye, A.; Metzger, R.; Pedrazzini, A. Use of LIDAR in Landslide Investigations: A Review. *Nat. Hazards* **2012**, *61*, 5–28. [CrossRef]
36. McKean, J.; Roering, J. Objective landslide detection and surface morphology mapping using high-resolution airborne laser altimetry. *Geomorphology* **2004**, *57*, 331–351. [CrossRef]
37. Glenn, N.F.; Streutker, D.R.; Chadwick, D.J.; Thackray, G.D.; Dorsch, S.J. Analysis of LiDAR-derived topographic information for characterizing and differentiating landslide morphology and activity. *Geomorphology* **2006**, *73*, 131–148. [CrossRef]
38. Van Den Eeckhaut, M.; Poesen, J.; Verstraeten, G.; Vanacker, V.; Nyssen, J.; Moeyersons, J.; Van Beek, L.P.H.; Vandekerckhove, L. Use of LIDAR-derived images for mapping old landslides under forest. *Earth Surf. Process. Landf.* **2006**, *32*, 754–769. [CrossRef]
39. Van Den Eeckhaut, M.; Poesen, J.; Gullentops, F.; Vandekerckhove, L.; Hervás, J. Regionall mapping and characterization of old landslides in hilly regions using LiDAR-based imagery in Southeren Flanders. *Quat. Res.* **2011**, *75*, 721–733. [CrossRef]
40. Amundsen, J.; Johnson, S.; Rouse, K.; Wang, H. Using LiDAR-Derived DEM's to Delineate and Characterize Landslides in Northern Kentucky and Hamilton County, Ohio. Available online: http://www.trishock.com/academic/pdf/lidar_landslides.pdf (accessed on 28 February 2022).
41. Ruszkiczay-Rüdiger, Z.; Fodor, L.; Horváth, E.; Telbisz, T. Discrimination of Fluvial; Eolian and Neotectonic Features in a Low Hilly Landscape: A DEM-based Morphotectonic Analysis in the Central Pannonian Basin; Hungary. *Geomorphology* **2009**, *104*, 203–217. [CrossRef]
42. Kokalj, Ž.; Somrak, M. Why Not a Single Image? Combining Visualizations to Facilitate Fieldwork and On-Screen Mapping. *Remote Sens.* **2019**, *11*, 747. [CrossRef]
43. Yokoyama, R.; Shlrasawa, M.; Pike, R.J. Visualizing topography by openness: A new application of image processing to digital elevation models. *Photogramm. Eng. Remote Sens.* **2002**, *68*, 251–266.
44. Zorn, M. Ali se zavedamo hitrosti erozijskih procesov—Primer iz Slovenske Istre. *Dela* **2007**, *28*, 183–195. [CrossRef]
45. Placer, L.; Jež, J.; Atanackov, J. Strukturni pogled na plaz Slano Blato. *Geologija* **2008**, *51*, 229–234. [CrossRef]
46. Van Den Eeckhaut, M.; Kerle, N.; Poesen, J.; Hervas, J. Identification of vegetated landslides using only a lidar-based terrain model and derivatives in an object-oriented environment. In Proceedings of the 4th GEOBIA, Rio de Janeiro, Brazil, 7–9 May 2012.
47. Van Den Eeckhaut, M.; Kerle, N.; Poesen, J. Object-oriented identification of forested landslides with derivatives of single pulse LiDAR data. *Geomorphology* **2012**, *173–174*, 30–42. [CrossRef]
48. Cruden, D.M.; Varnes, D.J. Landslides types and processes. In *Landslides: Investigation and Mitigation*; Transportation Research Board Special Report 247; Turner, A.K., Schuster, R.L., Eds.; National Academy Press: Washington, DC, USA.
49. Grohmann, C.H.; Smith, M.J.; Riccomini, C. Surface Roughness of Topography: A Multi-Scale Analysis of Landform Elements in Midland Valley, Scotland. In Proceedings of the Geomorphometry, Zurich, Switzerland, 31 August–2 September 2009.

50. Grohmann, C.H.; Smith, M.J.; Riccomini, C. Multiscale Analysis of Topographic Surface Roughness in the Midland Valley, Scotland. *IEEE Trans. Geosci. Remote Sens.* **2011**, *49*, 1200–1213. [CrossRef]
51. Shulz, W.H. Landslide susceptibility revealed by LIDAR imagery and historical records, Seattle, Washington. *Eng. Geol.* **2007**, *89*, 67–87. [CrossRef]
52. Habič, P. *Kraški Svet Mad Idrijco in Vipavo.: Prispevek k Poznavanju Razvoja Kraškega Reliefa*; Slovenska Akademija Znaosti in Umetnosti, Razred za Prirodoslovne in Medicinske Vede: Ljubljana, Slovenia, 1968.
53. Trček, B.; Veselič, M.; Pezdič, J. The vulnerability of karst springs—A case study of the Hubelj spring (SW Slovenia). *Environ. Geol.* **2006**, *49*, 865–874. [CrossRef]
54. Hungr, O.; Leroueil, S.; Picarelli, L. The Varnes classification of landslide types, an update. *Landslides* **2014**, *11*, 167–194. [CrossRef]
55. Varnes, D.J. Slope movement types and processes. In *Landslides: Analysis and Control*; Special Report 176. Transportation research board; Schuster, R.L., Krizek, R.J., Eds.; National Academy of Sciences: Washington, DC, USA, 1978.
56. Deline, P.; Alberto, W.; Broccolato, D.; Hungr, O.; Noetzli, J.; Ravanel, L.; Tamburini, A. The December 2008 Crammont rock avalanche; Mont Blanc massif area; Italy. *Nat. Hazards Earth Syst. Sci.* **2011**, *11*, 3307–3318. [CrossRef]
57. Placer, L.; Čar, J. Problem podzemeljske razvodnice Trnovskega gozda; Križne gore in Črnovške planote. *Acta Carsologica* **1974**, *6*, 81–93.
58. Placer, L. Vzroki nastajanja recentnih in subrecentnih plazov na Rebrnicah—Geohazard z vidika kompleksnosti geoloških procesov. In *Proceedings of the 18th Posvetovanje Slovenskih Geologov, Ljubljana, Slovenia*; Geološki Zbornik 19: Ljubljana, Slovenia, 2007.

remote sensing MDPI

Communication

EGMStream, a Desktop App for EGMS Data Downstream

Davide Festa and Matteo Del Soldato *

Earth Science Department, University of Firenze, 50121 Firenze, Italy; davide.festa@unifi.it
* Correspondence: matteo.delsoldato@unifi.it; Tel.: +39-055-2757551

Abstract: The recent release of European Ground Motion Service (EGMS) products implemented under the responsibility of the Copernicus Land Monitoring Service (CLMS) guarantees free and accessible Europe-wide ground motion data for ground deformation analysis at the local and regional scales. The need for value-adding services and tools for optimal dissemination of radar data from the Copernicus Sentinel-1 satellite mission urges the scientific community to find efficient solutions. A desktop R-based application with a user-friendly interface capable of automatically downloading and transforming EGMS products delivered as large .csv tiles, equivalent to a radar burst into geospatial databases, is presented here. EGMStream is a self-contained desktop app that enables users to systematically store, customize, and convert ground movement data into geospatial databases, burst per burst or for an area of interest directly selectable on the app interface.

Keywords: downstream; EGMS; InSAR; Europe; ground deformation

Citation: Festa, D.; Del Soldato, M. EGMStream, a Desktop App for EGMS Data Downstream. *Remote Sens.* **2023**, *15*, 2581. https://doi.org/10.3390/rs15102581

Academic Editor: Soe Myint

Received: 3 April 2023
Revised: 28 April 2023
Accepted: 12 May 2023
Published: 15 May 2023

1. Introduction

Copernicus is the Earth Observation Programme of the European Union, managed by the European Commission and implemented in partnership with the European Member States; the ESA (European Space Agency); the EUMETSAT (European Organization for the Exploitation of Meteorological Satellites); the ECMWF (European Centre for Medium-Range Weather Forecasts); and European Union (EU) agencies, such as the Environmental European Agency (EEA); and the Mercator Ocean. The program aims to have the planet and its environment benefit all European citizens by using in situ data collected from different sources, such as Earth Observation satellites. The program is divided into different services: (i) Copernicus Atmosphere Monitoring Service (CAMS); (ii) Copernicus Marine Service (or Copernicus Marine Environment Monitoring Service); (iii) Copernicus Land Monitoring Service (CLMS); (iv) Copernicus Climate Change Service (C3S); (v) Copernicus Service for Security; and (vi) Copernicus Emergency Management Services (EMS). The last addition in the CLMS program is the European Ground Motion Service (EGMS), launched in mid-2022 to provide consistent A-DInSAR (Advanced Differential Interferometric Synthetic Aperture Radar) data derived by satellite imagery in high resolution (Sentinel-1) at a continental scale and allow the ground motion analysis and monitoring [1]. At a national, or regional, scale, the ground motion service monitoring was sporadically adopted. In fact, starting from the first application launched in Italy in 2007 (Italian Special Plan of Remote Sensing of the Environment) and taking advantage of the ERS (European Remote Sensing), ENVISAT (Environmental Satellite), and partially the COSMO-SkyMed data [2,3], processed with the PSP (persistent scatterer pairs [4]), the PSInSAR [5,6], and the SqueeSAR [7] algorithms, other nations decided to develop the same service as Norway [8,9] and Germany in 2018 [10]. In addition, the Agency for Data Supply and Efficiency of Denmark is working on a WebGIS platform, not yet available for a full, free, and open to all user service based on the Sentinel-1 data processed providing InSAR data by the SqueeSAR algorithm data and information along LOS (Line of Sight) and GNSS (Global Navigation Satellite System)-calibrated velocity [11] and their components. Additionally, the Netherlands is implementing its GMS based on Sentinel-1 data at national level for investigating the height change obtained via A-DInSAR [1].

In Italy, taking advantage of the Sentinel-1 regularity of acquisition and coverage, in 2016, the Tuscany Region (central Italy) implemented the first continuous monitoring system according to the Prime Minister Decree 27 February 2004 [12,13]. Following the same approach, the northwestern Valle d'Aosta Region, in 2018, and the northeastern Veneto Region, in 2019, started with the near-real monitoring of the ground deformation [14]. These examples explain that the launch of the Sentinel-1 constellation opened the possibility to investigate several geohazards thanks to its short revisiting time, the worldwide coverage and the freely availability due to its scientific application purposes. Several examples can be found in the literature on landslide detection [15–18], characterization [19–21], and monitoring [22–24], subsidence phenomena analysis and monitoring [25–29], infrastructure monitoring [30–33], or mining instability analysis and monitoring [34–37].

On the same line, the EGMS was conceived following the direct request of many users for free InSAR data over Europe. The service is managed and being implemented under the responsibility of the EEA, and it is part of the CLMS portfolio. The EGMS provides an annual update of the ground motion over the whole European territory with the time series from February 2015 with full spatial resolution based on the Sentinel-1 radar data. The Sentinel-1 images are processed by four different algorithms [38] PSP-IFSAR [4,39,40], SqueeSAR [7], GSAR-GTSI [41,42], and PSI (Persistent Scatterer Interferometry), performed with an IWAP (Integrated Wide Area Processor) [43,44].

The processed data are available from November 2022 in visualization and download at three levels of processing [45], (i) basic (L2a), (ii) GNSS-calibrated (L2b), and (iii) Ortho (L3).

The basic data provides InSAR velocity and displacement information along the LoS (Line of Sight) with information about the geolocalization and quality measurements for every MP (Measurement Point). A time series is associated with each MP, while the velocity and displacement information are relative, thus referring to a stable virtual Reference Point chosen during processing of the Sentinel-1 SAR images stacks frame by frame. The L2a products are provided in full resolution as two discrete datasets referred to as the Sentinel-1 ascending (South to North) and descending (North to South) orbits. The basic product is a necessary first step to the more advanced products.

The calibrated data are an advanced product, considered the main EGMS product, consisting of a deformation map with LoS absolute velocity and displacement information corrected by a model derived from GNSS time series data across Europe. Since some isolated islands do not have GNSS data available, such as the L2b products, are calibrated, and "... *are produced by harmonizing Basic products with respect to each other, and then adjusting the mean ground velocity to zero.*" [22] The L2b products are provided in full resolution as two discrete ascending and descending datasets as the L2a.

The L3 data, named Ortho products, are the vertical and horizontal (East–West) components of velocity, completed for their time series, calculated from the L2b data. Differently to the previous products, the L3 data follow a regular grid since the L3 MP are synthetic points summarizing the L2b ascending and descending velocity in a cell of 100 m (coinciding to the Copernicus DEM). As for the L2a and L2b data, the time series, both for the vertical and horizontal components, are present in this level of product.

Prior to publication, all EGMS data underwent an extensive process quality control protocol involving the validation of several criteria, such as a suitable density of measurement points within the Corine Land Cover (CLC) categories and an appropriate displacement standard deviation [46]. This procedure of EGMS data velocity validation is still ongoing by an external consortium.

In this work, we introduce the R-based EGMStream tool developed with the Shiny package [47,48] and available as a desktop application for downloading, transforming, and exporting the EGMS products into customized geospatial databases storable in ESRI shapefiles or GeoPackage containers. The meaningfulness of such a value-adding service relies on offering an improved user experience for the management of A-DInSAR data covering Europe. In particular, the capability of cropping the EGMS products based on a personalized Area of Interest (AoI) is implemented within the EGMStream application. Within this framework, thanks to an easy-to-use interface, we foster the usage and the growth of the potential pool of users interested in the downstream applications of EGMS data and interferometric products.

2. Materials and Methods

EGMStream is a free app that does not require any previous software installation or the use of a third-party server, being a self-contained R-based application, which can be deployed to the desktop. The presented data downstream approach follows a precise workflow to download and convert the EGMS data, indifferently from the level of processing, by using a list of links available for download directly from the EGMS viewer [49]. The underlying flowchart of EGMStream can be split into two main parts, the upload of the input data required for the application to run and the design of the geospatial database prior to the final conversion (Figure 1).

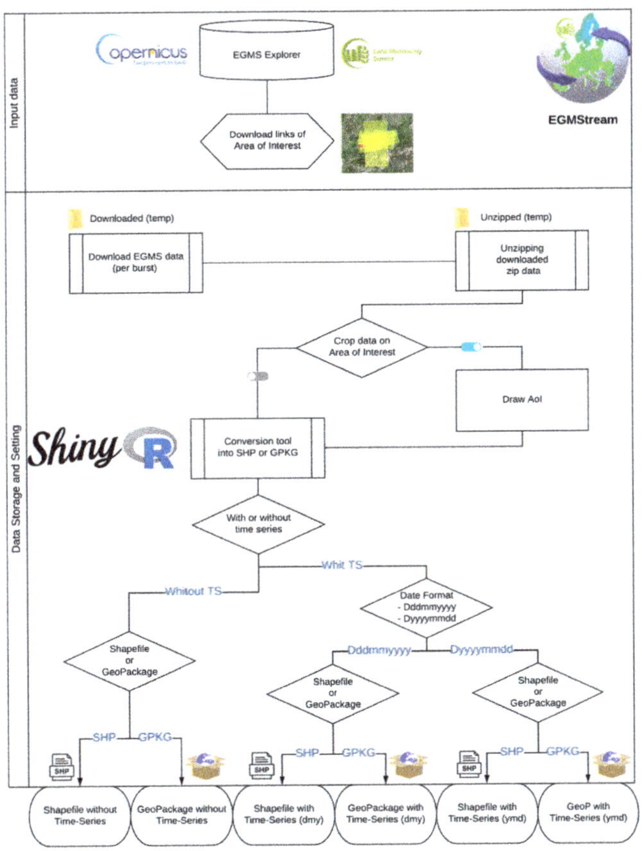

Figure 1. EGMStream concept workflow.

2.1. Input Data

The EGMStream app is designed to retrieve and manage EGMS ground deformation products at a pan-European level. The EGMS satellite-based land monitoring data are made available for download through the EGMS Product Archive and Dissemination System by accessing the EGMS Explorer [49].

Upon registration and authentication, users are enabled to locate and download multiple InSAR datasets (with a maximum of two simultaneously) related to a geographical territory. The only limit about the extension of the latter is a maximum width of 3 degrees. In addition, the EGMS Explorer system allows downloading the "Download links", which is an ASCII file (*.txt*) containing hyperlinks corresponding to a list of products that is queried for bulk download. Multiple EGMS product levels can be listed within the single ASCII file, where the provided hyperlinks come with a security token that keeps the data valid for downloading for one hour. Moreover, the token is refreshed if a download is in progress, allowing the hyperlinks to remain valid for another hour after a given download finishes [50]. When the token expires, the app will appear frozen, requiring a further re-load by the desktop shortcut or by the start menu. Therefore, new valid download links are then required to correctly deploy the app.

The EGMStream application is conceived to be fed with an ASCII file (*"download links"* from the EGMS viewer) containing one or several download-links via upload control. The selected file will serve as the input for the successive operations concerning data download, eventual data cropping over the AoI, and setting of the database attributes prior to InSAR data conversion.

2.2. *Data Storage Setting and Conversion*

The EGMStream can unscramble encoded URLs with a timeout limit set to 1 h to ensure an appropriate time window for starting the download of server-intensive files. In the first instance, the resulting EGMS products are temporarily stored as zipped files within an automatically created folder named 'Downloaded'. After this operation, EGMStream automatically proceed by unzipping the retrieved files and by creating the directory 'Unzipped' where the ground motion data are stored in the *.csv* format (standard download format from EGMS viewer). Both folders are automatically generated and then deleted at the end of the data conversion process. Pop-up notifications are extensively used within the app to inform the user about the processes achieved by the ongoing R session. In case the pop-up expires, all the processes made by the app are reported in the "Processing history" tab.

EGMStream's main functionalities concern EGMS data cropping based on the user's AoI and data conversion into geospatial databases. A major feature of the app regards the possibility to interactively draw a rectangular shape of the AoI through a map viewer panel created using the JavaScript 'Leaflet' library [51]. This functionality allows the user to manually derive the geographical area, which acts as a mask covering the underlying EGMS data products to be clipped. In the event that the AoI is on two different bursts (on the same track), the MP data will be merged into a unique shapefile or database. On the contrary, if the selected data overlay two or more tracks (which implies different acquisition dates from the satellite), the converted geodatabases will be kept separate and named after the track number. EGMStream ensures the possibility of choosing the folder location where the converted data will be saved locally. Moreover, several settings allow designing the data storage characteristics, which are selectable from the app interface. In particular, the geospatial database can be designed by the user by selecting:

- Inclusion ('With') or exemption ('Without') of the time series related columns;
- Two date formats, in case of time series inclusion;
- Shapefile or GeoPackage output data format.

Giving the possibility to exclude time series information would result in a lighter conversion process and a geospatial database with a reduced file size. This can be particularly helpful to adapt EGMS data to users whose need is to retrieve only ground deformation maps, requiring only mean velocity values of deformation. The default option is set to 'With Time Series'.

When retained, date column names related to the time series information can be adjusted to the *Dddmmyyyy* format or the *Dyyyymmdd* format. In this case, there is no default option; therefore, the date format selection needs to be explicitly selected.

Once all the parameters are set, the downloaded *.csv* files can be converted into two of the most common file formats for geospatial data, namely shapefiles (i.e., *.shp*) or GeoPackage (i.e., *.gpkg*), with the last being the default option. Shapefile is a native ESRI (Environmental Systems Research Institute) proprietary format which comes with a mandatory part of file collections and is especially designed for use in Geographic Information Systems (GIS) software. On the other hand, GeoPackage is an open and platform-independent format for storing geospatial information within an SQLite database in a unique file that supports its direct use. The required and supported content of a GeoPackage is entirely defined in the Open Geospatial Consortium (OGC®) standard document [52].

As a result of the selectable storage setting options, six different configurations are made available to convert EGMS products:

.shp without time series;
.gpkg without time series;
.shp with *Dddmmyyyy* time series format;
.gpkg with *Dddmmyyyy* time series format;
.shp with *Dyyyymmdd* time series format;
.gpkg with *Dyyyymmdd* time series format.

3. Results and Discussion

EGMStream is an open-source tool which is realized by using the Shiny R-package framework. To encourage distribution and make the program platform independent, EGMStream is shared as a Windows desktop application ready for immediate use without the need for installing any external software (i.e., R). To achieve this, the built framework uses R-Portable [53], while the app's primary package dependencies are loaded when the application is run for the first time. The presented application is bundled into an executable installation (i.e., a setup wizard), which guarantees control over the destination location and allows creating a program's desktop shortcut. The front-end interface of the app loads dynamically on the PC default web browser (tested on Google Chrome and Microsoft Edge) prior to installing EGMStream.

Leveraging on the Shiny's reactive framework, the user-driven draw toolbar can be used to reshape the AoI multiple times within the same session, where only the last drawn element will be kept. Additionally, the drawn AoI can be interactively deleted from the application interface. Additionally, EGMStream is designed to overwrite the output results when the same instructions are repeated. It should be noted that EGMStream tackles the conversion of very large EGMS files by limiting the process to 500,000 rows at a time; consequently, the targeted downloaded file would be split into different converted geospatial databases (which are suitably labeled with a progressive numbering for reference).

The EGMStream's current limit concerns the maximum amount of downloadable EGMS products within the same session. This amount varies according to the user's internet connection velocity and to the dimension of the requested file for conversion. Based on tests performed with an Intel Core i7-4790QM at 3.60 GHz, 4 cores, 8 threads, 8 MB cache, 16 GB RAM, 250 GB SSD disk, and Windows 10, 64 bits, we recommend to feed EGMStream with less than 50 download links.

3.1. Example

To demonstrate the added value of EGMStream for the rapid downstream and deployment of EGMS products, it briefly showcased the successful data conversion procedure of multiple levels of interferometric data covering part of the Rhenish coalfields of Germany, as shown in Figure 2. In particular, the converted database contains a geographical subset of the interferometric products overlaying the AoI drawn by the user.

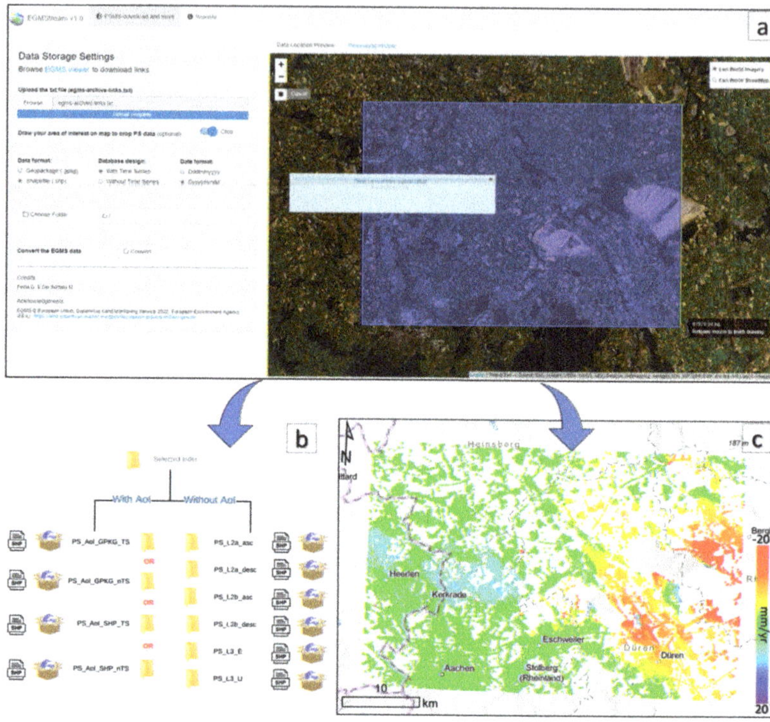

Figure 2. (**a**) Outline of the EGMStream app interface; (**b**) framework of EGMStream output products; and (**c**) visualization of the converted EGMS product via GIS platform.

EGMStream implements intuitive workflow routine tasks, such as (i) deployment of the ASCII input file obtained from the EGMS viewer; (ii) implementation of an arbitrary extraction mask to clip the input EGMS data to an AoI; (iii) data download-unzipping procedure; (iv) data conversion according to the specified storage settings; and (v) storage of converted data.

The output results, being vector data points, are stored following a precise framework; every level of EGMS data and every related acquisition geometry is stored within a dedicated folder (Figure 2b), enabling a direct reference to the queried products.

With the exploitation of the EGMStream functionalities, the downloaded and converted data result are particularly suited for visualization, data handling, and post-processing elaboration on GIS software (Figure 2c).

A deep understanding of the ongoing surface displacement can rely on the analysis of the spatial pattern of the various levels of EGMS data features. In Figure 3, a landslide is displayed, one of the most common geohazards that can be detected and analyzed via the available EGMS products (reference period: 2016–2021) and downloaded by the EGMStream tool for the area of interest. In particular, a joint use of the calibrated (Figure 3a,b) and ortho (Figure 3c,d) products enables the identification of clusters of points with comparable motion trends and deformation patterns (which can be identified according to the mean velocity). The displacement of the area of interest can be further evaluated by looking at the time series of every MP. Mapping and monitoring potentially risky and/or critical areas can greatly contribute to more resilient urban planning and prompt response from civil protection authorities.

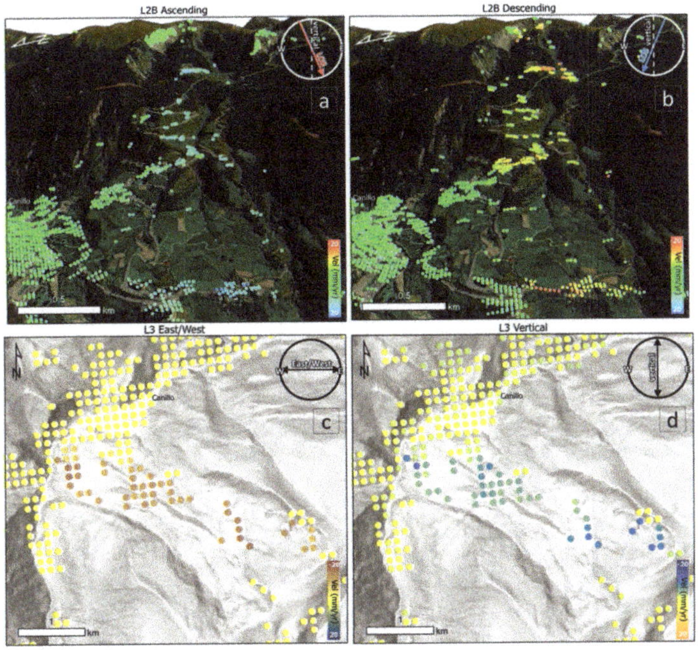

Figure 3. (**a**,**b**) Three-dimensional visualization of the calibrated ascending and descending EGMS products related to an ongoing slope displacement located near Canillo (Andorra); a more accurate picture of the deformation scenario can be obtained by consulting the EGMS (**c**,**d**) ortho data.

3.2. Future Developments

The EGMS foresees an annual update of the three levels of data, and the EGMStream tool is already set to process the new data.

In addition, in a further version of the EGMStream tool, the authors would like to implement several functions, to among:

- The possibility to select the information column to extract with or without displacement data, which are now are automatically chosen by the developers of the tool;
- The possibility to visualize on the map the downloaded and cropped data with different options for color-scales;
- The possibility to execute preliminary post-processing operations over the downloaded EGMS data, e.g., data mining of relevant ground deformations visible within the AoI;
- Possibility to automatically check the free space on the root chosen by the user for alerting if there is not enough space for the conversion.

In addition to the above-mentioned ideas in the pipeline, every future suggestion from the scientific community or end-users will be considered.

4. Conclusions

EGMStream is an open-source, interactive, and user-friendly Shiny/R desktop application designed to enhance the downstream of EGMS products by enabling the user to seamlessly download, customize, convert, and store radar-based geospatial databases. Distributed as a self-contained application, EGMStream deployment is tied to an initial installation procedure on a local Windows PC. Leveraging the app´s intuitive interface, no prior knowledge is required to obtain reliable and handy results.

The development of EGMStream is still ongoing in order to guarantee a more personalized experience.

Author Contributions: Conceptualization and methodology, M.D.S.; software, D.F. and M.D.S.; validation, D.F.; writing—original draft preparation, D.F.; writing—review and editing, M.D.S.; visualization, D.F. supervision and project administration, M.D.S. All authors have read and agreed to the published version of the manuscript.

Funding: This research received no external funding.

Data Availability Statement: EGMStream v1.0 is freely available as a standalone desktop app at http://cpc.unifi.it/EGMStream_v1.zip. By clicking on the link the download of the zip file containing a Windows installer (i.e., EGMStream_installer.exe) will start. To remain updated about future releases or to receive further support, please contact the authors at egmstream@dst.unifi.it. The code is freely available and can be found within the installed folders.

Acknowledgments: This work was developed within the project EGMS RASTOOL (ECHO–UCPM-2021-PP (Prevention and preparedness in civil protection and marine pollution), Contract No. 101048474). We would like to thank Lorenzo Solari for the support in the data comprehension and in the testing of the tool. In addition, we would like to thank the Copernicus Land Monitoring Service (CLMS) to provide the European Ground Motion Services (EGMS) data to users.

Conflicts of Interest: The authors declare no conflict of interest.

References

1. Crosetto, M.; Solari, L.; Mróz, M.; Balasis-Levinsen, J.; Casagli, N.; Frei, M.; Oyen, A.; Moldestad, D.A.; Bateson, L.; Guerrieri, L. The evolution of wide-area dinsar: From regional and national services to the European ground motion service. *Remote Sens.* **2020**, *12*, 2043. [CrossRef]
2. Costantini, M.; Ferretti, A.; Minati, F.; Falco, S.; Trillo, F.; Colombo, D.; Novali, F.; Malvarosa, F.; Mammone, C.; Vecchioli, F. Analysis of surface deformations over the whole Italian territory by interferometric processing of ERS, Envisat and COSMO-SkyMed radar data. *Remote Sens. Environ.* **2017**, *202*, 250–275. [CrossRef]
3. WebGIS of the PST-A Project. Available online: http://www.pcn.minambiente.it/viewer/ (accessed on 10 December 2022).
4. Costantini, M.; Falco, S.; Malvarosa, F.; Minati, F. A new method for identification and analysis of persistent scatterers in series of SAR images. In Proceedings of the IGARSS 2008 Geoscience and Remote Sensing Symposium, Boston, MA, USA, 7–11 July 2008; IEEE International: Boston, MA, USA, 2008; pp. II-449–II-452.
5. Ferretti, A.; Prati, C.; Rocca, F. Nonlinear subsidence rate estimation using permanent scatterers in differential SAR interferometry. *IEEE Trans. Geosci. Remote Sens.* **2000**, *38*, 2202–2212. [CrossRef]
6. Ferretti, A.; Prati, C.; Rocca, F. Permanent scatterers in SAR interferometry. *IEEE Trans. Geosci. Remote Sens.* **2001**, *39*, 8–20. [CrossRef]
7. Ferretti, A.; Fumagalli, A.; Novali, F.; Prati, C.; Rocca, F.; Rucci, A. A new algorithm for processing interferometric data-stacks: SqueeSAR. *IEEE Trans. Geosci. Remote Sens.* **2011**, *49*, 3460–3470. [CrossRef]
8. Dehls, J.F.; Larsen, Y.; Marinkovic, P.; Lauknes, T.R.; Stødle, D.; Moldestad, D.A. INSAR. No: A National Insar Deformation Mapping/Monitoring Service In Norway—From Concept To Operations. In Proceedings of the IGARSS 2019–2019 IEEE International Geoscience and Remote Sensing Symposium, Yokohama, Japan, 28 July 2019–2 August 2019; pp. 5461–5464.
9. InSAR Norway WebGIS. Available online: https://insar.ngu.no/ (accessed on 9 December 2022).
10. Kalia, A.; Frei, M.; Lege, T. A Copernicus downstream-service for the nationwide monitoring of surface displacements in Germany. *Remote Sens. Environ.* **2017**, *202*, 234–249. [CrossRef]
11. Bischoff, C.A.; Ferretti, A.; Novali, F.; Uttini, A.; Giannico, C.; Meloni, F. Nationwide deformation monitoring with SqueeSAR® using Sentinel-1 data. *Proc. Int. Assoc. Hydrol. Sci.* **2020**, *382*, 31–37. [CrossRef]
12. Del Soldato, M.; Solari, L.; Raspini, F.; Bianchini, S.; Ciampalini, A.; Montalti, R.; Ferretti, A.; Pellegrineschi, V.; Casagli, N. Monitoring ground instabilities using SAR satellite data: A practical approach. *ISPRS Int. J. Geo-Inf.* **2019**, *8*, 307. [CrossRef]
13. Raspini, F.; Bianchini, S.; Ciampalini, A.; Del Soldato, M.; Solari, L.; Novali, F.; Del Conte, S.; Rucci, A.; Ferretti, A.; Casagli, N. Continuous, semi-automatic monitoring of ground deformation using Sentinel-1 satellites. *Sci. Rep.* **2018**, *8*, 7253. [CrossRef]
14. Confuorto, P.; Del Soldato, M.; Solari, L.; Festa, D.; Bianchini, S.; Raspini, F.; Casagli, N. Sentinel-1-based monitoring services at regional scale in Italy: State of the art and main findings. *Int. J. Appl. Earth Obs. Geoinf.* **2021**, *102*, 102448. [CrossRef]
15. Barra, A.; Monserrat, O.; Mazzanti, P.; Esposito, C.; Crosetto, M.; Scarascia Mugnozza, G. First insights on the potential of Sentinel-1 for landslides detection. *Geomat. Nat. Hazards Risk* **2016**, *7*, 1874–1883. [CrossRef]
16. Intrieri, E.; Raspini, F.; Fumagalli, A.; Lu, P.; Del Conte, S.; Farina, P.; Allievi, J.; Ferretti, A.; Casagli, N. The Maoxian landslide as seen from space: Detecting precursors of failure with Sentinel-1 data. *Landslides* **2018**, *15*, 123–133. [CrossRef]
17. Mantovani, M.; Bossi, G.; Marcato, G.; Schenato, L.; Tedesco, G.; Titti, G.; Pasuto, A. New perspectives in landslide displacement detection using sentinel-1 datasets. *Remote Sens.* **2019**, *11*, 2135. [CrossRef]
18. Kalia, A.C. Landslide activity detection based on Sentinel-1 PSI datasets of the Ground Motion Service Germany—The Trittenheim case study. *Landslides* **2023**, *20*, 209–221. [CrossRef]

19. Lanari, R. From ERS-1 TO SENTINEL-1: A big data challenge for 25 years of DInSAR observations. In *2015 IEEE International Geoscience and Remote Sensing Symposium (IGARSS)*; IEEE: Milan, Italy, 2015; pp. 1460–1463.
20. Mirmazloumi, S.M.; Wassie, Y.; Navarro, J.A.; Palamà, R.; Krishnakumar, V.; Barra, A.; Cuevas-González, M.; Crosetto, M.; CuMonserrat, O. Classification of ground deformation using sentinel-1 persistent scatterer interferometry time series. *GIScience Remote Sens.* **2022**, *59*, 374–392. [CrossRef]
21. Confuorto, P.; Casagli, N.; Casu, F.; De Luca, C.; Del Soldato, M.; Festa, D.; Raspini, F. Sentinel-1 P-SBAS data for the update of the state of activity of national landslide inventory maps. *Landslides* **2023**, *20*, 1083–1097. [CrossRef]
22. Dai, K.; Li, Z.; Tomás, R.; Liu, G.; Yu, B.; Wang, X.; Cheng, H.; Chen, J.; Stockamp, J. Monitoring activity at the Daguangbao mega-landslide (China) using Sentinel-1 TOPS time series interferometry. *Remote. Sens. Environ.* **2016**, *186*, 501–513. [CrossRef]
23. Beladam, O.; Balz, T.; Mohamadi, B.; Abdalhak, M. Using ps-insar with sentinel-1 images for deformation monitoring in northeast Algeria. *Geosciences* **2019**, *9*, 315. [CrossRef]
24. Raspini, F.; Bianchini, S.; Ciampalini, A.; Del Soldato, M.; Montalti, R.; Solari, L.; Tofani, V.; Casagli, N. Persistent Scatterers continuous streaming for landslide monitoring and mapping: The case of the Tuscany region (Italy). *Landslides* **2019**, *16*, 2033–2044. [CrossRef]
25. Ferretti, A.; Colombo, D.; Fumagalli, A.; Novali, F.; Rucci, A. InSAR data for monitoring land subsidence: Time to think big. *Proc. Int. Assoc. Hydrol. Sci.* **2015**, *372*, 331–334. [CrossRef]
26. Solari, L.; Del Soldato, M.; Bianchini, S.; Ciampalini, A.; Ezquerro, P.; Montalti, R.; Raspini, F.; Moretti, S. From ERS 1/2 to Sentinel-1: Subsidence monitoring in Italy in the last two decades. *Front. Earth Sci.* **2018**, *6*, 149. [CrossRef]
27. Edalat, A.; Khodaparast, M.; Rajabi, A.M. Detecting Land Subsidence Due to Groundwater Withdrawal in Aliabad Plain, Iran, Using ESA Sentinel-1 Satellite Data. *Nat. Resour. Res.* **2019**, *29*, 1935–1950. [CrossRef]
28. Hakim, W.L.; Achmad, A.R.; Eom, J.; Lee, C.-W. Land Subsidence Measurement of Jakarta Coastal Area Using Time Series Interferometry with Sentinel-1 SAR Data. *J. Coast. Res.* **2020**, *102*, 75–81. [CrossRef]
29. Raspini, F.; Caleca, F.; Del Soldato, M.; Festa, D.; Confuorto, P.; Bianchini, S. Review of satellite radar interferometry for subsidence analysis. *Earth-Sci. Rev.* **2022**, *235*, 104239. [CrossRef]
30. Luo, Q.; Li, J.; Zhang, Y. Monitoring Subsidence over the Planned Jakarta–Bandung (Indonesia) High-Speed Railway Using Sentinel-1 Multi-Temporal InSAR Data. *Remote. Sens.* **2022**, *14*, 4138. [CrossRef]
31. Shami, S.; Azar, M.K.; Nilfouroushan, F.; Salimi, M.; Reshadi, M.A.M. Assessments of ground subsidence along the railway in the Kashan plain, Iran, using Sentinel-1 data and NSBAS algorithm. *Int. J. Appl. Earth Obs. Geoinf.* **2022**, *112*, 102898. [CrossRef]
32. Bakon, M.; Czikhardt, R.; Papco, J.; Barlak, J.; Rovnak, M.; Adamisin, P.; Perissin, D. remotIO: A Sentinel-1 multi-temporal InSAR infrastructure monitoring service with automatic updates and data mining capabilities. *Remote Sens.* **2020**, *12*, 1892. [CrossRef]
33. Ramirez, R.A.; Lee, G.J.; Choi, S.K.; Kwon, T.H.; Kim, Y.C.; Ryu, H.H.; Kim, S.; Bae, B.; Hyun, C. Monitoring of construction-induced urban ground deformations using Sentinel-1 PS-InSAR: The case study of tunneling in Dangjin, Korea. *Int. J. Appl. Earth Obs. Geoinf.* **2022**, *108*, 102721. [CrossRef]
34. Nam, B.X.; Van Anh, T.; Bui, L.K.; Long, N.Q.; Le Thu Ha, T.; Goyal, R. Mining-Induced Land Subsidence Detection by Persistent Scatterer InSAR and Sentinel-1: Application to Phugiao Quarries, Vietnam. *Lect. Notes Civ. Eng.* **2021**, *108*, 18–38.
35. Dang, V.K.; Nguyen, T.D.; Dao, N.H.; Duong, T.L.; Dinh, X.V.; Weber, C. Land subsidence induced by underground coal mining at Quang Ninh, Vietnam: Persistent scatterer interferometric synthetic aperture radar observation using Sentinel-1 data. *Int. J. Remote Sens.* **2021**, *42*, 3563–3582. [CrossRef]
36. Mehrabi, A.; Derakhshani, R.; Nilfouroushan, F.; Rahnamarad, J.; Azarafza, M. Spatiotemporal subsidence over Pabdana coal mine Kerman Province, central Iran using time-series of Sentinel-1 remote sensing imagery. *Epis. J. Int. Geosci.* **2023**, *46*, 19–33. [CrossRef]
37. Mura, J.C.; Gama, F.F.; Paradella, W.R.; de Oliveira, C.G.; Rodrigues, T.G. Ground displacements revealed by A-DInSAR analysis in the Germano iron mining complex before and after the Fundão Dam collapse using Sentinel-1 data. *J. Appl. Remote Sens.* **2021**, *15*, 034513. [CrossRef]
38. Ferreti, A.; Passera, E.; Capes, R. Algorithm Theoretical Basis Document. Available online: https://land.copernicus.eu/user-corner/technical-library/egms-algorithm-theoretical-basis-document (accessed on 2 October 2022).
39. Costantini, M.; Bignami, C.; Falco, S.; Malvarosa, F.; Moro, M.; Saroli, M.; Stramondo, S. New Results On Post-Seismic Deformations over L'aquila, Italy, by High Resolution Psp Sar Interferometry. In Proceedings of the IEEE International Geoscience and Remote Sensing Symposium (IGARSS), Melbourne, Australia, 21–26 July 2013; IEEE: Melbourne, Australia; pp. 1850–1852.
40. Costantini, M.; Falco, S.; Malvarosa, F.; Minati, F.; Trillo, F.; Vecchioli, F. Persistent scatterer pair interferometry: Approach and application to COSMO-SkyMed SAR data. *IEEE J. Sel. Top. Appl. Earth Obs. Remote Sens.* **2014**, *7*, 2869–2879. [CrossRef]
41. Sansosti, E.; Berardino, P.; Manunta, M.; Serafino, F.; Fornaro, G. Geometrical SAR image registration. *IEEE Trans. Geosci. Remote Sens.* **2006**, *44*, 2861–2870. [CrossRef]
42. Prats-Iraola, P.; Scheiber, R.; Marotti, L.; Wollstadt, S.; Reigber, A. TOPS interferometry with TerraSAR-X. *IEEE Trans. Geosci. Remote Sens.* **2012**, *50*, 3179–3188. [CrossRef]
43. Yagüe-Martínez, N.; Prats-Iraola, P.; Gonzalez, F.R.; Brcic, R.; Shau, R.; Geudtner, D.; Eineder, M.; Bamler, R. Interferometric processing of Sentinel-1 TOPS data. *IEEE Trans. Geosci. Remote Sens.* **2016**, *54*, 2220–2234. [CrossRef]
44. Eineder, M. Efficient simulation of SAR interferograms of large areas and of rugged terrain. *IEEE Trans. Geosci. Remote Sens.* **2003**, *41*, 1415–1427. [CrossRef]

45. Capes, R.; Passera, E. Product Description and Format Specification. Available online: https://land.copernicus.eu/user-corner/technical-library/egms-product-description-document (accessed on 2 October 2022).
46. Solari, L.; Crosetto, M. Validation of the EGMS Product Portfolio. Available online: https://land.copernicus.eu/user-corner/technical-library/validation-approach-of-the-egms-product-portfolio (accessed on 2 October 2022).
47. Racine, J.S. RStudio: A Platform-Independent IDE for R and Sweave. *J. Appl. Econom.* **2012**, *27*, 167–172. [CrossRef]
48. Team, R.C.; R: A Language and Environment for Statistical Computing. R Foundation for Statistical Computing, Vienna, Austria. 2013. Available online: http://www.R-project.org/ (accessed on 2 October 2022).
49. European Ground Motion Service Explorer. Available online: https://egms.land.copernicus.eu/ (accessed on 28 September 2022).
50. Larsen, Y.; Marinkovic, P.; Dehls, J.; Stødle, D. End User Interface Manual. Available online: https://land.copernicus.eu/user-corner/technical-library/egms-end-user-interface-manual (accessed on 2 October 2022).
51. Graul, C. LeafletR: Interactive Web-Maps Based on the Leaflet JavaScript Library. R-Package Version 0.4-0. 2016. Available online: https://github.com/chgrl/leafletR (accessed on 8 October 2022).
52. Van Rees, E. Open geospatial consortium (OGC). *Geoinformatics* **2013**, *16*, 28.
53. R-Portable Software. Available online: https://sourceforge.net/projects/rportable/ (accessed on 6 October 2022).

remote sensing MDPI

Technical Note

A High-Precision Remote Sensing Identification Method on Saline-Alkaline Areas Using Multi-Sources Data

Jingyi Yang [1,2,3], Qinjun Wang [1,2,3,4,*], Dingkun Chang [1,2,3], Wentao Xu [1,2,3] and Boqi Yuan [1,2,3,5]

1 Key Laboratory of Digital Earth Science, Aerospace Information Research Institute, Chinese Academy of Sciences (CAS), Beijing 100094, China
2 International Research Center of Big Data for Sustainable Development Goals, Beijing 100094, China
3 University of Chinese Academy of Sciences, Beijing 100049, China
4 Key Laboratory of the Earth Observation of Hainan Province, Hainan Aerospace Information Research Institute, Sanya 572029, China
5 College of Geoscience and Surveying Engineering, China University of Mining & Technology (Beijing), Beijing 100083, China
* Correspondence: wangqj@radi.ac.cn

Abstract: Soil salinization is a widespread and important environmental problem. We propose a high-precision remote sensing identification method for saline-alkaline areas using multi-source data, a method which is of some significance for improving ecological and environmental problems on a global scale which have been caused by soil salinization. Its principle is to identify saline-alkaline areas from remote sensing imagery by a decision tree model combining four spectral indices named $NDSI_{34}$ (Normalized Difference Spectral Index of Band 3 and Band 4), $NDSI_{25}$ (Normalized Difference Spectral Index of Band 2 and Band 5), $NDSI_{237}$ (Normalized Difference Spectral Index of Band 3 and Band 4) and $NDSI_{new}$ (New Normalized Difference Salt Index) that can distinguish saline-alkaline areas from other features. In this method, the complementary information within the multi-source data is used to improve classification accuracy. The main steps of the method include multi-source data acquisition, adaptive feature fusion of multi-source data, feature identification and integrated expression of the saline-alkaline area from multi-source data, fine classification of the saline-alkaline area, and accuracy verification. Taking Minqin County, Gansu Province, China as the study area, we use the method to identify saline-alkaline areas based on GF-2, GF-6/WFV and DEM data. The results show that the overall accuracy of the method is 88.11%, which is 7.69% higher than that of the traditional methods, indicating that it could effectively identify the distribution of saline-alkaline areas, and thus provide a scientific technique for the quick identification of saline-alkaline areas in large regions.

Keywords: remote sensing; saline-alkali areas; salinization identifying; high precision; multi-source data

Citation: Yang, J.; Wang, Q.; Chang, D.; Xu, W.; Yuan, B. A High-Precision Remote Sensing Identification Method on Saline-Alkaline Areas Using Multi-Sources Data. *Remote Sens.* **2023**, *15*, 2556. https://doi.org/10.3390/rs15102556

Academic Editor: Bas van Wesemael

Received: 27 March 2023
Revised: 10 May 2023
Accepted: 11 May 2023
Published: 13 May 2023

1. Introduction

Soil salinization is a major type of land degradation in arid and semi-arid areas [1,2], one which causes soil consolidation and crop yield decline, and thus results in huge losses in agricultural production. In addition, its mutual induction with soil desertification will cause more significant damage to the ecological environment and even cause serious geological disasters [3–6]. Soil salinization lasts for a long time, and the land encounters difficulties when it attempts to repair itself, which makes for a continuous impact on the human living environment and economic development [7–9]. More than 100 countries and 7% of land area on a global scale are affected by land salinization [10,11]. It has become a worldwide environmental issue of wide-ranging concern, thus leading many countries to pay high attention to the amelioration and development of saline-alkaline areas. China is one of the countries seriously affected by salinization [12]. Therefore, it is important to strengthen the dynamic monitoring of saline-alkaline areas to curb the source of land degradation, and to make rational use of land to improve the ecological environment.

The methods of saline-alkaline area monitoring can be currently divided into two types: instrument-measured soil data [13,14] and large-scale monitoring with remote sensing. With the development of spatial information technology, remote sensing has become the most widely used method in large-scale saline-alkaline area monitoring [5,15–17]. The methods for monitoring saline-alkaline areas based on remote sensing technology have mainly changed from visual interpretation to methods using computers to process image data and extract features [18].

The exploration of saline-alkaline area identification methods based on spectral features has been a subject of frequent scholarly discussion. In 1992, Dwivedi [19] performed experimental research on the best remote sensing bands combination for saline-alkaline areas monitoring, and concluded that the combination of bands 1, 3, and 5 of Landsat TM remote sensing images contained the largest amount of information, while the accuracy of a saline-alkaline area being identified was not proportional to the amount of information in the remote sensing data. Farifteh [20] found that soil reflectance had a good response to the salinity of a soil surface layer when using hyperspectral data for soil salinization classification, and concluded that there was a linear relationship between soil salinization and its spectral reflectance. By correlating the spectral parameters from MODIS images with salinization levels, Bouaziz et al. [21] constructed a linear spectral unmixing (LSU) model to examine the status of soil salinization in semi-arid areas. Xiao Dong [22] et al. obtained reflectance and salinity data by field sampling to construct an inversion model and a correction model. Yanhua Fu [23] constructed a model indicating the relationship of spectral data and salt content, and of organic matter content and PH level.

Research efforts using indirect features are mainly used to verify the saline-alkaline soil distribution with the help of some other auxiliary information. For example, the growth condition of vegetation can be affected by salinity; thus, vegetation is a good indirect indicator of salinity [24]. Some salt-tolerant vegetation can also be one of the salinization signs. On the ecological scale, soil salinity can adversely limit species diversity and species' ecological niches [25]. Salinity is especially associated with negative osmotic potential, which inhibits seed germination and debilitates cell turgidity [26]. R. L. Dehaan et al. [27] demonstrated that the growth and distribution of vegetation had a strong correlation with soil salinity. By developing the normalized difference vegetation index–salinity index (NDVI–SI) feature-space remote sensing model of soil salinization, Wang et al. [28] successfully monitored the change of saline soil in the Tarim Basin, Xinjiang.

Although these two methods have attained some achievements, how to effectively identify saline-alkaline areas with high accuracy is still the focus of present research.

Minqin County in Gansu Province, China is located at the junction of the Tengger Desert and the Badain Jaran Desert [29], where land degrades seriously. Since the middle of the 20th century, Qingtu Lake, which is located in the deepest part of the two deserts, has gradually dried up. In the 1970s, Minqin County started to use a large amount of groundwater, which caused soil salinization. If it continues this seriously, it will eventually lead to the merger of the Tengger Desert and the Badain Jaran Desert, which will directly affect the geomorphology, climate, and human environment of the northwest region and even threaten the survival of local peoples [30].

Although scholars have attained some achievements of quantitative monitoring of land cover using remote sensing, there is little research on the application of remote sensing data to the identification and monitoring of saline-alkaline areas at present. Traditional saline-alkaline area identification methods only rely on the selection of a single feature parameter, which is difficult to adapt to the optimal classification effect. To solve the problem of low accuracy of saline-alkaline area identification based on the traditional spectral indices, taking Minqin County as the study area, we propose a high-precision method of saline-alkaline area identification using multi-source data. By analyzing the trends and reasons of changes in saline-alkaline areas in the Minqin oasis, the objective of this paper is to provide a reference for the timely monitoring of saline-alkaline areas and ecological environment construction globally in arid areas.

2. Methodology and Experimental Application

2.1. Methodology

2.1.1. Identification Method of Saline-Alkaline Area

A decision tree [31] is a method for hierarchical processing of remote sensing images which is suitable for features with blurred boundaries and complex structures. Its main idea is to gradually mask and separate each feature as a layer from the imagery, avoiding any impact on the other features' identification. Therefore, it is possible to integrate various effective feature quantities, thus improving the identification accuracy of saline-alkaline areas.

Firstly, we use GF-6/WVF (Chinese satellite GaoFen-6/Wide Field View) image data, combining GF-2(Chinese satellite GaoFen-2) image and Google Earth high-resolution image data to select different types of samples, and find the best spectral index of band combinations for saline-alkaline areas. Secondly, the GF-2 image data is used to extract textures. Elevation and slope from DEM (Digital Elevation Model) data are used as elevation features to build a decision tree model for saline-alkaline area identification. Finally, the accuracy of the classification results of the constructed decision tree model are verified in ArcGIS using the random scattering function combined with visual interpretation. The technical flowchart is shown in Figure 1.

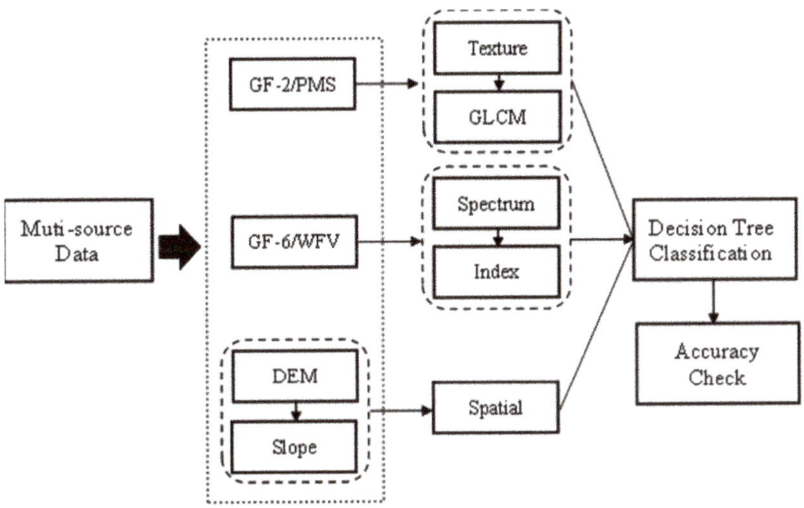

Figure 1. Technical flowchart (GLCM: grey-level co-occurrence matrix).

2.1.2. Accuracy Evaluation Method

Evaluation of feature classification is an important part of remote sensing monitoring, attempting to determine whether the results are credible. The most commonly-used evaluation method is the error matrix method, also called the confusion matrix method [32].

In this paper, the confusion matrix is calculated by comparing each actual measured image element with the corresponding classified one [33]. Each column of the confusion matrix represents the actual measured information, and each row of the confusion matrix represents the classified information of the remote sensing data (Table 1).

Table 1. Example of confusion matrix.

Classified Data	Truth Data				
	Class 1	**Class 2**	...	**Class n**	**Total**
Class 1	X_{11}	X_{12}	...	X_{1n}	$Cd_1 = \sum_{j=1}^{n} X_{1j}$
Class 2	X_{21}	X_{22}	...	X_{2n}	$Cd_2 = \sum_{j=1}^{n} X_{2j}$
...	$\vdots$	$\vdots$	$\ddots$	$\vdots$	$\vdots$
Class n	X_{n1}	X_{n2}	...	X_{nn}	$Cd_n = \sum_{j=1}^{n} X_{nj}$
Total	$Td_1 = \sum_{i=1}^{n} X_{i1}$	$Td_2 = \sum_{i=1}^{n} X_{i2}$	...	$Td_n = \sum_{i=1}^{n} X_{in}$	$\text{All} = \sum_{i=1,j=1}^{n} X_{ij}$

Various land type: Class 1, Class 2, ... , Class n.

User accuracy is the percentage of test points that fall on that category in that sub-category and are correctly classified as that category on the classification graph.

$$\text{UA (User's Accuracy)} = \frac{X_{nn}}{Cd_n} \quad (1)$$

Producer accuracy is the probability that the ground truth reference data for the category is correctly classified in this classification.

$$\text{PA(Producer accuracy)} = \frac{X_{nn}}{Td_n} \quad (2)$$

Overall accuracy is the percentage of check points of all correctly-classified land cover categories relative to the total number of check points.

$$\text{OA(Overall accuracy)} = \frac{\sum_{i=1}^{n} X_{ii}}{All} \quad (3)$$

The Kappa coefficient is a metric that indicates how much better the classification result is than random classification. The Kappa coefficient takes into account the difference between two kinds of consistency; one is the consistency between automatic classification and reference data, and the other is the consistency between sampling and reference classification. In general, the Kappa coefficient is between 0 and 1. A higher Kappa coefficient indicates a higher classification accuracy.

$$\text{Kappa} = \frac{OA - \frac{\sum_{i=1}^{n} Cd_i \times Td_i}{All}}{1 - \frac{\sum_{i=1}^{n} Cd_i \times Td_i}{All}} \quad (4)$$

From 5 to 8 March 2023, we collected 143 samples for verification at a depth of 0 to 5 cm from the surface. They were recorded, associated with information such as number, location, depth, personnel, date, and then brought back to the laboratory.

2.2. *Experimental Application*

2.2.1. Study Area

Minqin County is located in the downstream region of the Shiyang River Basin in eastern Gansu Province, China, with an altitude of 1200–1500 m (Figure 2). Tengger Desert is in the east, and Badain Jaran Desert is in the north [34] (Figure 3).

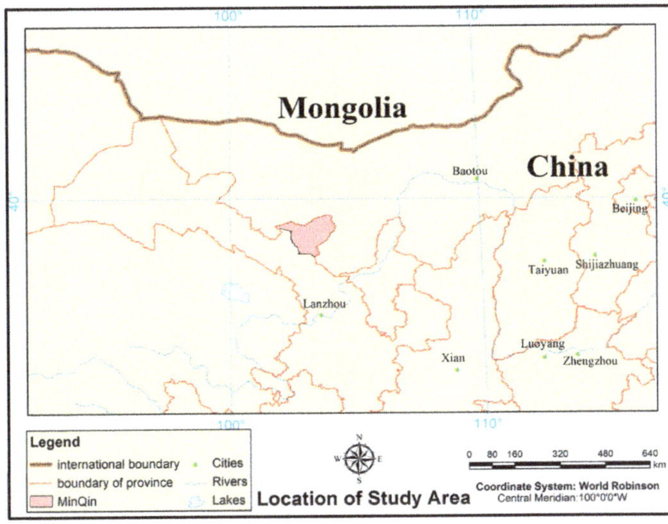

Figure 2. Location map of the study area.

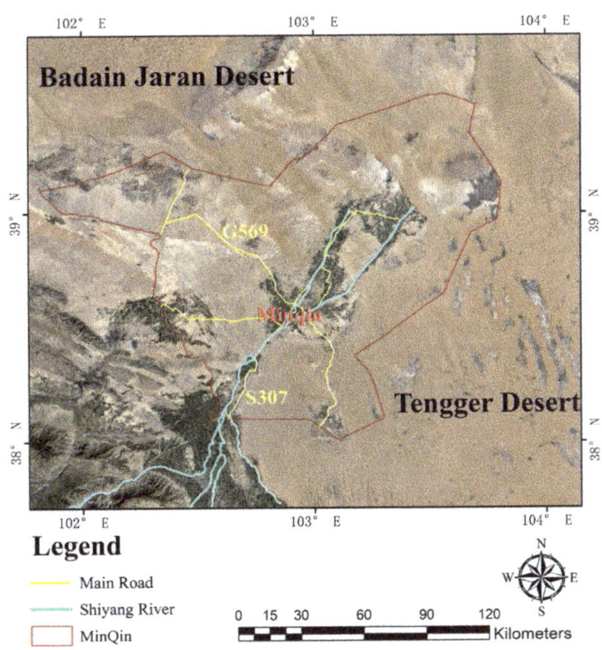

Figure 3. Map of Minqin.

As a temperate continental desert climate, the climate of the study area is characterized by cold winters and hot summers, and is dry, with little precipitation, as well as windy and sandy [35]. Its average annual temperature is 8.2 °C and the average annual precipitation is 115 mm [36]. The total area of oasis in this area is about 1352 km^2, which only accounts for 9% of the total area of Minqin County [37]. Due to environmental characteristics such as high temperatures, low precipitation, and high evaporation, the water resources in Minqin are lacking, which leads to soil desertification [30].

The main soil types are Fragic Arenosol, Solonchak, Solonetz, Plaggic Anthrosol, and Irragric Anthrosol [38], the first of which can be classified as Arenosol with sand content exceeding one-half [39].

2.2.2. Data

GF-6/WFV images and GF-2/PMS images, as well as DEM, slope, and vector boundary data, are used in this paper (Table 2). In this experiment, seasons of the remote sensing images were selected as being from June to July, because plants grow more luxuriantly and there is no snow and ice cover in this period, and thus, it is favorable for the identification of saline-alkaline areas.

Table 2. Data sources for identification.

Data	Date	Spatial Resolution	Parameter
GF-6/WFV	2022.6.13	16 m	Spectra
GF-2/PMS	2022.7.13	1 m(PAN)/4 m(MSS)	Texture
SRTM_DEM		30 m	Elevation
Slope			
Vector boundary	2020		Zone

The GF-6/WFV data were pre-processed for radiometric calibration, atmospheric correction, orthorectification correction, and vector cropping to obtain eight-band surface reflectance data for the study area (Figure 4). The atmospheric correction was implemented by the FLAASH (fast line-of-sight atmospheric analysis of spectral hypercubes) atmospheric correction module (Table 3).

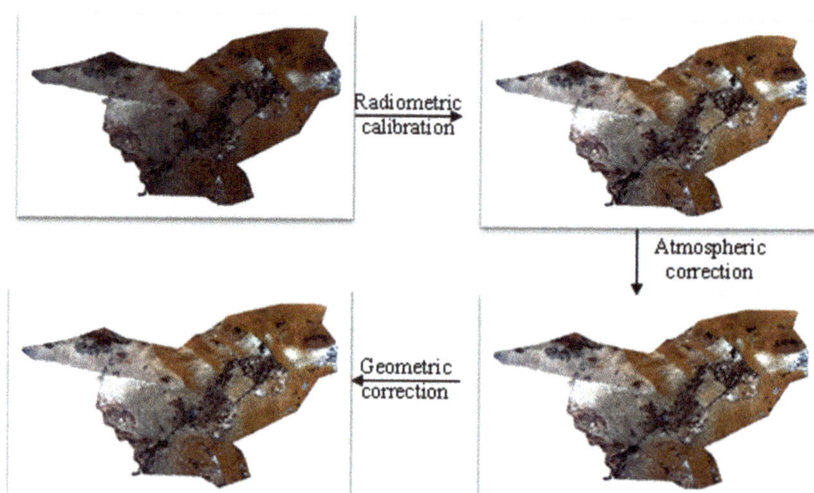

Figure 4. GF6 pre-processing flow chart.

Table 3. Parameters of the FLAASH atmospheric correction module.

Atmospheric Model	Aerosol Model	Aerosol Retrieval	Initial Visibility	Spectral Response Function
Mid-Latitude Summer	Rural	None	40 km	gf6_wfv.sli

The GF-2/PMS image has MSS (Multispectral) and PAN (Panchromatic) data. They have been pre-processed for radiometric calibration, atmospheric correction, geometric correction, image fusion, etc. to obtain a four-band fused image with a spatial resolution of 1 m for the study area (Figure 5).

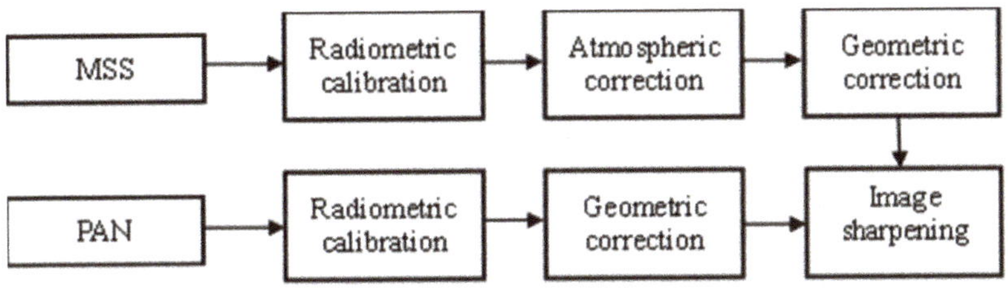

Figure 5. GF2 pre-processing process.

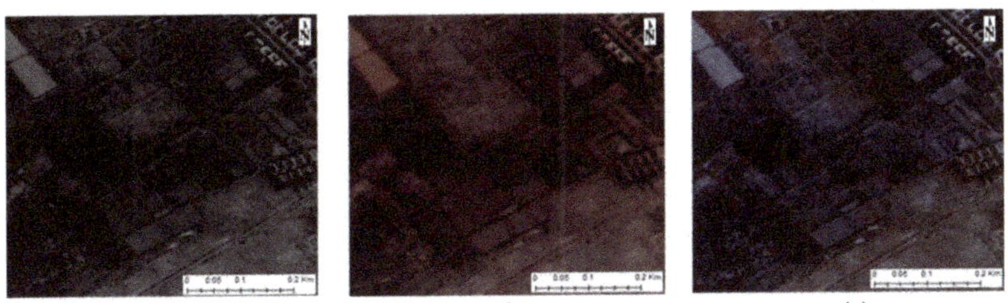

(**a**) (**b**) (**c**)

Figure 6. Comparison of image fusion: (**a**) image of PAN; (**b**) image of MSS; and (**c**) sharpened MSS.

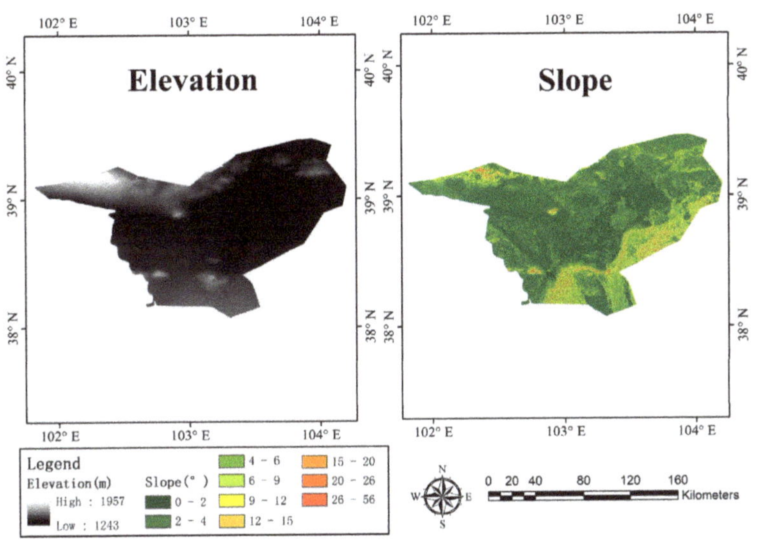

Figure 7. DEM with slope.

Image fusion is an image processing technique that resamples low-resolution multispectral images with a high-resolution pan image to generate a high-resolution multispectral image for remote sensing, enabling the processed image to have both high spatial resolution and multispectral characteristics. Here, we use the Gram–Schmidt pan sharpening (GS) fusion method (Figure 6). Its advantage is that it is not limited by the band, which is suitable for processing high spatial resolution images, and can better maintain the texture and spectral information.

The DEM data was cropped in ArcGIS using vector boundary files and then output to obtain the elevation data of the study area. Furthermore, slope was obtained from the cropped DEM in ArcGIS (Figure 7).

2.2.3. Feature Extraction

- Spectral features

The main land cover types in the study area include desert, saline-alkaline area, vegetation, urban, and water. After pre-processing the GF-6/WFV images, the original spectral characteristics of each type in the study area were analyzed (Table 4, Figure 8).

Table 4. GF-6/WFV band.

Band	Wavelength/μm	Name	Spatial Resolution/m	Scan Width/km
B01	0.45~0.52	Blue	16	800
B02	0.52~0.59	Green		
B03	0.63~0.69	Red		
B04	0.77~0.89	NIR		
B05	0.69~0.73	Red edge1		
B06	0.73~0.77	Red edge2		
B07	0.40~0.45	Violet		
B08	0.59~0.63	Yellow		

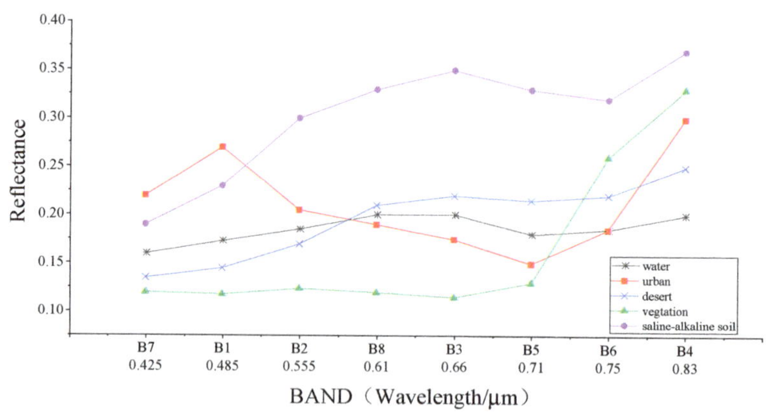

Figure 8. Spectra of features in the study area. (Bands are arranged by increasing wavelength).

From Figure 8, we can see that some spectral features of a saline-alkaline area and a desert are easily confused; more spectral indices are needed to improve the saline-alkaline areas' classification accuracy.

The $NDSI_{34}$ (Normalized Difference Spectral Index of Band 3 and Band 4) was constructed using band 3 (Red) and band 4 (NIR).

$$NDSI_{34} = (NIR - R)/(R + NIR) \tag{5}$$

The $NDSI_{25}$ (Normalized Difference Spectral Index of Band 2 and Band 5) was constructed using band 2 (Green) and band 5 (Red edge1: Re1).

$$NDSI_{25} = (Re1 - G)/(Re1 + G) \tag{6}$$

The NDSI 237 (Normalized Difference Spectral Index of Band 2, Band 3 and Band 7) was constructed using bands 2 (Green), 3 (Red), and 7 (Violet).

$$NDSI_{237} = (R + G - V)/(R + G + V) \tag{7}$$

The final composite salinity index NDSInew (New Normalized Difference Salt Index) was constructed as:

$$NDSI_{new} = NDSI_{25} + NDSI_{237} - NDSI_{34} \tag{8}$$

The spectral indices are mainly selected depending on the spectral characteristics of each feature. For example, $NDSI_{34}$ can sufficiently separate the vegetation in the image. The saline-alkaline area is associated with a large difference between the red edge1 band and the green band, therefore, $NDSI_{25}$ can distinguish the saline-alkaline areas from other features. For reflectance of saline-alkaline soil in the red and green bands, which are significantly higher than those in the violet band, $NDSI_{237}$ can sufficiently separate saline-alkaline soil from other features. Considering the three indices together, we finally construct the comprehensive index, $NDSI_{new}$, by which the saline-alkaline areas can be well distinguished.

- Texture features

When the spectra of the features are relatively similar, the spectral differentiability decreases and texture information can play an important role in distinguishing the features, raising the accuracy rates of classification [40].

Among the methods for computing image texture features, GLCM (grey-level co-occurrence matrix) is one of the most widely used statistical methods [41]. GLCM can describe the spatial distribution and structural characteristics of the image grayscale, which is advantageous in improving the classification of geological targets by using texture. There are eight main feature quantities commonly used for texture identification in remote sensing images: mean, variance, homogeneity, contrast, dissimilarity, entropy, angular second moment, and correlation.

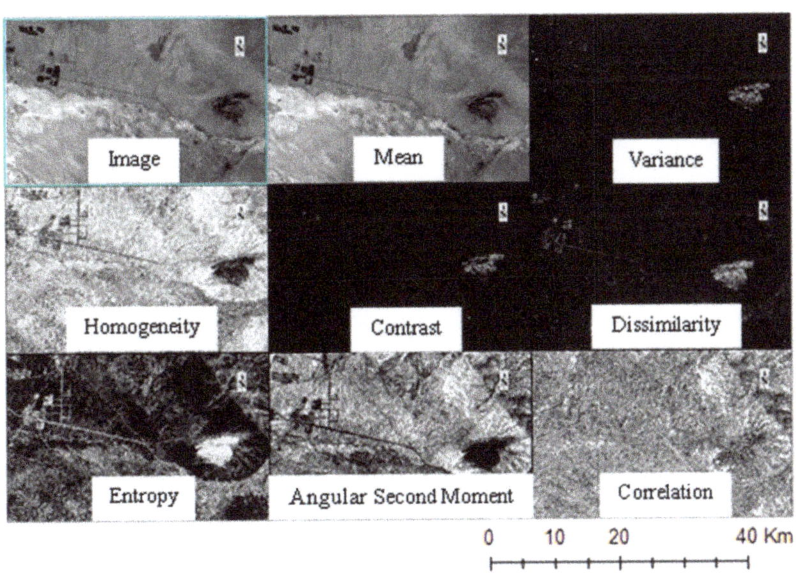

Figure 9. Textures.

We used the GLCM method to extract textures from GF-2 images and calculated eight textures on four bands with a 3 × 3 window (Figure 9). After that, we selected mean, dissimilarity and entropy as the parameters for classification.

- Elevation features

Height and slope information from DEM are introduced to carry out reclassification in ArcGIS. As shown in Figure 10, there are some differences in elevation among features. For example, vegetation and urban types are generally flatter.

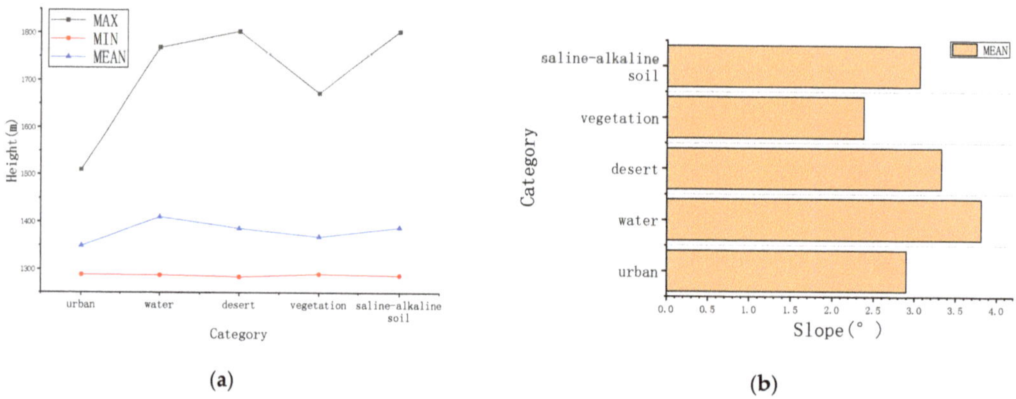

Figure 10. Elevation characteristics: (**a**) height of features; and (**b**) slope of features.

3. Results and Discussion

3.1. Classification and Verification

Based on multi-source data, the results of saline-alkaline area identification map in the study area using a decision tree classification method combining spectral features (NDSInew), texture features (mean, dissimilarity and entropy), and elevation features (height and slope) is shown in Figure 11.

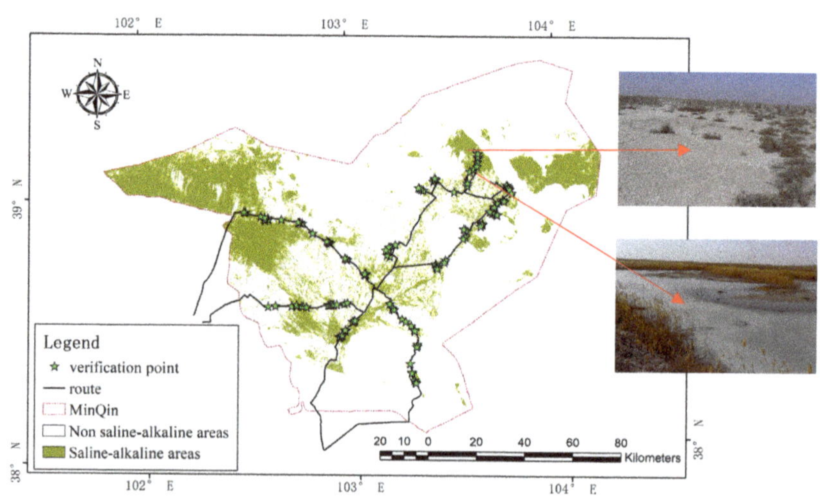

Figure 11. Saline-alkaline areas: map and sample points.

Through random distribution and considering the accessibility of each site, we travelled to Minqin County for a field survey (Figure 12). The land types of verification points were investigated and labeled, and 143 verification samples were obtained.

Figure 12. Field survey: (a) saline-alkaline areas; (b) handheld GPS recording; and (c) sampling.

Table 5. Confusion matrix of saline-alkaline area identification.

Classified Data	Checked Data			
	Non-Saline-Alkaline Area	Saline-Alkaline Area	Total	UA
Non-saline-alkaline area	64	4	68	94.12%
Saline-alkaline area	13	62	75	82.67%
Total	77	66	143	
PA	83.12%	93.94%		
OA		88.11%		
Kappa		0.76		

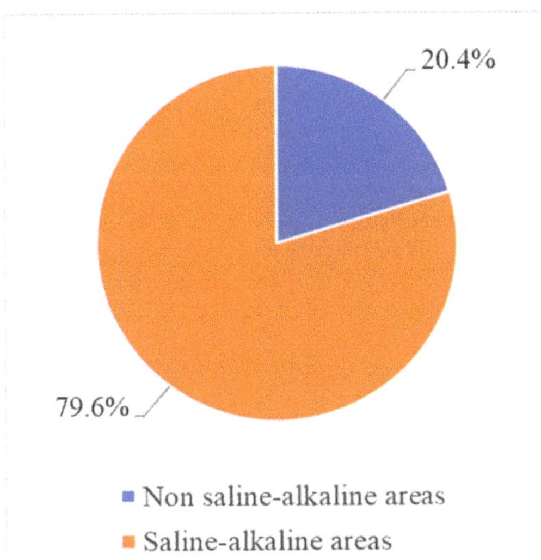

Figure 13. Percentage of saline-alkaline area in Minqin County.

The samples were established and verified by confusion matrix, using producer accuracy, user accuracy, total accuracy and Kappa coefficient. The results are shown in Table 5, and indicate that the accuracy of the proposed saline-alkaline area identification method is 88.11%.

Shown in Figure 13, the saline-alkaline area in Minqin County is 3385.17 km^2, accounting for 20.4% of the total area of Minqin County.

From its spatial distribution in Figure 11, the soil salinization in the northwest is the most serious, with a large area of saline-alkaline area implicated, followed by the eastern region, and finally the Minqin oasis area, where the saline-alkaline area is small, scattered, and distributed on both sides of the oasis.

3.2. Comparison of the Results of Different Indices

Based on the same data, the traditional salinity index $NDSI_{old}$ = (NIR − R)/(R + NIR) was used for salinity identification, and its accuracy was verified to be 80.42%. The comparison of the salinity identification results between these two methods is shown in Figure 14.

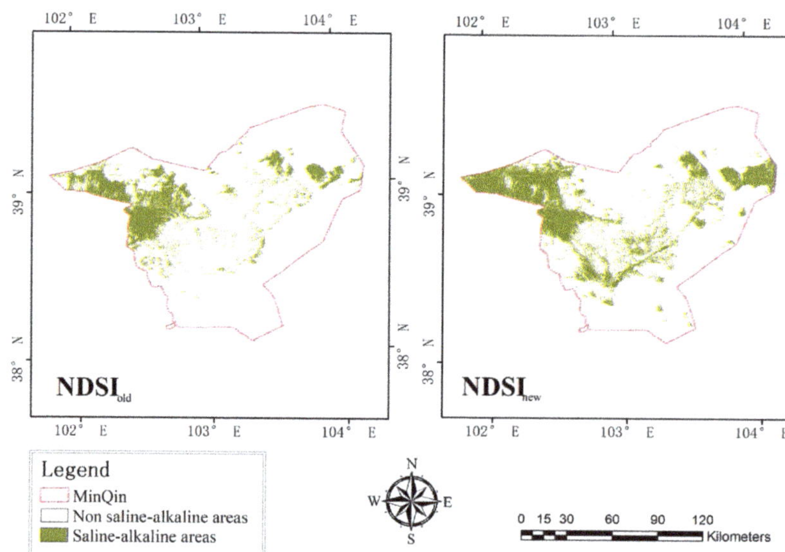

Figure 14. Comparison of the identification results of different indices.

From these results, we can see that the accuracy of the new salinity index $NDSI_{new}$ is improved by 7.69% compared with the traditional salinity index $NDSI_{old}$, indicating the effectiveness of the new spectral index in the identification of saline-alkaline areas.

3.3. Analysis of Saline-Alkaline Area Change

Three Landsat8 OLI remote sensing images were downloaded from https://www.gscloud.cn, accessed on 28 December 2022, in July and August (Table 6).

Table 6. Data sources for analysis.

Date	Satellite	Sensor	Number
2010.7.25 2015.7.22 2020.8.2	Landsat 8	OIL	131/33

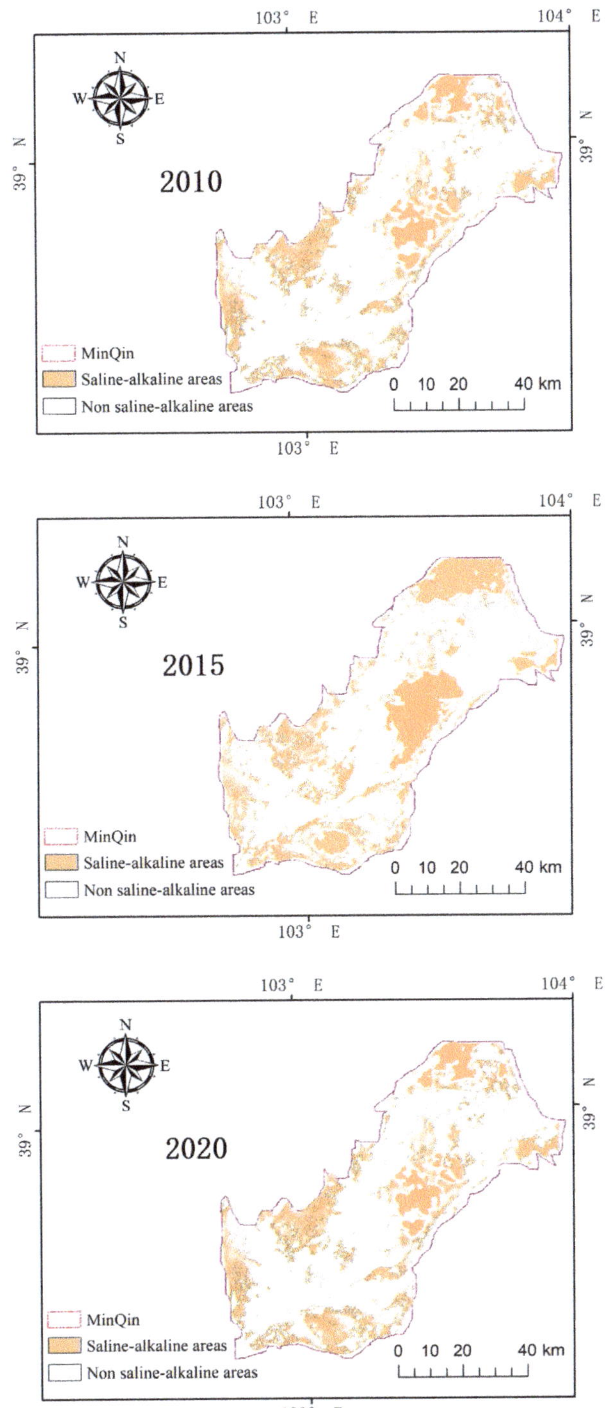

Figure 15. Saline-alkaline areas identified in the Minqin oasis in 2010, 2015, and 2020.

The remote sensing data were pre-processed with ENVI.

The results of the saline-alkaline area identification in 2010, 2015 and 2020 (Figure 15) were statistically analyzed in ArcGIS to classify the total areas of saline-alkaline land. The saline-alkaline areas in 2010, 2015 and 2020 were 2276.21 km^2, 2186.28 km^2 and 1922.93 km^2, respectively (Figure 16). From this, we can see that the saline-alkaline area decreased 353.28 km^2 from 2010 to 2020.

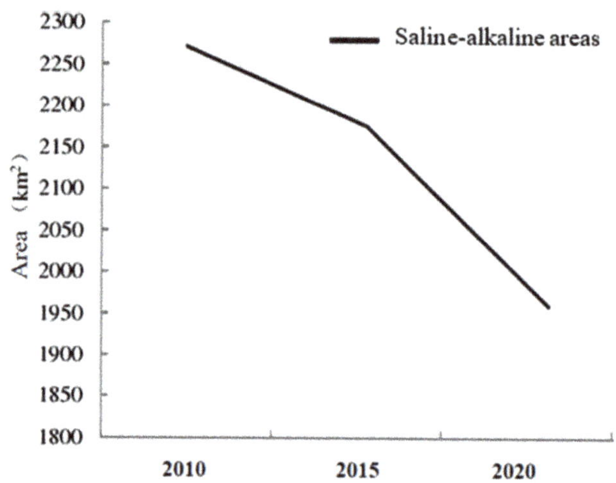

Figure 16. Change of saline-alkaline areas in the Minqin oasis.

There are many natural and human factors affecting the saline-alkaline area changes in the Minqin oasis:

(1) Climate change has brought many problems to the soil environment, such as a series of biological changes in the soil's physical composition (water content), chemical composition (various salt ion contents), and plant species. Climatic warming can not only cause microorganisms to rapidly decompose soil organic matter and soil nutrients' rapid decrement, but it can also cause soil moisture to evaporate, accelerating the upward movement of salt, and causing soil salinization.

According to the statistics of the Minqin meteorological station, Minqin has little rainfall but a high level of evaporation (Figure 17). Combined with the temperature rises, these promote the salinization of the soil.

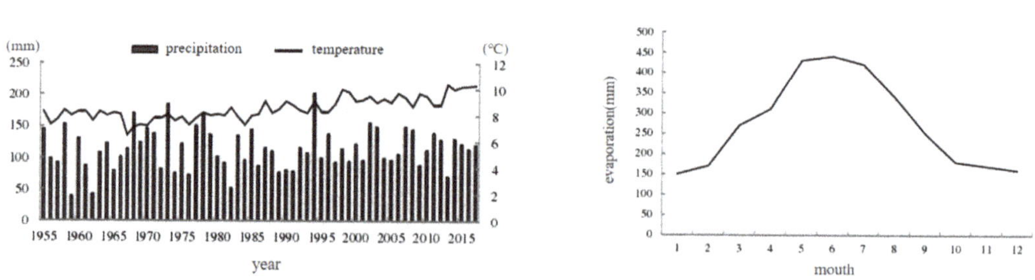

Figure 17. Climate in Minqin county: (**a**) curve of annual average temperature and precipitation; and (**b**) annual average monthly potential evaporation.

Coupled with the weathering effect of rocks, a large amount of salt is released in the soil's parent material of the northwestern remnant hills, and then carried to the lowlands through precipitation, resulting in serious salinization in the northwest [42].

(2) Historically, Minqin has been one of the important salt-producing areas, with many salt ponds [34]. With the gradual depletion of the Shiyang River, Minqin started to seek groundwater instead (Figure 18). Due to the adjustment of agricultural structure, water resources were redistributed spatially, salt was transferred with water, and the overuse of irrigation water also led to the transformation of some depressions at the edge of the oasis into saline-alkaline areas [43].

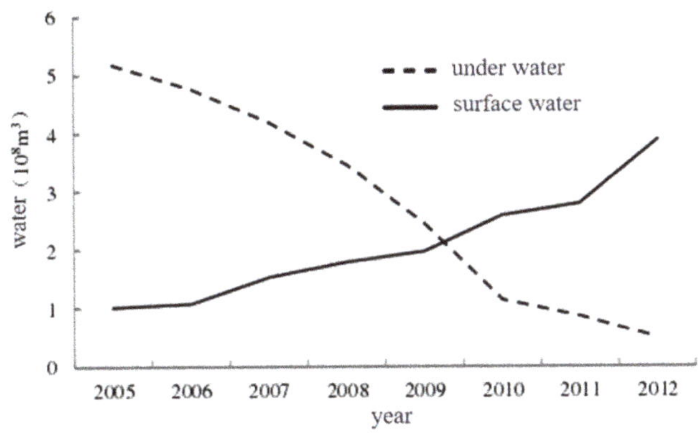

Figure 18. Changes of surface water inflow in Minqin County from 2005 to 2012.

(3) In 2007, the government implemented the "Key Control Plan of Shiyang River Basin" and began to transfer water to Minqin County at the lower reaches of the Shiyang River [44]. The surface water runoff into Minqin County has increased year by year (Figure 18). Since then, the soil salinization in Minqin has been improved to a certain extent.

Some studies [45–47] on saline-alkaline area identification in Minqin County are shown in Table 7. However, all only considered salinization as a type of desertification and did not conduct in-depth research on the fine classification of saline-alkaline areas. From this, it can be concluded that there are few studies on saline-alkaline area identification in Minqin County. At the same time, there have been precedents for the decision tree classification methods for land classification in this region, which proves the applicability of the method in Minqin.

Table 7. Comparison of studies on the identification of saline-alkaline areas in Minqin County.

Order Number	Contents	Date	Articles
1	Classification of unused land	2016	Yao, A. et al., 2014 [45]
2	Dynamic monitoring of land desertification	2004–2009	Chen, X. et al., 2014 [46]
3	Analysis of land desertification characteristics	2012–2013	Ma, J. et al., 2019 [47]

4. Conclusions

We use multi-source data for saline identification in Minqin County and draw the following conclusions:

(1) The proposed method is effective in saline identification.

Based on multi-source data, we use a decision tree classification method to extract saline-alkaline areas by constructing three features: spectral indexes, textures, elevations, and slopes. The results show that the accuracy of saline-alkaline area identification is

88.11%, which is 7.69% greater than the traditional salinity indices, indicating the effectiveness of the proposed method.

(2) The multi-source data can help to identify features and improve accuracy.

GF-6 data are beneficial to the improvement of the accuracy of saline-alkaline area identification. In which, band 3 and band 7 are important to the saline-alkaline area identification in the study area.

High spatial resolution of GF-2 data can provide rich texture information, thus reducing the mistakes of distinguishing or misclassifying between features due to "different features with the same spectrum" or "different features with the same spectrum".

The height and slope from DEM can quantify the topography of the study area, which is also helpful for identifying features and improving the classification accuracy.

(3) Monitoring and prevention of unused land in the study area are necessary.

With 20.4% of the land considered to be within a saline-alkaline area, soil salinization in Minqin County is a serious concern, especially in the northwestern areas. Therefore, we should strengthen the monitoring and prevention of unused land to prevent further soil salinization.

In summary, based on the previous studies, we proposed a high-precision saline-alkaline area identification method based on multi-source data. The results demonstrate the effectiveness of the method, thus solving the current problem of low accuracy of saline-alkaline area identification, a solution which may be applied to large-scale saline-alkaline area monitoring in the future. Meanwhile, it should be noted that, although the decision tree classification method achieved better classification results in this study, the significance of selected feature variables and grading criteria need to be further studied and improved to make the discriminative rules and classification results more realistic. Therefore, the research on the identification and classification of soil salinization in arid zones needs to be further developed.

Author Contributions: Conceptualization, J.Y. and Q.W.; methodology, J.Y.; validation, J.Y. and D.C.; formal analysis, J.Y. and D.C.; investigation, J.Y. and W.X.; resources, Q.W. and B.Y.; data curation, J.Y. and D.C.; writing—original draft preparation, J.Y.; writing—review and editing, Q.W.; visualization, J.Y. and W.X.; supervision, Q.W.; project administration, Q.W.; funding acquisition, Q.W. All authors have read and agreed to the published version of the manuscript.

Funding: This research was funded in part by the National Key R&D Program of China (grant No. 2021YFB3900503), the National Natural Science Foundation of China (grant number 42071312), the Hainan Hundred Special Project (grant number 31, JTT [2018]), the Innovative Research Program of the Foundation of China (grant number 42071312), the Innovative Research Program of the International Research Center of Big Data for Sustainable Development Goals (grant number CBAS2022IRP03), the National Key R&D Program (grant number CBAS2022IRP03), the National Key R&D Program (grant number 2021YFB3900503), the Special Project of Strategic Leading Science and Technology of the Chinese Academy of Sciences (grant number XDA19090139), the Second Tibetan Plateau Scientific Expedition and Research (STEP) (grant number 2019QZKK0806), and the Hainan Provincial Department of Science and Technology (grant number ZDKJ2019006).

Data Availability Statement: Not applicable.

Acknowledgments: Thanks are due for the GF data provided by CRESDA and AIRCAS.

Conflicts of Interest: The authors declare no conflict of interest.

References

1. Dennis, L.C. Climate change impacts on soil salinity in agricultural areas. *Eur. J. Soil Sci.* **2020**, *72*, 842–862.
2. Ivushkin, K.; Bartholomeus, H.; Bregt, A.K.; Pulatov, A.; Franceschini, M.H.D.; Kramer, H.; Loo, E.N.V.; Roman, V.J.; Finkers, R. UAV based soil salinity assessment of cropland. *Geoderma* **2018**, *338*, 501–512. [CrossRef]
3. Wu, D.; Jia, K.; Zhang, X.; Zhang, J.; Abd El Hamid, H.T. Remote Sensing Inversion for Simulation of Soil Salinization Based on Hyperspectral Data and Ground Analysis in Yinchuan, China. *Nat. Resour. Res.* **2021**, *30*, 4641–4656. [CrossRef]
4. Bani, A.; Daghari, I.; Hatira, A.; Chaabane, A.; Daghari, H. Sustainable management of a cropping system under salt stress conditions (Korba, Cap-Bon, Tunisia). *Environ. Sci. Pollut. Res. Int.* **2021**, *28*, 46469–46476. [CrossRef] [PubMed]

5. Masoud, A.; Koike, K. Arid land salinization detected by remotely-sensed landcover changes: A case study in the Siwa region, NW Egypt. *J. Arid Environ.* **2006**, *66*, 151–167. [CrossRef]
6. Arruda, G.P.D.; Demattê, J.A.M.; Chagas, C.D.S.; Fiorio, P.R.; Souza, A.; Fongaro, C.T. Digital soil mapping using reference area and artificial neural networks. *Sci. Agric.* **2015**, *73*, 266–273. [CrossRef]
7. Sheng, J.; Ma, L.; Jiang, P.; Li, B.; Huang, F.; Wu, H. Digital soil mapping to enable classification of salt-affected soils in desert agro-ecological zones. *Agric. Water Manag.* **2010**, *97*, 1944–1951. [CrossRef]
8. Peng, J.; Biswas, A.; Jiang, Q.; Zhao, R.; Hu, J.; Hu, B.; Shi, Z. Estimating soil salinity from remote sensing and terrain data in southern Xinjiang Province, China. *Geoderma* **2019**, *337*, 1309–1319. [CrossRef]
9. Sidike, A.; Zhao, S.; Wen, Y. Estimating soil salinity in Pingluo County of China using QuickBird data and soil reflectance spectra. *Int. J. Appl. Earth Obs. Geoinf.* **2014**, *26*, 156–175. [CrossRef]
10. Dong, F.; Tang, Y.; Xing, X.; Liu, Z.; Xing, L. Formation and Evolution of Soil Salinization in Shouguang City Based on PMS and OLI/TM Sensors. *Water* **2019**, *11*, 345. [CrossRef]
11. Fathizad, H.; Ardakani, M.A.H.; Sodaiezadeh, H.; Kerry, R.; Taghizadeh-Mehrjardi, R. Investigation of the spatial and temporal variation of soil salinity using random forests in the central desert of Iran. *Geoderma* **2020**, *365*, 114233. [CrossRef]
12. Wang, J.; Huang, X.; Zhong, T.; Chen, Z. Review on Sustainable Utilization of Salt-affected Land. *Acta Geol. Sin.* **2011**, *66*, 673–684.
13. Liu, L.; Zhang, J.; Zhang, K.; Jia, L.; Dou, L.; Nie, A. Effect of drip irrigation on the rapid growth and soil improvement of saline alkali ponies in coastal saline lands. *China Rural Water Conserv. Hydropower* **2021**, *12*, 209–215.
14. Cao, Y.; Bai, Y.; Li, J.; Chen, X.; Shen, G. Study on the spatial distribution characteristics of soil salinity in mudflats in the Yangtze River Delta. *Soil Crops* **2022**, *11*, 31–40.
15. Periasamy, S.; Shanmugam, R.S. Multispectral and Microwave Remote Sensing Models to Survey Soil Moisture and Salinity. *Land Degrad. Dev.* **2017**, *28*, 1412–1415. [CrossRef]
16. Metternicht, G.I.; Categorical, F. A comparison between crisp and fuzzy class boundary modelling for mapping salt-affected soils using Landsat TM data and a classification based on anion ratios. *Ecol. Model.* **2003**, *168*, 371–389. [CrossRef]
17. Mulder, V.L.; Bruin, S.D.; Schaepman, M.E.; Mayr, T.R. The use of remote sensing in soil and terrain mapping: A review. *Geoderma* **2011**, *162*, 1–19. [CrossRef]
18. Wang, X.L.; Hou, X.Y. Changes in normalized vegetation index (NDVI) in coastal areas of China from 1982–2014 and its effect on response of extreme climate. *Geogr. Stud.* **2019**, *38*, 807–821.
19. Dwivedi, R.S.; Rao, B.R.M. The selection of the best possible Landsat TM band combination for delineating salt-affected soils. *Int. J. Remote Sens.* **1992**, *13*, 2051–2058. [CrossRef]
20. Farifteh, J.; Farshad, A.; George, R.J. Assessing salt-affected soils using remote sensing, solute modelling, and geophysics. *Geoderma* **2006**, *130*, 191–206. [CrossRef]
21. Bouaziz, M.; Matschullat, J.; Gloaguen, R. Improved remote sensing detection of soil salinity from a semi-arid climate in northeast Brazil. *Comptes Rendus Geosci.* **2011**, *343*, 795–803. [CrossRef]
22. Xiao, D.; Wan, L.; Sun, X. Remote sensing retrieval of saline and alkaline land based on reflectance spectroscopy and RV-MELM in Zhenlai County. *Opt. Laser Technol.* **2021**, *139*, 106909. [CrossRef]
23. Fu, Y.; Wan, L.; Xiao, D.; Le, B. Remote sensing inversion of saline and alkaline land based on reflectance spectroscopy and D-TELM algorithm in Wuyuan areas. *Infrared Phys. Technol.* **2020**, *109*, 103367. [CrossRef]
24. Yang, Z.; Zhao, J.; Wen, Y. Study on vegetation and soil salinity response in Qingtu Lake area. *Agric. Res. Arid Areas* **2020**, *38*, 231–237.
25. Ashilenje, D.S.; Amombo, E.; Hirich, A.; Kouisni, L.; Devkota, K.P.; El Mouttaqi, A.; Nilahyane, A. Crop Species Mechanisms and Ecosystem Services for Sustainable Forage Cropping Systems in Salt-Affected Arid Regions. *Front. Plant Sci.* **2022**, *13*, 1518. [CrossRef] [PubMed]
26. Patel, P.R.; Kajal, S.S.; Patel, V.R.; Patel, V.J.; Khristi, S.M. Impact of salt stress on nutrient uptake and growth of cowpea. *Braz. J. Plant Physiol.* **2010**, *22*, 43–48. [CrossRef]
27. Dehaan, R.L.; Taylor, G.R. Field-derived spectra of salinized soils and vegetation as indicators of irrigation induced soil salinization. *Remote Sens. Environ.* **2002**, *80*, 406–417. [CrossRef]
28. Wang, F.; Ding, J.; Wu, M. Remote sensing monitoring models of soil salinization based on NDVI-SI feature space. *Trans. Chin. Soc. Agric. Eng.* **2010**, *26*, 168–173.
29. Yang, J.; Zhao, J.; Zhu, G.; Wang, Y.; Ma, X.; Wang, J.; Guo, H.; Zhang, Y. Soil salinization in the oasis areas of downstream inland rivers—Case Study: Minqin oasis. *Quat. Int.* **2020**, *537*, 69–78. [CrossRef]
30. Niu, Z.Y.; Ding, J.L. Soil salinization information extraction method based on gf-1 image. *Arid Land Geogr.* **2016**, *39*, 171–181.
31. Shan, J.; Yue, C. Research on decision tree classification method based on spectral features—A case study of northeastern Kunming city. *J. Northwest For. Acad.* **2010**, *25*, 222–226.
32. Yang, X. Overview of performance metrics of classification learning algorithms. *Comput. Sci.* **2021**, *48*, 209–219.
33. Barranco, C.I.; Carrillo, G.R.M. Techniques to Deal with Off-Diagonal Elements in Confusion Matrices. *Mathematics* **2021**, *9*, 3233. [CrossRef]
34. Sun, L.Q.; Zhang, F.H.; Yang, S.W. Historical change and evolution of sandy land in Minqin County (during 1991–2015). *Mapp. Spat. Geogr. Inf.* **2018**, *41*, 106–109.

35. Zhang, Z.; Li, Z.; Li, Q.; Hao, R. Analysis of dynamic changes of vegetation cover in Minqin Oasis from 2013-2015 based on GF-1 remote sensing images. *J. Southwest For. Univ. (Nat. Sci.)* **2017**, *37*, 163–170.
36. Li, W.; Lv, S.; Dong, Z.; Fan, G.; Chen, L. Temporal and spatial variation characteristics of precipitation around Badain Jardan Desert. *J. Desert Res.* **2015**, *35*, 94–105.
37. Lu, H.; Jie, Y.; Zhang, W.; Wei, J. Analysis of spatial and temporal changes of oasis in Minqin County from 1986–2015. *Arid Zone Res.* **2017**, *34*, 1410–1417.
38. Qian, T.; Tsunekawa, A.; Masunaga, T.; Wang,, T. Analysis of the Spatial Variation of Soil Salinity and Its Causal Factors in China's Minqin Oasis. *Math. Probl. Eng.* **2017**, *2017*, 1–9.
39. Chen, L.; Li, C.; Feng, Q.; Wei, Y.; Zhao, Y.; Zhu, M.; Ravinesh, C.D. Direct and indirect impacts of ionic components of saline water on irrigated soil chemical and microbial processes. *Catena* **2019**, *172*, 581–589. [CrossRef]
40. Yang, Y.; Tian, Q.; Jim, Y.; Tao, B.; Xu, K. Effects of spatial resolution and texture features on multispectral remote sensing classification. *J. Geoinf. Sci.* **2018**, *20*, 99–107.
41. Verma, H.; Vidyarthi, A.; Chitre, A.V.; Wanjale, K.H.; Anusha, M.; Majrashi, A.; Hinga, S.K. Local Binary Patterns Based on Neighbor Center Difference Image for Color Texture Classification with Machine Learning Techniques. *Wirel. Commun. Mob. Comput.* **2022**, *2022*. [CrossRef]
42. Wang, X.; Ren, T.; Dong, P. Research on climate change trend prediction technology in Minqin County. *Gansu Sci. Technol. Vert.* **2017**, *46*, 29–31+22.
43. Ma, J.; Chen, J.; Yang, X.; Zhang, X. The impact of land use change on social resilience of rural communities in arid zones: The case of Minqin Oasis. *Resour. Sci.* **2021**, *43*, 1615–1627.
44. Gansu Water Resources and hydropower Survey and Design Institute. Key Governance Planning of Shiyang River Basin. 2007. Available online: http://www.ndrc.gov.cn/fggz/fzzlgh/gjjzxgh/200806/P020191104623858217901.pdf (accessed on 27 December 2022).
45. Yao, A.; Che, T.; Jiang, L.; Feng, Y. Remote sensing classification of desertification in minqin county of gansu area unused study. *J. Res. For. Sci.* **2014**, *27*, 195–200.
46. Chen, X.; Gao, B.; Wang, X.; Zhang, C. Status and dynamics of land desertification in Minqin County, Gansu Province. *J. Desert Res.* **2014**, *34*, 970–974.
47. Ma, J.; Shan, L.; Sun, X.; Zhao, X.; Yan, H.; Zhang, Y.; Hu, S.; Zeng, X. Minqin county even the analysis of characteristics of the ancient city of national nature reserve of land desertification. *J. Gansu Agric. Univ.* **2019**, *54*, 99–107.

MDPI
St. Alban-Anlage 66
4052 Basel
Switzerland
www.mdpi.com

Remote Sensing Editorial Office
E-mail: remotesensing@mdpi.com
www.mdpi.com/journal/remotesensing